Lecture Notes in Computer Science 15760

The series Lecture Notes in Computer Science (LNCS), including its subseries Lecture Notes in Artificial Intelligence (LNAI) and Lecture Notes in Bioinformatics (LNBI), has established itself as a medium for the publication of new developments in computer science and information technology research, teaching, and education.

LNCS enjoys close cooperation with the computer science R & D community, the series counts many renowned academics among its volume editors and paper authors, and collaborates with prestigious societies. Its mission is to serve this international community by providing an invaluable service, mainly focused on the publication of conference and workshop proceedings and postproceedings. LNCS commenced publication in 1973.

Nathalie Bertrand · Clemens Dubslaff ·
Sascha Klüppelholz
Editors

Principles of Formal Quantitative Analysis

Essays Dedicated to Christel Baier on the
Occasion of Her 60th Birthday

 Springer

Editors
Nathalie Bertrand 🆔
Inria
University of Rennes
Rennes, France

Clemens Dubslaff 🆔
Eindhoven University of Technology
Eindhoven, The Netherlands

Sascha Klüppelholz 🆔
TU Dresden
Dresden, Germany

ISSN 0302-9743   ISSN 1611-3349 (electronic)
Lecture Notes in Computer Science
ISBN 978-3-031-97438-0   ISBN 978-3-031-97439-7 (eBook)
https://doi.org/10.1007/978-3-031-97439-7

Foto by Franziska Pilz

Preface

This Festschrift is dedicated to Christel Baier in recognition of her contributions to the field of theoretical computer science, particularly in formal methods, temporal logics, model checking, and probabilistic systems. Her work has shaped the foundations and practical applications of system verification. This volume honors Christel by celebrating her life, birthday, achievements, and, last but not least, her personal engagement.

After earning her doctorate from the University of Mannheim, Christel held research and professorial positions in Mannheim, Bonn, and today in Dresden, where she has been a full professor of Algebraic and Logical Foundations of Computer Science at Technische Universität Dresden since 2006. In 2011, Christel was elected to the Academia Europaea. She was the editor-in-chief of Acta Informatica from 2015 to 2022, and received an honorary doctorate from RWTH Aachen University in 2022. She currently serves as the Dean of the Faculty of Computer Science at Technische Universität Dresden. Among her many notable research achievements, she has made fundamental contributions to the field of model checking. She pioneered probabilistic model checking, advanced techniques for model checking of continuous-time Markov chains, and co-authored the standard textbook Principles of Model Checking with Joost-Pieter Katoen. Beyond her academic contributions, Christel is widely respected within the scientific community as a scientist as well as a person. She continuously contributes to the community with countless activities, including invited talks, membership in steering committees, program committee service, the organization of scientific events, and many other forms of academic service. Not only has she done groundbreaking research, but she has also mentored a generation of young scientists. Although Christel's group has always been comparatively small, it has over the decades welcomed many young researchers who greatly benefited from her boundless support. We, the three editors of this volume, are ourselves good examples of that species. Clemens Dubslaff began benefiting from Christel's supervision during his bachelor's studies, continued to do a Ph.D. with her, and has been an assistant professor at Eindhoven University of Technology since 2022. Nathalie Bertrand collaborated with Christel during her Ph.D. and held a postdoctoral position in her group before joining Inria Rennes as a researcher in 2007. Sascha Klüppelholz started as a student helper in Bonn in 2000, moved with Christel to Dresden for his Ph.D., and now holds a permanent position in her group. If we were to highlight a single trait that characterizes her most distinctly, it would be her unwavering consistency. She approaches every task with remarkable persistence, focus, and diligence—always driven by a deep sense of passion. We have yet to meet someone with a similar level of commitment and intrinsic drive. All three of us are very thankful to her and sincerely appreciate her continuous support, guidance, and collaboration.

May 2025

Nathalie Bertrand
Clemens Dubslaff
Sascha Klüppelholz

Foreword by Mila Majster-Cederbaum

I met Christel when she studied at the University of Mannheim majoring in mathematics with computer science as her minor. She attended some of the courses I was teaching at that time, in particular two courses on algorithms. Very soon it become clear that she was an outstanding student and it was interesting to deal with her questions on various subjects. And I am sure that she also excelled in her mathematics courses. So, when the end of her master studies came closer, I was convinced that some of the mathematics professors had offered her a possibility to continue her studies to obtain a Ph.D. At that time, I met her in the institute and asked her in what field of mathematics and with which professor she was going to work. It turned out that the mathematicians had not seized the opportunity but some professor in economics had offered her a position. And I am sure that she would have succeeded there, too. But as, by chance, I just had an open position to fill I reacted promptly and offered her to pursue a Ph.D. in my group. She accepted and a fruitful collaboration began. We first worked on models for concurrency and later Christel got interested in probabilistic model checking which was a starting point for her future career. Interestingly, for this work she adapted a concept that was dealt with years before in my advanced course on algorithms, i.e. augmenting paths for network flow algorithms. This transfer of ideas demonstrates once more her talents and clarity of mind. It was a great pleasure working with you, Christel, Happy Birthday!

Mila

Foreword by Thomas A. Henzinger

I am glad for this opportunity to join the line of congratulants for Christel Baier's 60th birthday celebration. Christel has been a pioneering figure in German informatics over the past three decades. At the end of the last century, more so than elsewhere, the German academic landscape in system verification was dominated by deductive approaches such as automated and interactive theorem proving and term rewriting. Christel was one of the first to popularize model checking in Germany. At the same time, internationally she was one of the first to bring model checking to probabilistic systems, which embody a perfect target for the method: theoretically challenging and practically important. Thirdly, the textbook of Baier and Katoen on model checking has educated more than a generation of young researchers and remains unmatched in clarity and scope after almost twenty years. So much for the facts. The reason I find Christel's work even more appealing than these impressive facts suggest is perhaps a matter of personal taste. It is her choice of problems: while many others focus on particular syntactic formalisms, Christel has repeatedly identified and solved essential problems that apply to entire semantic classes of formalisms, such as continuous-time Markov chains. I am sure that this will cause her work to be referenced for a long time to come. Happy birthday, Christel, and I wish you many more productive years!

Tom

Foreword by Prakash Panangaden

Few things in a research career are as enjoyable as celebrating a long friendship with a fellow researcher. I met Christel at a workshop in the Netherlands in 1992 and somehow we bonded right away. I had never heard of her and did not suspect that she was going to be the star that she later became. At the time she was a Ph.D. student working with Mila Majster-Cederbaum in Mannheim. I think at that first workshop she was quite quiet, at least during the day. If I remember correctly, we spoke in French because at that point her French was better than her English. In the evenings, however, she was a spark plug of fun and sociability. I will never forget being dragged out onto the dance floor by her where I did a very good illustration of randomness in action. It was the first of many fun evenings that we spent at conferences.

Of course, her main claim to fame is her outstanding contributions to verification, model checking and temporal logics. Her textbook with Joost-Pieter Katoen, another famous researcher whom I am delighted to count among my friends, is the absolute best book on the topic. It is impossible to summarize the nearly 200 great papers she has written or co-authored; a glance at her Google Scholar page reveals very impressive bibliometrics indices. But I have valued a few great papers more than quantity and she has produced outstanding papers that have had a great impact.

The work that has influenced me the most is the work on probabilistic logics and probabilistic verification. That is a topic close to my own interests. She had some classic papers on model checking continuous-time Markov chains which inspired my own work with Josée Desharnais on a logical characterization of bisimulation for CTMCs. Those papers are over 20 years old now, but still extremely relevant and important.

I do not want to give the impression that her important work was all done 20 years ago. What is amazing is how she has managed to sustain a high level of top quality research over a 30 year span and shows no signs of slowing down. A glance at her DBLP page showed me papers from the last 12 months that are at the centre of the subject. She has diversified her interests over the years, for example, she has a recent paper on positivity hardness of decision problems for MDPs which I was unaware of. She has joined forces with Joël Ouaknine, Ben Worrell and their collaborators and written important papers on linear dynamical systems. The list of different things she has done is nearly endless.

Why did we never collaborate? Certainly we had occasions to do so. I even found an old list of "things to do together" which she and I prepared after she visited me in Montreal in the early 2000's. It is just chance I suppose, but whether we were direct collaborators or not we were certainly very close in interest and had a high degree of mutual respect.

So let me end with "Alles Gute zum Geburtstag" which, unless Google Translate is playing games with me, is Happy Birthday Christel!

Prakash

Contents

The Optimal Strategy to Meet the Deadline: Ask Christel!

Joost-Pieter Katoen[1,2]([✉])(iD)

[1] Software Modelling and Verification, RWTH Aachen University, Aachen, Germany
katoen@cs.rwth-aachen.de
[2] Formal Methods and Tools, University of Twente, Enschede, The Netherlands

Abstract. Do we want to meet deadlines? Most probably, yes. Several of Christel Baier's celebrated scientific contributions are concerned with verifying time-bounded reachability queries on continuous-time Markov models. The central question is to compute the (extremal) probability to meet the deadline. This note provides a historical perspective on Christel's contributions and addresses the (practical and theoretical) relevance of these results.

Keywords: bisimilarity · continuous-time Markov chain · continuous-time Markov decision process · rewards · temporal logic · time-bounded reachability probabilities · uniformisation · Volterra integral

1 A Pub in Birmingham

Birmingham, UK, 1998. Christel and I participated in a meeting of the ARC project "Stochastic Modelling and Verification" funded by the British Council. The project involved amongst others LFCS Edinburgh (Jane Hillston's group), the University of Birmingham (Marta Kwiatkowska's group), as well as the universities we were employed: Mannheim (Mila Majster-Cederbaum's group) and Erlangen-Nürnberg (Ulrich Herzog's group), respectively. Those days, Christel was intensively working on model checking of Markov decision processes (MDPs) with her influential contributions on fairness, symbolic model checking, and bisimilarity, to mention a few. After finishing a PhD in Twente on true concurrent models for real-time and probabilistic systems, I had started early 1997 as post-doctoral researcher in the research group of Ulrich Herzog—a specialist on classical performance evaluation such as queueing networks—and just developed a course on model checking. The Birmingham meeting was fruitful and very informative. Hot topics were process algebras such as PEPA [23] and TIPP [18] to describe continuous-time Markov chains in a modular and compositional fashion and the model checking of discrete-time probabilistic models such as MDPs and Markov chains.

N. Bertrand et al. (Eds.): Christel Baier Festschrift, LNCS 15760, pp. 1–14, 2026.
https://doi.org/10.1007/978-3-031-97439-7_1

Context. To understand the context, let me shortly describe the state of affairs in probabilistic model checking around that time. Those days, the verification of probabilistic models was completely focused on *discrete-time* Markov models, such as finite Markov chains and MDPs. This included the seminal works on reachability and ω-regular properties by renowned researchers such as Moshe Vardi (and Pierre Wolper), Amir Pnueli (with Hart and Sharir, as well as with Lenore Zuck), and the duo Costas Courcoubetis and Mihailis Yannakakis. This linear-time setting was extended by Hans Hansson and Bengt Jonsson in 1989 [19]. They came up with a probabilistic version of the branching-time temporal logic CTL for Markov chains. The key was to replace the universal and existential path-quantifiers by a probabilistic quantifier and to introduce a bounded until-operator. Their probabilistic CTL enabled to incorporate thresholds on probabilities of path-formulae in a flexible and nested manner. Their polynomial-time model-checking algorithm gave practical perspective and initiated a first prototypical implementation of a PCTL model checker by Lars Fredlund. Luca de Alfaro and Andrea Bianco extended model checking to PCTL*, whereas Christel with Marta Kwiatkowska considered fairness for MDP model checking [14], and developed with Edmund Clarke et al. the idea to use extensions of binary decision diagrams for PCTL model checking [5]. Another very relevant development was the work by Luca de Alfaro on verifying long-run properties on MDP-like models [1].

Markov chains are conceptually very simple. They are basically finite-state transition systems whose transitions are equipped with probabilities. They are pivotal operational models for randomised algorithms; MDPs (and their extension by Roberto Segala allowing equally labelled outgoing distributions) extend Markov chains with non-determinism and are essential to describe concurrent systems with randomisation such as randomised distributed algorithms at the operational level. These discrete-time probabilistic models address important application areas. Randomised algorithms can be simpler and more efficient than deterministic solutions, and randomisation is a key mechanism to circumvent impossibility results in distributed computing. The ability to verify the correctness of (finite instances of) such algorithms and protocols in an automated manner was a milestone. Despite their conceptual simplicity, reasoning about such algorithms and protocols is very tricky! Any support to verify properties was—and still is—highly relevant.

Continuous-Time Markov Chains. But what about continuous-time models? In Herzog's group in Erlangen, the primary focus was on performance evaluation for which continuous-time Markov chains (CTMCs) are a natural fit. At least half of the aforementioned ARC project worked on process algebras for CTMCs. What about model checking CTMCs? This question seemed natural, but had not been addressed yet. At least as far as we were aware of back then. The CAV 1996 paper by Adnan Aziz *et al.* [4]—to my knowledge the shortest CAV paper ever—proposed a continuous-time version of PCTL and showed its model checking problem for finite CTMCs to be decidable. Their proof heavily depends on transcendental number theory. At the Birmingham meeting, Christel and I were unaware of this paper.

After the project meeting, we spent a weekend in Birmingham and at some point ended up in a pub in the city center. While enjoying some good glasses of English ale—every time again you wonder about its (too warm?) temperature— we ended up discussing about timed reachability probabilities in CTMCs. After all, this could not be so difficult, as CTMCs are Markov chains equipped with just one extra piece of information: a real number associated to each state which is the parameter of an exponential distribution. That is to say, the state residence time in a CTMC is governed by an exponential distribution. So what about timed reachability, the probability to reach a goal state within a real deadline t? As reachability probabilities were the central element for model checking MDPs and Markov chains, timed reachability probabilities could have a similar importance for verifying CTMCs. (And indeed they have.)

How Likely to Reach the Goal? Reachability probabilities can be captured as a unique solution of a linear equation system. One assigns to each state s in the Markov model a real-valued variable x_s that is intended to be the probability to reach the goal state from the state s. This can be done in a backward fashion: if state s is a goal state, x_s is set to one; and otherwise equals the probability to reach a direct successor state t, say, multiplied by x_t, the probability from t to reach the goal state. As the goal state can be reached from state s via different direct successors, one has to take the sum over all direct successors of s. As the Markov chain is finite, there are only finitely many such direct successors rendering the sum to be finite. If one deletes all states from the Markov chain that almost surely never reach the goal state upfront—a simple graph analysis on the graph of the Markov chain suffices—then the reachability probability is the unique solution of this linear equation system. Simple, efficient, and can also be nicely captured as unique least fixed point of a higher-order function enabling e.g., iterative computational methods.

Characterising reachability probabilities in finite Markov chains.
Remove all states that cannot reach the goal set G. Let S be the remaining state space. Introduce for each state s a real-valued variable $x_s \in [0, 1] \cap \mathbb{R}$. Set $x_s = 1$ if $s \in G$, and for all remaining states, i.e., the states not in G but that can reach G with positive probability, we set:

$$x_s = \underbrace{\sum_{t \in S} \mathbf{P}(s, t) \cdot x_t}_{\text{reach } G \text{ via } t}$$

where $\mathbf{P}(s, u)$ denotes the one-step transition probability from state s to u.

How Likely to Reach the Goal on Time? Can we obtain something similar for timed reachability probabilities in CTMCs? For instance, the expected time until reaching a goal state from a state s in a CTMC can be solved by a linear equation system, in fact a small twist of the equation system for reachability probabilities in Markov chains. We started scribbling some equations on our beer mats. The beer mats quickly turned out to be too small. Luckily, Christel—prepared as always—carried a sufficient amount of paper with her. After quite a couple of failed attempts, at some point the idea was to split the problem of characterising timed reachability probabilities—reaching a goal state within d time units from state s—into the probability to reach some direct successor u, say at some exact time point y, multiplied with the probability to reach the goal state from that state u with the remaining time period $d-y$. The probability to go from state s to u at time point y amounts to the density at y of the exponential residence time in state s, the state to-be-left. As for reachability probabilities, one has to consider all possible direct successors of state s, yielding a finite sum over all these successors. All good. But what about y? There are uncountably many possibilities for y as any value in the real interval $[0, d]$ would do. This can be captured by an integral over $[0, d]$. All in all, we equipped each state s with a function x_s—rather than a variable—where $x_s(d)$ for any real d intends to be the probability to reach the goal state starting from s within d time units. The remaining time to reach the goal from state u after staying y time units in s is thus $x_u(d-y)$.

Characterising timed reachability probabilities in finite CTMCs.
Remove all states that cannot reach the goal set G. Let S be the remaining state set. Introduce for each state s, a function $x_s : \mathbb{R}_{\geq 0} \to [0, 1]$. Set $x_s(d) = 1$ for all $d \in \mathbb{R}$ if $s \in G$, and for all remaining states, i.e., the states not in G but that can reach G with positive probability, we set:

$$x_s(d) = \int_0^d \sum_{u \in S} \underbrace{\mathbf{R}(s, u) \cdot e^{-r(s) \cdot y}}_{\substack{\text{probability to go from } s \\ \text{to state } u \text{ at time } y}} \cdot \underbrace{x_u(d-y)}_{\substack{\text{prob. to reach } G \\ \text{within } d-y \text{ from } u}} dy$$

where $r(s)$ is the exit rate of state s and $\mathbf{R}(s, u) = \mathbf{P}(s, u) \cdot r(s)$ denotes the rate of going from state s to state u.

What a fruitful pub visit! Back in Germany, we worked out all the details and formalised all this by a unique least fixed point characterisation of timed reachability probabilities in finite CTMCs. Mission accomplished. Whereas infinite horizon until-properties could be characterised as unique least fixed points of a linear equation system, time-bounded until-probabilities were unique solutions of a Volterra integral equation system. Using Christel's background on using symbolic data structures for probabilistic model checking, we also proposed a generalisation of multi-terminal binary decision diagrams to enable symbolic numerical integration.

How Likely to be in a Goal State on the Long Run? Mission completed, we thought. The fixed-point characterisation could nicely be used as mathematical building block to verify a PCTL-like logic where the discrete time step-bounded until is replaced by a real-valued time-bounded variant. This allows to express that the probability to reach a goal state while avoiding certain bad states on the way meets a certain threshold. Such real-time bounded until-operator was not new; real-time until-modalities existed already in timed CTL, a logic for timed automata. And for the stochastic setting by Adnan Aziz *et al.*, though they even allowed for nested timed-until modalities which in our view was unnecessarily complicated. We wanted to keep things simple though, and just considered timed-until as in timed CTL. Back home, Holger Hermanns convinced us that expressing timed reachability probabilities—the natural pendant to transient probabilities—is nice but is not the holy grail. What about addressing the most relevant measures in performance and dependability evaluation: stationary (aka: long-run) distributions? An excellent idea to also add this to our logic. In fact, Luca de Alfaro considered a similar idea about a year earlier using a kind of observer automata for MDP-like processes. We thus added a long-run operator to the logic and showed that it could be algorithmically tackled using a combination of computing reachability probabilities of terminal strongly connected components (SCCs that once entered can almost surely never be left) combined with computing stationary distributions within such SCCs (which can be computed by solving a linear equation system in size linearly proportional to the size of the SCC). We submitted the results to CONCUR.

2   Eindhoven—Aachen—Chicago—Geneva

2.1   No One Cares once the Deadline is over

We presented the paper at CONCUR 1999 in Eindhoven [12], the Netherlands. Moshe Vardi asked us about the numerical accuracy (rounding errors etc.), while Luca de Alfaro congratulated us to our results. We had however no idea on how to solve these Volterra integral equation systems in an efficient and numerically stable manner! This is also witnessed by the final (handwritten) slide of the CONCUR talk, cf. Fig. 1. We naively experimented with Simpson and Romberg integration. No success. Extremely slow, and numerically instable—even obtaining values far beyond one as probabilities. After I presented the CONCUR results at a seminar in Aachen[1], Boudewijn Haverkort observed that once a goal state

---

[1] During this visit to Aachen, Wolfgang Thomas confessed that he was using my model checking script that I wrote in Erlangen for his course in Aachen. He encouraged me to write a book on this topic. This tremendous challenge I took up with Christel who around the same time had a syllabus (in German) on model checking. Around 2002–2003 we decided to jointly write a book, which in 2008 finally was published [11]. Meeting the deadline was a serious issue for this endeavour. I remember postponing the imposed deadline by the publisher (MIT Press) at least twice. And I have to confess that even the published version was not ready; e.g., Chap. 10 on probabilistic model checking in our repository is called ch13 (sic).

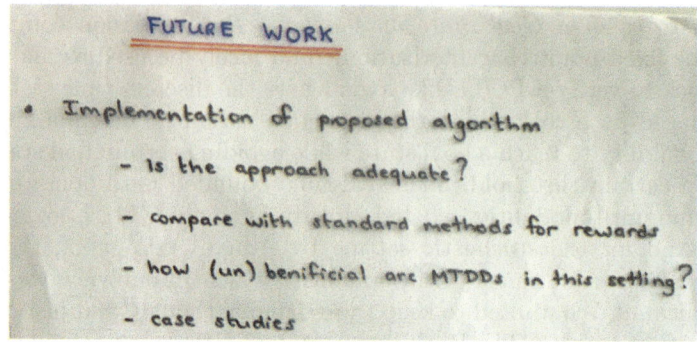

Fig. 1. Future work as mentioned in the slides of CONCUR 1999

is successfully reached, it does not matter what happens thereafter. A simple but perfectly true insight. Goal states can thus be made absorbing, replacing all their outgoing transitions with a self-loop with probability one[2]. One simply waits until the deadline d, and then measures the total probability mass of being in a goal state. If a run arrives earlier than d in the goal state, it remains there forever.

This simple insight enabled us to reduce computing timed reachability probabilities—indeed, the unique solutions of a Volterra integral equation system—by computing the probability to be *exactly* after d time units in a goal state. These so-called transient probabilities are mathematically described as solutions of a linear differential equation system, see the shaded box.

Characterising transient probabilities in finite CTMCs.
The transient probability vector of a CTMC on state set s_1 through s_n at time point $d \in \mathbb{R}_{\geq 0}$, denoted $\underline{p}(d) = (p_{s_1}(d), \ldots, p_{s_n}(d))$, is defined by:

$$\underbrace{\underline{p}'(d)}_{derivative} = \underline{p}(d) \cdot (\mathbf{R} - \mathbf{r}) \quad \text{given} \quad \underline{p}(0)$$

where $\mathbf{r}$ is the $n \times n$-diagonal matrix of the exit rates $r(s_i)$.
Solving this equation using standard analysis techniques yields:

$$\underline{p}(d) = \underline{p}(0) \cdot e^{(\mathbf{R} - \mathbf{r}) \cdot d}$$

Thus, transient probabilities are given as a matrix exponential. This poses a serious problem, as solving matrix exponentials is a notoriously difficult problem. Especially if the matrix such as $\mathbf{R} - \mathbf{r}$ contains negative as well as positive values that can be arbitrarily large (and small). Computing the matrix exponential is

---

[2] In fact, a similar trick was used by Hansson and Jonsson [20, pp. 521] to treat step-bounded reachability probabilities in discrete-time Markov chains. We only realised this much later.

described in detail in the paper by Cleve Moler[3] and Charles van Loan from 1987 entitled "Nineteen dubious ways to compute the exponential of a matrix". The abstract of this paper [28] says it all:

"In practice, consideration of computational stability and efficiency indicates that some of the methods are preferable to others, but that none are completely satisfactory."

25 years later, the authors reported on advances in the field, but the conclusion is still the same [29].

Luckily, the matrix exponential for transient probabilities of CTMCs can be numerically approximated efficiently and in a numerically stable manner. The key is to use a technique from the early 1950s: uniformisation, aka: Jensen's method [24][4]. The beauty of uniformisation is its simplicity. One takes the largest exit rate, r, say in the CTMC at hand. One then accelerates all states by setting their exit rate to r. To compensate for this acceleration, the outgoing transition probabilities of each state are scaled by r^{-1}, and the probability of self-loops is increased by the missing transition probability. Thus,the frequency of transitions was fixed to r^{-1} and the self-loops avoided to leave states too early. Uniformisation in fact yields a CTMC in which all states have the same exit rate, r. As a result, the problematic matrix exponential $e^{(\mathbf{R}-\mathbf{r})\cdot d}$ reduces to $e^{r\cdot d\cdot\mathbf{P}'}$ (multiplied with the constant $e^{-r\cdot d}$). Still a matrix exponential but now of $\mathbf{P}'$, the transition probability matrix of the uniformised CTMC. As this matrix is stochastic, approximating its exponential can be done efficiently and numerically stable.

We presented the reduction of computing timed reachability probabilities to transient probabilities at the CAV 2000 conference in Chicago [6]. Its realisation in a prototypical software tool built by the Ph.D. student Joachim Meyer-Kayser in Erlangen (jointly supervised by Markus Siegle) showed promising results. The first CTMC model checker, called the Erlangen-Twente Markov Chain Checker, with logo $E \vdash M\cdot C^2$, resulted, and was presented at TACAS 2000 in Berlin.

How Likely to Reach a Goal State Within the Budget? Now things started to evolve rapidly with the gang-of-four, see Fig. 2. We discovered that similar techniques as for timed reachability in CTMCs could be used to consider cost-bounded probabilities in CTMCs with rewards. Here, the question was to determine the probability to reach a goal state within a given budget. Costs can model e.g., power consumption, the number opf steps, etc. This culminated in an ICALP paper [7] in 2000 (held in Geneva). We did a small tour in the US to present our results to research groups in performance evaluation and to NASA, and we gave a tutorial at the IFIP Performance conference in Rome, 2002.

Another relevant result was to show that formulas in our timed version of PCTL were preserved by probabilistically bisimilar states. This connection to probabilistic bisimilarity—the coarsest lumpable partition of the CTMC— enabled us to minimise CTMCs prior to carrying out the necessary numerical

---

[3] Moler is the developer of MATLAB and co-founder of MathWorks.

[4] Unfortunately, Wikipedia contributes this technique to Grassmann, 1977.

Christel Baier
University of Bonn

Boudewijn Haverkort
RWTH Aachen University

Holger Hermanns
University of Twente

Joost-Pieter Katoen
University of Twente

Fig. 2. The gang of four around the year 2000

calculations for model checking[5]. We also showed that almost surely, paths in a CTMC are non-Zeno. That is to say, the accumulated delay along any infinite timed path through a CTMC diverges with probability one. This means that Zeno paths, paths in which the total time converges occur, but with probability zero. This fact simplifies the analysis. We submitted a collection of our results to Journal of the ACM. Unsuccessful. In 2003, our journal paper on the model checking of CTMCs was accepted in IEEE Transactions of Software Engineering [8]. Later it turned out that the paper belongs to the top-100 of most cited papers (among 70,000 papers) in software engineering [17]. Quite an achievement for a paper that does not directly has something to do with software engineering....

3   The Starbucks-in-San-Francisco Paper

What about non-determinism in CTMCs? The rationale behind this question was not only the natural pendant to model checking of MDPs. Whereas reachability probabilities in discrete-time Markov chains can be obtained by solving a linear equation system, maximal, or dually, minimal reachability probabilities in MDPs can be obtained by solving a linear inequation system. This also results in a polynomial-time verification procedure for finite MDPs. As a side result, memoryless deterministic policies for MDPs exist that either maximise or minimise the reachability probabilities. How about continuous-time MDPs? Another

---

[5] With Verena Wolf, we later extended these results substantially by considering weak and strong versions of bisimilarity and simulation relations on Markov chains, and investigating logical fragments of (continuous-time) PCTL that characterise these behavioural comparison relations [13]. Elementary ingredients of this work emerged during a joint meeting in Saarbrücken. As a matter of fact, this meeting would have a rather decisive character, as during that meeting I was invited for a job interview at RWTH Aachen. And from December 2004 on I was (and still am) a full professor at that university.

important reason to consider a merger of CTMCs with non-determinism was that parallel composition of CTMCs—a key operator in process algebras—naturally gives rise to non-determinism [15]. This is nicely reflected in Hermanns' interactive Markov chains [22] that were later equipped with discrete probabilistic branching [16]. If complex real-time stochastic systems can be described in a modular way using non-determinism, why can't such models be verified yet?

After some failed attempts, we came up with a greedy algorithm in the summer of 2003. We decided to submit to TACAS 2004, and a full draft with all technical results was—exceptionally—ready a couple of weeks before the submission deadline. Early October 2003, Holger Hermanns and I, together with David N. Jansen, were in California to attend a UML (sic, yes) conference and pay a visit to Mariëlle Stoelinga, those days a post-doctoral researcher at Santa Cruz. After a short visit to the UML conference, we drove along Highway One from SF to Santa Cruz. As the sunny weather was extremely appealing, half way we decided to take a swim in the Pacific. Pure ignorance. Due to a gulf stream from Alaska, the water is simply freezing. We barely made it until our knees, and then decided to go out. David, the only wise person who did not go into the sea, must have found the entire situation quite hilarious. The next day, we visited the University of Santa Cruz, and gave a talk about our greedy approach for CTMDPs. Luca de Alfaro posed a question. In fact, it was more a remark than a question. At first sight, it looked like an innocent remark, but the more we thought about it, we realised that there could be a flaw in our approach! After Mariëlle confronted us with the nicest dessert in Santa Cruz—Chocolate Madness[6]—next day we drove back to SF, fully disillusioned.

Back in SF, we contacted Christel and Boudewijn and started discussing about Luca's remark. And indeed, we were able to construct a counterexample to our approach. What was the problem? Our greedy approach was based on lifting the concept of uniformisation, the key ingredient to computing transient probabilities in CTMCs, to CTMDPs. As mentioned before, uniformisation equalises the exit rates of states by the fastest state, and compensates the acceleration of states by increasing the self-loop probabilities. But in case of non-determinism, it means that a policy in a state can count the number of self-loops taken, and thus can obtain a good guess of the amount of time already spent in that state. This knowledge can be utilised in optimising its decisions. As this information however is not at the disposal of the policies in the original, i.e., non-uniformised CTMDP, it means that after uniformisation, policies are more powerful. As the optimal policies for maximal (or minimal) timed reachability in CTMDPs are the ones that know precisely how much time is spent so far [27], it meant that uniformisation influences the optimality—the uniformised CTMDP admits better policies than the original model.

There was no way to fix this while keeping the greedy algorithmic approach. We were at the point to not submit the paper. In the absence of any good WiFi-connection, we spent our days at the Starbucks coffee corner close to our hotel

---

[6] Although I love chocolate, this was madness indeed! There was no way to finish this enormous amount of chocolate cake with chocolate sauce and melted chocolate.

in SF. We spent entire days at Starbucks in order to figure out whether a greedy approach could be applicable to an interesting subset of CTMDPs. Clearly our approach works for models in which the exit rate for each state is the same. We found this class however quite restrictive. After plenty of discussions whether we should submit the remaining result, we finally decided to do so. (Holger and I celebrated the submission with an enjoyful dinner in the restaurant The Stinking Rose where all courses are made with lots of garlic.) The paper was accepted at TACAS 2004 in Barcelona [9]. A full version of the paper was published in the journal Theoretical Computer Science in 2005 [10] and interestingly enough some years later we received a certificate for the paper to be one of the top-cited papers in the journal in the period 2005–2010. Several other authors then established approximate algorithms for computing maximal timed reachability probabilities in CTMDPs, primarily based on discretising the time horizon.

Nowadays it is known that timed reachability probabilities in CTMDPs is in fact a hard problem. Whereas the decision problem "Does the probability in a CTMC to reach a goal state within deadline d exceed a given rational probability p?" is decidable by Adnan Aziz *et al.*'s achievement of CAV 1996, recent results by Rupak Majumdar *et al.* indicate that the situation for similar questions on CTMDPs is significantly harder. Decision problems such as "Does the maximal probability in a CTMDP to reach a goal state within deadline d exceed a given rational probability p?" and "Can this maximal probability be achieved by a stationary policy?" have recently been shown [26] to be decidable *provided* Schanuel's conjecture holds. The conjecture by Stephen Schanuel about tran-scedental number theory was first published in 1966 and is still open. Proving this conjecture is an enormous challenge and would imply, amongst others, that the decidable theory of the real numbers—a famous result by Tarski—can be extended with exponential and trigonometric functions while remaining decidable.

4   Impact: 25 Years Later

It is fair to say that the results on timed reachability probabilities have had quite some impact. The algorithms have been implemented in various software tools such as the probabilistic model checkers PRISM and Storm. One of the reasons is that the approach is very scalable. In particular for the resulting approximate algorithm to compute timed reachability probabilities in CTMCs scales up to models of millions of states. This is illustrated in Fig. 3 that plots some run-times (in log-scale) for various CTMCs of varying size (in log-scale). As a result, this approach has been adapted by other communities, most notably the research community on performance evaluation, reliability engineering, and systems biology. In performance evaluation, for instance, various stochastic Petri net tools have adopted CTMC model checking in their catalogue of analysis techniques [2]. On the one hand, the flexibility of expressing various relevant properties such as nested combinations of long-run and transient properties in temporal logic, and on the other hand, the efficiency of the algorithms have led to

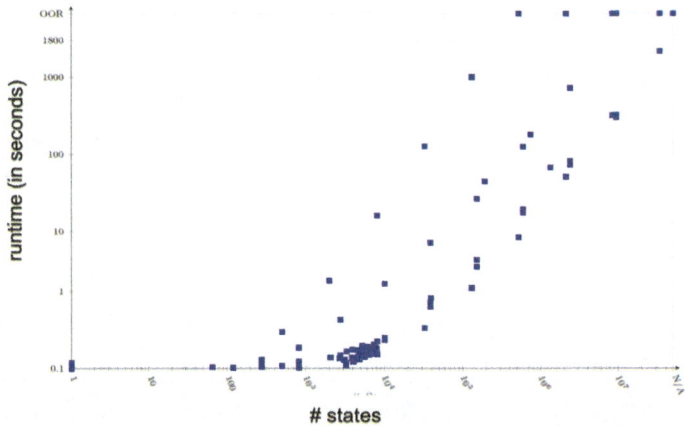

Fig. 3. Run-times (log scale) versus state space size (log size) for computing timed reachability probabilities on CTMC benchmarks from the QVBS benchmark suite using the Storm model checker [21]

this adoption. Also analysis tools for genetic circuits have adapted the approach to compute time to reachability probabilities [25]. And finally, it has quite some impact in reliability engineering, in particular in fault tree analysis. The central question in fault tree analysis is what is the probability that the system fails, or phrased dually, stays operational, within a certain time frame. This question is naturally phrased as a timed reachability query on CTMCs and this has given rise to various tools for (dynamic) fault trees that have adapted CTMC model checking [3, 30]. Experimental evaluations show the superiority of this technique over existing methods in the literature. The commercialised probabilistic risk analysis tool SAFEST uses CTMC model checking as one of its key fault tree analysis techniques [31].

It is fascinating to see that CTMC model checking both received recognition from a theory conference, as well as from a more practical conference in computer science. At the CONCUR 2022 conference in Warsaw, we received the test-of-time award for the 1999 paper presented in Eindhoven[7]. At the conference DSN (Dependable Systems and Networks) in Porto, 2023, we received the Jean-Claude Laprie Award; an award that recognises outstanding papers that have significantly influenced the theory and/or practice of dependable computing. As the jury writes

> "This paper profoundly impacted the research and practice in applying the theoretical formulation of model checking in engineering practice to design and evaluate dependable and mission-critical systems."

---

[7] https://processalgebra.blogspot.com/2022/05/orna-kupfermans-interview-with-christel.html.

5 Epilogue

In this note, I've tried to give a personal reflection on the work that we (Holger, Boudewijn, and myself) carried out together with Christel about 25 years ago and equipped it with some of the (many) anecdotes. It has been a great journey! It is great to see that our collaboration still continues. Let's continue this for many many more years.

Christel Baier   Boudewijn Haverkort   Holger Hermanns   Joost-Pieter Katoen
TU Dresden   University of Twente   Saarland University   RWTH Aachen University

Fig. 4. The gang of four, 25 years later

References

1. de Alfaro, L.: How to specify and verify the long-run average behavior of probabilistic systems. In: LICS, pp. 454–465. IEEE Computer Society (1998)
2. Amparore, E.G., Beccuti, M., Donatelli, S.: (Stochastic) model checking in Great-SPN. In: Ciardo, G., Kindler, E. (eds.) PETRI NETS 2014. LNCS, vol. 8489, pp. 354–363. Springer, Cham (2014). https://doi.org/10.1007/978-3-319-07734-5_19
3. Arnold, F., Belinfante, A., Van der Berg, F., Guck, D., Stoelinga, M.: DFTCalc: a tool for efficient fault tree analysis. In: Bitsch, F., Guiochet, J., Kaâniche, M. (eds.) SAFECOMP 2013. LNCS, vol. 8153, pp. 293–301. Springer, Heidelberg (2013). https://doi.org/10.1007/978-3-642-40793-2_27
4. Aziz, A., Sanwal, K., Singhal, V., Brayton, R.: Verifying continuous time Markov chains. In: Alur, R., Henzinger, T.A. (eds.) CAV 1996. LNCS, vol. 1102, pp. 269–276. Springer, Heidelberg (1996). https://doi.org/10.1007/3-540-61474-5_75
5. Baier, C., Clarke, E.M., Hartonas-Garmhausen, V., Kwiatkowska, M., Ryan, M.: Symbolic model checking for probabilistic processes. In: Degano, P., Gorrieri, R., Marchetti-Spaccamela, A. (eds.) ICALP 1997. LNCS, vol. 1256, pp. 430–440. Springer, Heidelberg (1997). https://doi.org/10.1007/3-540-63165-8_199
6. Baier, C., Haverkort, B., Hermanns, H., Katoen, J.-P.: Model checking continuous-time markov chains by transient analysis. In: Emerson, E.A., Sistla, A.P. (eds.) CAV 2000. LNCS, vol. 1855, pp. 358–372. Springer, Heidelberg (2000). https://doi.org/10.1007/10722167_28
7. Baier, C., Haverkort, B., Hermanns, H., Katoen, J.-P.: On the logical characterisation of performability properties. In: Montanari, U., Rolim, J.D.P., Welzl, E. (eds.) ICALP 2000. LNCS, vol. 1853, pp. 780–792. Springer, Heidelberg (2000). https://doi.org/10.1007/3-540-45022-X_65

8. Baier, C., Haverkort, B.R., Hermanns, H., Katoen, J.: Model-checking algorithms for continuous-time Markov chains. IEEE Trans. Softw. Eng. **29**(6), 524–541 (2003)
9. Baier, C., Haverkort, B., Hermanns, H., Katoen, J.-P.: Efficient computation of time-bounded reachability probabilities in uniform continuous-time markov decision processes. In: Jensen, K., Podelski, A. (eds.) TACAS 2004. LNCS, vol. 2988, pp. 61–76. Springer, Heidelberg (2004). https://doi.org/10.1007/978-3-540-24730-2_5
10. Baier, C., Hermanns, H., Katoen, J., Haverkort, B.R.: Efficient computation of time-bounded reachability probabilities in uniform continuous-time Markov decision processes. Theor. Comput. Sci. **345**(1), 2–26 (2005)
11. Baier, C., Katoen, J.: Principles of Model Checking. MIT Press, Cambridge (2008)
12. Baier, C., Katoen, J.-P., Hermanns, H.: Approximative symbolic model checking of continuous-time markov chains. In: Baeten, J.C.M., Mauw, S. (eds.) CONCUR 1999. LNCS, vol. 1664, pp. 146–161. Springer, Heidelberg (1999). https://doi.org/10.1007/3-540-48320-9_12
13. Baier, C., Katoen, J., Hermanns, H., Wolf, V.: Comparative branching-time semantics for Markov chains. Inf. Comput. **200**(2), 149–214 (2005)
14. Baier, C., Kwiatkowska, M.Z.: Model checking for a probabilistic branching time logic with fairness. Distrib. Comput. **11**(3), 125–155 (1998)
15. Brinksma, E., Hermanns, H.: Process algebra and Markov chains. In: Brinksma, E., Hermanns, H., Katoen, J.-P. (eds.) EEF School 2000. LNCS, vol. 2090, pp. 183–231. Springer, Heidelberg (2001). https://doi.org/10.1007/3-540-44667-2_5
16. Eisentraut, C., Hermanns, H., Zhang, L.: On probabilistic automata in continuous time. In: LICS, pp. 342–351. IEEE Computer Society (2010)
17. Garousi, V., Fernandes, J.M.: Highly-cited papers in software engineering: the top-100. Inf. Softw. Technol. **71**, 108–128 (2016)
18. Götz, N., Herzog, U., Rettelbach, M.: TIPP - introduction and application to protocol performance analysis. In: FBT, pp. 105–125. K. G. Saur Verlag (1992)
19. Hansson, H., Jonsson, B.: A framework for reasoning about time and reliability. In: RTSS, pp. 102–111. IEEE Computer Society (1989)
20. Hansson, H., Jonsson, B.: A logic for reasoning about time and reliability. Formal Aspects Comput. **6**(5), 512–535 (1994)
21. Hensel, C., Junges, S., Katoen, J., Quatmann, T., Volk, M.: The probabilistic model checker Storm. Int. J. Softw. Tools Technol. Transf. **24**(4), 589–610 (2022)
22. Hermanns, H., Katoen, J.-P.: The how and why of interactive Markov chains. In: de Boer, F.S., Bonsangue, M.M., Hallerstede, S., Leuschel, M. (eds.) FMCO 2009. LNCS, vol. 6286, pp. 311–337. Springer, Heidelberg (2010). https://doi.org/10.1007/978-3-642-17071-3_16
23. Hillston, J.: A compositional approach to performance modelling. Ph.D. thesis, University of Edinburgh, UK (1994)
24. Jensen, A.: Markov chains as an aid in the study of Markov processes. Skand. Aktuarietidskrift. **3**, 87–91 (1953)
25. Madsen, C., Zhang, Z., Roehner, N., Winstead, C., Myers, C.J.: Stochastic model checking of genetic circuits. ACM J. Emerg. Technol. Comput. Syst. **11**(3), 23:1–23:21 (2014)
26. Majumdar, R., Salamati, M., Soudjani, S.: On decidability of time-bounded reachability in CTMDPs. In: ICALP. LIPIcs, vol. 168, pp. 133:1–133:19. Schloss Dagstuhl - Leibniz-Zentrum für Informatik (2020)
27. Miller, B.: Finite state continuous time Markov decision process with a finite planning horizon. SIAM J. Control **6**(2), 266–279 (1968)

28. Moler, C.B., Van Loan, C.: Nineteen dubious ways to compute the exponential of a matrix. SIAM Rev. **20**(4), 801–836 (2003)
29. Moler, C.B., Van Loan, C.: Nineteen dubious ways to compute the exponential of a matrix, twenty-five years later. SIAM Rev. **45**(1), 3–49 (2003)
30. Volk, M., Junges, S., Katoen, J.: Fast dynamic fault tree analysis by model checking techniques. IEEE Trans. Ind. Inf. **14**(1), 370–379 (2018)
31. Volk, M., Sher, F., Katoen, J., Stoelinga, M.: SAFEST: fault tree analysis via probabilistic model checking. In: RAMS, pp. 1–7. IEEE (2024)

Nondeterminism in Interactive Markov Chains, with Application to the Erlangen Mainframe

Hubert Garavel[1] and Holger Hermanns[2]

[1] Univ. Grenoble Alpes, INRIA, CNRS, Grenoble INP, LIG, F-38000 Grenoble, France
[2] Saarland University, Saarland Informatics Campus, Saarbrücken, Germany
hubert.garavel@inria.fr, hermanns@cs.uni-saarland.de

Abstract. To formally describe and numerically analyse the performance of systems consisting of concurrent stochastic processes, there exist two orthogonal approaches: "compound" models, each transition of which carries both an event (as in process algebra) and a rate (as in exponential distributions), and Interactive Markov Chains, each transition of which carries either an event or a rate. We investigate the subtle differences between both approaches, with a particular focus on the presence of nondeterminism in Interactive Markov Chains, and the ways nondeterminism can be either handled (resulting in Markov automata) or eliminated (resulting in Continuous-Time Markov Chains). We apply these ideas to the Erlangen mainframe, a challenging problem that exhibits the limitations of classical queueing theory by featuring multiple processors, job queues of different priorities, parallel composition with multiway synchronisation between two or more concurrent processes, and aperiodic failure/repair events.

1 Introduction

The present article was written in honour of Christel Baier and included in a collective *Festschrift* book offered to her on the occasion of her 60th birthday.

The work presented here builds upon the solid foundations established by Christel Baier for modelling and analysing complex systems that combine functional behaviour and quantitative aspects. To model such systems, a difficult challenge lies in the proper integration of process calculi and (discrete-time or continuous-time) Markov chains. Mainstream process calculi are inherently nondeterministic (e.g., due to their choice operators and to the interleaving semantics of their parallel composition operators), whereas the concept of nondeterminism is absent in Markov chains. Thus, it is not straightforward to define conservative extensions of a process calculi with Markov chains. A beautiful, though not sufficiently well known approach can be found in Christel Baier's habilitation thesis [1, Chap. 4], which provides a blueprint, by moving to Markov decision processes, for incorporating discrete time in process calculi.

The work presented here faces another challenge relating to the efficient verification of large quantitative models. Christel Baier also pioneered the development of verification algorithms for Markovian models. On the discrete-time

© The Author(s), under exclusive license to Springer Nature Switzerland AG 2026
N. Bertrand et al. (Eds.): Christel Baier Festschrift, LNCS 15760, pp. 15–69, 2026.
https://doi.org/10.1007/978-3-031-97439-7_2

side, she contributed substantially to model-checking algorithms for Markov decision processes and discrete-time Markov chains [1, Chap. 9–10]. On the continuous-time side, which is at the core of our work, Christel Baier (together with Holger Hermanns, Joost-Pieter Katoen and, later, Boudewijn Haverkort) designed major model-checking algorithms for continuous-time Markov chains and continuous-time Markov decision processes. Many of her papers are acknowledged as reference publications, e.g., [4], which received a CONCUR Test-of-Time Award (2022), [2], which received the IFIP Jean-Claude Laprie Award in Dependable Computing (2022), or [3], which is a most-cited TCS paper.

These results have been transferred to other continuous-time models, such as interactive Markov chains and Markov automata, and implemented in modern verification and performance-evaluation tools. The work presented here substantially reuses these results, while exploring less known facets of the modelling and analysis of stochastic systems. Specifically, the present article addresses three related issues:

1. Two types of state-transition models have been proposed for the formal description of continuous-time stochastic systems, in which delays are governed by exponential distributions of known rate parameters. In the first type, each transition consists of an event and a rate glued together, while in the second type, each transition carries either an event or a rate. Both models are theoretically different but have been proposed to describe the same classes of systems. This raises compatibility issues (to which extent are both models interchangeable, in the sense that any of them could be used for a given system?), as well as choice issues (which model should be used preferably?).
2. The next issues concern the possibility of automated conversions between both types of models. Specifically, we define a translation function that converts models of the first type into models of the second type. We study how faithful this translation is with respect to crucial properties such as deadlocks, determinism, and the preservation of steady-state and transient probabilities.
3. Finally, we study the issues related to the presence of nondeterminism in the models of the second type produced by our translation function. We analyse the reasons why such nondeterminism appears and discuss how to cope with it or to eliminate it using successive transformations of the model.

We investigate these issues on a concrete case study, the "Erlangen mainframe", for which models of the first type already exist, and for which we develop a new model of the second type. This example is concise but involved enough to raise all major issues that may arise with continuous-time stochastic systems.

The present article is organised as follows. Section 2 introduces the models of the first type ("compound models") and the models of the second type ("interactive Markov chains"); it defines a translation function between both types of models and compares them with respect to essential properties; it also addresses related questions about state spaces, such as the proper access to information contained in states and transitions. Section 3 introduces the Erlangen mainframe, its already existing formal models, and the quantitative properties to be measured on this system. Section 4 presents a new model, based on interactive Markov chains, for the Erlangen mainframe and shows how this model must be extended to express the quantitative properties of interest. Section 5 reports about the (nondeterministic) state spaces generated from this formal model and the various correctness checks performed during and after generation. Section 6 presents the results of the quantitative analyses of these nondeterministic state spaces using two state-of-the-art tools for Markov automata. Section 7 suggests an alternative approach in which nondeterminism is gradually removed from the model by using two types of transformations that preserve either the atomicity or the chronology of sequences of event/rate transitions. Finally, Section 8 gives concluding remarks and suggests directions for future work. The set of model files and scripts mentioned in the present article is publicly available on Zenodo[3].

2   Events-and-Rates versus Event-or-Rates

2.1   Events-and-Rates: Compound Models

A first approach to formally describe the functional behaviour and quantitative performance of concurrent systems with stochastic time is to adopt *compound models*. These are state-transition models, the transitions of which are pairs (a, λ), where a is an *event*, possibly synchronised with other events of processes executing concurrently, and where λ is a *rate* that defines a delay exponentially distributed with parameter λ (mean: $1/\lambda$). Compound models are also called *integrated models* elsewhere, e.g., in [5].

The theory of compound models has been explored throughout various process calculi, among which MTIPP [25] [39], PEPA [42], and EMPA [6]. The essential difference lies in the semantics chosen for the synchronisation of two compound transitions (a, λ) and (a, μ). MTIPP defines the result of this synchronisation as a transition with the product of rates $(a, \lambda\mu)$. PEPA computes the maximum of mean delays while incorporating the individual synchronisation capacities of processes. Finally, EMPA forbids this type of synchronisation and requires one process to impose the rate, e.g., (a, λ), while the other process(es) must accept any rate, written $(a, \mathbb{1})$, where $\mathbb{1}$ denotes a neutral element semantically different from the constant 1.0.

These ideas have been implemented in many dedicated software tools (compilers, model checkers, etc.), a few of which are actively maintained. It was also shown recently [20, Sect. 4] that compound models obeying the EMPA restriction on synchronisation (which we call the "*one-to-many*" property) can easily be described in LNT [13] and, by extrapolation, in any value-passing concurrent language that supports CSP-like rendezvous synchronisation, even without specific provisions for modelling stochastic aspects, provided that the compiler for this language preserves multiple identical transitions (i.e., transitions are stored in a multiset, not a set).

---

[3] https://doi.org/10.5281/zenodo.15243150

The compound models given below as examples satisfy the one-to-many property. They are written in a small process calculus, using only a few operators having their usual CSP/LOTOS/LNT semantics (see [18] for details): "(a, λ)" denotes an *active* compound transition that imposes rate λ; "$(a, \mathbb{1})$" denotes a *passive* compound transition that accepts any rate; "$P^{\bullet}$" denotes the infinite repetition of process P; "$P\,; Q$" denotes the sequential composition of process P followed by process Q; "$P \square Q$" denotes the nondeterministic choice between processes P and Q; "$P \parallel Q$" denotes the parallel composition without synchronisation (i.e., full interleaving) of processes P and Q; "$P \parallel_E Q$" the parallel composition of processes P and Q with synchronisation on the set E of events. Operator ";" has higher syntactic precedence than all other binary operators.

Table 1 summarizes the semantics of this small process calculus. In this table, "$\checkmark$" denotes a special event indicating the termination; "ϵ" denotes the process that terminates immediately; C denotes a compound event that is either (a, λ) or $(a, \mathbb{1})$; the (overloaded) predicate $C \in E$ is true iff C is (a, λ) or $(a, \mathbb{1})$ and $a \in E$; L denotes an event that is either (a, λ), or $(a, \mathbb{1})$, or $\checkmark$; the predicate $sync(L, L_1, L_2, E)$ is true iff $L = L_1 = L_2 = \checkmark$, or $L = L_1 = L_2 = (a, \mathbb{1})$ and $a \in E$, or $L = L_1 = (a, \lambda)$ and $L_2 = (a, \mathbb{1})$ and $a \in E$, or $L = L_2 = (a, \lambda)$ and $L_1 = (a, \mathbb{1})$ and $a \in E$ — thus, any violation of the one-to-many property results in a local deadlock, as it is impossible to synchronise (a, λ) with (a, μ). The operator $P \parallel Q$ is defined as a shorthand for $P \parallel_{\varnothing} Q$.

$$
\frac{}{\epsilon \xrightarrow{\checkmark}} \qquad \frac{}{C \xrightarrow{C} \epsilon} \qquad \frac{P \xrightarrow{C} P'}{P^{\bullet} \xrightarrow{C} P'\,; P^{\bullet}} \qquad \frac{P \xrightarrow{C} P'}{P\,; Q \xrightarrow{C} P'\,; Q} \qquad \frac{P \xrightarrow{\checkmark} \quad Q \xrightarrow{L} Q'}{P\,; Q \xrightarrow{L} Q'}
$$

$$
\frac{P \xrightarrow{L} P'}{P \square Q \xrightarrow{L} P'} \qquad \frac{Q \xrightarrow{L} Q'}{P \square Q \xrightarrow{L} Q'} \qquad \frac{P \xrightarrow{C} P' \quad C \notin E}{P \parallel_E Q \xrightarrow{C} P' \parallel_E Q} \qquad \frac{Q \xrightarrow{C} Q' \quad C \notin E}{P \parallel_E Q \xrightarrow{C} P \parallel_E Q'}
$$

$$
\frac{P \xrightarrow{L_1} P' \quad Q \xrightarrow{L_2} Q' \quad sync(L, L_1, L2, E)}{P \parallel_E Q \xrightarrow{L} P' \parallel_E Q'}
$$

Table 1. Structural Operational Semantics rules for the small "compound" calculus

2.2   Events-or-Rates: Interactive Markov Chains

Compound models are not entirely satisfactory for, at least, two reasons:

- Compound transitions (a, λ) can be introduced in a process calculus, but they tend to modify the intuitive meaning of the choice and parallel composition operators, relative to the original process calculus.
- When modelling "real" systems, compound transitions force the modeller to attach a rate to every event. In practice, the concrete values of rates are not known for all events in a system, and rates are often useful for only a few,

well-chosen events. One can always replace an unknown or irrelevant rate by the neutral rate 𝟙, but this approach is not fully convincing.

For these and other reasons, Holger Hermanns and Joost-Pieter Katoen proposed *Interactive Markov Chains* (IMCs, for short), an alternative model in which events a and rates λ are dissociated transitions [32] [38]. Among other qualities, such dissociated transitions (which are called *orthogonal* in [5]) allow the introduction of rates only where actually needed and deliver conservative extensions of the process calculi they are incorporated to.

IMCs have been equipped with model-checking facilities by adapting and extending analysis tools already available for untimed concurrent systems [15], and also by developing dedicated algorithms for timed reachability [43] [51] [53].

IMCs (expressed in LOTOS, Statemate or, more recently, LNT) have been used in many case studies, including an access control mechanism of ATM networks [32, Chap. 4.6], a plain old telephony system [37] [32, Chap. 6.2], a reliability model for the Hubble space telescope[4] [31], the arbitration mechanism of the SCSI-2 bus protocol[5] [19], a turntable system for drilling products[6] [47], a signalling system for high-speed trains [7], a distributed cache-coherence protocol [14], a multiprocessor data-flow architecture for multimedia streams [16], a comprehensive collection of mutual exclusion protocols for shared-memory systems[7] [49], etc.

From a methodological point of view, the typical use of IMCs for modelling and analysing a system can be decomposed into successive steps:

1. The system under study is formally modelled in a language that supports the concepts of IMCs. There is no language dedicated to IMCs, but they can easily be described in value-passing process calculi, such as LOTOS or LNT. In the formal model, rates are described in the same way as events, but given symbolic names (usually, λ, μ, etc.). A key principle of IMCs is that synchronising rates is forbidden, even if they are equal.
2. From the formal model, a Labelled Transition System (LTS, for short) is generated. Its transitions are labelled either with rates, or with visible events, or with τ, which is the CCS notation for hidden events [50]. Duplicated rate transitions having the same source state, the same target state, and the same label should be preserved (that is, rate transitions are stored in a multiset).
3. On the rate transitions of the LTS, all symbolic rates are replaced by their concrete values (e.g., all rates π are replaced by 3.14).
4. On the event transitions of the LTS, all visible events are renamed to τ, which makes them both invisible and urgent.
5. The LTS is modified using *maximal progress* to enforce the urgency of τ-transitions: any rate transition going out of some state s is removed if there exists a τ-transition also going out of state s.

---

[4] https://cadp.inria.fr/ftp/demos/demo_30
[5] https://cadp.inria.fr/ftp/demos/demo_31
[6] https://cadp.inria.fr/ftp/demos/demo_39
[7] https://cadp.inria.fr/ftp/demos/demo_10

6. The LTS is minimised according to an equivalence relation compatible with the IMC semantics. Throughout the present article, we only consider *stochastic branching bisimulation*, which combines branching bisimulation [24] and lumpability [44]. This minimisation preserves self-loops labelled with τ when they are attached to states having no other outgoing transition (contrary to branching bisimulation). We also call IMC the state-transition model obtained after minimisation.

7. If the IMC contains only rate transitions, it is said to be *deterministic* and falls in the category of Continuous-Time Markov Chains (CTMCs, for short).

8. If the IMC contains at least one τ-transition, it is said to be *nondeterministic* and falls in the category of Markov Automata (MAs, for short).

This methodology is not inflexible. Certain steps could be permuted (e.g., 3 and 4) or merged with other steps (e.g., 5 and 6, or even 2 and 5). More complex scenarios are also possible, especially with *compositional verification* of IMCs [37] [19], which we do not address in the present article.

Like compound models, our IMC examples given below are written in a small LNT-like process calculus having the following operators: "λ" (or any other Greek letter but τ) denotes a rate transition; "a" denotes a visible event; "τ" denotes the hidden event; "$P^{\bullet}$" denotes infinite repetition; "$P\,;Q$" denotes sequential composition; "$P \,\square\, Q$" denotes nondeterministic choice; "$P \parallel Q$" denotes parallel composition without synchronisation; "$P \parallel_E Q$" denotes parallel composition with synchronisation. Again, operator "$;$" has higher syntactic precedence than all other binary operators.

Table 2 summarizes the semantics of this small process calculus. In this table, "$\checkmark$" denotes a special event indicating the termination; "ϵ" denotes the process that terminates immediately; C denotes either a visible event a, or the hidden event τ, or a rate transition λ; L denotes an event that is either C or $\checkmark$; in any parallel operator $\parallel_E$, the synchronisation set E should contain only visible events; yet, our semantics also covers the extended case where E contains rates, which is normally not accepted in the usual IMC theory. Again, the operator $P \parallel Q$ is defined as a shorthand for $P \parallel_\varnothing Q$.

$$
\begin{array}{cccccc}
& & P \xrightarrow{C} P' & P \xrightarrow{C} P' & P \xrightarrow{\checkmark} & Q \xrightarrow{L} Q' \\[4pt]
\hline
\epsilon \xrightarrow{\checkmark} & C \xrightarrow{C} \epsilon & P^{\bullet} \xrightarrow{C} P'\,;P^{\bullet} & P\,;Q \xrightarrow{C} P'\,;Q & P\,;Q \xrightarrow{L} Q'
\end{array}
$$

$$
\begin{array}{cc}
P \xrightarrow{L} P' & Q \xrightarrow{L} Q' \\[4pt]
\hline
P\,\square\,Q \xrightarrow{L} P' & P\,\square\,Q \xrightarrow{L} Q'
\end{array}
\qquad
\begin{array}{cc}
P \xrightarrow{C} P' \quad C \notin E \\[4pt]
\hline
P \parallel_E Q \xrightarrow{C} P' \parallel_E Q
\end{array}
\qquad
\begin{array}{cc}
Q \xrightarrow{C} Q' \quad C \notin E \\[4pt]
\hline
P \parallel_E Q \xrightarrow{C} P \parallel_E Q'
\end{array}
$$

$$
\frac{P \xrightarrow{L} P' \quad Q \xrightarrow{L} Q' \quad L \in E \cup \{\checkmark\}}{P \parallel_E Q \xrightarrow{L} P' \parallel_E Q'}
$$

Table 2. Structural Operational Semantics rules for the small IMC calculus

2.3   Standard Translation from Compound Models to IMCs

Any compound model S that satisfies the one-to-many property can be translated to an IMC denoted $[\![S]\!]$ by applying the rewrite rules given in Tab. 3. Rule (1) is the most significant one: it expresses that any compound transition (a, λ) translates to a sequence of two transitions, in which event a occurs after a random delay of rate λ. Rule (2) expresses that any compound transition $(a, \mathbb{1})$ translates to a single transition a. The remaining rules (3)–(7) state that the translation function $[\![\cdot]\!]$ is a morphism with respect to the repetition, sequence, choice, and parallel operators.

$$
\begin{array}{ll}
[\![(a, \lambda)]\!] = \lambda; a & (1) \\
[\![(a, \mathbb{1})]\!] = a & (2) \\
[\![P^\bullet]\!] = [\![P]\!]^\bullet & (3) \\
[\![P; Q]\!] = [\![P]\!] ; [\![Q]\!] & (4) \\
[\![P \square Q]\!] = [\![P]\!] \square [\![Q]\!] & (5) \\
[\![P \parallel Q]\!] = [\![P]\!] \parallel [\![Q]\!] & (6) \\
[\![P \parallel_E Q]\!] = [\![P]\!] \parallel_E [\![Q]\!] & (7)
\end{array}
$$

Table 3. Rules defining the standard translation from compound models to IMCs

Although Rule (2) looks surprising at first, it is justified by the fact that the fundamental equality $(a, \lambda) \parallel_{\{a\}} (a, \mathbb{1}) = (a, \lambda)$ is preserved by $[\![\cdot]\!]$, as both terms of the equation translate to $(\lambda; a)$.

This translation function is quite similar to the encoding from MTIPP to IML [32] given by Marco Bernardo et al. in [5, Def. 5.1], with two differences concerning rules (1) and (2): (i) MTIPP uses action prefix, whereas our calculus models use sequential composition; (ii) MTIPP defines the synchronisation of rates as a product of rates, so that $\mathbb{1}$ corresponds to the numeric rate 1.0 in MTIPP; thus, the encoding of [5] does not have rule (2) and uses instead rule (1) with $\lambda = 1.0$.

Despite these differences, we proudly call our function $[\![.]\!]$ the *standard translation* from compound models to IMCs — especially to distinguish it clearly from the other translations and translators presented in Sect. 6.2 below.

Alternative translations could be considered instead. On the one hand, we are not aware of any approach that would translate (a, λ) to a sequence $(a; \lambda)$ — conventionally, λ represents a preliminary delay, which should be spent before event a, rather than after it. On the other hand, one could imagine an approach that would translate (a, λ) to a sequence $(a'; \lambda; a'')$ and $(a, \mathbb{1})$ to a sequence $(a'; a'')$, where the events a' and a'' respectively correspond to the start and the end of the original event a. We have not explored this approach and only rely, throughout the present article, on the standard translation.

2.4   Compound Models vs IMCs

Much could be written about the comparison of compound models and IMCs, although some questions are still open, especially in the presence of synchronisation and nondeterminism. We focus here on three key points:

1. IMCs are theoretically more expressive than compound models because they can express both probabilistic choice and nondeterministic choice, whereas compound models can only express probabilistic choice.
2. There exist results establishing that strong Markovian bisimilarity is preserved for processes that do not contain synchronisation [5]. Precisely, if two MTIPP compound models are bisimilar, their translations to IML by the aforementioned encoding of [5, Def. 5.1] are bisimilar. Conversely, if two IML models are bisimilar, their translations to MTIPP using the reverse encoding of [5, Def. 6.1] are also bisimilar. These results are valuable but cannot be reused directly in the present article, since rule (2) of our standard translation differs from the encoding of [5, Def. 5.1], and since our case study intensively uses synchronisations, as shown below in Sect. 3.2.
3. A crucial issue is the preservation of steady-state and transient probabilities between compound models and IMCs. When strong Markovian bisimilarity is preserved, e.g., under the assumptions of [5], both steady-state and transient probability distributions over states are preserved. However, the situation is not always so propitious.

   If the IMC contains nondeterminism, then one gets [min, max] intervals instead of single probabilities. This is a major difference, which raises a new question: are the single probabilities computed for a compound model S within the bounds of the [min, max] intervals computed for $[\![S]\!]$?

   Without nondeterminism, but in the presence of synchronisations, transient probabilities are not preserved, as shown by the following example.

Example 1. Consider the compound system $S_1 := (a, \lambda) \ \|_{\{a\}} \ (b, \mu); (a, \mathbb{1})$. Because of the synchronisation on event a, the left-hand side operand cannot spend time (i.e., λ) before the right-hand side has fully executed (b, μ). The corresponding CTMC is, thus, the sequence "$\mu; \lambda$". The standard translation of this system is $[\![S_1]\!] = \lambda; a \ \|_{\{a\}} \ \mu; b; a$. In this IMC, the rates λ and μ start elapsing from the initial state and the corresponding CTMC is the diamond "$\lambda \| \mu$". Hence, $[\![\cdot]\!]$ does not preserve transient probabilities.   ∎

The next example (suggested by Marco Bernardo) shows that steady-state probabilities are also not preserved in the presence of synchronisations.

Example 2. Consider the system $S_2 := (a, \lambda)^\bullet \ \|_{\{a\}} \ ((b, \mu); (a, \mathbb{1}))^\bullet$, which is the compound system S_1 of Example 1 modified by adding repetitions to ensure that both concurrent processes never terminate, so as to obtain an ergodic CTMC. The corresponding CTMC has two states and can be expressed as $(\mu; \lambda)^\bullet$. The steady-state probability of being in the initial state of this CTMC is $\lambda/(\lambda + \mu)$. The standard translation of this system is $[\![S_2]\!] =$

$(\lambda; a)^\bullet \parallel_{\{a\}} (\mu; b; a)^\bullet$. The corresponding CTMC has three states and can be expressed as $(\lambda; \mu \square \mu; \lambda)^\bullet$. The steady-state probability of being in the initial state of this CTMC is $\lambda\mu/(\lambda^2 + \mu^2 + \lambda\mu)$. Thus, $[\![\cdot]\!]$ does not preserve steady-state probabilities. ∎

In the next sections, we extend to other kinds of properties this comparison between compound models and their IMCs obtained by standard translation.

2.5   Alternation in IMCs

In compound models, events and rates alternate on all execution paths: every event is preceded by a rate, and vice versa (except in the initial state). In IMCs, this alternation property does not hold in general: rates are only introduced where they are actually needed, so that several events not separated by rates may occur in sequence. It is possible, however, to consider only *strictly alternating* IMCs, whose semantics was shown to coincide with that of Continuous-Time Markov Decision Processes (CTMDPs) [43]. One may wonder whether the IMCs generated from compound models are strictly alternating or not. The answer is negative in the general case, and difficult to predict at intermediate steps:

- The standard translation clearly breaks alternation by removing all $\mathbb{1}$ rates, as each compound transition $(a, \mathbb{1})$ translates to a.
- Parallel composition may restore alternation, as in $(a; b \parallel_{\{a\}} \lambda; a)$, or break it, as in $(\lambda; a \parallel \mu; b)$, where interleaving enables λ and μ to occur in sequence.
- The successive application of event hiding and maximal progress may also restore alternation by removing sequences of rates, as in $(\lambda; a \parallel \mu; b)$, where maximal progress cuts a λ-transition that follows a μ-transition and a μ-transition that follows a λ-transition.
- Finally, stochastic branching bisimulation breaks alternation by removing τ-transitions. If all of them are removed, the minimised IMC is a CTMC with only rate transitions, making the notion of alternation meaningless. If some τ-transitions remain, the minimised IMC is a MA, a model that does not impose any constraint concerning alternation.

2.6   Deadlocks in IMCs

Another point to be mentioned is that the standard translation does not preserve the absence of deadlocks, and may indeed produce IMCs containing "artefact" deadlocks that did not exist in the original compound models.

Example 3. Consider a system $S := P \parallel_{\{a,b\}} Q$ consisting of two concurrent processes P and Q that synchronise on events a and b. Assume that both P and Q are infinitely looping processes that offer a choice between a $\mathbb{1}$ rate and a non-$\mathbb{1}$ rate: $P := ((a, \lambda) \square (b, \mathbb{1}))^\bullet$ and $Q := ((a, \mathbb{1}) \square (b, \mu))^\bullet$. This compound model has no deadlock. Its standard translation to IMCs is $[\![S]\!] = [\![P]\!] \parallel_{\{a,b\}} [\![Q]\!]$, where $[\![P]\!] = (\lambda; a \square b)^\bullet$ and $[\![Q]\!] = (a \square \mu; b)^\bullet$. The LTS of $[\![S]\!]$ contains a deadlock, which is reached when P does a λ-transition and Q does a μ-transition, in any order. This deadlock is not removed later by applying maximal progress and stochastic branching bisimulation. ∎

Such deadlocks can be explained as follows: the standard translation converts any compound transition (a, λ) into a sequence "$\lambda; a$" of two transitions. Following the IMC methodology, the λ-transition should not be synchronised with anything else. It can thus evolve freely (exactly like τ-transitions) and may cause deadlocks if it appears in a context of nondeterministic choice.

It is worth noticing that maximal progress may not introduce deadlocks, as it only cuts rate transitions that are in choice with τ-transitions, leaving such τ-transitions unchanged. Stochastic branching bisimulation does not introduce deadlocks either since, unlike "standard" branching bisimulation, it does not remove τ-loops.

It has been suggested by Pedro d'Argenio that deadlocks could be avoided if all concurrent processes in the compound model would be either *active* (i.e., do not use the rate $\mathbb{1}$) or *passive* (i.e., only use the rate $\mathbb{1}$). The following example illustrates this idea.

Example 4. Consider a system $S' := P' \|_{\{a,b\}} Q'$ where $P' := ((a, \lambda) \,\Box\, (b, \mu))^{\bullet}$ and $Q' := ((a, \mathbb{1}) \,\Box\, (b, \mathbb{1}))^{\bullet}$. This compound model is bisimilar to system S of Example 3. Its standard translation to IMCs is $[\![S']\!] = [\![P']\!] \|_{\{a,b\}} [\![Q']\!]$, where $[\![P']\!] = (\lambda; a \,\Box\, \mu; b)^{\bullet}$ and $[\![Q]\!] = (a \,\Box\, b)^{\bullet}$. The LTS of $[\![S']\!]$ contains no deadlock; this property is preserved by maximal progress and stochastic branching bisimulation. ∎

Two remarks can be drawn from the comparison of these examples. First, bisimilarity is not preserved by the standard translation, as the two compound systems S and S' are bisimilar, while their translations $[\![S]\!]$ and $[\![S']\!]$ are not (the LTS and IMC of $[\![S]\!]$ both have deadlocks, and the IMC contains τ-transitions, which is not the case for $[\![S']\!]$).

Second, these examples suggest a strategy to remove deadlocks by carefully permuting, in the compound model before translation, (a, λ) and $(a, \mathbb{1})$ transitions across concurrent processes, still preserving bisimilarity at the compound-model level. However, such transformations might not be always possible, e.g., when several active processes compete to synchronise with, and impose their rates to, a passive process.

2.7 Nondeterminism in IMCs

As mentioned before, IMCs can genuinely express nondeterminism, if it is an explicit intention of the specifier. But nondeterminism may also arise unexpectedly when translating a compound model to an IMC.

Example 5. Consider the system $S_1 := ((a, \lambda)^{\bullet} \| (a, \mu)^{\bullet}) \|_{\{a\}} ((b, \varphi); (a, \mathbb{1}))^{\bullet}$. Its standard translation is $[\![S_1]\!] = ((\lambda; a)^{\bullet} \| (\mu; a)^{\bullet}) \|_{\{a\}} (\varphi; b; a)^{\bullet}$. In this

example, repetitions are intended to avoid deadlocks. The IMC of $[\![S_1]\!]$ is nondeterministic in the case both delays specified by λ and μ expire before the delay
specified by φ. Indeed, from the initial state of the IMC, there are two sequences
"$(\lambda \parallel \mu); \varphi; (\tau \square \tau)$" leading to a nondeterministic state.

Incidentally, this scenario is referred to as the "lunch with Christel" problem.
Christel invites two friends, with the intention of going for lunch with the one
who arrives first. Her plan is safe, because they both need different exponentially
distributed delays (λ and μ) to join her, so that the probability that they arrive
at the same moment is null. But Christel needs a cigarette and goes to the
smoking area. Smoking also takes an exponentially distributed delay (φ). If it
is too long (especially if φ is much smaller than λ and μ), the two friends may
be already there when she returns from the smoking area, and then, she faces a
choice problem.

There is a similar situation in queueing theory. When a user has the choice
between two waiting queues, the strategy "join the shortest queue" avoids nondeterminism in many cases. But if both queues are equally filled, the choice is
necessarily nondeterministic.   ■

One might argue that Example 5 is peculiar and that its nondeterminism primarily comes from the interleaving between two occurrences of the same event a
on the left-hand side of the $\parallel_{\{a\}}$ operator, meaning that nondeterminism was
already present and intended in the compound model. The next example refutes
this objection.

Example 6. Consider $S_2 := ((a,\lambda)^\bullet \parallel (b,\mu)^\bullet) \parallel_{\{a,b\}} ((c,\varphi);((a,\mathbb{1}) \square (b,\mathbb{1})))^\bullet$.
Its standard translation is $[\![S_2]\!] = ((\lambda;a)^\bullet \parallel (\mu;b)^\bullet) \parallel_{\{a,b\}} (\varphi;c;(a \square b))^\bullet$.
The IMC of $[\![S_2]\!]$ is nondeterministic since, from its initial state, there are two
sequences "$(\lambda \parallel \mu); \varphi; (\tau \square \tau)$" leading to nondeterminism.   ■

In Examples 5 and 6, and in the more complex situations presented in
Sect. 7.5, 7.7, and 7.8 below, the occurrences of nondeterminism arising in IMCs
generated from compound models seem to follow a common scenario. Two processes P and Q execute in parallel, without being synchronised together. Maximal progress ensures the natural alternation of their rates and events. It also
preserves some form of "atomicity", in the sense that a rate λ originating from
a given compound transition (a,λ) is immediately followed by its corresponding event a. But processes P and Q are synchronised on certain events a and
b, respectively, with a third process R. At some point, R blocks both events a
and b, but cannot block their corresponding rate transitions λ and μ — since
rate transitions are never synchronised in classical IMC theory. This breaks the
alternation of rates and events, and also breaks atomicity, as occurrences of λ
and μ are no longer followed by occurrences of a and b, respectively. The system
continues to evolve and, eventually, R unlocks both a and b, which suddenly
become enabled at the same moment, hence causing nondeterminism.

Examples 5 and 6 are simple, but we tried hard to make them even simpler. Doing so, we experimented with dozens of variations and observed that

nondeterminism in IMCs is "fragile", like unstable physical particles or chemical compounds, and tends to disappear in most cases, unless certain precise conditions are met. As a result, it does not occur so often statistically, which may explain why former case studies based on IMCs did not face this issue. Yet, nondeterminism remains a possibility that must be considered.

2.8   Building and Debugging State Spaces

A practical difference between compound models and IMCs lies in their respective state spaces. Because the standard translation replaces each compound transition (a, λ) by a sequence of two transitions "$\lambda; a$", the LTSs generated from IMCs are likely to be much larger (possibly by several orders of magnitude, as shown in Sect. 5.3 below) than those produced from compound models.

Such an increased risk of state explosion has two implications. First, IMCs cannot be used in the absence of robust tools capable of handling large state spaces. Second, to avoid generating huge LTSs that will be considerably reduced later by stochastic branching bisimulation, one should use *compositional verification* techniques [23], which have been advocated quite early for IMCs [34] [37] [19].

Models of stochastic systems, like any concurrent program, may be wrong, e.g., contain unexpected behaviour, unreachable code, deadlocks, etc. Thus, it is often necessary to debug them so as to understand and correct the mistakes. In the case of compound models, one can apply the standard debugging techniques to the LTSs generated from compound models, the transitions of which are labelled with pairs (a, λ). In the case of IMCs, debugging is more difficult, as one must examine models at two different levels:

- The IMCs themselves contain little information, as all the visible transitions have been hidden and, later, sequences of τ-transitions have been compacted by stochastic branching bisimulation. Most transitions are labelled with numeric rates, and of some these rates may correspond to different symbolic rates, or be the sum of several other rates as a consequence of lumpability. There may also be τ-transitions, if the model contains nondeterminism, and self-loop transitions, as explained in Sect. 2.10 below. From such a low-level information, it is often tedious to trace down the correspondence with the high-level formal model from which the IMC was produced.
- The LTSs take place half-way between the high-level formal models and the low-level IMCs. LTSs are easier to understand, since they contain all event transitions, with their visible names, and all rate transitions, with their symbolic rates. However, LTSs may contain large parts of unreachable code eliminated later by maximal progress.

Therefore, debugging of large IMCs is not feasible using visual inspection only. Dedicated tools that extract and combine information from both LTSs and IMCs must be developed for this task — an example is given in Sect. 7.4 below.

2.9   Rate Tagging

IMCs create additional difficulties for expressing and computing throughputs of rate transitions, state probabilities, and throughputs of events. Let us start by the former point.

The throughput of a rate transition λ is the sum, for all states s having an outgoing transition λ, of the probability, multiplied by λ, of being in state s.

Normally, λ is a symbolic rate that corresponds to a situation of interest in the system under study. For instance, in the Hubble space telescope case study [31], the throughput of rate μ expresses the probability of entering the so-called "sleep mode".

Two difficulties may occur: (i) there exist several symbolic rates having the same concrete value as λ, making it difficult to distinguish between λ and these other rates in the IMC; (ii) lumpability may remove some rate transitions λ and create new rate transitions where λ is summed with other rates.

These issues can be avoided as follows: when substituting λ with its concrete value, e.g., 100, one may attach a *tag* to the concrete value to remember that it comes from the replacement of λ. For instance, instead of replacing λ with 100, one may replace it with "a; rate 100" [36]. Notice that "a; rate λ" neither denotes a compound transition (a, λ) nor a sequence of two transition "a; λ"; it denotes a rate transition, in which tag a is merely a decoration.

Then, maximal progress and stochastic branching minimisation are applied. Notice that the presence of tags (i.e., when two rate transitions have different tags, or when one has a tag and the other not) prevents states to be merged by bisimulation and rate transitions to be merged by lumpability.

This *rate tagging* approach has been intensively used in prior applications of IMCs: mutual exclusion[8], Hubble[9], SCSI-2[10], drilling unit[11], etc. It has two limitations: (i) it only applies to rate transitions, but not to event transitions, the throughput of which may also be of interest; (ii) it is not supported by all tools and, thus, not "portable" across different tools.

2.10   State Probes

Another difficulty lies in the expression of state probabilities in IMCs. More generally, this difficulty is faced with all event-based formalisms, e.g., process calculi, where information is not attached to states, but to transitions. In such formalisms, state variables are not visible and, thus, their values cannot be used to characterise sets of states, the probabilities of which is of interest for computing steady-state or transient quantitative properties.

The solution to this problem is well-known: it consists in exporting state-based information (e.g., the value of state variables) through additional transitions (called *state probes*) introduced in the model. Such extra-transitions usually

---

[8] https://cadp.inria.fr/ftp/demos/demo_10/demo.svl (search for "rate_")
[9] https://cadp.inria.fr/ftp/demos/demo_30/demo.svl (search for "rate_")
[10] https://cadp.inria.fr/ftp/demos/demo_31/demo.svl (search for "rate_")
[11] https://cadp.inria.fr/ftp/demos/demo_39/demo.svl (search for "rate_")

have the form of *self-loops* attached to certain states, the information relevant to these states being given by the transition labels. The concept of probes is not specific to IMCs: probes are also used in compound models, as well as in non-quantitative models of concurrent systems.

Traditionally, in the latter case (e.g., models written in untimed process calculi), any probe should respect three "neutrality" conditions: (i) it is a visible transition, i.e., not renamed to τ; (ii) it is not synchronised with any other transition; and (iii) it only displays information present in the state it is attached to, but does not modify this state, e.g., by assigning state variables.

Hence, the introduction of probes in a model does not increase the number of states, but only the number of transitions. Moreover, probes are preserved when minimising the model for strong or branching bisimulation.

In the case of IMCs, however, the situation is less simple. One may wonder about the nature of probes: are they event transitions or rate transitions?

- If probes are event transitions (and assuming they remain visible, rather than being hidden), they are preserved by maximal progress (since only rate transitions are cut by τ-transitions), but they may prevent stochastic branching minimisation from merging certain states, potentially increasing the numbers of states and τ-transitions in the resulting IMCs. A frequent situation is the case of a transition $s_1 \xrightarrow{\tau} s_2$, where state s_1 has only a self-loop probe and where state s_2 has only outgoing rate transitions. If probes are events, minimisation will neither merge states s_1 and s_2, nor remove the τ-transition, thus leaving nondeterminism in the IMC.

- If probes are rate transitions (i.e., if a probe with label z is transformed into "z; $\mathtt{rate}$ λ" by rate tagging), they may be cut by maximal progress if they are in choice with a τ-transition. In the previous example $s_1 \xrightarrow{\tau} s_2$, the probe attached to s_1 is cut by the τ-transition, allowing minimisation to merge s_1 and s_2 and to remove the τ-transition. This is not systematic: because of concurrency and synchronisations, many probes are not deleted by maximal progress after visible transitions have been hidden. Yet, cutting probes may be annoying, as they convey relevant information about states (e.g., being in a critical section, etc.). However, state s_1 here is the source state of a τ-transition, which is considered to be immediate; thus, the probability of s_1 should be zero, so that this state plays no role in calculations of state probabilities and event throughputs.

In the remainder of the present article, we consider probes as rate transitions: each probe z is changed to "z; $\mathtt{rate}$ λ" before applying maximal progress and branching stochastic minimisation. All the experiments reported below are based on this assumption.

Notice that adding such probes does not alter the state probabilities, because self-loops, which correspond to diagonal elements, are discarded when building the generator matrix of a CTMC. We give to λ the concrete value 1.0, so that the sum of state probabilities for all the states s having the same self-loop z can be computed as the throughput of "z; $\mathtt{rate}$ 1.0".

The appropriate placement of state probes in IMCs obtained by standard translation of compound models is discussed in Sect. 4.4 below.

2.11 Event Probes

One sometimes needs to compute the throughput of an event transition a, rather than the throughput of a rate transition as in Sect. 2.9. This is difficult in IMCs, since all visible transitions are hidden and later compacted by minimisation, making it impossible to recognise transitions labelled with a. To address this issue:

- A first option would be not to hide the event a if one needs to compute its throughput, but this would prevent minimisation to be done completely. Consequently, we discard this approach.
- A second option would be to identify, if it exists, some rate λ closely related to the event a of interest, so that the throughput of a can be expressed in terms of the throughput of λ. This is not always easy, nor even possible but, if the IMC has been generated by standard translation from a compound model, one may consider choosing the rate λ associated to a in the compound transitions (a, λ).
- A third option would be to modify the LTS and the IMC by adding self-loop probes z (hereafter called *event probes*) to well-chosen states, so that the throughput of event a can be expressed in terms of the throughput of z.

A combination of the second and third options is presented in Sect. 4.5 below, so as to compute, in an IMC generated by standard translation, the throughput of a transition (a, λ) present in the original compound model.

3 The Erlangen Mainframe

3.1 Origin and History of the Problem

The mainframe example was designed in the mid-1990s by the IMMD-7 research team at the University of Erlangen. The IMMD-7 group leader, Ulrich Herzog, had identified three major issues in the design of modern distributed systems: (i) the manageability problems due to increasing complexity; (ii) the fundamental uncertainty in timing and state caused by randomly occurring transmission errors, leading to varying traffic intensities and communication delays; and (iii) the issues of partitioning and allocating tasks of proper granularity on computational resources for optimal performance and dependability. He was convinced that neither classical queueing theory, nor the informal and ad-hoc extensions to queueing networks proposed at that time, were sufficient to handle complex systems and address the resulting challenges.

The publications made by IMMD-7 in the 1990s, e.g. [40] [25] [35], provide pioneering ideas about formal methods and tools, incremental and compositional design, validation, and maintenance, with a particular focus on process-algebraic extensions for discrete- and continuous-time — a topic to which Christel Baier also greatly contributed during her PhD thesis. These ideas were illustrated with many interesting case studies, among which the Erlangen mainframe [41, Sect. 4] [34, Sect. 4] seems to be the oldest example proposed by IMMD-7.

3.2   System Description

This case study represents a multiprocessor computer that is designed to serve two purposes: (i) it has to maintain an important database and, therefore, has to process transactions submitted by a number of *users*, and (ii) it is used for program development and has to provide computing capacity to *programmers* for compiling and testing their programs. In addition, two interesting features, failures and priorities, are present. *Failures* may cause system downtimes, by making the mainframe become unavailable until it is repaired. *Priorities* of two types are built into the system. On the one hand, database users need immediate reaction, so they *explicitly* have priority over the jobs issued by programmers. On the other hand, failures cannot be preempted, which implies that they are neither buffered nor delayed; thus, they *implicitly* have the highest priority and take down the system immediately, until repair.

The description of the Erlangen mainframe is modular and hierarchical. It consists in ten processes that execute concurrently and synchronise together using n-party ("multiway") rendezvous, i.e., simultaneous events involving n processes, where n can be as large as 6. The various interactions between these processes are shown in Fig. 1.

The Erlangen mainframe is a compound model, as each event is associated with a rate. Table 4 lists the nine possible events and their corresponding rates.

event	synchronisation pattern	rates
c	3-party rendezvous	$\varphi = 0.00334$
prog_job	3-party rendezvous	$\lambda_1 = 0.01667$ or $\lambda_2 = 0.16$
user_job	3-party rendezvous	$\mu_1 = 0.033$ or $\mu_2 = 2$
fail	6-party rendezvous	$\delta_1 = 0.00035$ or $\delta_2 = 0.0007$
repair	5-party rendezvous	$\beta = 0.01$
get_prog_job	4-party rendezvous (3 queues, 1 processor)	$\alpha = 48$
get_user_job	3-party rendezvous (2 queues, 1 processor)	$\alpha = 48$
prog_job_ready	no rendezvous (interleaving)	$\xi = 0.3$
user_job_ready	no rendezvous (interleaving)	$\nu = 12$

Table 4. Events, synchronisations, and rates in the Erlangen mainframe

On the topmost level, the Erlangen mainframe can be seen as the parallel composition of three principal modules, namely, the *loads*, the *queues*, and the *processors*:

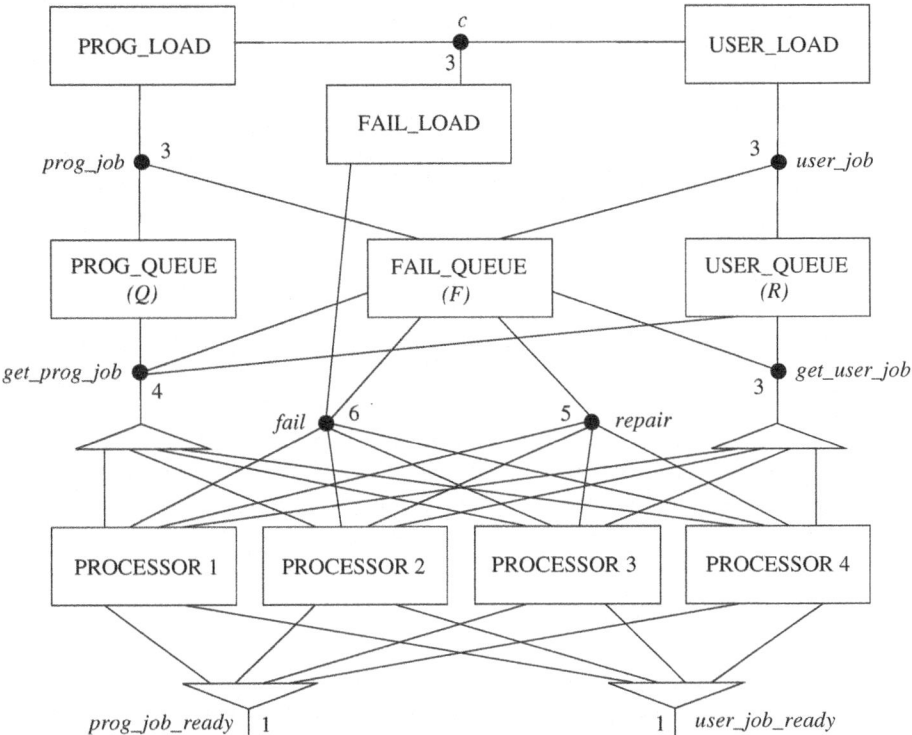

Fig. 1. Architecture of the Erlangen mainframe (black bullets represent n-ary synchronisations and white triangles represent 1-among-4 competitions for synchronisation)

- *Loads*: There are three different arrival streams that put load on the system, namely the database *users*, the *programmers*, and *failures*. Each of these arrival streams produces events according to a given arrival rate. This rate however is not constant, but is instead modelled to vary according to a so-called Markov-Modulated Poisson Process (MMPP) [17]. Each arrival stream has three phases, which come with different arrival rates for the streams: extended working hours (low load: λ_1, μ_1, and δ_1), normal working hours (high load: λ_2, μ_2, and δ_2), and night hours (no activity). The phase changes are governed by yet another rate φ and happen simultaneously across the streams: this is achieved by synchronising the three load processes, which otherwise run independently in parallel, on a common event c that indicates a phase change.
- *Queues*: The mediation between the events arriving from the loads and the processors is handled by three queues. The *prog_queue* buffers the jobs generated by programmers; the *user_queue* buffers the jobs generated by database users; the *fail_queue* reacts to failure events by triggering repairs. The priority mechanisms discussed above are implemented using clever synchronisations,

ensuring that programmer jobs are only served if no user jobs are pending. Both types of jobs are, of course, processed only if the mainframe is not in a failure state.

- *Processors*: The mainframe possesses four identical processors running in parallel. Each processor receives from the queues, at rate α, programmers' jobs and users' jobs (via the *get_prog_job* and *get_user_job* events) and delivers these jobs (via the *prog_job_ready* and *user_job_ready* events) at rates ξ and ν. The processors synchronise altogether on failures and repairs, meaning that failures affect the entire system, halting all processors, until repair. When a failure occurs, the jobs being executed by the processors are silently discarded and not delivered, but the contents of queues are preserved.

The combination of all these features (multiple queues, multiple processors, priorities, failures/repairs, three-phase Markov-Modulated Poisson Processes, and multiway synchronisations) makes the Erlangen mainframe significantly more involved and challenging than the standard models studied by queueing theory.

3.3   Quantitative Properties

The original paper [41, Sect. 4.3] provides eight figures corresponding to various analysis scenarios. These figures are numbered from $\boxed{3}$ to $\boxed{10}$ — we surround their numbers by square boxes to distinguish them from the figure numbers of the present article. These eight figures have been carefully designed to exercise diverse kinds of quantitative properties, as summarised in Tab. 5:

- The first line gives the type of property: "S" for steady-state probabilities (i.e., in the long term, after an equilibrium has been reached) or "T" for "transient" probabilities (i.e., at specific time instants).
- The second line gives the kind of property: "s" for state probabilities (i.e., the probability to be in a certain set of states characterised by a predicate over variables of the mainframe) or "t" for event throughputs.
- The third line gives the parameters that may vary in each figure, e.g., the concrete values of certain rates (β, δ_2, and μ_2), the current size of the *prog_queue* or *user_queue*, the current time instant for transient properties, or the initial phase of the arrival streams.
- The fourth line gives the maximal sizes of the *prog_queue* and the *user_queue*. These values are those used in [20].
- The fifth line gives the number of different state spaces that need to be generated or explored for each figure. These state spaces differ by the values of the corresponding parameters. We call *instances* these variants of the same model, following the terminology of the Model Checking Contest [45]. Certain pairs of figures (namely, Fig. $\boxed{5}$–$\boxed{6}$, Fig. $\boxed{7}$–$\boxed{8}$, and Fig. $\boxed{9}$–$\boxed{10}$) are constructed using the same set of instances; this is why their columns in Tab. 5 are (partially) merged.

- The sixth line gives the number of properties to be evaluated on each instance. The number of plots to be drawn for this figure is the product of the number of instances by the number of properties. The indication "(+1)" for Fig. 3 and 4 means that, although one does not need to compute the probabilities that the queues are empty (these probabilities would be much too large compared to the other values in these figures), one needs to compute the extra properties mentioned in Sect. 6.3.

figure	3	4	5	6	7	8	9	10
type	S	S	S	S	T	T	T	T
kind	s	s	s	t	s	t	s	t
parameters	prog queue, μ_2	user queue, μ_2	β, δ_2		β, time		phase, time	
queue sizes	(40, 4)	(10, 10)	(4, 4)		(4, 4)		(4, 4)	
nb of instances	10	10	36		3		3	
nb of properties per instance	40(+1)	10(+1)	1	1	15	15	16	16

Table 5. Overview of the quantitative properties

3.4 Compound Models of the Erlangen Mainframe

The first formal model of the mainframe [41, Sect. 4] appeared in 1994. It was specified using the TIPP (Timed Processes and Performance Evaluation) process algebra and analysed with the TIPPtool software developed at Erlangen [33], which is no longer maintained. A few variants (also expressed in TIPP) of this model have been explored, e.g. in [34, Sect. 4], where the four processors are replaced by one sequential process.

Thirty years later, two novel models of the Erlangen mainframe have been designed [20]. The first model can be processed by either the PRISM[12] [46] or Storm[13] [30] model checkers; it is written in the PRISM language and its files can be found in the PRISM library of models [14]. The second model was developed using the CADP[15] [22] verification toolbox; it is written in the LNT language and its files are available from the list of CADP demo examples[16].

This work brought a few corrections and simplifications to the original TIPP model of the Erlangen mainframe, e.g., by pointing out that the size of queues can be greatly reduced without noticeably affecting steady-state and transient probabilities. It was shown in [20] that the PRISM and LNT models can closely

---

[12] https://prismmodelchecker.org
[13] https://www.stormchecker.org
[14] https://www.prismmodelchecker.org/files/erlangen
[15] https://cadp.inria.fr
[16] https://cadp.inria.fr/demos.html (see "demo_15")

reproduce the eight Figures ③ to ⑩ originally produced using the TIPP model. It was also shown that the various CTMCs generated by PRISM and CADP are pairwise equivalent modulo strong stochastic bisimulation. For these reasons, we adopt the PRISM and LNT models of [20] as reference specifications to build upon in the present work.

3.5 IMC Models of the Erlangen Mainframe

The three aforementioned specifications of the Erlangen mainframe (in TIPP, PRISM, and LNT languages) all are compound models. So far, there exists no IMC model, and it is tempting to know whether the Erlangen mainframe could be described using IMCs.

To progress this question, the mainframe problem was given in 2022, as a homework exercise, to seven master students of Univ. Grenoble Alpes. Starting from the TIPP specification given in [41], their mission was to develop an LNT specification and reproduce the eight figures ③ to ⑩. The students were offered the possibility to use either compound models or IMCs; all of them chose IMCs — likely because prior lectures had put emphasis on teaching IMC concepts.

Although we were expecting that modelling the Erlangen mainframe using IMCs would be straightforward, it turned out that this was not the case. Most students managed to produce, using the standard translation (see Sect. 2.3), an LNT specification and generate the corresponding IMC using the CADP tools, but they did not succeed in reproducing the eight figures. Indeed, the generated IMCs were subsequently rejected by the CADP tools BCG_STEADY[17] and BCG_TRANSIENT[18] [36] because these IMCs contained (i) unreachable states, (ii) deadlocks, and/or (iii) τ-transitions (i.e., nondeterminism).

Issue (i) has been understood and solved: the presence of unreachable states in the generated IMCs was due to the BCG_MIN[19] tool, which implements branching stochastic minimisation. Indeed, BCG_MIN may cut certain transitions if their rate is close to zero or if they are of low priority according to maximal progress. This might create unreachable states, even if, in the model given as input to BCG_MIN, all states are reachable from the initial state. Although this feature was agreed upon by experts and properly documented in the manual page of BCG_MIN, it remained potentially confusing for novice users. To address this issue, Frédéric Lang extended, in September 2023, the probabilistic and stochastic bisimulation algorithms of BCG_MIN to ensure that the output model is truly minimal, i.e., contains only reachable states.

Concerning issue (ii), it appeared that most deadlocks arose from incorrect parallel compositions, which is no surprise given the complex synchronisation patterns (see Fig. 1 and Tab. 4) between the ten processes of the Erlangen mainframe.

---

[17] https://cadp.inria.fr/man/bcg_steady.html

[18] https://cadp.inria.fr/man/bcg_transient.html

[19] https://cadp.inria.fr/man/bcg_min.html

Concerning issue (iii), a preliminary analysis concluded that the causes of nondeterminism were unclear and that the Erlangen mainframe was substantially more involved than all prior case studies done using IMCs, where nondeterminism would usually vanish after application of maximal progress and stochastic branching bisimulation. This raised rekindled interest in the Erlangen mainframe and prompted a deeper analysis of this problem.

4   Formal Modelling of the Erlangen Mainframe

4.1   Choice of the Modelling Language

The compound models developed for the Erlangen mainframe are either based on process calculi (TIPP and LNT languages) or on automata (PRISM language). For modelling the mainframe using IMCs, we chose LNT, for two reasons: (i) since their inception, IMCs have been closely associated to process calculi, successively TIPP, LOTOS, and LNT; (ii) IMCs in general, and IMC models of the mainframe in particular, may contain nondeterminism and, thus, result in Markov automata, which are not supported by the PRISM tool.

To develop an IMC model in LNT of the mainframe, we started from the compound model in LNT [20] and applied the standard translation rules (see Sect. 2.3). This transformation was straightforward. The main difference between the input and output LNT models is that, in the IMC model, rates are expressed as events whereas, in the compound model, they are expressed as data variables. The transformation thus converted, in all LNT processes, most value parameters into event parameters. Yet, a few value parameters remain in certain processes, e.g., to represent queue sizes or initial phases of Markov-Modulated Poisson Processes.

In the remainder of this section, we detail specific modelling aspects that are not covered by the standard translation.

4.2   Modelling φ-rates

In the Erlangen mainframe, each change of phase (between extended working hours, normal working hours, and night) for the three arrival streams is modelled using a three-party synchronisation on an event c with the rate $\varphi = 0.00334$. This situation can be expressed in various ways. In the PRISM compound model (see [20, Fig. 2]), the three queue processes all propose $(c, \mathbb{1})$ and synchronise together with a fourth additional process that proposes (c, φ) and, thus, imposes rate φ. In the LNT compound model (see [20, Fig. 3]), there is no such additional process: each queue process proposes (c, φ) and the three-way synchronisation of these transitions results in a (c, φ) transition, according to the parallel composition rules of LNT — unlike the parallel composition rules of TIPP, which would give (c, φ^3) in such case (see [20, Sect. 2.3(2)] for a discussion).

For the IMC version of the mainframe, both solutions could have been reused. Yet, we did not retain the PRISM solution, because we preferred not to introduce an extra process, and we did not retain the LNT solution, because it would have required to synchronise the three queue processes on both c and φ, although synchronisation of rates is discouraged in IMCs. Instead, we opted for a third solution, in which the *fail_queue* process proposes $[\![(c, \varphi)]\!]$, i.e., "$\varphi; c$", while the two other processes *prog_queue* and *user_queue* both propose $[\![(c, \mathbb{1})]\!]$, i.e., "c". These three processes are synchronised on c (but not on φ) and a Boolean parameter *delayed* is introduced to distinguish *fail_queue* from *prog_queue* and *user_queue*.

4.3   Modelling α-rates

In all compound models, two different events have the same rate, namely (get_prog_job, α) and (get_user_job, α), with $\alpha = 48$ (see Tab. 4). Given that these events respectively appear in two concurrent processes *prog_queue* and *user_queue*, this entails, in the LTSs obtained after translation, nondeterminism between two sequences of transitions "$\alpha; get\_prog\_job$" and "$\alpha; get\_user\_job$".

Such an "overloading" of rates α is likely to make nondeterminism harder to explain. Therefore, in our LNT model, we distinguish both rates α by giving them different names α_1 (for get_prog_job) and α_2 (for get_user_job), but keeping the same value $\alpha_1 = \alpha_2 = 48$. However, when searching for the shortest sequences leading to nondeterministic states (see Sect. 7.4 below), we give to α_2 a slightly different value (e.g., 49) to unambiguously back-translate rate sequences to event sequences.

4.4   Introducing State Probes

The compound model in LNT of the Erlangen mainframe contains three state probes, each defined in one of the three queue processes [20, Sect. 4]:

- The probe of the *fail_queue* process attaches a self-loop labelled "z_avail" to each global state in which the mainframe is available (i.e., is not in failed state). This state probe is needed for Fig. 5, 7, and 9 only.
- The probe of the *prog_queue* process attaches a self-loop labelled "$z\_prog\_queue(n)$" to each global state in which the *prog_queue* contains n programmer jobs ($n \geq 0$). This state probe is needed for Fig. 3 only.
- Similarly, the probe of the *user_queue* process attaches a self-loop labelled "$z\_user\_queue(n)$" to each global state in which the *user_queue* contains n user jobs ($n \geq 0$). This state probe is needed for Fig. 4 only.

These three probes must also be present in the IMC model of the mainframe. Specifying the z_avail probe in the IMC setting is easy. The LNT code for this probe (see Fig. 2, left) is the same as that of the compound model in LNT.

Specifying the two other probes is more involved. We only consider here the z_prog_queue probe — the case of z_user_queue being fully symmetric. In the compound model, z_prog_queue is straightforward, but the IMC model is different, as the standard translation splits the compound transition (α_1, get_prog_job)

in two successive transitions $s_1 \xrightarrow{\alpha_1} s_2 \xrightarrow{get\_prog\_job} s_3$. If the probe is inserted in the IMC model exactly in the same way as in the compound model, a self-loop "$z\_prog\_queue(n)$" will appear on state s_1 and another self-loop "$z\_prog\_queue(n-1)$" will appear on state s_3. This correctly expresses the fact that event get_prog_job removes one element from the queue.

But what about the intermediate state s_2 created by the IMC semantics? Should it have also a self-loop "$z\_prog\_queue(n)$" or not? There are conflicting arguments on this matter.

- For the self-loop: after the delay of rate α_1, the queue still contains n elements until event get_prog_job occurs, which may take time if all the processors are busy; it is thus logical to have a self-loop "$z\_prog\_queue(n)$" on s_2; moreover, in this particular case, the numeric value of α_1 is so large that little time will be spent in state s_1.
- Against the self-loop: the event get_prog_job will be ultimately hidden, i.e., renamed to τ; if the intermediate state s_2 has no self-loop, then stochastic branching bisimulation will remove this τ-transition and merge both states s_2 and s_3 in one state; this would be impossible if s_2 had a self-loop with a different label than that of s_3, and such a residual τ-transition could introduce nondeterminism as an artefact in the model.

To decide this question, we devised a probabilistic criterion that the correct modelling solution should satisfy and we experimented both approaches using Storm. The results (see Sect. 6.3 below) convinced us that a self-loop should be put on state s_2.

The last difficulty was to specify this self-loop properly in LNT, taking into account that this probe should be enclosed between α_1 and get_prog_job. One must explicitly specify a finite loop, which can be interrupted at any time to let get_prog_job happen (see Fig. 2, right), and not an infinite loop running forever.

4.5   Introducing Event Probes

To reproduce Fig. [6], [8], and [10], one needs to compute the throughput of high-priority jobs (which is displayed on the y-axis of these figures), i.e., the throughput of event get_user_job. In the compound model, this event is visible and its throughput is the sum, for all states s having an outgoing transition (get_user_job, α_2), of the probability, multiplied by α_2, of being in state s.

In the IMC model of the mainframe, all events are hidden (and possibly merged by bisimulation minimisation), which prevents one from referring to get_user_job directly. We decided not to use rate tagging (see Sect. 2.9) on α_2, for two reasons:

- As pointed out in Sect. 2.9, rate tagging may prevent lumpability from merging rate transitions if they carry different tags, or if some of them have tags and others not. But lumpability certainly plays a role in the Erlangen mainframe, especially nearby the four processors. A posteriori, one can indeed

```
process PROG_QUEUE [...] (L: nat) is
    var N: nat in
        N := 0;
        loop
            alt
                -- self-loop on states (s1) and (s3)
                Z_PROG_QUEUE (N)
            []
                only if N < L then
                    PROG_JOB;
                    N := N + 1
                end if
            []
                only if N > 0 then
                    ALPHA1;
                    -- self-loop on state (s2)
                    loop SELF in
                        alt
                            Z_PROG_QUEUE (N)
                        []
                            break SELF
                        end alt
                    end loop;
                    GET_PROG_JOB;
                    N := N - 1
                end if
            end alt
        end loop
    end var
end process
```

```
process FAIL_QUEUE [...] is
    loop
        alt
            -- self-loop
            Z_AVAIL
        []
            FAIL;
            BETA;
            REPAIR
        []
            GET_USER_JOB
        []
            GET_PROG_JOB
        []
            USER_JOB
        []
            PROG_JOB
        end alt
    end loop
end process
```

Fig. 2. LNT model fragments illustrating the placement of state probes

observe that the IMCs generated for the mainframe contain rate transitions 0.6, 0.9, 1.2, 24, 36, and 48 that presumably correspond to the lumped rates 2ξ, 3ξ, 4ξ, 2ν, 3ν, and 4ν (the latter rate might also correspond to α_1 or α_2), where $\xi = 0.3$ and $\nu = 12$ are the rates attached to the output events of the processors. The situation could very well have been similar for the input events *get_user_job* and *get_prog_job* of the processors. Today, we know retrospectively that it is not the case, as the IMCs contain no rates larger than α_1 and α_2 but, in doubt, we adopted a cautious modelling approach that avoids rate tagging.

— Rate tagging is implemented in the various tools of CADP that generate or manipulate CTMCs, but it is not supported by other tools capable of handling MAs. Rather than using this convenient, CADP-specific extension, we opted for event probes, which remain in the standard framework of IMCs (without rate tags) and are more likely to be "portable" across various analysis tools.

We thus decided to use event probes (see Sect. 2.11). The LNT specification of the mainframe was extended with an additional self-loop probe *z_get_user_job* intended to measure the throughput of the hidden event *get_user_job*.

The compound transition (α_2, get_user_job) is split, when translated to IMCs, in two successive transitions $s_1 \xrightarrow{\alpha_2} s_2 \xrightarrow{get\_user\_job} s_3$. Given that a through-put measures the probability of states before a given transition, the self-loop $z\_get\_user\_job$ must appear before event get_user_job, which excludes state s_3. But should this self-loop be attached to state s_1 only, to state s_2 only, or to both states s_1 and s_2? This problem is similar to the one of Sect. 4.4, yet different as it deals with throughputs instead of state probabilities.

At first sight, placing the self-loop $z\_get\_user\_job$ on state s_2 seems to be the most natural option, given that the throughput of get_user_job is defined using the probabilities of the states from which this event is enabled. However, this option should be excluded, since the self-loops attached to s_2 may very well disappear completely from the IMC because of maximal progress.

Example 7. Consider a compound system $S := ((a, \lambda); (b; \mu))^\bullet$ and its standard translation $[\![S]\!] = (\lambda; a; \mu; b)^\bullet$. Any self-loop probe inserted between λ and a, or between μ and b, will be cut by maximal progress after a and b are hidden. ∎

However, the self-loops attached to s_2 may also remain in the global state space after minimisation by stochastic branching bisimulation, and the states to which these self-loops are attached may have non-null probabilities. This was confirmed by Storm on the various IMCs generated for Fig. 6.

Contrary to Sect. 4.4, where state probes had to be attached to both s_1 and s_2, so as to total the probabilities of these states, event probes should be attached to s_1 only, and not to both s_1 and s_2, because they express flows, not stocks.

Example 8. Consider a compound system $S := (a, \lambda)^\bullet \;\|_{\{a\}}\; ((b; \mu); (a, \mathbb{1}))^\bullet$ and its standard translation $[\![S]\!] = (\lambda; a)^\bullet \;\|_{\{a\}}\; (\mu; b; a)^\bullet$. Assume that the goal is to compute, on the IMC, the throughput of the compound transition (a, λ). To do so, three self-loops are inserted in $[\![S]\!]$: probe z is put before λ, probe z' is put between λ and a, and probe z'' is put between b and a. After minimisation, the resulting IMC has three states, four rate transitions, and four self-loops. Probe z has the intended meaning, as it is attached to all (and only those) states having an outgoing rate transition λ. Probe z' is irrelevant, as it is attached to one state having only one outgoing rate transition μ. Probe z'' is irrelevant as well, but for a different reason: it is attached to one state having an outgoing rate transition λ, but not to the initial state, which also has an outgoing rate transition λ; thus, probe z'' does not capture the entire mass of probability needed to compute the throughput of (a, λ). ∎

So, our solution to compute the throughput of event get_user_job is to put a self-loop on state s_1 only, and to multiply the obtained probability by $\alpha_2 = 48$ in order to obtain values comparable with the compound model.

5   State-Space Generation

5.1   Models, Instances, and SVL Scripts

We now indicate how the LNT model of the Erlangen mainframe can be translated to IMCs, and how this is implemented concretely. Actually, there are several LNT models to be considered: the model of Sect. 4 will be hereafter referred to as V1 (for version 1), but there will be other models V2, V3, V4, etc. For each model, each of the eight figures Fig. 3 to 10 requires to generate between 3 and 36 instances (see Tab. 5).

All these models and instances create a notable complexity, with more than 1,800 files containing LNT code, temporal-logic properties, execution logs, probability/throughput results, as well as LTSs, MAs, and CTMCs in multiple formats for the various tools. Consequently, experiments cannot be carried out manually (the risk of mistakes would be too high) and must be properly automated.

To this aim, we developed a verification script written in the SVL language[20] [21]. This script (560 lines, 380 lines if not counting comments and blank lines) reuses and extends the SVL script formerly developed for the compound model of the Erlangen mainframe — see [20, Sect. 5.8] for a detailed presentation of this former script that, starting from the LNT compound model, performs all the steps needed to automatically produce the eight picture files for Fig. 3–10.

The SVL script developed for the IMC approach also starts from a given LNT model, generates the corresponding LTS, hides all visible events but probes, renames labels to replace symbolic rates by their concrete values, and invokes BCG_MIN to apply maximal progress and stochastic branching minimisation. It performs the required iterations on the aforementioned variables and rate parameters and seeks to avoid duplicated calculations by building intermediate models that mix symbolic and concrete rates.

After stochastic branching minimisation, the SVL script checks, using BCG_INFO[21], whether the resulting IMC is deterministic, i.e., free from τ-transitions. If so, the IMC is a CTMC: the script invokes BCG_STEADY or BCG_TRANSIENT to compute probability/throughput results and then converts these numbers into PNG pictures by reusing the eight Gnuplot scripts developed for the compound version of the mainframe. If not, the IMC is a MA: this case will be detailed in Sect. 6 below.

5.2   Sanity Checks

To make sure that our various models and instances of the Erlangen mainframe contain no obvious mistakes, we devised another SVL script that performs sanity checks about deadlocks, alternation, and probes.

Although deadlocks may appear when translating a deadlock-free compound model to an IMC (see Sect. 2.6), we found no deadlocks in the LTSs and IMCs

---

[20] https://cadp.inria.fr/man/svl-lang.html
[21] https://cadp.inria.fr/man/bcg_info.html

generated for the various instances. This is fortunate, given that the Erlangen mainframe does not satisfy the sufficient condition mentioned in Sect. 2.6 for avoiding deadlocks: six out of the ten mainframe processes (namely, *prog_queue*, *user_queue*, and the four processors) are both active and passive, i.e., contain both $\mathbb{1}$ and non-$\mathbb{1}$ rates. This also confirms that the deadlocks reported by some students (see Sect 3.5) were not specific to the Erlangen mainframe itself, but rather caused by synchronisation mistakes.

The IMCs obtained after stochastic branching minimisation are not CT-MDPs, because they do not respect the mandatory alternation between events and rates. Indeed, for each of the eight figures, we checked that at least one instance enables, from its initial state, a sequence of two consecutive rate transitions labelled with $\varphi = 0.00334$. More generally, when the mainframe is in its initial state, one can perform an infinite sequence "$\varphi; c; \varphi; c; \varphi; c...$" that corresponds to the elapse of nights and days while the mainframe remains inactive. When the c events are hidden, the sequence of rates φ remain. This confirms the point of Sect. 2.5: alternation is not necessarily preserved when translating compound models to IMCs.

Concerning probes, the SVL script checks, using XTL[22] [48], that each transition labelled with a probe (namely, *z_avail*, *z_get_user_job*, *z_prog_queue*, and *z_user_queue*) is a self-loop, i.e., has identical source and target states. The SVL script also checks, using BCG_CMP[23], that the LTS obtained from a modified LNT model, in which all the probes have been removed, is strongly bisimilar to the LTS obtained from the original LNT model after deleting all probe transitions from the latter LTS.

5.3   Sizes of State Spaces

Table 6 gives a comparative overview of our state-space generation experiments. It is worth noticing that, in each column, all the instances for a given figure have the same number of states and transitions, as they only differ by the values of the varying parameters.

The first four lines give the sizes of the LTSs and CTMCs generated for the compound model of [20]. The LTSs have not been minimised and the CTMCs (which contain no τ-transitions) have been minimised for stochastic strong bisimulation — essentially to make sure that the various CTMCs generated by CADP, PRISM, and Storm are comparable, i.e., have similar sizes.

The last four lines give the sizes of the LTSs and MAs generated for the IMC model of the Erlangen mainframe. The LTSs have not been minimised and the MAs have been minimised for stochastic branching bisimulation after applying maximal progress and hiding all visible transitions.

These numbers confirm the point of Sect. 2.8: state spaces are much larger with the IMC approach than with compound models. Indeed, the numbers of states and transitions are 85 and 120 times larger for the LTSs, and 7.4 and 7 times larger for the CTMCs or MAs.

---

[22] https://cadp.inria.fr/man/xtl.html
[23] https://cadp.inria.fr/man/bcg_cmp.html

figures	[3]	[4]	[5] and [6]	[7] and [8]	[9] and [10]
nb of states compound/LTS	414,387	238,323	48,195	48,195	48,195
nb of transitions compound/LTS	3,188,415	1,826,679	361,389	361,389	361,389
nb of states compound/CTMC	9,840	5,808	1,200	1,200	1,200
nb of transitions compound/CTMC	55,785	33,069	6,570	6,570	6,570
nb of states IMC/LTS	38,834,935	23,274,055	4,213,159	4,213,159	4,213,159
nb of transitions IMC/LTS	382,382,187	230,112,423	41,065,683	41,065,683	41,065,683
nb of states IMC/MA	74,800	43,708	8,560	8,560	8,560
nb of transitions IMC/MA	387,515	227,243	47,060	47,060	47,060

Table 6. Statistics on state-spaces for the compound and IMC models

5.4   Execution Times

Similarly, the IMC approach takes more time than compound models. For the compound model of [20], generating the eight figures takes 10 min. 35 sec. on a Linux laptop with an Intel core i5 processor and 16 GB RAM. For the IMC approach, generating the eight figures for model V4 (see Sect. 7.8 below) that produces CTMCs and invokes BCG_STEADY and BCG_TRANSIENT (thus, having comparable workload with the compound model) takes nearly five hours on the same laptop.

In practice, execution times could be reduced in three ways: (i) potential optimisations have been identified in some CADP tools, but not implemented yet; (ii) experiments have been done sequentially to give an idea of the total time required, but could easily be run in parallel, given that the calculations needed for the various figures (and even for the multiple instances of a given figure) are largely independent one from the other; (iii) so far, compositional verification [19] [23] has not been used, since execution times, although long, were not prohibitive, but it would surely be effective given the large difference in size between the generated LTSs and the minimised MAs or CTMCs.

6   Nondeterministic Analyses

6.1   Analysis Tools for Markov Automata

All the IMCs generated in Sect. 5 for model V1 of the Erlangen mainframe are not CTMCs, as they contain nondeterministic (i.e., τ-transitions not eliminated by stochastic branching minimisation). Thus, they cannot be analysed using PRISM or the BCG_STEADY and BCG_TRANSIENT tools of CADP. One must use tools that genuinely support Markov automata. To this aim, we selected Modest[24] version v3.1.273 (Oct. 2024) [29] [10] and Storm version 1.9.0 (Nov. 2024). These tools come with various constraints, so that our IMC-based approach must be adapted to fit the possibilities of each tool:

- The LNT model of the Erlangen mainframe extensively uses both parallel composition operators $\|$ and $\|_E$ to describe 1-among-n synchronisations (see the white triangles in Fig. 1). Currently, the native input languages of Modest and Storm only provide a CSP-like $\|$ operator that synchronises processes on their common events. To address this issue, one option would be to combine this CSP-like $\|$ operator with relabellings, so as to express 1-among-n synchronisations, while duplicating certain events in the IMCs to give them unique names — as can be seen in the PRISM compound model of the mainframe[25], in the three queue processes of which the *get_prog_job* and *get_user_job* events are replicated four times, one per processor. Another option would be to use the JANi interchange format [9], which has a $\|_E$ operator and translates automatically in the input languages of Modest and Storm. Eventually, we found it simpler to let CADP expand parallel composition and provide Modest and Storm with explicit state-transition models, from which concurrency has been eliminated.
- It does not seem that Modest and Storm implement the stochastic branching bisimulation needed for the IMC approach — although Storm has an option to quotient the state space using strong or weak bisimulation. It is thus appropriate to rely on the BCG_MIN tool of CADP and give to Modest and Storm models that have been minimised beforehand.
- The property languages of Modest and Storm do not allow to compute event throughputs, which are needed for Fig. 6, 8, and 10. However, in Sect. 4.5, we have shown how to represent event throughputs in terms of event probes. But the property languages of Modest and Storm support neither state probes nor event probes, as they cannot directly express the presence or absence of transitions entering or leaving a state — except perhaps by extending the model with the introduction of rewards [12]. Consequently, we must find a way to encode the concept of probes, so essential in the IMC approach, in terms of state variables, state labels, and state probabilities, which are the concepts understood by Modest and Storm.

6.2   Translators from IMCs to Modest and Storm

Given the large number of models and instances to analyse, the sole solution was to develop translators for automatically converting the explicit-state, minimised IMCs with probes into the input formats supported by Modest and Storm.

Modest accepts the Modest language [28] and the JANi interchange format, which is also supported by many other tools. Storm accepts Markov automata

---

[24] https://www.modestchecker.net
[25] https://www.prismmodelchecker.org/files/erlangen

in various formats: DRN, IMCA/TRA, JANi, PRISM-MA, etc. Faced with such a multiplicity, we were unsure about the right format to choose, especially with respect to the proper way of encoding probes. We finally opted for the Modest language, which is the native input language of Modest, and for PRISM-MA, which is a Storm extension for Markov automata of PRISM's input language we were already familiar with. We felt that both languages would be easy to generate and proofread.

We developed two translators, one for Modest, another for PRISM-MA, which follow the same principles:

- Each generated program contains a single process (or "module" in PRISM-MA terminology), without concurrency nor recursion.
- Assuming that the IMC has N states, a bounded-integer variable s in the range $[0, N-1]$ encodes the current state in the generated program.
- There is also one variable per state probe and one variable per event probe. Such additional *probe variables* store, in the generated program, the information conveyed by the presence or absence of self-loops on IMC states.
- Precisely, the state probe z_avail (resp. the event probe $z\_get\_user\_job$) is translated to a Boolean probe variable $avail$ (resp. get_user_job), which is true in any state iff the corresponding self-loop is attached to this state.
- Similarly, the state probe "$z\_prog\_queue(n)$" (resp. "$z\_user\_queue(n)$") is translated to a bounded-integer probe variable $prog\_queue$ (resp. $user\_queue$) in the range $[-1, L]$, where L is the maximal size of the $prog\_queue$ (resp. $user\_queue$). This probe variable equals $n \geq 0$ in any state iff a corresponding self-loop holding parameter n is attached to this state, or -1 if the state does no have such a self-loop.
- Each rate transition of the IMC is straightforwardly translated to a rate transition in the generated program.
- Each τ-transition (if any) of the IMC is translated to a nondeterministic transition in the generated program.
- Each probe self-loop of the IMC is translated, by its presence or absence, into assignments to the corresponding probe variable. Such a self-loop does not create any transition in the generated program and its rate is not used during the translation.
- In the generated program, each transition starts with a Boolean guard of the form "$s = n$" that checks if the current state is n. Probe variables are not used in these guards and, thus, do not change the number of states or transitions; they only decorate states with extra information.
- Each transition terminates with assignments that modify the value of the current-state variable s and, possibly, of the probe variables. The translation avoids generating useless assignments of the form "$x := v$" if variable x already has value v in the current state.
- With Storm, our translator generates, after the Markov automaton, a list of definitions for *state labels*, which express predicates on the probe variables and are used in the properties to evaluate.

– With Modest, the properties to evaluate are part of the model — whereas, with Storm, the Markov automaton and its properties are put in two separate files. For this reason, our translator for Modest gets an extra command-line argument, which is the name of a file containing the properties, and inserts these properties in the generated Modest model.

Technically, our translators are written in C (230 lines of code each, 180 lines if not counting comments and blank lines) and rely on the BCG_READ[26] programming interface for reading the input IMC, which is stored in a binary file using the BCG format of the CADP toolbox.

Our translators also perform various checks that confirm or extend those of Sect. 5.2, namely: (i) each transition labelled with a probe name is a self-loop; (ii) no state has exactly one outgoing τ-transition; (iii) each state has either no outgoing rate transition and at least one outgoing τ-transition, or no outgoing τ-transition and at least one outgoing rate transition.

6.3   Assumption Checking for State Probes

As discussed above in Sect. 4.4, it is unclear whether, in process *prog_queue*, a self-loop "$z\_prog\_queue(n)$" must be attached or not to the intermediate state s_2 located after rate α_1 and before event *get_prog_job*. The same question holds for process *user_queue* by replacing α_1, *get_prog_job*, and *z_prog_queue* with α_2, *get_user_job*, and *z_user_queue*, respectively.

To decide between compelling arguments for and against these self-loops, we formulate the following criterion that a correct modelling of the Erlangen mainframe should satisfy: any reachable global state should have a self-loop "$z\_prog\_queue(n)$", because the *prog_queue* is active at any time and always contains zero or more elements. Given that this probe is only needed for Fig. 3, which deals with steady-state probabilities, this criterion can be refined and generalised as follows: the long-run probability of being in a global state that does not have a self-loop "$z\_prog\_queue(n)$" should be zero. Obviously, a similar criterion should hold for "$z\_user\_queue(n)$" and Fig. 4.

We modified our Storm translator to generate, in the PRISM-MA models, an additional state label *prog_queue_no_loop* that identifies all states in which the *prog_queue* variable equals -1. We added two Storm formulas that compute the minimal and maximal long-run probabilities for this state label. Similar changes were also done for the *user_queue*.

We used Storm to analyse the Markov automata generated for Fig. 3 and Fig. 4 on two variants of the mainframe model, with and without self-loops on the intermediate states s_2. Notice that the ten instances generated for Fig. 3 (resp. Fig. 4) all give identical results for the present matter.

For the variant with self-loops: there are many states without self-loops (5,088 states without *z_prog_queue* probes, out of 74,800 states for Fig. 3, 2,940 states without *z_user_queue* probes, out of 43,708 states for Fig. 4). Our Storm translator indicates that each of these states without self-loops has no outgoing rate

---

[26] https://cadp.inria.fr/man/bcg_read.html

transition and more than one outgoing τ-transition. Given that τ-transitions are assumed to be immediate, the sojourn time in these states should be null. This is confirmed by Storm, which reports that the minimal and maximal long-run probabilities for these states are (exactly) zero.

For the variant without self-loops: interestingly, the number of τ-transitions in the Markov automata for Fig. [3] and Fig. [4] remains unchanged, compared to their variants with self-loops, thus showing that having self-loops on states s_2 does not necessarily increase nondeterminism. Also, Storm reports much larger numbers of states without self-loops (48,368 states without z_prog_queue probes, out of 74,800 states for Fig. [3], 19,520 states without z_user_queue probes, out of 43,708 states for Fig. [4]). Our Storm translator indicates that each of these states without self-loops has (between 1 and 7) outgoing rate transitions, but no outgoing τ-transition — this may be explained by stochastic branching bisimulation merging the τ-transition arriving to state s_3 with the various rate transitions leaving this state s_3. Consequently, the sojourn time in the states without self-loops cannot be null, since only rate transitions are leaving these states. This is confirmed by Storm, which reports that the long-run probabilities for these states are greater than a minimal value ranging from 0.0002 to 0.001 for the ten instances of Fig. [3], and from 0.0008 to 0.01 for the ten instances of Fig. [4]. These values are all but negligible compared to the other probabilities computed by Storm for these figures.

These results thus indicate that self-loops should be present on the intermediate states s_2 of processes $prog\_queue$ and $user\_queue$.

6.4 Results of Steady-State Analyses

We experimented with Modest and Storm in order to obtain the steady-state probabilities needed for Fig. [3]-[6]. We wrote, in the specific format required by each tool, properties that compute these probabilities and, given that all instances of model V1 contain nondeterminism, we asked each tool to compute both the minimal and maximal long-term probabilities, which had the effect of doubling the numbers of properties given on the last line of Tab. 5.

All the properties to evaluate on a given instance are put in the same file, which speeds up the analysis by ensuring that the instance is parsed only once.

For convenience, all our experiments with Modest and Storm were done in a VirtualBox virtual machine running Debian Linux 12 with two processors and 8 MB RAM, hosted on the same laptop as in Sect. 5.4. Therefore, the execution times reported below should not be taken too strictly, as (i) executing the tools natively on a high-end server would certainly give better numbers, and (ii) the tools, especially Storm, have so many options that we could not try them all, so that we are unsure whether we selected the optimal ones for our experiments.

Experiments with Fig. [3] of Model V1. Modest was found too slow, as it took more than 50 hours to build the state space of the first instance (we halted the experiment). Storm took 100 hours to evaluate the 820 formulas needed for

Fig. [3]. At first sight, the results obtained seem close to those reported by [20] for the compound models, but various observations can be made: (i) in some cases, the "min" probabilities are slightly larger than the corresponding "max" probabilities; (ii) all probabilities that the queue contains a large number L of elements remain close to 5.10^{-7} (the default precision of Storm being 10^{-6}), whereas the probabilities computed by CADP for the compound model in LNT decrease from 10^{-33} to 10^{-38} regularly; (iii) the most meaningful cases seem to be $L = 1$ and $L = 2$, where probabilities are not too small, but we observe that the probabilities for the compound model are, with $L = 1$, always smaller than the corresponding "min" probabilities and, with $L = 2$, always larger than the corresponding "max" probabilities. We consider that these results are too uncertain for drawing valid conclusions.

Experiments with Fig. [4] of Model V1. Modest was found too slow: 18 hours to build the state space of the first instance, and more than 6 hours to evaluate the first formula (we halted the experiment). Storm took 12 hours to evaluate the 220 formulas needed for Fig. [4] and produced results that, at first sight, look similar to those of compound models. Yet, in the case $L = 1$, the probability computed by Storm is twice as large (0.03 on compound models vs 0.05 when using Storm on MAs). Like for Fig. [3], the probabilities computed by Storm for larger values of L stagnate around 5.10^{-7} and some "min" probabilities are larger than the corresponding "max" probabilities. We tried without success to vary the precision of calculations and the LP solver used by Storm. Hence, we consider that these results are not conclusive.

Experiments with Fig. [5] of Model V1. On the first five instances, Modest reported floating-point errors and displayed probabilities that were all equal to 1 or very close to 1, e.g., 0.99999999... (we halted the experiment). Storm took 2 hours to produce the 36 "min" and 36 "max" probabilities needed for Fig. [5] — together with those of Fig. [6]. These results are plausible: (i) the "min" probabilities are always smaller than the corresponding "max" probabilities; (ii) the "min" and "max" probabilities are very close (the most distant pair being 0.818153 vs 0.818168), so that the differences are not likely to be visible on the picture generated by Gnuplot; (iii) interestingly, the probabilities computed for the compound models [20] are not in the ["min", "max"] interval, as one would expect, but are all above "max"; (iv) yet, they are close to "max" except for the lowest values of β, for which discrepancy becomes visible, the most distant pair being 0.818168 vs 0.858825 (absolute difference: 0.0407, relative difference: 4.97%). Figure 3 compares the results obtained on compound models and MAs.

Experiments with Fig. [6] of Model V1. The observations for Fig. [6] are similar to those made for Fig. [5], given that both experiments have been done together. The main changes are: (i) the discrepancy between the "min" and "max" probabilities computed by Storm is larger, the most distant pair being

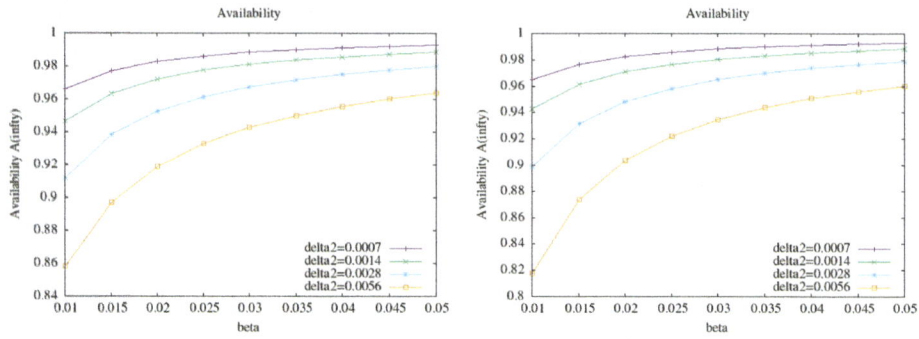

Fig. 3. Figure ⑤ generated by CADP for the compound model (left) and by Storm for the IMC-MA of model V1 (right)

0.440619 vs 0.441169; (ii) the discrepancy between the "max" probabilities and the probabilities computed for the compound models is also larger, the most distant pair being 0.441169 vs 0.474733 (absolute difference: 0.0336, relative difference: 7.61%). Nevertheless, the shapes of the curves remain comparable, as shown by Fig. 4.

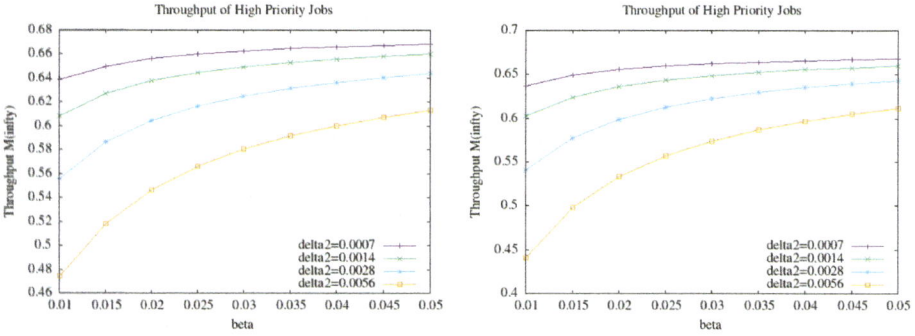

Fig. 4. Figure ⑥ generated by CADP for the compound model (left) and by Storm for the IMC-MA of model V1 (right)

Thus, IMC modelling and Markov-automata tools can be used to describe a complex system such as the Erlangen mainframe and, when the state space is not too large, produce results close to those obtained with compound models.

6.5   Results of Transient Analyses

We experimented with Modest and Storm in order to obtain the transient probabilities needed for Fig. ⑦–⑩. Our experiments were similar to those of Sect. 6.4, but the properties to evaluate were meant to compute state probabilities or event throughputs at particular instants in time $t = 1$, $t = 100$, $t = 200$, ..., and $t = 1500$.

Modest emitted, for each property π, a warning that π was unsupported and skipped it.

Storm warned that it would use the IMCA method rather the Unif+ method and announced, for the first property ($t = 100$) of Fig. [7], that it would perform 106,819,632,945,608 iterations. For the last property ($t = 1500$), the number of iterations was even 225 times larger. Of course, we halted the execution, which had not converged after 8 hours. We then tried to lower the accuracy of results by setting Storm's time-bounded precision option to 0.1: this decreased to 1,068,196,330 the number of iterations planned for $t = 100$, but this reduction was still not sufficient and we had to halt the execution after 18 hours.

Observing that Modest and Storm fail to compute transient "time-point" properties of the form $t = k$ (where k is a constant) when the Markov automaton gets large, and knowing that they better handle transient "time-bounded" properties of the form $t \leq k$, we tried to compute the former using the latter by expressing $p(t = k)$ either as $p(t \leq k) - p(t < k)$ or as $\big(p(t \leq k+\epsilon) - p(t \leq k-\epsilon)\big)/2\epsilon$, but these attempts did not succeed.

6.6   Possible Enhancements of Analysis Tools

The availability of Modest and Storm to analyse Markov automata is, in itself, a great achievement, considering the rare combination of theoretical breakthroughs and skilful development over many years that was needed to produce these tools. Clearly, our experiments with the Erlangen mainframe are pushing these tools to their current limits, but we are confident that these could be overcome in a near future. To this aim, we can make five suggestions:

1. The main limitation arising from our experiments is the current impossibility of obtaining transient time-point probabilities for Fig. [7]–[10]. On the one hand, Storm's early algorithm [53] for computing these probabilities, which is based on discretisation and inherited from the IMCA tool [27], faces problems scaling to large MAs. On the other hand, the Unif+ algorithm [11] [26], which is implemented in both Modest and Storm, scales much better but only supports time-bounded properties. Extending Unif+ to address time-point properties would be desirable. A discussion in [8] (just before Lemma 2) suggests that this could be done by omitting the initial preprocessing step that turns all target states into absorbing states, prior to the main Unif+ computation.

2. Most of the complexity of the translators from IMCs to Modest and Storm (see Sect. 6.2) lies in the conversions of probes to state variables. Such conversions (and their consecutive introduction of additional probe variables in the models) could be avoided if the property language of Modest and the state labels of Storm and PRISM, which, at present, only refer to state variables, would also refer to transitions. We suggest to extend their syntax with

a predicate *enable(a)* that is true for all states having an outgoing transition labelled with *a*. This is an old idea [52], which is easy to implement. This way, the states in which the mainframe is available could be directly characterised as *enable("z_avail")*. One could also introduce predicates, e.g., *enable(z_prog_queue(n) where n < 2)*, but this is not needed for the Erlangen mainframe and such properties may often be expressed in a different manner, e.g., *enable("z_prog_queue_0")* ∨ *enable("z_prog_queue_1")*.

3. Although the various instances of model V1 produced by our translators have a simple, explicit-state structure and are not so large, Modest and Storm are surprisingly slow in parsing them and building their state spaces. Indeed, the time spent grows more than linearly as the number of states increases. On the three sizes of instances shown in Tab. 6 for Fig. 5, 4, and 3 (namely 8,560, 43,708, and 74,800 states — the corresponding numbers of transitions being in a linear ratio, i.e., 5.2–5.5 times larger), Modest takes more than 15 seconds, 18 hours, and 50 hours, respectively, while Storm takes more than 25 seconds, 10 minutes, and 29 minutes, respectively. A similar observation can be made with PRISM on model V4 (see Sect. 7.8), which, given CTMCs having 9,972, 53,640, and 90,396 states, takes more than 10 seconds, 6 minutes, and 17 minutes, respectively. This suggests that the tools are currently not able to recognise explicit-state models and handle them linearly, and that the user should help them to do so. This could easily be implemented by a command-line option or a pragma in the input file to inform the tool that the model, which should contain no parallel composition, is an explicit-state one, with an indication of the particular variable *s* chosen to store the current state.

4. We observed that the time taken by Storm on model V1 to compute each of the 800 "min"/"max" probabilities of Fig. 3 is always larger than 300 seconds per property. For the 200 "min"/"max" probabilities of Fig. 4, this time is always larger than 141 seconds. Similarly, PRISM takes on model V4 at least 21 seconds per property of Fig. 3 and at least 14 seconds per property of Fig. 4. This suggests that these tools just evaluate properties as they come, one after the other, without examining the set of properties as a whole to avoid redundant calculations. However, the properties of Fig. 3 (resp. 4) are all alike: each of them computes the steady-state probability for a given subset of states, the collection of these subsets forming a partition of the entire state space. Therefore, computing the stationary solution vector first (the tools have algorithms for this) would allow to answer all the properties at once. Alternatively, the tools could implement a simple form of memoisation by storing in a cache the stationary solution vector computed for the first property and reusing it for the next ones; this way, evaluating the first property would take time, whereas the evaluation of subsequent properties would be instantaneous.

5. Finally, the output format of the tools should be enhanced. At present, this format (visibly inspired by the output of the PRISM tool) consists in human-readable text, in which the essential results (probabilities and throughputs) come together with a lot of auxiliary information about progress of calcula-

tions, models and properties analysed, size of state spaces, timing, etc. To produce the eight figures of the Erlangen mainframe, 778 probabilities and throughputs must be evaluated (see Tab. 5); this number is even multiplied by two when dealing with Markov automata, since one computes both the minimal and maximal probabilities. To produce the eight figures using Gnuplot, each of these numerical results must be dispatched to a precise file, where it must be inserted in a matrix, at a precise line and column. Such a task must be done by the user, since it is really application-specific and beyond the mission of tools for Markov automata. To do so, we developed five scripts (160 lines of Bourne shell) that make intensive use of Awk. This proved to be a time-consuming activity, since it required to finely parse the output files produced by the tools. We noticed that the formats of these files are not convenient for automated exploitation, since the name μ of the model being analysed, the name π of the property being evaluated, and the numerical result p all appear on different lines, intertwined with other information. Gathering all important information on the same line (e.g., "Result of property π on model μ is p") would enable users to exploit the results faster and more easily.

7   Deterministic Analyses

7.1   Removing Nondeterminism

Considering the results of the two analyses that succeeded using tools for Markov automata, the "min" and "max" probabilities are very close, with a largest difference of 1.5×10^{-5} for Fig. [5] and 5.5×10^{-4} for Fig. [6]. One may wonder whether nondeterminism is really meaningful here, and worth the difficulties it causes, when its impact on the final results seems negligible.

Taking this line of thought further, would it be possible to slightly modify the IMC model in order to eliminate nondeterminism (and, thus, to obtain a CTMC instead of a MA), still preserving the properties needed to build the eight figures?

From a conceptual point of view, the presence of nondeterminism in a formal model, unless it was purposely intended by the specifier, raises questions. In the IMC models of the Erlangen mainframe, nondeterminism was surely not intended, and its causes were totally unclear when we undertook the present study. In such case, one hesitates between two possible explanations. Is nondeterminism simply an artefact of the standard translation (see Sect. 2.7)? Or is it rather the consequence of specification mistakes that were originally present in the compound model, in some latent form that compound semantics did not reveal, but that IMCs, being a more demanding formalism, bring to light?

From a practical point of view, nondeterminism is currently a nuisance for the analysis, so that replacing MAs with CTMCs would offer many advantages: (i) CTMCs deliver exact probabilities while MAs only produce intervals of "min" and "max" probabilities; (ii) there are many more tools for CTMCs than tools for MAs; (iii) algorithms for solving CTMCs are older, simpler, and probably faster than those for MAs.

We conjecture that a model having a nondeterministic IMC can be turned into another model whose IMC is a CTMC, by using step-by-step transformations that progressively eliminate all occurrences of nondeterminism and should, hopefully, preserve the numerical solutions (probabilities and throughputs) of the original model.

We additionally require that each transformation from a model M to a model M' ensures that the LTS of M' is (strictly) included in the LTS of M according to the strong bisimulation preorder, which we denote $M' \sqsubseteq M$. This condition means: (i) that M' does only a part of what M can do, with the expectation that some nondeterministic parts of M are absent from M', and (ii) that everything M' can do is already doable by M, i.e., M' does not introduce novel possibilities.

We now present two such transformations, which we then apply to the Erlangen mainframe, starting from model V1 (see Sect. 5.1) to produce successive models V2, V3, and V4, from which nondeterminism is gradually eliminated.

7.2   Removal of Nondeterminism by Enforcing Atomicity

The presence of nondeterminism in many IMC models, including the Erlangen mainframe, arises from broken atomicity (see Sect. 2.7), i.e., the presence of concurrent sequences "$\lambda; a$" and "$\mu; b$", where λ and μ occur first, then a and b get blocked due to impossible synchronisations with a third process, before being unlocked simultaneously by the third process.

One way of avoiding such nondeterminism between a and b is to relax the synchronisation patterns to prevent a (resp. b) from being both blocked by the third party. We will not use this approach, as it violates our requirement of preserving inclusion between LTSs for the strong bisimulation preorder — indeed, the transformed model would have states in which a (resp. b) is enabled, whereas it is not enabled in the corresponding states of the original model.

Notice that it is not mandatory to address both concurrent sequences "$\lambda; a$" and "$\mu; b$": addressing only one of them might be sufficient, in some cases, to remove nondeterminism.

Another way of avoiding nondeterminism is to delay the occurrence of λ (resp. μ) as long as there is a risk that a (resp. b) becomes blocked afterwards. This requires either that (i) the process(es) executing "$\lambda; a$" (resp. "$\mu; b$") must observe the system globally to predict situations in which a (resp. b) is likely to be blocked, or that (ii) an additional *bouncer* process is added to the system, with the capability of blocking λ (resp. μ) in certain situations and unlocking them at other moments.

Solution (ii) automatically guarantees the preorder inclusion $M' \sqsubseteq M$ if M denotes the original model and if the transformed model M' has the form $M' := M \parallel_L B$, where B is a bouncer that does not contain τ-transitions, and where L is the set of all transition labels in B, including (symbolic) rate transitions. Notice that synchronising concurrent processes on rate transitions violates a tenet of IMCs, but this is a meaningful derogation from this principle, especially when solution (ii) can be expressed equivalently using solution (i), which only requires synchronisations on non-rate transitions — a detailed example is shown in Sect. 7.6 below.

Globally, all these approaches aim at enforcing atomicity, in the sense they try to ensure that either λ and a (resp. μ and b) will occur "together" (i.e., without too long interval in between), or none of them will.

7.3   Removal of Nondeterminism by Enforcing Chronology

In the presence of broken atomicity for two concurrent sequences "$\lambda; a$" and "$\mu; b$", there is an alternative approach to those preserving atomicity. Rather than delaying or blocking λ and/or μ to prevent posterior nondeterminism between a and b, one may let λ and μ happen without restriction, and wait until the point of nondeterminism to resolve it, by preventing the occurrence of either a or b.

Obviously, the choice between a and b cannot be decided using fixed priorities (e.g., $a \gg b$), otherwise it would be unfair. We instead propose the following criterion based on chronology: if λ occurred first, then a will be enabled and b blocked, whereas if μ occurred first, then a will be blocked and b enabled.

This is implemented by adding to the system an additional *resolver* process, which synchronises on the transitions λ, μ, a, and b. This process records the chronology of λ and μ in its internal state and uses this information to decide between a and b.

Example 9. For instance, the model $M := (\lambda; a; z_a^{\bullet} \parallel \mu; b; z_b^{\bullet}) \parallel_{\{a,b\}} \nu; (a \square b)$, where $z_a^{\bullet}$ and $z_b^{\bullet}$ are two self-loop probes, generates an IMC that is nondeterministic due to a choice between a and b. Among various possibilities, an effective resolver for M is $R := \lambda; (a \parallel \mu) \square \mu; (b \parallel \lambda)$. Indeed, the parallel composition $M' := (M \parallel_{\{\lambda,\mu,a,b\}} R)$ generates an IMC that has the same number of states as the IMC of M but one transition less (actually, the two τ-transitions are replaced by one rate transition ν).   ■

Again, if R is a resolver that contains no τ-transition and if L is the set of all transition labels in R, including rate transitions, then defining $M' := M \parallel_L R$ ensures that $M' \sqsubseteq M$. Again, synchronising M and R on rates goes against an established principle of IMCs, but notice that any meaningful resolver should only observe rate transitions and never block them.

7.4   Tools for Understanding Nondeterminism

Explaining the presence of nondeterminism is the dual problem of explaining the presence of deadlocks caused by synchronisation mistakes. Such deadlocks arise when two events are not present as expected at the same instant. Nondeterminism arises when two events are unexpectedly present at the same instant. Thus, the difficulty is similar in both cases, although nondeterminism is perhaps simpler, as it is easier to study the presence of an event rather than its absence.

Yet, in the case of IMCs, an extra difficulty comes from the fact that all visible transitions have been hidden.

Given the difficulty of analysing IMCs (see Sect. 2.8) and the large sizes of IMC instances (see Tab. 6), visual inspection would not be feasible. So, we reused various software components of CADP to develop a *debugging toolchain* that automatically locates and explains nondeterminism in IMCs. This toolchain works in successive steps described as follows:

1. It first generates the LTS and the IMC corresponding to a given instance of a model (see Sect. 5).
2. It identifies, using an XTL script, all *conflict states* in the IMC, i.e., all states having outgoing τ-transitions.
3. It computes, using the BCG_INFO tool, the shortest paths leading to each of these conflict states in the IMC. Each of these paths is a sequence of transitions labelled with concrete rates.
4. It transforms these paths by replacing all concrete rates with symbolic rates, e.,g., 0.00334 by φ, 0.01667 by λ_1, etc. We call *rate sequences* the resulting sequences of symbolic rates, dropping out all information about states.
5. It detects and removes, using the FDUPES[27] tool, duplicated rate sequences, i.e., identical rate sequences leading to different conflict states.
6. It converts, using the EXHIBITOR[28] tool, each rate sequence into one or many corresponding sequences in the LTS, by retrieving and inserting the visible events that have been hidden and then abstracted away by stochastic branching minimisation. We call *full sequences* the resulting sequences, each of which is such that, by keeping only its symbolic rates, one obtains the rate sequence it was generated from.
7. Finally, it orders full sequences by increasing length, with the expectation that a user wanting to remove nondeterminism will first study the shortest sequences, which are easier to replay and analyse.

7.5   Understanding Nondeterminism in Model V1

We applied our debugging toolchain to model V1, which was obtained by applying the standard translation (see Sect. 2.3) to the compound model of the Erlangen mainframe [20]. For our experiments, we chose one instance among those needed to produce Fig. 5 and 6, because these are the smallest instances of model V1 (the other instances differing mostly because of larger queue sizes) and because Storm successfully handles these instances.

The toolchain reports the existence of 552 conflict states. All the 552 rate sequences leading to these states (there are no duplicate sequences) end with a rate transition β, which suggests that the remaining nondeterminism arises from failures and repairs. The two shortest rate sequences (of length 4) are $s_1 := \text{``}\delta_1; \lambda_1; \delta_1; \beta\text{''}$ and $s_2 := \text{``}\delta_1; \mu_1; \delta_1; \beta\text{''}$. The full sequence corresponding

---

[27] https://github.com/adrianlopezroche/fdupes
[28] https://cadp.inria.fr/man/exhibitor.html

to s_1 is "$\delta_1; fail; \lambda_1; \delta_1; \beta; repair; (fail \,\square\, prog\_job)$", of which the full sequence corresponding to s_2 is dual, by replacing λ_1 with μ_1 and $prog\_job$ with $user\_job$.

The scenario of sequence s_1 can be explained as follows: while the mainframe is in its initial state, the $fail\_load$ process triggers a failure (δ_1 followed by $fail$). The $prog\_load$ process initiates a new job (λ_1) but the next event $prog\_job$ is blocked by the $fail\_queue$ process. The $fail\_load$ process then initiates another failure (δ_1), but the next event $fail$ is also blocked by the $fail\_queue$ process. Eventually, the mainframe restarts (β followed by $repair$). The $fail\_queue$ process resumes, which simultaneously unlocks both pending events $prog\_job$ and $fail$. When all events are hidden, this creates a conflict state with two outgoing τ-transitions.

7.6   Model V2 of the Erlangen Mainframe

The analysis of nondeterminism in model V1 shows that the $fail\_load$ process may initiate a new failure by spending rate δ_1 (resp. δ_2) while the mainframe is already failed. This questionable scenario cannot happen in the compound model, where δ_1 and δ_2 are, each, glued with $fail$ and, thus, cannot occur before the next $repair$. This is clearly a broken atomicity issue, caused by the splitting of compound transitions $(fail, \delta_1)$ and $(fail, \delta_2)$.

To address the issue, we tried various modifications of model V1 in order to enforce atomicity (see Sect. 7.2) or chronology (see Sect. 7.3), before concluding that atomicity would be more effective at this point. We thus present two possible modifications of model V1 that, respectively, illustrate the approaches (i) and (ii) mentioned in Sect. 7.2.

Our first approach consists in modifying the $fail\_load$ process so that it does not initiate failures while the mainframe is already failed, which means that, in such case, $fail\_load$ has to wait until the next $repair$ before spending δ_1 or δ_2. Consequently, model V1 must be modified to allow $fail\_load$ to synchronise on event $repair$ — this restores some broken symmetry, considering that $fail$ was involved in six-party synchronisations while $repair$ was involved in only five-party ones (see Fig. 1). Process $fail\_load$ itself must also be deeply modified (see its LNT code on the left of Fig. 5). This process, which already slightly differed from the two other processes $prog\_load$ and $user\_load$ due to the Boolean parameter $delayed$ (see Sect. 4.2), becomes even more different with the introduction of $repair$ as a new event parameter. The behaviour of $fail\_load$ is further modified to ensure that, between $fail$ and $repair$, events c (which express phase changes) are not blocked. Conversely, the process must also accept a potential occurrence of $repair$ between φ and c, not to cause deadlocks. This first approach does not synchronise on rate transitions, thus respecting an IMC principle, but it is involved and error-prone.

Our second approach is much simpler: the $fail\_load$ process of model V1 is kept unchanged, while the mainframe is composed in parallel with a small bouncer (see its LNT code at the top right of Fig. 5) that prevents rates δ_1 and δ_2 from occurring between $fail$ and $repair$. The mainframe and the bouncer synchronise on $\delta_1, \delta_2, fail$, and $repair$. We checked, using the BCG_CMP tool, that the LTSs generated by both approaches are strongly bisimilar. Hence, we adopt as model V2 the LNT program of the second approach and will use it as a basis for the next steps.

```
process FAIL_LOAD [...] (in var P: PHASE,
                        in var FAILED: bool) is
  loop
    alt
        if FAILED then
            REPAIR;
            FAILED := false
        else
            case P in
                1 ->
                    -- low-load phase
                    DELTA1;
                    FAIL;
                    FAILED := true
              | 2 ->
                    -- high-load phase
                    DELTA2;
                    FAIL;
                    FAILED := true
              | 3 ->
                    -- idle phase
                    null
            end case
        end if
    []
        PHI;
        if FAILED then
            alt
                    null
            []
                REPAIR;
                FAILED := false
            end alt
        end if;
        C;
        case P in
            1 -> P := 2
          | 2 -> P := 3
          | 3 -> P := 1
        end case
    end alt
  end loop
end process
```

```
process BOUNCER [...] is
  loop
    alt
        DELTA1
    []
        DELTA2
    []
        FAIL;
        REPAIR
    end alt
  end loop
end process
```

```
process BOUNCER [...] is
  loop
    alt
        DELTA1
    []
        DELTA2
    []
        MU1
    []
        MU2
    []
        FAIL;
        REPAIR
    end alt
  end loop
end process
```

Fig. 5. LNT model fragments documenting the changes in models V2 and V3

Experiments with model V2 give encouraging results: (i) we checked, using the BISIMULATOR[29] tool, that V2 ⊑ V1; (ii) the number of conflict states drops from 552 in model V1 to 56 in model V2; (iii) the discrepancy between the "min" and "max" probabilities computed by Storm on the MAs of model V2 is at most 4.10^{-9} for Fig. 5 and 4.10^{-8} for Fig. 6; (iv) the discrepancy

---
[29] https://cadp.inria.fr/man/bisimulator.html

between the probabilities of the compound model and those computed by Storm on model V2 is smaller than with model V1 (maximal relative differences: 0.39% for Fig. $\boxed{5}$ and 2.19% for Fig. $\boxed{6}$).

7.7   Model V3 of the Erlangen Mainframe

We applied our debugging toolchain to model V2 to better know about the 56 conflict states. All the 56 rate sequences leading to these states (there are no duplicate sequences) end with a rate transition β. The shortest rate sequence (of length 5) is "$\lambda_1; \delta_1; \mu_1; \alpha_1; \beta$" and the corresponding full sequence is "$\lambda_1; prog\_job; \delta_1; fail; \mu_1; \alpha_1; \beta; repair; (user\_job \,\square\, get\_prog\_job)$". The scenario is the following: the $prog\_load$ process sends a job to the $prog\_queue$ (λ_1 followed by $prog\_job$). The $fail\_load$ process triggers a failure (δ_1 followed by $fail$). While the mainframe is in failed state, the $user\_load$ initiates a job (μ_1), but the next event $user\_job$ is blocked by the $fail\_queue$ process. The $prog\_queue$ initiates a job transfer to the processors (α_1), but the next event get_prog_job is also blocked by the failure. Eventually, the mainframe restarts (β followed by $repair$). The $fail\_queue$ unlocks the inputs and outputs of the two other queues, simultaneously enabling both pending events $user\_job$ and get_prog_job, hence causing nondeterminism.

The root cause of the problem is that, in the compound model, only (c, φ) transitions may occur between $(fail, \delta_i)$ and $(repair, \beta)$, i.e., while the mainframe is in failed state. But, in the IMC model, many other transitions may happen between $fail$ and $repair$: occurrences of c, β, and φ are normal; occurrences of δ_1 and δ_2 are already prohibited by the bouncer of model V2; occurrences of α_i, λ_i, μ_i, etc. should be examined carefully, as they might break atomicity and introduce nondeterminism.

We first extended the bouncer of model V2 to block, beyond δ_1 and δ_2, all rate transitions α_1, α_2, λ_1, λ_2, μ_1, μ_2, ν, and ξ when the mainframe is in failed state, so as to replicate the behaviour of the compound model in this situation. Using this bouncer, the number of conflict states was reduced from 56 to 28 and Storm produced a Fig. $\boxed{5}$ similar to that of model V2. However, Storm also produced a Fig. $\boxed{6}$ radically different from the one of model V2 (see Fig. 6 for a comparative view), making it clear that, if this bouncer is effective in further removing nondeterminism, it is overly restrictive concerning synchronisations.

Undertaking a systematic study, we tried to remove one by one the rate transitions α_1, α_2, λ_1, λ_2, μ_1, μ_2, ν, and ξ from the bouncer, while watching the outcome of such removals on the number of conflict states. Changes were only observed with μ_1 and μ_2, the other rates having no impact. When removing either μ_1 or μ_2, the number of conflict states was reduced from 56 to 44. When removing both μ_1 and μ_2, it dropped down to 28. Consequently, we opted for a new model V3 derived from model V2 by adding μ_1 and μ_2 (together with δ_1 and δ_2) to the bouncer (see the LNT code at the bottom right of Fig. 5) and to the list of rates on which the mainframe and the bouncer synchronise.

Our experiments with this model V3 gave results corroborating this decision: (i) we checked, using the BISIMULATOR tool, that V3 $\sqsubseteq$ V2; (ii) the discrep-

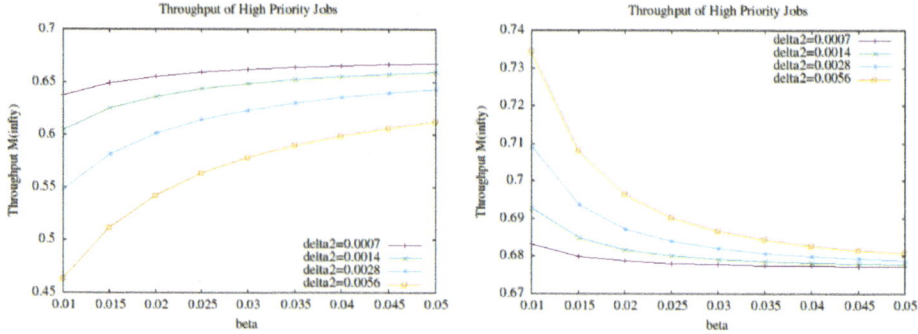

Fig. 6. Figure ⑥ for model V2 (left) and for model V2 equipped with a restrictive bouncer blocking all rate transitions but β and φ (right)

ancy between the "min" and "max" probabilities computed by Storm on the MAs of model V3 is at most 2.10^{-10} for Fig. ⑤ and 2.10^{-9} for Fig. ⑥; (iii) the discrepancy between the probabilities of the compound model and those computed by Storm on model V3 is comparable to that of model V2 (maximal relative differences: 0.37% for Fig. ⑤ and 2.37% for Fig. ⑥).

7.8   Model V4 of the Erlangen Mainframe

Again, we applied our debugging toolchain to understand the 28 conflict states in model V3. Again, all the 28 rate sequences leading to these states terminate with a rate transition β. The shortest rate sequence (of length 15) is[30] "$(\lambda_1; \alpha_1)^5; \mu_1; \alpha_2; \mu_1; \delta_1; \beta$", where s^n means n consecutive repetitions of the sub-sequence s. The corresponding full sequence (of length 29) is "$(\lambda_1; prog\_job\,; \alpha_1; get\_prog\_job)^4\,; \lambda_1; prog\_job\,; \alpha_1; \mu_1; user\_job\,; \alpha_2; \mu_1; \delta_1; fail; \beta;$ $repair; get\_user\_job\,; (get\_prog\_job \,\square\, user\_job)$". The scenario can be explained as follows: the $prog\_load$ process sends four jobs, which are received by the $prog\_queue$ and delivered to the four processors. The $prog\_load$ process sends a fifth job to the $prog\_queue$, which initiates a transfer to the processors (α_1) that cannot complete since all processors are busy. The $user\_load$ does the same, sending a job to the $user\_queue$, which also initiates a transfer (α_2) that cannot complete. The $user\_load$ initiates another job (μ_1). The $fail\_load$ process triggers a failure (δ_1) that does not last long, since the mainframe gets immediately repaired within the expected delay (β). When the mainframe restarts, the four jobs formerly running on the processors are cleared, so that each processor can now accept a new job. The job stored in the $user\_queue$ is taken first, because of the priority mechanism. This queue becomes empty, so that the lower-priority job stored in the $prog\_queue$ becomes eligible. At the same time, the $user\_queue$

---

[30] For the clarity of this section, we replace the sequences found by our toolchain by "equivalent" sequences having the same lengths, but simpler to explain.

may receive the pending job sent by the *user_load*. Hence, nondeterminism ensues between the output of the *prog_queue* and the input of the *user_queue*.

We did all kinds of attempts to eliminate such nondeterminism from model V3. For instance, if rate transitions α_2 (or both α_1 and α_2) are removed from model V3, based on the assumption that the corresponding numeric rates are so large that these transitions can be considered as immediate, the number of conflict states drops from 28 to zero (interestingly, if rate transitions α_1 only are removed, the number of conflict states increases from 28 to 37). Yet, we discarded this idea, as the resulting model M does not satisfy $M \sqsubseteq V3$ and the figures ③–⑩ obtained were much too different from the expected ones.

We tried hard to reuse our atomicity-preserving approach that had been fruitfully applied to reduce nondeterminism in models V1 and V2, but we had no success with model V3. Indeed, the *prog_load* and *user_load* processes are independent, and the *prog_queue* and the *user_queue* processes are only weakly correlated. Due to the complexity of scenarios leading to conflict states (their length ranges between 29 and 39), it would be difficult for the *user_load* or *prog_queue* processes to acquire a global knowledge of the system precise enough to decide whether transferring a job to the *user_queue* or from the *prog_queue* may cause nondeterminism.

In such a case, our chronology-preserving approach seems more appropriate for avoiding, at the very last moment, nondeterminism between far-remote events, a problem that would otherwise be difficult to anticipate and prevent by blocking rate transitions in advance. We designed a new model V4 obtained by equipping model V3 with an additional resolver process. Given that the conflict is between *get_prog_job* and *user_job*, model V3 (i.e., the mainframe and its bouncer) is synchronised with the resolver on these two events and on their associated rates α_1, μ_1, and μ_2. The resolver (see its LNT code in Fig. 7) maintains an internal FIFO queue that stores the arrival order of α_1, on the one hand, and μ_1 or μ_2, on the other hand. Depending of this order, the resolver enables either *get_prog_job* or *user_job*. We observed that, in any reachable state of the LTSs, this FIFO queue contains at most two elements. Notice that this resolver retrospectively gives one more justification to our decision of splitting α into α_1 and α_2 (see Sect. 4.3), as the resolver would not work with nondeterministic choices "α; *get_prog_job* $\square$ α; *get_user_job*".

Experiments with model V4 gave the following results: (i) we checked, using the BISIMULATOR tool, that V4 $\sqsubseteq$ V3; (ii) the number of conflict states in model V4 is zero, meaning that all instances of this model are CTMCs and can be analysed using tools such as CADP or PRISM; (iii) the discrepancy between the probabilities of the compound model and those computed by BCG_STEADY on model V4 are nearly identical to those observed with Storm on model V3 (maximal relative differences: 0.37% for Fig. ⑤ and 2.38% for Fig. ⑥); (iv) we checked that BCG_STEADY and Storm produce similar results for these CTMCs (maximal absolute differences: 3.10^{-5} for Fig. ⑤ and 3.10^{-6} for Fig. ⑥); (v) noticing that a CTMC in PRISM-MA format can easily be converted to PRISM format by replacing "`ma`" by "`ctmc`" and all symbols "`<>`" by "`[]`",

```
process RESOLVER [...] is
   var F: FIFO in
      F := {};
      loop
         assert length (F) <= 2;
         alt
            ALPHA1;
            F := append ('a', F)
         []
            MU1;
            F := append ('m', F)
         []
            MU2;
            F := append ('m', F)
         []
            GET_PROG_JOB where head (F) = 'a';
            F := tail (F)
         []
            USER_JOB where head (F) = 'm';
            F := tail (F)
         end alt
      end loop
   end var
end process
```

```
type FIFO is
   list of char
   with head, tail,
        length, append
end type
```

Fig. 7. LNT model fragments documenting the changes in model V4

we checked that BCG_STEADY and PRISM produce similar results (maximal absolute differences: 8.10^{-7} for Fig. [5] and 9.10^{-6} for Fig. [6]).

7.9 Assessment of Deterministic Analyses

An overview of the experiments with our four successive models V1, V2, V3, and V4 is given in Tab. 7. As mentioned before, the LTSs of these models are included in each other according to the strong bisimulation preorder, i.e., $V4 \sqsubseteq V3 \sqsubseteq V2 \sqsubseteq V1$. Notice that $M' \sqsubseteq M$ does not imply that the LTS of M' has fewer states or transitions than the LTS of M (consider, e.g., $M := a^{\bullet}$ and $M' := a; a; a$). From the 4th and 5th columns of Tab. 7, one may conjecture that bouncers decrease the numbers of states and transitions in the generated CTMC, while resolvers increase them. The 6th and 7th column show the discrepancy (maximal absolute difference and maximal relative difference) for Fig. [5] and [6] between the compound model and our models V1–V4. Although there are too few values for drawing firm conclusions, these columns suggest that progressively removing nondeterminism while maintaining LTS inclusion preserves the steady-state probabilities and throughputs of Fig. [5] and [6]. This was confirmed in contrast by other experiments that exhibited larger discrepancies when applying transformations that did not maintain LTS inclusion.

Given that the instances of model V4 are CTMCs, one can now use CADP, PRISM, and/or Storm to generate the other figures than Fig. [5]–[6] and compare them with those produced for the compound model:

model	conflicts	inclusion	states CTMC	transitions CTMC	max.difference for Fig. [5]	max.difference for Fig. [6]
V1	552	—	8,560	36,427	0.0407 (4.97%)	0.0336 (7.61%)
V2	56	V2 ⊑ V1	7,416	32,803	0.00335 (0.39%)	0.0104 (2.19%)
V3	28	V3 ⊑ V2	7,024	31,195	0.0032 (0.37%)	0.0113 (2.37%)
V4	0	V4 ⊑ V3	9,972	41,619	0.0032 (0.37%)	0.0113 (2.38%)

Table 7. Overview of results obtained with models V1, V2, V3, and V4

– Concerning Fig. [3] and [4], the maximal differences between the compound model and model V4 are small in absolute values (9.10^{-4} for Fig. [3] and 3.10^{-2} for Fig. [4]) but large enough in relative values (172% for Fig. [3] and 221% for Fig. [4]) to produce visible differences that can be observed in Fig. 8 and 9. The right parts of these figures (those for model V4) have been computed independently using CADP, PRISM, and Storm, which all give nearly identical results (maximal absolute differences between CADP and PRISM: 1.10^{-8} for Fig. [3] and 4.10^{-7} for Fig. [4], and between CADP and Storm: 9.10^{-6} for Fig. [3] and 3.10^{-5} for Fig. [4]).
– The four remaining figures obtained from model V4 are very close to those of the compound model (maximal relative differences: 0.014% for Fig. [7], 0.43% for Fig. [8], 0.015% for Fig. [9], and 1.70% for Fig. [10]). Again, CADP, PRISM, and Storm give close results on these four figures.

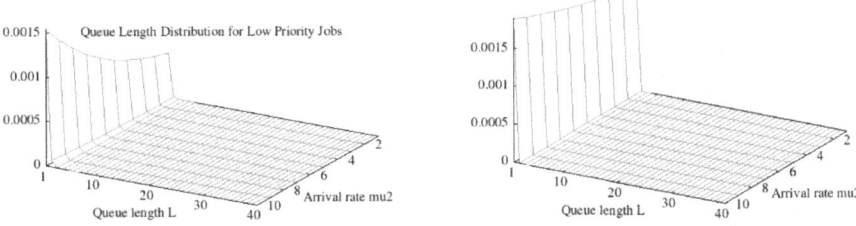

Fig. 8. Figure [3] for the compound model (left) and for model V4 (right)

Thus, our deterministic analysis approach, which is based on the progressive removal of nondeterminism through successive transformations, has two merits:

– The final model V4 is simply the initial model V1 composed in parallel with two small processes, a bouncer and a resolver. No involved modification of model V1 was needed to produce model V4.

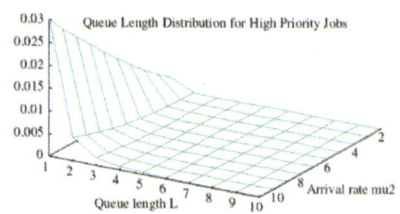 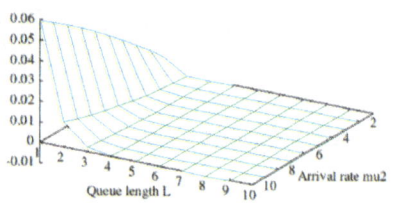

Fig. 9. Figure ④ for the compound model (left) and for model V4 (right)

- Despite the fact that standard translation does not preserve steady-state and transient probabilities in the general case (see Sect. 2.4), it was eventually possible to reproduce quite closely the eight figures of the Erlangen mainframe — except a few probabilities in Fig. ③ and ④ that visibly differ by their relative values.

So, transforming progressively a nondeterministic IMC into a CTMC seems promising. But one may wonder whether this approach is truly a methodology, in the sense that it can be reused and applied by others to solve similar problems. The results obtained should not hide the efforts and difficulties:

- There are many ways to transform a model. Our debugging toolchain may suggest plausible directions, but deciding which changes should be made requires a deep understanding of the system under analysis, as well as intuition, intense brainstorming, and lively discussions.
- This is a trial-and-error approach. Even if a change seems rational and promising, it is difficult to predict whether it will be effective in removing nondeterminism while preserving the expected numerical results.
- Designing bouncers and resolvers is a narrow path. If these processes are too permissive, they will not reduce nondeterminism. If they are too restrictive (e.g., bouncers synchronising certain transitions without necessity, resolvers blocking unforeseen rate transitions, etc.), they may introduce deadlocks or, worse, slow down the system in a way that alters the expected throughputs.
- To find our way in the search space, we used a combination of three guidelines: (i) the number of conflict states, which had to decrease strictly after each transformation; (ii) the inclusion of LTSs, which had to be preserved; and (iii) the probabilities and throughputs computed for Fig. ⑤ and ⑥ of the compound model, which we used as a trigger warning to discard those transformations that would diverge too much from these numbers.

8   Conclusion

The present article explored the hidden side of IMCs to get new insights about this formalism. In a nutshell, our contributions fall into three categories:

- We provided a stepwise methodology explaining how IMCs can be applied to an involved example such as the Erlangen mainframe. We set out the concept of "rate tagging", which has been used in most case studies so far, but never documented in theoretical papers. We introduced the concepts of "state probes" and "event probes", which are required to express state probabilities and throughput measures on IMCs. We exposed the inherent difficulties that may arise when using IMCs (proper placement of probes, larger state spaces, unexpected nondeterminism) and indicated solutions to these problems, thus paving the way for a wider dissemination of IMCs.
- We progressed the comparison between compound models and IMCs by assessing, for the first time, both approaches on a common case study through the prism of our "standard translation" function that converts compound models to IMCs. We established that, in the general case, this translation preserves neither the alternation of events and rates, nor the absence of deadlocks, nor the determinism property, nor the steady-state and transient probabilities. We however observed that this same translation, when applied to the Erlangen mainframe, preserved six figures out of eight, two other figures being quite similar except for corner-case values. Our experiments also indicate that, although IMCs have been designed for systems with only a few rates, they can also be used for compound-like models, in which (almost) every event has a rate.
- We studied how to handle nondeterminism, an issue not considered so far in publications related to IMCs. We proposed two different approaches: either (i) accepting nondeterminism as a fact, converting nondeterministic IMCs to Markov automata, and using state-of-the-art tools for Markov automata, or (ii) progressively eliminating nondeterminism by means of IMC transformations that preserve either atomicity ("bouncers") or chronology ("resolvers"), still enforcing model inclusion according to the strong bisimulation preorder.

This work could be pursued in three different directions:

- This study would not have been possible without the availability of advanced software tools, such as CADP, PRISM, Modest, and Storm. Our experiments sometimes pushed these tools to their current limits, which led us to suggest (see Sect. 6.6) five possible enhancements to these tools: two about the expressiveness of quantitative properties, two about the performance (speed) of model checking, and one about the user-friendliness of output formats.
- The ability to specify nondeterministic choices gives IMCs a greater expressiveness than compound models. Yet, in our study, nondeterminism was rather a nuisance than a desirable feature: (i) it popped up where it was not expected, but vanished when we tried to generate it purposely; (ii) it caused

problems to Markov automata tools; (iii) it turned single probabilities into [min, max] intervals, although we observed in our experiments that the difference between "min" and "max" was always negligible, thus questioning the significance of nondeterminism; (iv) the scenarios causing nondeterminism were often difficult to understand. In all the case studies published so far, with the exception of [14,54] and [49], IMCs are deterministic; one would thus welcome new IMC-based case studies, in which nondeterminism would be present and undoubtedly useful.

– Accurately comparing compound models and IMCs is still an open problem. Our results are paradoxical: our translation function does not preserve steady-state and transient probabilities on tiny examples, but widely preserves them on the much more involved Erlangen mainframe. Having two perfectly reasonable modelling formalisms that, when applied to the same system, might give different numeric results is annoying. Just stating that they are incomparable in the general case is not satisfactory, because this fails to explain which formalism should be chosen in practice. More work is needed to understand the situation, namely by extending the theoretical results of [5] to the case of synchronised processes. This may require to modify our translation function to better take synchronisations into account and to perform other case-studies to observe if our results can be reproduced.

Apologies and Acknowledgements

We owe an apology to the master students in Grenoble who tried to model the Erlangen mainframe using IMCs (see Sect. 3.5). A number of them produced valid LNT models, but nondeterminism popped up unexpectedly, making steady-state and transient analyses intractable with CTMC tools. Solving this problem, which did not appear in prior case studies tackled using IMCs, required substantial research and an amount of effort out of proportion to a lab exercise.

We would like to thank Pedro d'Argenio, Marco Bernardo, and Frédéric Lang for sharing their expertise in helpful discussions, as well as Radu Mateescu and the anonymous reviewers for proofreading and commenting the present article. We are also grateful to Arnd Hartmanns (Modest), Dave Parker (PRISM), Tim Quatmann (Storm), and Matthias Volk (Storm) for their most useful answers and advises; we hope, but are not 100% sure, that we used their tools in the most effective way and look forward to get the next versions.

References

1. Baier, C.: On Algorithmic Verification Methods for Probabilistic Systems (1998), Habilitation thesis, Universität Mannheim, Germany
2. Baier, C., Haverkort, B.R., Hermanns, H., Katoen, J.P.: Model-Checking Algorithms for Continuous-Time Markov Chains. IEEE Transactions on Software Engineering **29**(6), 524–541 (2003). https://doi.org/10.1109/TSE.2003.1205180
3. Baier, C., Hermanns, H., Katoen, J.P., Haverkort, B.R.: Efficient Computation of Time-Bounded Reachability Probabilities in Uniform Continuous-Time Markov Decision Processes. Theoretical Computer Science **345**(1), 2–26 (2005)

4. Baier, C., Katoen, J.P., Hermanns, H.: Approximate Symbolic Model Checking of Continuous-Time Markov Chains. In: Baeten, J.C.M., Mauw, S. (eds.) Proceedings of the 10th International Conference on Concurrency Theory (CONCUR'99), Eindhoven, The Netherlands. Lecture Notes in Computer Science, vol. 1664, pp. 146–161. Springer (Aug 1999). https://doi.org/10.1007/3-540-48320-9_12

5. Bernardo, M., Corradini, F., Tesei, L.: Timed Process Calculi with Deterministic or Stochastic Delays: Commuting between Durational and Durationless Actions. Theoretical Computer Science **629**, 2–39 (2016). https://doi.org/10.1016/J.TCS.2016.02.022

6. Bernardo, M., Gorrieri, R.: A Tutorial on EMPA: A Theory of Concurrent Processes with Nondeterminism, Priorities, Probabilities and Time. Theoretical Computer Science **202**(1–2), 1–54 (1998). https://doi.org/10.1016/S0304-3975(97)00127-8

7. Böde, E., Herbstritt, M., Hermanns, H., Johr, S., Peikenkamp, T., Pulungan, R., Rakow, J.H., Wimmer, R., Becker, B.: Compositional Dependability Evaluation for STATEMATE. IEEE Transactions on Software Engineering **35**(2), 274–292 (2009). https://doi.org/10.1109/TSE.2008.102

8. Buchholz, P., Hahn, E.M., Hermanns, H., Zhang, L.: Model Checking Algorithms for CTMDPs. In: Gopalakrishnan, G., Qadeer, S. (eds.) Proceedings of the 23rd International Conference on Computer Aided Verification (CAV'11), Snowbird, UT, USA. Lecture Notes in Computer Science, vol. 6806, pp. 225–242. Springer (Jul 2011). https://doi.org/10.1007/978-3-642-22110-1_19

9. Budde, C.E., Dehnert, C., Hahn, E.M., Hartmanns, A., Junges, S., Turrini, A.: JANI: Quantitative Model and Tool Interaction. In: Legay, A., Margaria, T. (eds.) Proceedings of the 23rd International Conference on Tools and Algorithms for the Construction and Analysis of Systems (TACAS'17), Part II, Uppsala, Sweden. Lecture Notes in Computer Science, vol. 10206, pp. 151–168 (Apr 2017). https://doi.org/10.1007/978-3-662-54580-5_9

10. Butkova, Y., Hartmanns, A., Hermanns, H.: A Modest Approach to Markov Automata. ACM Transactions on Modeling and Computer Simulation **31**(3), 14:1–14:34 (2021). https://doi.org/10.1145/3449355

11. Butkova, Y., Hatefi, H., Hermanns, H., Krcál, J.: Optimal Continuous Time Markov Decisions. In: Finkbeiner, B., Pu, G., Zhang, L. (eds.) Proceedings of the 13th International Symposium on Automated Technology for Verification and Analysis (ATVA'15), Shanghai, China. Lecture Notes in Computer Science, vol. 9364, pp. 166–182. Springer (Oct 2015). https://doi.org/10.1007/978-3-319-24953-7_12

12. Butkova, Y., Wimmer, R., Hermanns, H.: Long-Run Rewards for Markov Automata. In: Legay, A., Margaria, T. (eds.) Proceedings of the 23rd International Conference on Tools and Algorithms for the Construction and Analysis of Systems (TACAS'17), Part II, Uppsala, Sweden. Lecture Notes in Computer Science, vol. 10206, pp. 188–203 (Apr 2017). https://doi.org/10.1007/978-3-662-54580-5_11

13. Champelovier, D., Clerc, X., Garavel, H., Guerte, Y., McKinty, C., Powazny, V., Lang, F., Serwe, W., Smeding, G.: Reference Manual of the LNT to LOTOS Translator (Version 7.5) (Feb 2025), https://cadp.inria.fr/publications/Champelovier-Clerc-Garavel-et-al-10.html, INRIA, Grenoble, France

14. Chehaibar, G., Zidouni, M., Mateescu, R.: Modeling Multiprocessor Cache Protocol Impact on MPI Performance. In: Proceedings of the IEEE International Workshop on Quantitative Evaluation of Large-Scale Systems and Technologies

(QuEST'09), Bradford, UK. pp. 1073–1078. IEEE Computer Society Press (May 2009)

15. Coste, N., Garavel, H., Hermanns, H., Lang, F., Mateescu, R., Serwe, W.: Ten Years of Performance Evaluation for Concurrent Systems Using CADP. In: Margaria, T., Steffen, B. (eds.) Proceedings of the 4th International Symposium on Leveraging Applications of Formal Methods, Verification and Validation ISoLA 2010 (Amirandes, Heraclion, Crete), Part II. Lecture Notes in Computer Science, vol. 6416, pp. 128–142. Springer (Oct 2010)

16. Coste, N., Hermanns, H., Lantreibecq, E., Serwe, W.: Towards Performance Prediction of Compositional Models in Industrial GALS Designs. In: Bouajjani, A., Maler, O. (eds.) Proceedings of the 21th International Conference on Computer Aided Verification (CAV'09), Grenoble, France. Lecture Notes in Computer Science, vol. 5643, pp. 204–218. Springer (Jul 2009)

17. Fischer, W., Meier-Hellstern, K.S.: The Markov-Modulated Poisson Process (MMPP) Cookbook. Performance Evaluation **18**(2), 149–171 (Sep 1993). https://doi.org/10.1016/0166-5316(93)90035-S

18. Garavel, H.: Revisiting Sequential Composition in Process Calculi. Journal of Logical and Algebraic Methods in Programming **84**(6), 742–762 (Nov 2015). https://doi.org/10.1016/j.jlamp.2015.08.001

19. Garavel, H., Hermanns, H.: On Combining Functional Verification and Performance Evaluation using CADP. In: Eriksson, L.H., Lindsay, P.A. (eds.) Proceedings of the 11th International Symposium of Formal Methods Europe (FME'02), Copenhagen, Denmark. Lecture Notes in Computer Science, vol. 2391, pp. 410–429. Springer (Jul 2002), full version available as INRIA Research Report 4492

20. Garavel, H., Hermanns, H., Parker, D.: Revisiting a Pioneering Concurrent Stochastic Problem: The Erlangen Mainframe. In: Jansen, N., Junges, S., Kaminski, B.L., Matheja, C., Noll, T., Quatmann, T., Stoelinga, M., Volk, M. (eds.) Principles of Verification: Cycling the Probabilistic Landscape (Part II) – Essays Dedicated to Joost-Pieter Katoen on the Occasion of His 60th Birthday. Lecture Notes in Computer Science, vol. 15261, pp. 46–74. Springer (Nov 2024). https://doi.org/10.1007/978-3-031-75775-4

21. Garavel, H., Lang, F.: SVL: a Scripting Language for Compositional Verification. In: Kim, M., Chin, B., Kang, S., Lee, D. (eds.) Proceedings of the 21st IFIP WG 6.1 International Conference on Formal Techniques for Networked and Distributed Systems (FORTE'01), Cheju Island, Korea. pp. 377–392. Kluwer Academic Publishers (Aug 2001). https://doi.org/10.1007/0-306-47003-9_24, full version available as INRIA Research Report RR-4223

22. Garavel, H., Lang, F., Mateescu, R., Serwe, W.: CADP 2011: A Toolbox for the Construction and Analysis of Distributed Processes. Springer International Journal on Software Tools for Technology Transfer (STTT) **15**(2), 89–107 (Apr 2013). https://doi.org/10.1007/s10009-012-0244-z

23. Garavel, H., Lang, F., Mounier, L.: Compositional Verification in Action. In: Howar, F., Barnat, J. (eds.) Proceedings of the 23rd International Conference on Formal Methods for Industrial Critical Systems (FMICS'18), Maynooth, Ireland – Essays Dedicated to Susanne Graf at the Occasion of Her 60th Birthday. Lecture Notes in Computer Science, vol. 11119, pp. 189–210. Springer (Sep 2018)

24. van Glabbeek, R.J., Weijland, W.P.: Branching Time and Abstraction in Bisimulation Semantics. Journal of the ACM **43**(3), 555–600 (1996)

25. Götz, N., Herzog, U., Rettelbach, M.: Multiprocessor and Distributed System Design: The Integration of Functional Specification and Performance Analysis

Using Stochastic Process Algebras. In: Donatiello, L., Nelson, R.D. (eds.) Performance Evaluation of Computer and Communication Systems, Joint Tutorial Papers of Performance'93 and Sigmetrics'93, Santa Clara, CA, USA. Lecture Notes in Computer Science, vol. 729, pp. 121–146. Springer (May 1993). https://doi.org/10.1007/BFB0013851

26. Gros, T.P.: Markov Automata Taken by Storm. Master's thesis, Saarland University, Germany (Jan 2018)

27. Guck, D., Han, T., Katoen, J.P., Neuhäußer, M.R.: Quantitative Timed Analysis of Interactive Markov Chains. In: Goodloe, A., Person, S. (eds.) Proceedings of the 4th International NASA Formal Methods Symposium (NFM'12), Norfolk, VA, USA. Lecture Notes in Computer Science, vol. 7226, pp. 8–23. Springer (Apr 2012). https://doi.org/10.1007/978-3-642-28891-3_4

28. Hahn, E.M., Hartmanns, A., Hermanns, H., Katoen, J.P.: A Compositional Modelling and Analysis Framework for Stochastic Hybrid Systems. Formal Methods in System Design **43**(2), 191–232 (2013). https://doi.org/10.1007/S10703-012-0167-Z

29. Hartmanns, A., Hermanns, H.: The Modest Toolset: An Integrated Environment for Quantitative Modelling and Verification. In: Ábrahám, E., Havelund, K. (eds.) Proceedings of the 20th International Conference on Tools and Algorithms for the Construction and Analysis of Systems (TACAS'14), Grenoble, France. Lecture Notes in Computer Science, vol. 8413, pp. 593–598. Springer (Apr 2014). https://doi.org/10.1007/978-3-642-54862-8_51

30. Hensel, C., Junges, S., Katoen, J.P., Quatmann, T., Volk, M.: The Probabilistic Model Checker Storm. International Journal on Software Tools for Technology Transfer **24**(4), 589–610 (2022). https://doi.org/10.1007/S10009-021-00633-Z

31. Hermanns, H.: Construction and Verification of Performance and Reliability Models. Bulletin of the EATCS **74**, 135–154 (2001)

32. Hermanns, H.: Interactive Markov Chains: The Quest for Quantified Quality, Lecture Notes in Computer Science, vol. 2428. Springer (2002)

33. Hermanns, H., Herzog, U., Klehmet, U., Mertsiotakis, V., Siegle, M.: Compositional Performance Modelling with the TIPPtool. Performance Evaluation **39**(1-4), 5–35 (Feb 2000). https://doi.org/10.1016/S0166-5316(99)00056-5

34. Hermanns, H., Herzog, U., Merksiotakis, V.: Stochastic Process Algebras as a Tool for Performance and Dependability Modelling. In: Iyer, R.K. (ed.) Proceedings of the International Computer Performance and Dependability Symposium (IPDS'95), Erlangen, Germany. pp. 102–111. IEEE (Apr 1995). https://doi.org/10.1109/IPDS.1995.395813

35. Hermanns, H., Herzog, U., Mertsiotakis, V.: Stochastic Process Algebras – Between LOTOS and Markov Chains. Computer Networks **30**(9-10), 901–924 (1998). https://doi.org/10.1016/S0169-7552(97)00133-5

36. Hermanns, H., Joubert, C.: A Set of Performance and Dependability Analysis Components for CADP. In: Garavel, H., Hatcliff, J. (eds.) Proceedings of the 9th International Conference on Tools and Algorithms for the Construction and Analysis of Systems (TACAS'03), Warsaw, Poland. Lecture Notes in Computer Science, vol. 2619, pp. 425–430. Springer (Apr 2003)

37. Hermanns, H., Katoen, J.P.: Automated compositional markov chain generation for a plain-old telephone system. Sci. Comput. Program. (2000)

38. Hermanns, H., Katoen, J.P.: The How and Why of Interactive Markov Chains. In: de Boer, F.S., Bonsangue, M.M., Hallerstede, S., Leuschel, M. (eds.) Revised

Selected Papers of the 8th International Symposium on Formal Methods for Components and Objects (FMCO'09), Eindhoven, The Netherlands. Lecture Notes in Computer Science, vol. 6286, pp. 311–337. Springer (Nov 2009)

39. Hermanns, H., Rettelbach, M.: Syntax, Semantics, Equivalences, and Axioms for MTIPP. In: Herzog, U., Rettelbach, M. (eds.) Proceedings of the 2nd Workshop on Process Algebras and Performance Modelling (PAPM'94), Erlangen, Germany. Lecture Notes in Computer Science, vol. 1601, pp. 71–88. University of Erlangen-Nürnberg, Germany (Jul 1994)

40. Herzog, U.: Formal Description, Time and Performance Analysis: A Framework. In: Härder, T., Wedekind, H., Zimmermann, G. (eds.) Entwurf und Betrieb verteilter Systeme, Proceedings Fachtagung des Sonderforschungsbereiche 124 und 182 (Dagstuhl, Germany). Informatik-Fachberichte, vol. 264, pp. 172–190. Springer (Sep 1990). https://doi.org/10.1007/978-3-642-76309-0_10

41. Herzog, U., Merksiotakis, V.: Stochastic Process Algebras Applied to Failure Modelling. In: Herzog, U., Rettelbach, M. (eds.) Proceedings of the 2nd Workshop on Process Algebras and Performance Modelling (PAPM'94), Regensberg/Erlangen, Germany. pp. 107–126 (Jul 1994), https://www.researchgate.net/publication/2731331

42. Hillston, J.: A Compositional Approach to Performance Modelling. Cambridge University Press (1996)

43. Johr, S.: Model Checking Compositional Markov Systems. Ph.D. thesis, Saarland University (Aug 2007)

44. Kemeny, J.G., Snell, J.L.: Finite Markov Chains. Undergraduate Texts in Mathematic, Springer (1976)

45. Kordon, F., Garavel, H., Hillah, L.M., Paviot-Adet, E., Jezequel, L., Rodríguez, C., Hulin-Hubard, F.: MCC'2015 – The Fifth Model Checking Contest. Transactions on Petri Nets and Other Models of Concurrency **XI**, 262–273 (2016)

46. Kwiatkowska, M.Z., Norman, G., Parker, D.: PRISM 4.0: Verification of Probabilistic Real-Time Systems. In: Gopalakrishnan, G., Qadeer, S. (eds.) Proceedings of the 23rd International Conference on Computer Aided Verification (CAV'11), Snowbird, UT, USA. Lecture Notes in Computer Science, vol. 6806, pp. 585–591. Springer (Jul 2011). https://doi.org/10.1007/978-3-642-22110-1_47

47. Mateescu, R.: Specification and Analysis of Asynchronous Systems using CADP. In: Merz, S., Navet, N. (eds.) Modeling and Verification of Real-Time Systems – Formalisms and Software Tools, chap. 5, pp. 141–170. ISTE publishing / John Wiley (2008), http://hal.inria.fr/inria-00264235

48. Mateescu, R., Garavel, H.: XTL: A Meta-Language and Tool for Temporal Logic Model-Checking. In: Margaria, T. (ed.) Proceedings of the International Workshop on Software Tools for Technology Transfer (STTT'98), Aalborg, Denmark. pp. 33–42. BRICS (Jul 1998)

49. Mateescu, R., Serwe, W.: Model Checking and Performance Evaluation with CADP Illustrated on Shared-Memory Mutual Exclusion Protocols. Science of Computer Programming **78**(7), 843–861 (Jul 2013)

50. Milner, R.: A Calculus of Communicating Systems, Lecture Notes in Computer Science, vol. 92. Springer (1980). https://doi.org/10.1007/3-540-10235-3

51. Neuhäußer, M.R.: Model Checking Nondeterministic and Randomly Timed Systems. Ph.D. thesis, RWTH Aachen University, Germany (2010)

52. Queille, J.P., Sifakis, J.: Specification and Verification of Concurrent Systems in CESAR. In: Dezani-Ciancaglini, M., Montanari, U. (eds.) Proceedings of the 5th International Symposium on Programming, Torino, Italy. Lecture Notes in Computer Science, vol. 137, pp. 337–351. Springer (1982)

53. Zhang, L., Neuhäußer, M.R.: Model Checking Interactive Markov Chains. In: Esparza, J., Majumdar, R. (eds.) Proceedings of the 16th International Conference on Tools and Algorithms for the Construction and Analysis of Systems (TACAS'10), Paphos, Cyprus. Lecture Notes in Computer Science, vol. 6015, pp. 53–68. Springer (Mar 2010). https://doi.org/10.1007/978-3-642-12002-2_5
54. Zidouni, M.: Modélisation et analyse des performances de la bibliothèque MPI en tenant compte de l'architecture matérielle. Ph.D. thesis, Université de Grenoble (May 2010), http://hal.inria.fr/tel-00526164/en

Decisiveness for Countable MDPs and Insights for NPLCSs and POMDPs

Nathalie Bertrand[1], Patricia Bouyer[2], Thomas Brihaye[3],
Paulin Fournier[1,2,3], and Pierre Vandenhove[3](✉)

[1] IRISA – Inria, CNRS, Univ. Rennes, Rennes, France
[2] Université Paris-Saclay, CNRS, ENS Paris-Saclay, Laboratoire Méthodes Formelles,
91190 Gif-sur-Yvette, France
[3] UMONS – Université de Mons, Mons, Belgium
`pierre.vandenhove@umons.ac.be`

Abstract. Markov chains and Markov decision processes (MDPs) are well-established probabilistic models. While finite Markov models are well-understood, analyzing their infinite counterparts remains a significant challenge. Decisiveness has proven to be an elegant property for countable Markov chains: it is general enough to be satisfied by several natural classes of countable Markov chains, and it is a sufficient condition for simple qualitative and approximate quantitative model-checking algorithms to exist.

In contrast, existing works on the formal analysis of countable MDPs usually rely on *ad hoc* techniques tailored to specific classes. We provide here a general framework to analyze countable MDPs by extending the notion of decisiveness. Compared to Markov chains, MDPs exhibit extra non-determinism that can be resolved in an adversarial or cooperative way, leading to multiple natural notions of decisiveness. We show that these notions enable the approximation of reachability and safety probabilities in countable MDPs using simple model-checking procedures.

We then instantiate our generic approach to two concrete classes of models inducing countable MDPs: non-deterministic probabilistic lossy channel systems and partially observable MDPs. This leads to an algorithm to approximately compute safety probabilities in each of these classes.

Keywords: Markov decision processes · Reachability · Decisiveness · Lossy channel systems · Partially observable Markov decision processes

1 Introduction

Formal methods for systems with random or unknown behaviours call for models with probabilistic aspects, and appropriate automated verification techniques. One of the simplest classes of probabilistic models is the one of Markov chains. The verification of finite-state Markov chains has been thoroughly studied in the literature and is supported by multiple mature tools such as PRISM [31] and STORM [18].

N. Bertrand et al. (Eds.): Christel Baier Festschrift, LNCS 15760, pp. 70–98, 2026.
https://doi.org/10.1007/978-3-031-97439-7_3

Countable Markov Chains. In some cases, *finite* Markov chains fall short at providing an appropriate modelling formalism, and *infinite* Markov chains must be considered. There are two general directions for the model checking of infinite-state Markov chains. One option is to focus on Markov chains generated in a specific way; for instance, when the underlying transition system is the configuration graph of a lossy channel system [2,25], a pushdown automaton [30], or a one-counter system [19]. In this case, *ad hoc* model-checking techniques have been developed for the qualitative and quantitative analysis. The second option is to establish general criteria on infinite Markov chains that are sufficient for their qualitative and/or quantitative model checking to be feasible.

Abdulla et al. explored the latter direction and proposed the elegant notion of *decisive* Markov chains [1]. Intuitively, decisive countably infinite Markov chains exhibit certain desirable properties of finite-state Markov chains. For instance, one such property is that if a state is continuously reachable with a positive probability, then it will almost surely be reached. Precisely, a Markov chain is decisive (with respect to a target state ☺, from a given initial state s_0) if almost all runs from s_0 either reach ☺ or end in states from which ☺ is no longer reachable. This is convenient to deal with *reachability objectives*, i.e., the event of reaching a specified set of states. Assuming decisiveness, the qualitative model checking of reachability objectives reduces—as in the finite case—to simple graph analysis. Moreover, decisiveness is the property that allows for approximating the probability of reachability objectives up to any desired error margin and for sampling trajectories towards statistical model-checking of infinite Markov chains [8]. While certain decisive classes have been exhibited [1], decidability of the decisiveness property has been shown in some other classes [22]. A stronger property for countable Markov chains is the existence of a *finite attractor*, i.e., a finite set of states that is reached almost surely from any state of the Markov chain. Sufficient conditions for the existence of a finite attractor are given in [5].

Markov Decision Processes. Purely probabilistic models are too limited to represent features such as, e.g., the lack of any assumption regarding scheduling policies or relative speeds (in concurrent systems), the lack of information regarding values that have been abstracted away (in abstract models), or the latitude left for delayed implementation decisions (in early designs). In such situations, it is not desirable to assume the choices to be resolved probabilistically, and non-determinism is needed. *Markov decision processes* (MDPs) are an extension of Markov chains with nondeterministic choices; they exhibit both nondeterminism and probabilistic phenomena. In MDPs, the nondeterminism is resolved by a *scheduler*, which can either be adversarial or cooperative, so that for a given event it is relevant to consider both the infimum and supremum probabilities that it occurs, ranging over all schedulers.

Similarly to the case of Markov chains, when considering infinite systems, one can either opt for *ad hoc* model-checking algorithms for classes of infinite-state MDPs, or derive generic results under appropriate assumptions. In the first scenario, one can mention MDPs which are generated by lossy channel systems [2,6], with nondeterministic action choices and probabilistic message losses. Up to our knowledge, only qualitative verification algorithms—based on the finite-attractor property—have been developed. In particular, the existence of a scheduler that ensures a reachability objective with probability 1 (or with positive probability) is decidable for lossy channel systems [6]; however, the existence of a scheduler ensuring a Büchi objective with positive probability is undecidable [2]. More generally, there are also examples of *games* on infinite arenas with underlying tractable model for which decidability results exist: recursive concurrent stochastic games [14,20], one-counter stochastic games [12,13] or lossy channel systems [3,11]. In the second scenario, general countable MDPs have been considered with the aim of characterizing the value function for various quantitative objectives [35] or identifying the resources (memory requirements, randomness) needed by optimal or ε-optimal schedulers [27,33,35]. Up to our knowledge, there are however no generic approaches to provide quantitative model-checking algorithms. This is the purpose of this paper.

Contributions. In this paper, we address the design of generic algorithms for the quantitative model checking of reachability objectives in countable MDPs. To do so, we first build on the seminal work on decisive Markov chains [1] and explore how the notion of decisiveness can be extended to Markov decision processes. We propose two notions of decisiveness, called *inf-decisiveness* and *sup-decisiveness*, which differ on whether the resolution of nondeterminism is adversarial or cooperative. These notions are natural extensions of the existing decisiveness for Markov chains. Second, we provide approximation schemes for the infimum and supremum probabilities of reachability objectives. These schemes provide a non-decreasing sequence of lower bounds, as well as a non-increasing sequence of upper bounds, for the probability one wishes to compute. Third, we identify sufficient conditions related to decisiveness for the two sequences to converge towards the same limit, which is necessary for the scheme to terminate for any given error margin. We obtain that for inf-decisive MDPs, one can approximate the infimum reachability probability up to any error, and for sup-decisive MDPs, one can approximate the supremum reachability probability up to any error.

We end the paper by instantiating our generic approach to two concrete classes of models inducing countably infinite MDPs of very different nature: *nondeterministic probabilistic lossy channel systems* and finite *partially observable MDPs*. Using decisiveness, we show in both classes that the infimum reachability probabilities can be approximated up to any desired precision. To the best of our knowledge, this is the first time that *quantitative* model-checking algorithms are provided for these classes. As we will discuss, existing algorithms often focus on the *qualitative* problems (e.g., whether there is a scheduler reaching a state almost surely) due to the undecidability of most other quantitative problems.

For consistency, we mostly discuss *reachability* objectives throughout the paper. However, note that minimizing the probability of a reachability objective is equivalent to maximizing the probability of the dual *safety* objective (consisting of avoiding a specified set of states). All results regarding the infimum probability of a reachability objective can therefore be thought of as results about the supremum probability of a safety objective (and vice versa).

Due to space constraints, we omit most proofs in this conference version. Where proofs are omitted, we provide references to the full version of the paper [9].

2   Preliminaries

2.1   Markov Decision Processes

Definition 1. *A* Markov decision process *(MDP) is a tuple* $\mathcal{M} = (S, \mathsf{Act}, \mathsf{P})$ *where S is a countable set of states,* Act *is a countable set of actions,* $\mathsf{P}: S \times \mathsf{Act} \times S \to [0,1] \cap \mathbb{Q}$ *is a probabilistic transition function satisfying $\sum_{s' \in S} \mathsf{P}(s, a, s') \in \{0, 1\}$ for all $(s, a) \in S \times \mathsf{Act}$.*

An MDP $\mathcal{M}$ is *finite* if S is finite. Let $\mathcal{M} = (S, \mathsf{Act}, \mathsf{P})$ be an MDP. Given $(s, a) \in S \times \mathsf{Act}$, we say that the action a is *enabled* at state s whenever $\sum_{s' \in S} \mathsf{P}(s, a, s') = 1$. We write $\mathsf{En}(s)$ for the set of actions enabled at s. We assume that each state has at least one enabled action. A state s is *absorbing* if for all enabled actions $a \in \mathsf{En}(s)$, $\mathsf{P}(s, a, s) = 1$. The MDP $\mathcal{M}$ is *finitely action-branching* if for every $s \in S$, $\mathsf{En}(s)$ is finite. It is *finitely prob-branching* if for every $(s, a) \in S \times \mathsf{Act}$, the support of $\mathsf{P}(s, a, \cdot)$ is finite. It is *finitely branching* if it is both finitely action-branching and finitely prob-branching.

A *history* (resp. *path*) in $\mathcal{M}$ is an element $s_0 s_1 s_2 \cdots$ of S^+ (resp. S^ω) such that for every relevant $i \geq 0$, there is $a_i \in \mathsf{Act}$ such that $\mathsf{P}(s_i, a_i, s_{i+1}) > 0$ (in particular, a_i is enabled at s_i). We write $\mathsf{Hist}(\mathcal{M})$ for the set of histories in $\mathcal{M}$ and $\mathsf{Paths}(\mathcal{M})$ for the set of paths in $\mathcal{M}$. We define the *length* of a history $h = s_0 s_1 \cdots s_k$ as k, and denote its last state by $\mathsf{last}(h) = s_k$. We sometimes write $h \cdot s$ for a history ending in a state s, to emphasize its last state.

We consider the σ-algebra generated by cylinders in $\mathsf{Paths}(\mathcal{M})$: for a history $h \in \mathsf{Hist}(\mathcal{M})$, the *cylinder generated by h* is

$$\mathsf{Cyl}(h) = \{\rho \in \mathsf{Paths}(\mathcal{M}) \mid h \text{ is a prefix of } \rho\} \ .$$

Definition 2. *A* scheduler *in $\mathcal{M}$ is a function* $\sigma: \mathsf{Hist}(\mathcal{M}) \to \mathsf{Dist}(\mathsf{Act})$ *which assigns a probability distribution over actions to any history, with the constraint that for every $h \in \mathsf{Hist}(\mathcal{M})$, the support of $\sigma(h)$ is included in $\mathsf{En}(\mathsf{last}(h))$. We write $\mathsf{Sched}(\mathcal{M})$ for the set of schedulers in $\mathcal{M}$.*

Schedulers are sometimes called *strategies* or *policies* in the literature. We fix a scheduler σ in $\mathcal{M}$. If σ only depends on the last state of histories, i.e., if $\mathsf{last}(h) = \mathsf{last}(h')$ implies $\sigma(h) = \sigma(h')$, then it is called *positional*. If for every history h, $\sigma(h)$ is a Dirac probability measure, it is said *pure*. A pure and positional scheduler can alternatively be described as a function $\sigma: S \to \mathsf{Act}$. We write $\mathsf{Sched}_{\mathsf{pp}}(\mathcal{M})$ for the set of pure and positional schedulers in $\mathcal{M}$, and $\mathsf{Sched}_{\mathsf{ph}}(\mathcal{M})$ for the set of pure (*a priori* not positional, that is, *history-dependent*) schedulers.

Given a scheduler σ in $\mathcal{M}$ and an initial state $s_0 \in S$, one can define a probability measure $\mathbb{P}^\sigma_{\mathcal{M}, s_0}$ on $\mathsf{Paths}(\mathcal{M})$ inductively as follows:

– $\mathbb{P}^{\sigma}_{\mathcal{M},s_0}(\mathsf{Cyl}(s_0)) = 1$;
– if $h = s_0 \cdots s_k \in \mathsf{Hist}(\mathcal{M})$ and $h \cdot s_{k+1} \in \mathsf{Hist}(\mathcal{M})$, then

$$\mathbb{P}^{\sigma}_{\mathcal{M},s_0}(\mathsf{Cyl}(h \cdot s_{k+1})) = \mathbb{P}^{\sigma}_{\mathcal{M},s_0}(\mathsf{Cyl}(h)) \cdot \sum_{a \in \mathsf{En}(\mathsf{last}(h))} \sigma(h)(a) \cdot \mathsf{P}(s_k, a, s_{k+1}) \ .$$

Equivalently, it is the probability measure in the (infinite) Markov chain $\mathcal{M}_\sigma$ induced by the scheduler σ on $\mathcal{M}$.

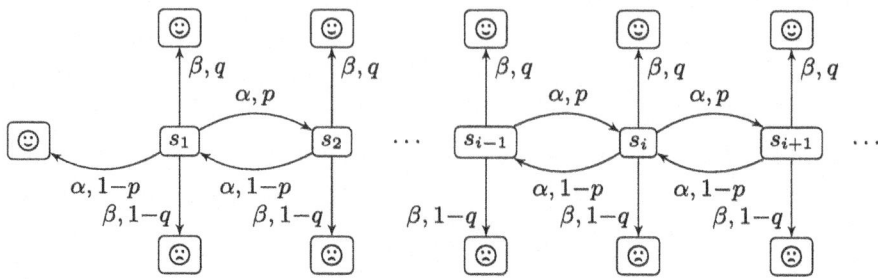

Fig. 1. Example of a finitely branching MDP with infinite state space. For readability, the absorbing states ☺ and ☹ are duplicated in the figure. Self-loops on absorbing states are omitted.

Figure 1 presents an example of a countably infinite MDP, which is finitely branching. Under a scheduler which always selects α, this yields a random walk [36, Section 3.1]. It is "diverging" if $p > \frac{1}{2}$, which entails that the probability λ_p *not* to reach ☺ is positive from every state (except ☺). In particular, in this case, the infimum probability of reaching ☺ depends on the relative values of q and λ_p.

2.2 Optimum Reachability Probabilities

Depending on the application, the non-determinism in Markov decision processes can be thought of as adversarial or as cooperative. For the probability of a given event, it thus makes sense to consider both the infimum and supremum probabilities when ranging over all schedulers.

We describe path properties using the standard LTL operators **F** and **G**, and their step-bounded variants $\mathbf{F}_{\leq n}$ and $\mathbf{G}_{\leq n}$. Let $\rho = s_0 s_1 \cdots \in \mathsf{Paths}(\mathcal{M})$ be a path in $\mathcal{M}$. If ψ is a state property, the path property $\mathbf{F}\psi$ holds on ρ if there is some index $k \in \mathbb{N}$ such that s_k satisfies ψ. Given $n \in \mathbb{N}$, $\mathbf{F}_{\leq n}$ holds on ρ if there is some index $k \leq n$ such that s_k satisfies ψ. Dually, ρ satisfies $\mathbf{G}\psi$ if all indices $k \in \mathbb{N}$ are such that s_k satisfies ψ, and ρ satisfies $\mathbf{G}_{\leq n}\psi$ if for all indices $k \leq n$, s_k satisfies ψ. Now, given a path property ϕ, we write $[\![\phi]\!]_{\mathcal{M},s_0}$ for the set of paths from s_0 in $\mathcal{M}$ that satisfy ϕ.

In this paper, we focus on the optimization of the probability of reachability objectives, and thus aim at computing or approximating the following values: given an MDP $\mathcal{M}$, an initial state s_0, a set of target states T, and $\mathsf{opt} \in \{\inf, \sup\}$,

$$\mathbb{P}^{\mathsf{opt}}_{\mathcal{M},s_0}(\mathbf{F}\,T) \stackrel{\text{def}}{=} \mathsf{opt}_{\sigma \in \mathsf{Sched}(\mathcal{M})} \mathbb{P}^{\sigma}_{\mathcal{M},s_0}(\mathbf{F}\,T) \ .$$

Without loss of generality, one can assume that T consists of a single absorbing state which we denote ☺ in the sequel.

Remark 1. The literature often considers *safety objectives*, which correspond to events $\mathbf{G}\neg T$ for T a set of states. Note that by the duality of reachability and safety objectives, all results below also hold for safety objectives by inverting inf and sup.

For finite MDPs, the computation of the above values for $\mathsf{opt} = \inf$ and $\mathsf{opt} = \sup$ is well-known (see e.g. [7, Chap. 10]). It reduces to solving a linear program (of linear size), resulting in a polynomial-time algorithm. Moreover, the infimum and supremum values are attained by pure and positional schedulers, as stated below.

Lemma 1. *Let $\mathcal{M}$ be a finite MDP, s_0 be an initial state, and ☺ be a target state. Then, for $\mathsf{opt} \in \{\inf, \sup\}$, there exists a pure and positional scheduler $\sigma^{\mathsf{opt}} \in \mathsf{Sched}_{\mathsf{pp}}(\mathcal{M})$ such that $\mathbb{P}^{\sigma^{\mathsf{opt}}}_{\mathcal{M},s_0}(\mathbf{F}\,☺) = \mathbb{P}^{\mathsf{opt}}_{\mathcal{M},s_0}(\mathbf{F}\,☺)$.*

Alternatively to solving a linear program, value-iteration techniques can also be used and often turn out to be more efficient in practice; see [24]. They rely on a fixed-point characterization (the *Bellman equations*) of the values $\mathsf{val}^{\mathsf{opt}}_{\mathcal{M}}(s) \stackrel{\text{def}}{=} \mathbb{P}^{\mathsf{opt}}_{\mathcal{M},s}(\mathbf{F}\,☺)$, where $\mathsf{opt} \in \{\inf, \sup\}$. This characterization also holds for finitely action-branching countable MDPs [35], and can even be extended to stochastic turn-based two-player games with reachability objectives [14,29]. Yet, the convergence of the fixed point does not imply the existence of a *stopping criterion* that can be used to identify when the computed value is sufficiently close to the actual value.

We recall existing results about the complexity of optimal schedulers for reachability objectives in MDPs, which we will use in later sections. The two items below are implied respectively by [35, Theorem 7.3.6] and [33, Theorem B]. The latter was also discussed more recently in [27].

Lemma 2. *Let $\mathcal{M} = (S, \mathsf{Act}, \mathsf{P})$ be a countable MDP and ☺ $\in S$ be a target state.*

1. *Assume $\mathcal{M}$ is finitely action-branching. There exists $\sigma \in \mathsf{Sched}_{\mathsf{pp}}(\mathcal{M})$ s.t. for all $s \in S$, $\mathbb{P}^{\sigma}_{\mathcal{M},s}(\mathbf{F}\,☺) = \mathbb{P}^{\inf}_{\mathcal{M},s}(\mathbf{F}\,☺)$.*
2. *For all $\varepsilon > 0$, there exists $\sigma \in \mathsf{Sched}_{\mathsf{pp}}(\mathcal{M})$ s.t. for all $s \in S$, $\mathbb{P}^{\sigma}_{\mathcal{M},s}(\mathbf{F}\,☺) \geq \mathbb{P}^{\sup}_{\mathcal{M},s}(\mathbf{F}\,☺) - \varepsilon$.*

A couple of remarks are of interest:

– The finite action-branching assumption is needed for the first item. Optimal schedulers for infimum reachability probabilities may not exist for infinitely branching MDPs, and ε-optimal schedulers may even require memory [28, Theorem 3].
– For supremum reachability probabilities, optimal schedulers may not exist, even in finitely branching MDPs; such an example is provided in [27, Figure 1]. This is why we only consider ε-optimal schedulers in the second item. Interestingly, item 2 fails to hold in MDPs with an uncountable state space [33, Theorem A]. As per the definition above, all MDPs in this paper are assumed countable.

Approximation Schemes and Algorithms. Even if characterizations of the values exist in infinite MDPs [35], no general algorithm is known to compute $\mathbb{P}^{\inf}_{\mathcal{M},s_0}(\mathbf{F} \; \text{☺})$ and $\mathbb{P}^{\sup}_{\mathcal{M},s_0}(\mathbf{F} \; \text{☺})$, or to decide whether these values exceed a threshold. Of course, such algorithms would very much depend on the representation of infinite MDPs.

In this paper, we aim at providing generic *approximation schemes* for infimum and supremum reachability probabilities in countable MDPs.

Definition 3. *An* approximation algorithm *takes as an input an MDP $\mathcal{M}$, an initial state s_0, a target state ☺, an optimization criterion* opt $\in \{\inf, \sup\}$, *and a precision $\varepsilon > 0$, and returns a value v such that $|v - \mathbb{P}^{\text{opt}}_{\mathcal{M},s_0}(\mathbf{F} \; \text{☺})| \leq \varepsilon$.*

In this paper, we provide generic *approximation schemes*, defined by two sequences $(r_n^-)_n$ and $(r_n^+)_n$, respectively non-decreasing and non-increasing, such that for every $n \in \mathbb{N}$, $r_n^- \leq \mathbb{P}^{\text{opt}}_{\mathcal{M},s_0}(\mathbf{F} \; \text{☺}) \leq r_n^+$. An approximation scheme is *converging* on $\mathcal{M}$ from s_0 if for every precision $\varepsilon > 0$, there exists $n \in \mathbb{N}$ such that $|r_n^+ - r_n^-| \leq \varepsilon$ (which means that any v in the interval $[r_n^-, r_n^+]$ is a solution to our problem). An approximation scheme yields an *approximation algorithm* if it is converging and the values r_n^- and r_n^+ can be effectively computed for arbitrarily large n.

Converting a converging approximation scheme into an algorithm requires hypotheses on the MDPs considered (e.g., finitely representable, restrictions on branching). In Sect. 4, we focus on approximation schemes; in Sect. 5, we investigate when these schemes are converging; in Sect. 6, we show on specific classes of models that these converging schemes can be made into algorithms. All these results will be enabled by the notion of *decisiveness* for MDPs, discussed in Sect. 3.

3   Decisiveness for MDPs

In this section, we define several flavors of *decisiveness* for MDPs, inspired by the notion of decisiveness defined for Markov chains [1]. We fix an MDP $\mathcal{M} = (S, \text{Act}, \text{P})$ and an absorbing target state ☺ $\in S$ for the rest of this section.

3.1   Avoid Sets

For Markov chains, the first ingredient to define decisiveness is the notion of *avoid set*, which is the set of states from which one can no longer reach ☺ (the avoid set was denoted $\widetilde{☺}$ in [1]). We extend this notion in several directions.

If $\sigma \in \mathsf{Sched_{pp}}(\mathcal{M})$, we define the *avoid set of $\mathcal{M}$ w.r.t. σ* as:

$$\mathsf{Avoid}_{\mathcal{M}}^{\sigma}(☺) = \left\{ s \in S \mid \mathbb{P}_{\mathcal{M},s}^{\sigma}(\mathbf{F}\ ☺) = 0 \right\}\ .$$

This is the avoid set of the Markov chain (as defined in [1]) induced by the pure and positional scheduler σ on $\mathcal{M}$.

We also define two other notions of *avoid set*, depending on whether one considers the infimum or supremum value over schedulers. For $\mathsf{opt} \in \{\inf, \sup\}$, we let:

$$\mathsf{Avoid}_{\mathcal{M}}^{\mathsf{opt}}(☺) = \left\{ s \in S \mid \mathsf{opt}_{\sigma \in \mathsf{Sched}(\mathcal{M})} \mathbb{P}_{\mathcal{M},s}^{\sigma}(\mathbf{F}\ ☺) = 0 \right\}\ .$$

Note that

$$\sup_{\sigma \in \mathsf{Sched}(\mathcal{M})} \mathbb{P}_{\mathcal{M},s}^{\sigma}(\mathbf{F}\ ☺) = 0 \quad \text{iff} \quad \forall \sigma \in \mathsf{Sched}(\mathcal{M}),\ \mathbb{P}_{\mathcal{M},s}^{\sigma}(\mathbf{F}\ ☺) = 0$$

$$\text{iff} \quad \forall \sigma \in \mathsf{Sched_{pp}}(\mathcal{M}),\ \mathbb{P}_{\mathcal{M},s}^{\sigma}(\mathbf{F}\ ☺) = 0\ ,$$

where the second equivalence can be shown using Lemma 2, item 2. We deduce that:

$$\mathsf{Avoid}_{\mathcal{M}}^{\sup}(☺) = \bigcap_{\sigma \in \mathsf{Sched_{pp}}(\mathcal{M})} \mathsf{Avoid}_{\mathcal{M}}^{\sigma}(☺)\ .$$

In contrast, it may happen that $\inf_{\sigma \in \mathsf{Sched}(\mathcal{M})} \mathbb{P}_{\mathcal{M},s}^{\sigma}(\mathbf{F}\ ☺) = 0$, yet there is no $\sigma \in \mathsf{Sched}(\mathcal{M})$ such that $\mathbb{P}_{\mathcal{M},s}^{\sigma}(\mathbf{F}\ ☺) = 0$. For instance, on the MDP $\mathcal{M}^{\mathtt{L}}$ in Fig. 2 (left), when choosing action α_i from s_0, the probability of $\mathbf{F}\ ☺$ is $\frac{1}{2^i}$. Recall that given Lemma 2 (item 1), this behaviour requires infinite action-branching: when $\mathcal{M}$ is finitely action-branching, we have that there exists a pure and positional scheduler $\sigma_{\inf}$ such that $\mathsf{Avoid}_{\mathcal{M}}^{\sigma_{\inf}}(☺) = \mathsf{Avoid}_{\mathcal{M}}^{\inf}(☺)$.

Following the definitions, for every $\sigma \in \mathsf{Sched_{pp}}(\mathcal{M})$, we have

$$\mathsf{Avoid}_{\mathcal{M}}^{\sup}(☺) \subseteq \mathsf{Avoid}_{\mathcal{M}}^{\sigma}(☺) \subseteq \mathsf{Avoid}_{\mathcal{M}}^{\inf}(☺)\ .$$

We show two examples to illustrate various kinds of avoid sets and when they can differ.

Example 1. Consider the three-state MDP $\mathcal{M}^{\mathtt{R}}$ on the right of Fig. 2. We have that $\mathsf{Avoid}_{\mathcal{M}^{\mathtt{R}}}^{\sup}(☺) \neq \mathsf{Avoid}_{\mathcal{M}^{\mathtt{R}}}^{\sigma}(☺)$ for some scheduler σ: indeed, $\mathsf{Avoid}_{\mathcal{M}^{\mathtt{R}}}^{\sup}(☺) = \{☺\}$, but for the pure and positional scheduler σ_{α} that chooses α in s_0, we have $\mathsf{Avoid}_{\mathcal{M}^{\mathtt{R}}}^{\sigma_{\alpha}}(☺) = \{☺, s_0\}$.

Consider again the infinitely branching MDP $\mathcal{M}^{\mathtt{L}}$ given in Fig. 2 (left). We have that $\mathsf{Avoid}_{\mathcal{M}^{\mathtt{L}}}^{\sigma}(☺) \neq \mathsf{Avoid}_{\mathcal{M}^{\mathtt{L}}}^{\inf}(☺)$ for all pure and positional schedulers σ: indeed, $\mathsf{Avoid}_{\mathcal{M}^{\mathtt{L}}}^{\inf}(☺) = \{s_0, ☺\}$, but $\mathsf{Avoid}_{\mathcal{M}^{\mathtt{L}}}^{\sigma}(☺) = \{☺\}$ for all such σ.

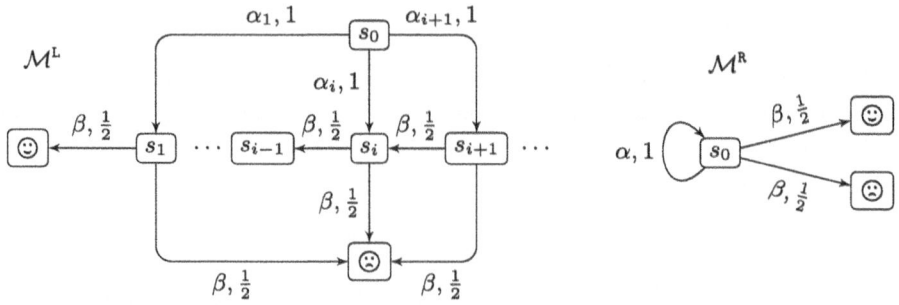

Fig. 2. Left: MDP $\mathcal{M}^{\mathrm{L}}$ for which $\mathbb{P}^{\mathrm{inf}}_{\mathcal{M}^{\mathrm{L}},s_0}(\mathbf{F} \ \odot) = 0$, yet for every scheduler σ, $\mathbb{P}^{\sigma}_{\mathcal{M}^{\mathrm{L}},s_0}(\mathbf{F} \ \odot) > 0$. Right: MDP $\mathcal{M}^{\mathrm{R}}$ such that $\mathsf{Avoid}^{\mathrm{sup}}_{\mathcal{M}^{\mathrm{R}}}(\odot) \neq \mathsf{Avoid}^{\sigma}_{\mathcal{M}^{\mathrm{R}}}(\odot)$ for some scheduler σ.

3.2 Decisiveness Properties

We now define several notions of decisiveness for MDPs, which are natural extensions of the decisiveness for Markov chains [1].

Definition 4 (Decisiveness). *Let $\mathcal{M} = (S, \mathsf{Act}, \mathrm{P})$ be an MDP, $\odot \in S$ be an absorbing target state, and $s \in S$ be a state.*

- *Let $\sigma \in \mathsf{Sched}_{\mathrm{pp}}(\mathcal{M})$. MDP $\mathcal{M}$ is said σ-decisive w.r.t. $\odot$ from s whenever*

$$\mathbb{P}^{\sigma}_{\mathcal{M},s}\big(\mathbf{F} \ \odot \vee \mathbf{F} \ \mathsf{Avoid}^{\sigma}_{\mathcal{M}}(\odot)\big) = 1 \ .$$

- *MDP $\mathcal{M}$ is univ-decisive w.r.t. $\odot$ from s whenever for all $\sigma \in \mathsf{Sched}_{\mathrm{pp}}(\mathcal{M})$, $\mathcal{M}$ is σ-decisive w.r.t. $\odot$ from s; that is,*

$$\forall \sigma \in \mathsf{Sched}_{\mathrm{pp}}(\mathcal{M}), \ \mathbb{P}^{\sigma}_{\mathcal{M},s}\big(\mathbf{F} \ \odot \vee \mathbf{F} \ \mathsf{Avoid}^{\sigma}_{\mathcal{M}}(\odot)\big) = 1 \ .$$

- *Let $\mathsf{opt} \in \{\inf, \sup\}$. MDP $\mathcal{M}$ is opt-decisive w.r.t. $\odot$ from s whenever*

$$\forall \sigma \in \mathsf{Sched}_{\mathrm{pp}}(\mathcal{M}), \ \mathbb{P}^{\sigma}_{\mathcal{M},s}\big(\mathbf{F} \ \odot \vee \mathbf{F} \ \mathsf{Avoid}^{\mathsf{opt}}_{\mathcal{M}}(\odot)\big) = 1 \ .$$

In the case of Markov chains, all these notions are equivalent and coincide with the notion of decisiveness defined in [1]. In the case of MDPs, these notions are different. Since $\mathsf{Avoid}^{\mathrm{sup}}_{\mathcal{M}}(\odot) \subseteq \mathsf{Avoid}^{\sigma}_{\mathcal{M}}(\odot) \subseteq \mathsf{Avoid}^{\mathrm{inf}}_{\mathcal{M}}(\odot)$ (for all pure and positional schedulers σ), sup-decisiveness is a stronger condition than univ-decisiveness, which is itself stronger than inf-decisiveness. We show examples distinguishing these notions.

Example 2. To distinguish sup-decisiveness from univ-decisiveness, we go back to the three-state MDP $\mathcal{M}^R$ from Example 1 (Fig. 2, right). Recall that $\mathsf{Avoid}^{\sup}_{\mathcal{M}^R}(\odot) = \{\odot\}$. Hence, for the scheduler σ_α that chooses α in s_0, we have $\mathbb{P}^{\sigma_\alpha}_{\mathcal{M}^R,s_0}(\mathbf{F}\,\odot \vee \mathbf{F}\,\mathsf{Avoid}^{\sup}_{\mathcal{M}^R}(\odot)) = 0$. Hence, $\mathcal{M}^R$ is not sup-decisive w.r.t. $\odot$ from s_0. On the other hand, we can show it is univ-decisive by considering the only two pure and positional schedulers σ_α and σ_β. We have $\mathsf{Avoid}^{\sigma_\alpha}_{\mathcal{M}^R}(\odot) = \{\odot, s_0\}$, so $\mathbb{P}^{\sigma_\alpha}_{\mathcal{M}^R,s_0}(\mathbf{F}\,\odot \vee \mathbf{F}\,\mathsf{Avoid}^{\sigma_\alpha}_{\mathcal{M}^R}(\odot)) = 1$. We have $\mathsf{Avoid}^{\sigma_\beta}_{\mathcal{M}^R}(\odot) = \{\odot\}$, so $\mathbb{P}^{\sigma_\beta}_{\mathcal{M}^R,s_0}(\mathbf{F}\,\odot \vee \mathbf{F}\,\mathsf{Avoid}^{\sigma_\beta}_{\mathcal{M}^R}(\odot)) = 1$. Hence, $\mathcal{M}^R$ is univ-decisive w.r.t. $\odot$ from s_0.

To distinguish univ-decisiveness from inf-decisiveness, consider the MDP $\mathcal{M}$ that was depicted in Fig. 1, and assume that $p > \frac{1}{2}$ (i.e., the random walk when choosing α repeatedly is *diverging*). Add to this MDP $\mathcal{M}$ an initial state s_0 from which one can go to any state s_i with an action α_i. We have that $\mathsf{Avoid}^{\inf}_{\mathcal{M}}(\odot) = \{s_0, \odot\}$, since the probability to reach $\odot$ can be made arbitrarily small by choosing α_i for a sufficiently large i. Hence, $\mathbb{P}^\sigma_{\mathcal{M},s_0}(\mathbf{F}\,\odot \vee \mathbf{F}\,\mathsf{Avoid}^{\inf}_{\mathcal{M}}(\odot)) = 1$ for all schedulers σ, so $\mathcal{M}$ is inf-decisive w.r.t. $\odot$ from s_0. However, for all fixed schedulers σ, we have $\mathsf{Avoid}^\sigma_{\mathcal{M}}(\odot) = \{\odot\}$, so $\mathbb{P}^\sigma_{\mathcal{M},s_0}(\mathbf{F}\,\odot \vee \mathbf{F}\,\mathsf{Avoid}^\sigma_{\mathcal{M}}(\odot)) < 1$. So $\mathcal{M}$ is not univ-decisive w.r.t. $\odot$ from s_0.

We finally show an example which is not inf-decisive (and thus, not univ-decisive or sup-decisive either). Consider again the MDP $\mathcal{M}$ in Fig. 1, also with $p > \frac{1}{2}$, but this time without the extra state s_0. It is such that $\mathsf{Avoid}^{\inf}_{\mathcal{M}}(\odot) = \{\odot\}$ since there is a positive probability to visit $\odot$ from every state (except from $\odot$), no matter the scheduler. The MDP $\mathcal{M}$ is not inf-decisive from s_0 w.r.t. $\odot$, since the scheduler which always selects α avoids $\odot$ and $\odot$ with positive probability λ_p.

Remark 2. Observe that the definitions of avoid sets and decisiveness only quantify over *pure and positional* schedulers. This will turn out to be sufficient for our purposes, notably thanks to the scheduler complexity results from Lemma 2.

Also, intuitively, quantifying over arbitrary schedulers would allow the cause for non-decisiveness to arise from the scheduler rather than the structure of the MDP. This would make the properties harder to check and less commonly satisfied. To see why, consider again the three-state MDP $\mathcal{M}^R$ in Fig. 2 (right). Consider the (infinite-memory) scheduler σ that, as long as s_0 is not left, chooses α with probability $1 - \frac{1}{2^{i+1}}$ and β with probability $\frac{1}{2^{i+1}}$ at step i. This scheduler avoids $\odot$ with probability $\prod_i (1 - \frac{1}{2^{i+1}}) > 0$. Yet, there is always a non-zero probability to reach $\odot$. Fixing σ induces an infinite Markov chain whose avoid set is $\{\odot\}$, but we do not have that $\{\odot, \odot\}$ is reached with probability 1. If we were to consider such schedulers, the MDP $\mathcal{M}^R$ would not be univ-decisive.

3.3 Decisiveness Criteria

We show how to adapt two existing criteria for the decisiveness of Markov chains [1, Lemmas 3.4 & 3.7] to MDPs. In both cases, we generalize the definition of a property to MDPs and show that this property implies some form of decisiveness. The proofs are in [9, Appendix A].

The first criterion relates to the existence of a *finite attractor*. It will be used in Sect. 6.1 to show that a class of infinite MDPs (*NPLCSs*) is inf-decisive.

Definition 5. *Let $\mathcal{M} = (S, \mathsf{Act}, \mathsf{P})$ be an MDP. We say that $\mathcal{M}$ has a* finite attractor *if there exists a finite set $A \subseteq S$ such that from all states $s \in S$, for all schedulers $\sigma \in \mathsf{Sched}_{pp}(\mathcal{M})$, $\mathbb{P}^\sigma_{\mathcal{M},s}(\mathbf{F}\,A) = 1$.*

Remark 3. Quantifying only over *pure and positional* schedulers in the definition of a finite attractor is sufficient for our purposes (such as the upcoming result). It would be stronger to require that $\mathbb{P}^\sigma_{\mathcal{M},s}(\mathbf{F}\,A) = 1$ for all $\sigma \in \mathsf{Sched}(\mathcal{M})$, as witnessed, e.g., by [28, Figure 3a] with $A = \{t\}$.

Proposition 1. *Let $\mathcal{M} = (S, \mathsf{Act}, \mathsf{P})$ be an MDP and $\odot \in S$ be an absorbing target state. If $\mathcal{M}$ has a finite attractor, then $\mathcal{M}$ is univ-decisive (hence also inf-decisive) w.r.t. $\odot$ from every state.*

Observe that, in particular, all finite MDPs are univ-decisive and inf-decisive (as for finite MDPs, the full state space S is a finite attractor). However, not all finite MDPs are sup-decisive; a counterexample was given in Example 2.

In finite MDPs, we can relate the notion of sup-decisiveness to the notion of *end component* [4]. An end component of an MDP $\mathcal{M} = (S, \mathsf{Act}, \mathsf{P})$ is a pair (R, A) where $R \subseteq S$ and $A \colon R \to 2^{\mathsf{Act}}$ such that for all $s \in R$, $A(s) \subseteq \mathsf{En}(s)$ and for all $a \in A(s)$, the support of $P(s, a, \cdot)$ is included in R, and the graph induced by (R, A) is strongly connected. As end components are commonly studied in MDPs, we formally state the relation here; however, we will use neither this result nor the notion of end component in the sequel.

Proposition 2. *Let $\mathcal{M} = (S, \mathsf{Act}, \mathsf{P})$ be a finite MDP and $\odot \in S$ be an absorbing target state. We have that $\mathcal{M}$ is sup-decisive w.r.t. $\odot$ from every state if and only if for all end components (R, A) of $\mathcal{M}$, either $R = \{\odot\}$ or $R \subseteq \mathsf{Avoid}^{\sup}_{\mathcal{M}}(\odot)$.*

This result gives another reason why the three-state MDP $\mathcal{M}^{\mathrm{R}}$ from Example 1 is not sup-decisive, as $(\{s_0\}, \{\alpha\})$ is an end component which is neither $\{\odot\}$ nor contained in $\mathsf{Avoid}^{\sup}_{\mathcal{M}}(\odot)$.

For finite MDPs, the property of end components used in Proposition 2 already appears in various works as a necessary property for the value-iteration algorithm to converge [15,24]. However, the notion of end components and its related results do not carry over straightforwardly to infinite MDPs; we believe that sup-decisiveness is a natural candidate for a property that is both well-defined on infinite MDPs and happens to coincide with this known property of finite MDPs.

We now extend a second decisiveness criterion by generalizing the concept of *globally coarse* Markov chains [1, Lemma 3.7]. Here, this extension yields a criterion for sup-decisiveness in MDPs.

Definition 6. *Let $\mathcal{M} = (S, \mathsf{Act}, \mathsf{P})$ be an MDP with an absorbing target state $\odot$ and a distinct absorbing state $\odot$ (in particular, $\odot \in \mathsf{Avoid}^{\sup}_{\mathcal{M}}(\odot)$). The MDP $\mathcal{M}$ is* semantically stopping *w.r.t $\odot$ and $\odot$ if there exists $p > 0$ such that from every state s, for all schedulers $\sigma \in \mathsf{Sched}_{pp}(\mathcal{M})$, $\mathbb{P}^\sigma_{\mathcal{M},s}(\mathbf{F}\,\odot \vee \mathbf{F}\,\odot) \geq p$.*

Proposition 3. *Let $\mathcal{M} = (S, \mathsf{Act}, \mathsf{P})$ be an MDP, ☺ be an absorbing target state, and ☹ be an absorbing state. If $\mathcal{M}$ is semantically stopping w.r.t. ☺ and ☹, then $\mathcal{M}$ is sup-decisive w.r.t. ☺ from every state.*

We can immediately deduce a natural syntactic class of MDPs that are sup-decisive. We say that an MDP $\mathcal{M} = (S, \mathsf{Act}, \mathsf{P})$ is *stopping* if there exists $p > 0$ from every state s, for every action $a \in \mathsf{En}(s)$, $\mathsf{P}(s, a, \{☺, ☹\}) \geq p$. It means that there is a uniformly bounded probability that a path "terminates" *at every step*. This is a natural adaptation to MDPs of the concept of *stopping* introduced by Shapley for stochastic games in 1953 [37] and used, e.g., in [17].

4 Generic Approximation Schemes

The objective of this section is to provide generic approximation schemes for optimum reachability probabilities, and to understand under which conditions they are converging. For conciseness, most proofs are omitted from this section; they can be found in [9, Appendix B and C].

For the rest of this section, we let $\mathcal{M} = (S, \mathsf{Act}, \mathsf{P})$ be an MDP, $s_0 \in S$ be an initial state, and ☺ $\in S$ be an absorbing target state.

4.1 Collapsing Avoid Sets and First Approximation Scheme

For opt $\in \{\inf, \sup\}$, we build a new MDP $\mathcal{M}^{\mathsf{opt}} = (S^{\mathsf{opt}}, \mathsf{Act}, \mathsf{P}^{\mathsf{opt}})$ by merging states in $\mathsf{Avoid}_{\mathcal{M}}^{\mathsf{opt}}(☺)$ into a fresh absorbing state $☺^{\mathsf{opt}}$.

Formally, $\mathcal{M}^{\mathsf{opt}} = (S^{\mathsf{opt}}, \mathsf{Act}, \mathsf{P}^{\mathsf{opt}})$ with

- $S^{\mathsf{opt}} = \left(S \backslash \mathsf{Avoid}_{\mathcal{M}}^{\mathsf{opt}}(☺)\right) \cup \{☺^{\mathsf{opt}}\}$;
- for every $s, s' \in S^{\mathsf{opt}} \backslash \{☺^{\mathsf{opt}}\}$, for every $a \in \mathsf{Act}$, $\mathsf{P}^{\mathsf{opt}}(s, a, s') = \mathsf{P}(s, a, s')$;
- for every $s \in S^{\mathsf{opt}} \backslash \{☺^{\mathsf{opt}}\}$, $\mathsf{P}^{\mathsf{opt}}(s, a, ☺^{\mathsf{opt}}) = \sum_{s' \in \mathsf{Avoid}_{\mathcal{M}}^{\mathsf{opt}}(☺)} \mathsf{P}(s, a, s')$;
- for every $a \in \mathsf{Act}$, $\mathsf{P}^{\mathsf{opt}}(☺^{\mathsf{opt}}, a, ☺^{\mathsf{opt}}) = 1$.

In both cases (when opt = inf or when opt = sup), notice that ☺ $\in S^{\mathsf{opt}}$. W.l.o.g., we assume that the initial state is preserved in the collapsed MDP (i.e., $s_0 \in S \cap S^{\mathsf{opt}}$); otherwise, by definition of $\mathsf{Avoid}_{\mathcal{M}}^{\mathsf{opt}}(☺)$, $\mathsf{opt}_\sigma \mathbb{P}_{\mathcal{M}, s_0}^\sigma(\mathbf{F} ☺) = 0$ and the value to be computed is trivially 0.

Note also the following two properties:

- for every $s \in S^{\inf} \backslash \{☺^{\inf}\}$, for every $\sigma \in \mathsf{Sched}(\mathcal{M}^{\inf})$, $\mathbb{P}_{\mathcal{M}^{\inf}, s}^\sigma(\mathbf{F} ☺) > 0$;
- for every $s \in S^{\sup} \backslash \{☺^{\sup}\}$, there is $\sigma \in \mathsf{Sched}(\mathcal{M}^{\sup})$ s.t. $\mathbb{P}_{\mathcal{M}^{\sup}, s}^\sigma(\mathbf{F} ☺) > 0$.

The above constructions collapsing avoid sets preserve optimum probabilities (with no prior assumption on $\mathcal{M}$; proof in [9, Appendix C]):

Lemma 3. $\mathbb{P}_{\mathcal{M}, s_0}^{\mathsf{opt}}(\mathbf{F} ☺) = \mathbb{P}_{\mathcal{M}^{\mathsf{opt}}, s_0}^{\mathsf{opt}}(\mathbf{F} ☺)$.

According to Lemma 3, computing the supremum probability (resp. infimum probability) in $\mathcal{M}$ can equivalently be done in $\mathcal{M}^{\mathrm{sup}}$ (resp. $\mathcal{M}^{\mathrm{inf}}$).

To do so, for every integer n, we define the following events in $\mathcal{M}^{\mathrm{opt}}$:

$$\begin{cases} R_n = \mathbf{F}_{\leq n}\odot \\ H_n^{\mathrm{opt}} = \mathbf{G}_{\leq n}(\neg\odot \wedge \neg\odot^{\mathrm{opt}}) \end{cases}.$$

In words, R_n expresses that the target is *reached* within n steps, and H_n^{opt} denotes that the target has not been reached within n steps, but that we are still in a region from which the probability of reaching $\odot$ is bounded away from 0 (in the case $\mathsf{opt} = \inf$) or from which reaching $\odot$ is possible with positive probability (in the case $\mathsf{opt} = \sup$). Note that $R_n \vee H_n^{\mathrm{opt}} = \mathbf{F}_{\leq n}\odot \vee \mathbf{G}_{\leq n}\neg\odot^{\mathrm{opt}}$.

We use these events to find lower and upper bounds on the desired probability $p = \mathbb{P}_{\mathcal{M}^{\mathrm{opt}},s_0}^{\mathrm{opt}}(\mathbf{F}\ \odot)$. The aim is that, thanks to the step bound n, these bounds are easier to compute than p in many classes of MDPs. A lower bound for p is trivially given by $\mathbb{P}_{\mathcal{M}^{\mathrm{opt}},s_0}^{\mathrm{opt}}(R_n)$: reaching $\odot$ *within n steps* naturally implies reaching $\odot$. An upper bound is given by $\mathbb{P}_{\mathcal{M}^{\mathrm{opt}},s_0}^{\mathrm{opt}}(R_n \vee H_n^{\mathrm{opt}})$: to reach $\odot$, it is necessary to either reach $\odot$ within n steps or to be in a state from which reaching $\odot$ is still possible after n steps. We state these observations formally.

Lemma 4. *For every initial state $s_0 \in S$ and every $n \in \mathbb{N}$,*

$$\mathbb{P}_{\mathcal{M}^{\mathrm{opt}},s_0}^{\mathrm{opt}}(R_n) \leq \mathbb{P}_{\mathcal{M}^{\mathrm{opt}},s_0}^{\mathrm{opt}}(\mathbf{F}\ \odot) \leq \mathbb{P}_{\mathcal{M}^{\mathrm{opt}},s_0}^{\mathrm{opt}}(R_n \vee H_n^{\mathrm{opt}})$$
$$\leq \mathbb{P}_{\mathcal{M}^{\mathrm{opt}},s_0}^{\mathrm{opt}}(R_n) + \mathbb{P}_{\mathcal{M}^{\mathrm{opt}},s_0}^{\mathrm{sup}}(H_n^{\mathrm{opt}}).$$

Thanks to Lemma 4, it is natural to define an approximation scheme with $\mathbb{P}_{\mathcal{M}^{\mathrm{opt}},s_0}^{\mathrm{opt}}(R_n)$ as a lower bound, and $\mathbb{P}_{\mathcal{M}^{\mathrm{opt}},s_0}^{\mathrm{opt}}(R_n \vee H_n^{\mathrm{opt}})$ as an upper bound, as formalised in Scheme 1. If the input is a Markov chain, this corresponds exactly to the path enumeration algorithm from [1, Algorithm 1].

Input : An MDP $\mathcal{M}$, $s_0 \in S$, $\odot \in S$, and $\varepsilon \in (0,1)$.
Output: A value $v \in [0,1]$.

$n := 0$;
repeat
  $\quad n := n + 1$;
  $\quad p_n^{\mathrm{opt},-} := \mathbb{P}_{\mathcal{M}^{\mathrm{opt}},s_0}^{\mathrm{opt}}(\mathbf{F}_{\leq n}\odot)$;
  $\quad p_n^{\mathrm{opt},+} := \mathbb{P}_{\mathcal{M}^{\mathrm{opt}},s_0}^{\mathrm{opt}}(\mathbf{F}_{\leq n}\odot \vee \mathbf{G}_{\leq n}(\neg\odot \wedge \neg\odot^{\mathrm{opt}}))$;
until $|p_n^{\mathrm{opt},+} - p_n^{\mathrm{opt},-}| \leq \varepsilon$;
return $p_n^{\mathrm{opt},-}$

Scheme 1: Approx_Scheme$_1^{\mathrm{opt}}$

Thanks to the last inequality of Lemma 4, if we prove that if in some MDP, for all schedulers, the probability of H_n^{opt} becomes negligible as n grows, then this ensures the convergence of the scheme for this MDP.

Theorem 1. *Let $\mathcal{M} = (S, \mathsf{Act}, \mathsf{P})$ be an MDP, $s_0 \in S$ be an initial state and $\odot \in S$ be a target state. Assume that $\lim_{n\to\infty} \mathbb{P}^{\sup}_{\mathcal{M}^{\mathsf{opt}},s_0}(H_n^{\mathsf{opt}}) = 0$. Then* `Approx_Scheme`$_1^{\mathsf{opt}}$ *provides a converging approximation scheme for $\mathbb{P}^{\mathsf{opt}}_{\mathcal{M},s_0}(\mathbf{F} \odot)$.*

Proof. The sequence $(p_n^{\mathsf{opt},-})_n$ is non-decreasing and the sequence $(p_n^{\mathsf{opt},+})_n$ is non-increasing. Assuming they converge to the same value (which is the case when $\lim_{n\to\infty} \mathbb{P}^{\sup}_{\mathcal{M}^{\mathsf{opt}},s_0}(H_n^{\mathsf{opt}}) = 0$ thanks to Lemma 4), then `Approx_Scheme`$_1^{\mathsf{opt}}$ converges, which means that it returns an ε-approximation of $\mathbb{P}^{\mathsf{opt}}_{\mathcal{M}^{\mathsf{opt}},s_0}(\mathbf{F} \odot)$. By Lemma 3, this corresponds to an ε-approximation of $\mathbb{P}^{\mathsf{opt}}_{\mathcal{M},s_0}(\mathbf{F} \odot)$. $\square$

Scheme `Approx_Scheme`$_1^{\mathsf{opt}}$ is based on unfoldings of the MDP to deeper and deeper depths. Precisely, the lower bound $p_n^{\mathsf{opt},-}$ is the probability in the unfolding up to depth n of histories that reach $\odot$; $p_n^{\mathsf{opt},+}$ is the probability in the same unfolding of histories that either reach $\odot$ or end in a state from which there is a path to $\odot$ in $\mathcal{M}^{\mathsf{opt}}$.

For completeness, we further clarify the relevance of the sequences $(p_n^{\mathsf{opt},-})_n$ and $(p_n^{\mathsf{opt},+})_n$ with respect to our aim. Focusing on $(p_n^{\mathsf{opt},-})_n$, observe that what the scheme computes is (an approximation) of the limit of this sequence, i.e., $\lim_n \mathsf{opt}_\sigma \mathbb{P}^\sigma_{\mathcal{M}^{\mathsf{opt}},s_0}(\mathbf{F}_{\leq n} \odot)$. Yet, the actual value we want to approximate is $\mathbb{P}^{\mathsf{opt}}_{\mathcal{M}^{\mathsf{opt}},s_0}(\mathbf{F} \odot)$, which is equal to $\mathsf{opt}_\sigma \lim_n \mathbb{P}^\sigma_{\mathcal{M}^{\mathsf{opt}},s_0}(\mathbf{F}_{\leq n} \odot)$. The convergence of the scheme is a sufficient condition for $\lim_{n\to\infty} p_n^{\mathsf{opt},-} = \mathbb{P}^{\mathsf{opt}}_{\mathcal{M}^{\mathsf{opt}},s_0}(\mathbf{F} \odot)$: indeed, given Lemma 4, this is the only possible limit value. Independently of the convergence of the scheme, these two values also always coincide in finitely action-branching MDPs. This statement is proved in [9, Appendix B].

Lemma 5. *Let $\mathcal{M}$ be a finitely action-branching MDP. Then,*

$$\lim_{n\to\infty} p_n^{\mathsf{opt},-} = \mathbb{P}^{\mathsf{opt}}_{\mathcal{M}^{\mathsf{opt}},s_0}(\mathbf{F} \odot) \quad and \quad \lim_{n\to\infty} p_n^{\mathsf{opt},+} = \mathbb{P}^{\mathsf{opt}}_{\mathcal{M}^{\mathsf{opt}},s_0}(\mathbf{F} \odot \vee \mathbf{G}(\neg\odot \wedge \neg\odot^{\mathsf{opt}})) \ .$$

However, this fails to hold in some infinitely branching MDPs.

Example 3. Consider the infinitely branching MDP $\mathcal{M}$ from Fig. 3. Note that $\mathsf{Avoid}^{\inf}_{\mathcal{M}}(\odot) = \emptyset$, so $\mathcal{M}^{\inf} = \mathcal{M}$. From s_0, there is a single choice α_i ($i \geq 1$) to make, determining that $\odot$ will be reached in exactly i steps. This means that for all n, it is possible to avoid seeing $\odot$ within n steps (e.g., by choosing α_{n+1}). Hence, for all n, $\mathbb{P}^{\inf}_{\mathcal{M},s_0}(\mathbf{F}_{\leq n} \odot) = 0$. We deduce that $\lim_n p_n^{\inf,-} = 0$.

However, $\mathbb{P}^{\inf}_{\mathcal{M},s_0}(\mathbf{F} \odot) = 1$, as any action leads surely to $\odot$. We conclude that, unlike the finitely branching case, we have

$$0 = \liminf_n \inf_\sigma \mathbb{P}^\sigma_{\mathcal{M},s_0}(\mathbf{F}_{\leq n} \odot) < \inf_\sigma \lim_n \mathbb{P}^\sigma_{\mathcal{M},s_0}(\mathbf{F}_{\leq n} \odot) = 1 \ .$$

Using Lemma 4, we also have that $1 = \mathbb{P}^{\inf}_{\mathcal{M},s_0}(\mathbf{F} \odot) \leq \lim_n p_n^{\inf,+}$. Hence, the scheme `Approx_Scheme`$_1^{\inf}$ does not converge on that particular MDP.

When $\mathsf{opt} = \sup$, `Approx_Scheme`$_1^{\mathsf{opt}}$ may not converge, even on finite MDPs. Consider the three-state MDP $\mathcal{M}^{\mathbb{R}}$ from Example 1: we have that for every $n \geq 1$, $p_n^{\sup,-} = \frac{1}{2}$ and $p_n^{\sup,+} = 1$. Hence, the scheme does not terminate when $\varepsilon < \frac{1}{2}$.

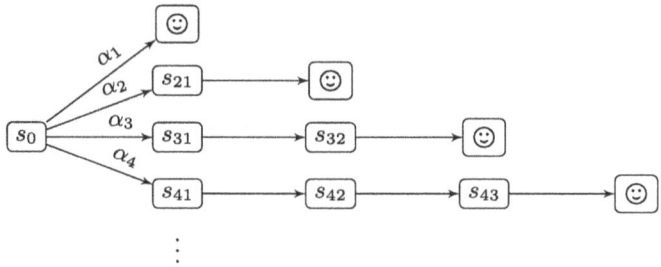

Fig. 3. An infinitely branching MDP $\mathcal{M}$ such that $0 = \lim_n \inf_\sigma \mathbb{P}^\sigma_{\mathcal{M},s_0}(\mathbf{F}_{\leq n}\smiley) < \inf_\sigma \lim_n \mathbb{P}^\sigma_{\mathcal{M},s_0}(\mathbf{F}_{\leq n}\smiley) = 1$.

4.2   Sliced MDP and Second Approximation Scheme

To overcome the above-mentioned shortcoming of $\mathtt{Approx\_Scheme}_1^{\mathrm{sup}}$, we propose a refined approximation scheme. Intuitively, instead of unfolding the MDP up to a fixed depth, as implicitly done in $\mathtt{Approx\_Scheme}_1^{\mathrm{opt}}$, we consider slices of the MDP consisting of the restrictions to all states that are reachable from s_0 within a fixed number of steps. Doing so, the convergence on finite MDPs is ensured.

Let $\mathcal{M} = (S, \mathsf{Act}, \mathsf{P})$ be an MDP, $s_0 \in S$ be an initial state, $\mathsf{opt} \in \{\inf, \sup\}$, and $\mathcal{M}^{\mathrm{opt}}$ be as defined in Sect. 4.1. For every $n \in \mathbb{N}$, we define the *sliced MDP* $\mathcal{M}_n^{\mathrm{opt}}$ as the restriction of $\mathcal{M}^{\mathrm{opt}}$ to states that can be reached within n steps from s_0. This construction is illustrated in Fig. 4.

For $n \geq 0$, let $\mathsf{Reach}_{s_0}^{\leq n}$ be the set of states reachable from s_0 with a positive probability in at most n steps. Formally, $\mathsf{Reach}_{s_0}^{\leq 0} = \{s_0\}$ and for $n \geq 0$,

$$\mathsf{Reach}_{s_0}^{\leq n+1} = \mathsf{Reach}_{s_0}^{\leq n} \cup \{s' \in S^{\mathrm{opt}} \mid \exists s \in \mathsf{Reach}_{s_0}^{\leq n}, \exists a \in \mathsf{Act}, \mathsf{P}^{\mathrm{opt}}(s,a,s') > 0\}\ .$$

For $n \geq 0$, the sliced MDP $\mathcal{M}_n^{\mathrm{opt}} = (S_n^{\mathrm{opt}}, \mathsf{Act}, \mathsf{P}_n^{\mathrm{opt}})$ is defined as follows:

- $S_n^{\mathrm{opt}} = \mathsf{Reach}_{s_0}^{\leq n} \cup \{s_\perp^n\}$;
- for all $s, s' \in \mathsf{Reach}_{s_0}^{\leq n}$, for all $a \in \mathsf{Act}$, $\mathsf{P}_n^{\mathrm{opt}}(s,a,s') = \mathsf{P}^{\mathrm{opt}}(s,a,s')$;
- for all $s \in \mathsf{Reach}_{s_0}^{\leq n}$, for all $a \in \mathsf{Act}$, $\mathsf{P}_n^{\mathrm{opt}}(s,a,s_\perp^n) = \sum_{s' \notin \mathsf{Reach}_{s_0}^{\leq n}} \mathsf{P}^{\mathrm{opt}}(s,a,s')$;
- for all $a \in \mathsf{Act}$, $\mathsf{P}_n^{\mathrm{opt}}(s_\perp^n, a, s_\perp^n) = 1$.

The state spaces of $\mathcal{M}^{\mathrm{opt}}$ and $\mathcal{M}_n^{\mathrm{opt}}$ coincide on $\mathsf{Reach}_{s_0}^{\leq n}$, and all transitions going out of $\mathsf{Reach}_{s_0}^{\leq n}$ in $\mathcal{M}^{\mathrm{opt}}$ are directed to $s_\perp^n$ in $\mathcal{M}_n^{\mathrm{opt}}$. Any path in $\mathcal{M}^{\mathrm{opt}}$ induces a unique path in $\mathcal{M}_n^{\mathrm{opt}}$ which either stays in the common state space $\mathsf{Reach}_{s_0}^{\leq n}$ or reaches $s_\perp^n$. Moreover, any path in $\mathcal{M}_n^{\mathrm{opt}}$ that reaches $\smiley$ corresponds to a path in $\mathcal{M}^{\mathrm{opt}}$ that also reaches $\smiley$. In the sequel, we use transparently the correspondence between paths in $\mathcal{M}^{\mathrm{opt}}$ that only visit states reachable within n steps, and paths in $\mathcal{M}_n^{\mathrm{opt}}$ that avoid $s_\perp^n$. Observe that the sliced MDPs of finitely branching MDPs are all finite.

We use events on the sliced MDP to find lower and upper bounds on the desired probability $p = \mathbb{P}^{\mathrm{opt}}_{\mathcal{M}^{\mathrm{opt}},s_0}(\mathbf{F}\ \smiley)$. A lower bound can be obtained through

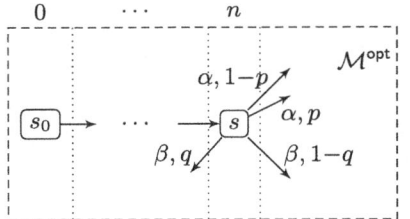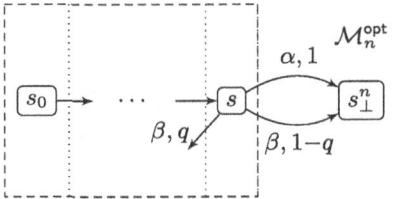

Fig. 4. Construction of the sliced MDP $\mathcal{M}_n^{\text{opt}}$ (right) from $\mathcal{M}^{\text{opt}}$ (left).

$\mathbb{P}_{\mathcal{M}_n^{\text{opt}},s_0}^{\text{opt}}(\mathbf{F} \odot)$: reaching $\odot$ in $\mathcal{M}_n^{\text{opt}}$ (only through states in $\mathsf{Reach}_{s_0}^{\leq n}$) implies reaching $\odot$ in $\mathcal{M}^{\text{opt}}$. An upper bound is given by $\mathbb{P}_{\mathcal{M}_n^{\text{opt}},s_0}^{\text{opt}}(\mathbf{F}(\odot \vee s_\perp^n))$: a path that reaches $\odot$ in $\mathcal{M}^{\text{opt}}$ would either reach $\odot$ or $s_\perp^n$ in $\mathcal{M}_n^{\text{opt}}$. We state these bounds formally, along with relations with the sequences $(p_n^{\text{opt},-})_n$ and $(p_n^{\text{opt},+})_n$ from `Approx_Scheme`$_1^{\text{opt}}$, in the following lemma (proved in [9, Appendix C]).

Lemma 6. *The sliced MDP $\mathcal{M}_n^{\text{opt}}$ enjoys the following inequalities:*

1. $\mathbb{P}_{\mathcal{M}_n^{\text{opt}},s_0}^{\text{opt}}(\mathbf{F} \odot) \leq \mathbb{P}_{\mathcal{M}^{\text{opt}},s_0}^{\text{opt}}(\mathbf{F} \odot) \leq \mathbb{P}_{\mathcal{M}_n^{\text{opt}},s_0}^{\text{opt}}(\mathbf{F}(\odot \vee s_\perp^n))$,
2. $\mathbb{P}_{\mathcal{M}_n^{\text{opt}},s_0}^{\text{opt}}(\mathbf{F}(\odot \vee s_\perp^n)) \leq \mathbb{P}_{\mathcal{M}_n^{\text{opt}},s_0}^{\text{opt}}(\mathbf{F} \odot) + \mathbb{P}_{\mathcal{M}_n^{\text{opt}},s_0}^{\sup}(\mathbf{F} s_\perp^n)$,
3. $p_n^{\text{opt},-} = \mathbb{P}_{\mathcal{M}^{\text{opt}},s_0}^{\text{opt}}(\mathbf{F}_{\leq n} \odot) \leq \mathbb{P}_{\mathcal{M}^{\text{opt}},s_0}^{\text{opt}}(\mathbf{F} \odot)$,
4. $\mathbb{P}_{\mathcal{M}_n^{\text{opt}},s_0}^{\text{opt}}(\mathbf{F}(\odot \vee s_\perp^n)) \leq \mathbb{P}_{\mathcal{M}^{\text{opt}},s_0}^{\text{opt}}(\mathbf{F}_{\leq n} \odot \vee \mathbf{G}_{\leq n}(\neg \odot \wedge \neg \odot^{\text{opt}})) = p_n^{\text{opt},+}$.

Thanks to Lemma 6 (item 1), it is natural to define an approximation scheme with $\mathbb{P}_{\mathcal{M}_n^{\text{opt}},s_0}^{\text{opt}}(\mathbf{F} \odot)$ as a lower bound, and $\mathbb{P}_{\mathcal{M}_n^{\text{opt}},s_0}^{\text{opt}}(\mathbf{F}(\odot \vee s_\perp^n))$ as an upper bound. It is formalised in Scheme 2.

Input : An MDP $\mathcal{M}$, $s_0 \in S$, $\odot \in S$, and $\varepsilon \in (0,1)$.
Output: A value $v \in [0,1]$.

$n := 0$;
repeat
$\quad | \quad n := n+1$;
$\quad | \quad q_n^{\text{opt},-} := \mathbb{P}_{\mathcal{M}_n^{\text{opt}},s_0}^{\text{opt}}(\mathbf{F} \odot)$;
$\quad | \quad q_n^{\text{opt},+} := \mathbb{P}_{\mathcal{M}_n^{\text{opt}},s_0}^{\text{opt}}(\mathbf{F}(\odot \vee s_\perp^n))$;
until $|q_n^{\text{opt},+} - q_n^{\text{opt},-}| \leq \varepsilon$;
return $q_n^{\text{opt},-}$

Scheme 2: `Approx_Scheme`$_2^{\text{opt}}$

Through Lemma 6 (items 3 and 4), we learn that for every n, $p_n^{\text{opt},-} \leq q_n^{\text{opt},-}$ and $q_n^{\text{opt},+} \leq p_n^{\text{opt},+}$. Thus, `Approx_Scheme`$_2^{\text{opt}}$ is a refinement of `Approx_Scheme`$_1^{\text{opt}}$, which we can state as follows.

Theorem 2. *Let $\mathcal{M} = (S, \mathsf{Act}, \mathsf{P})$ be an MDP, $s_0 \in S$ be an initial state, and $\odot \in S$ be a target state. Assume that `Approx_Scheme`$_1^{\text{opt}}$ provides a converging approximation scheme for $\mathbb{P}_{\mathcal{M},s_0}^{\text{opt}}(\mathbf{F} \odot)$. Then so does `Approx_Scheme`$_2^{\text{opt}}$.*

We give below a criterion for ensuring that $\texttt{Approx\_Scheme}_2^{\text{opt}}$ is an approximation scheme. It refines Theorem 1, as we have $\mathbb{P}^{\text{opt}}_{\mathcal{M}_n^{\text{opt}},s_0}(\mathbf{F}\, s_\perp^n) \leq \mathbb{P}^{\text{opt}}_{\mathcal{M}^{\text{opt}},s_0}(H_n^{\text{opt}})$: indeed, any path in $\mathcal{M}_n^{\text{opt}}$ that reaches $s_\perp^n$ (which takes at least $n+1$ steps) corresponds to a path that reaches neither ☺ nor ☹ within n steps in $\mathcal{M}^{\text{opt}}$.

Theorem 3. *Let* $\mathcal{M} = (S, \text{Act}, \mathsf{P})$ *be an MDP,* $s_0 \in S$ *be an initial state, and* ☺ $\in S$ *be a target state. Assume that* $\lim_{n\to\infty} \mathbb{P}^{\text{sup}}_{\mathcal{M}_n^{\text{opt}},s_0}(\mathbf{F}\, s_\perp^n) = 0$*. Then,* $\texttt{Approx\_Scheme}_2^{\text{opt}}$ *provides a converging approximation scheme for* $\mathbb{P}^{\text{opt}}_{\mathcal{M},s_0}(\mathbf{F}\,☺)$*.*

Proof. The sequences $(q_n^{\text{opt},-})_n$ and $(q_n^{\text{opt},+})_n$ are respectively non-decreasing and non-increasing. When they converge to the same limit, $\texttt{Approx\_Scheme}_2^{\text{opt}}$ terminates for all $\varepsilon > 0$. Moreover, thanks to Lemma 6 (item 1), for every $n \in \mathbb{N}$, $q_n^{\text{opt},-} \leq \mathbb{P}^{\text{opt}}_{\mathcal{M}^{\text{opt}},s_0}(\mathbf{F}\,☺) \leq q_n^{\text{opt},+}$ so that upon termination, $\texttt{Approx\_Scheme}_2^{\text{opt}}$ returns an ε-approximation of $\mathbb{P}^{\text{opt}}_{\mathcal{M}^{\text{opt}},s_0}(\mathbf{F}\,☺)$. This is an ε-approximation of $\mathbb{P}^{\text{opt}}_{\mathcal{M},s_0}(\mathbf{F}\,☺)$ by Lemma 3.

Under the assumption that $\lim_{n\to\infty} \mathbb{P}^{\text{sup}}_{\mathcal{M}_n^{\text{opt}},s_0}(\mathbf{F}\, s_\perp^n) = 0$, item 2 of Lemma 6 implies that $\lim_{n\to\infty} \mathbb{P}^{\text{opt}}_{\mathcal{M}_n^{\text{opt}},s_0}(\mathbf{F}\,(☺ \vee s_\perp^n)) = \lim_{n\to\infty} \mathbb{P}^{\text{opt}}_{\mathcal{M}_n^{\text{opt}},s_0}(\mathbf{F}\,☺)$. We obtain that the two sequences $(q_n^{\text{opt},-})_n$ and $(q_n^{\text{opt},+})_n$ converge towards $\mathbb{P}^{\text{opt}}_{\mathcal{M}^{\text{opt}},s_0}(\mathbf{F}\,☺)$, and $\texttt{Approx\_Scheme}_2^{\text{opt}}$ converges. $\qquad\square$

Remark 4 (The case of finite MDPs). $\texttt{Approx\_Scheme}_2^{\text{opt}}$ converges for finite MDPs (and stops at the latest after a number of iterations equal to the number of reachable states). In contrast, recall that approximating the supremum probability of a reachability objective with $\texttt{Approx\_Scheme}_1^{\text{sup}}$ may not converge on some finite MDPs (see Example 2 and Proposition 2).

5   When Do These Schemes Converge?

In this section, we give criteria related to decisiveness that ensure convergence of the approximation schemes. We start with criteria that ensure convergence of $\texttt{Approx\_Scheme}_1^{\text{opt}}$ (hence of $\texttt{Approx\_Scheme}_2^{\text{opt}}$ by Theorem 2). We then show that, for finitely branching MDPs, the convergence of $\texttt{Approx\_Scheme}_2^{\text{inf}}$ implies the convergence of $\texttt{Approx\_Scheme}_1^{\text{inf}}$. Finally, we give conditions on the MDPs that ensure the convergence of $\texttt{Approx\_Scheme}_2^{\text{sup}}$ (but not necessarily $\texttt{Approx\_Scheme}_1^{\text{sup}}$). Missing proofs for this section are in [9, Appendix D].

5.1   Convergence of $\texttt{Approx\_Scheme}_1^{\text{opt}}$

We give conditions related to decisiveness which ensure the convergence of the approximation schemes (recall that the convergence of $\texttt{Approx\_Scheme}_1^{\text{opt}}$ implies the convergence of $\texttt{Approx\_Scheme}_2^{\text{opt}}$ by Theorem 2).

Theorem 4. *Let* $\mathcal{M} = (S, \text{Act}, \mathsf{P})$ *be an MDP,* $s_0 \in S$ *be an initial state, and* ☺ *be a target state. Let* $\text{opt} \in \{\inf, \sup\}$*. Assume that* $\mathcal{M}$ *is finitely action-branching and* opt*-decisive w.r.t.* ☺ *from* s_0*. Then* $\texttt{Approx\_Scheme}_1^{\text{opt}}$ *converges on* $\mathcal{M}$ *from* s_0*.*

To highlight the role of the decisiveness hypotheses in Theorem 4, we show on some examples that without decisiveness, the approximation schemes may not converge. We first show that for some non-inf-decisive MDPs, the approximation schemes do not converge. Consider indeed the MDP $\mathcal{M}$ from Fig. 1, which has $\mathsf{Avoid}_{\mathcal{M}}^{\inf}(\smiley) = \{\smiley\}$, so that $\mathcal{M}^{\inf} = \mathcal{M}$. In case $p > \frac{1}{2}$, $\mathcal{M}$ is not inf-decisive from s_1 w.r.t. $\smiley$ since the pure and positional scheduler σ that always picks action α has a positive probability, say λ_p, to never reach $\smiley$ nor $\frownie$. For every $n \geq 1$,

$$p_n^{\inf,+} = \mathbb{P}_{\mathcal{M},s_1}^{\inf}(\mathbf{F}_{\leq n}\smiley \vee \mathbf{G}_{\leq n}(\neg\smiley \wedge \neg\frownie)) = 1 - \mathbb{P}_{\mathcal{M},s_1}^{\sup}(\mathbf{F}_{\leq n}\smiley) = q \ .$$

This is achieved by choosing β straight away; any other scheduler runs the risk of reaching $\smiley$. On the other hand, $p_n^{\inf,-} \leq \mathbb{P}_{\mathcal{M},s_1}^{\inf}(\mathbf{F}\,\smiley) \leq 1 - \lambda_p$ (which is the value obtained by the scheduler always choosing α). Hence, by picking q and p such that $q > 1 - \lambda_p$, $\mathtt{Approx\_Scheme}_1^{\inf}$ does not converge on $\mathcal{M}$ from s_1.

Similar arguments show that $\mathtt{Approx\_Scheme}_2^{\inf}$ does not converge on $\mathcal{M}$ from s_1. First, $q_n^{\inf,+} = \mathbb{P}_{\mathcal{M}_n^{\inf},s_1}^{\inf}(\mathbf{F}\,(\smiley \vee s_\perp^n)) = q$—this is achieved by choosing β straight away; any other scheduler runs the risk of reaching $\smiley$ or $s_\perp^n$. Second, $q_n^{\mathrm{opt},-} \leq \mathbb{P}_{\mathcal{M},s_1}^{\inf}(\mathbf{F}\,\smiley) \leq 1 - \lambda_p$. Thus, $\mathtt{Approx\_Scheme}_2^{\inf}$ does not converge either if $q > 1 - \lambda_p$.

Observe also that $\mathtt{Approx\_Scheme}_2^{\sup}$ (and thus $\mathtt{Approx\_Scheme}_1^{\sup}$) do not converge on the MDP $\mathcal{M}$ from Fig. 1 from s_1. This MDP is not sup-decisive (as it is not inf-decisive) w.r.t. $\smiley$ from s_1. We have that for every $n \in \mathbb{N}$, $q_n^{\sup,+} = 1$ (achieved by only choosing α), and yet $\mathbb{P}_{\mathcal{M},s_1}^{\sup}(\mathbf{F}\,\smiley) < 1$.

Finally, the finitely action-branching hypothesis is also critical. Recall that for the infinitely branching MDP in Example 3, the $\mathtt{Approx\_Scheme}_1^{\inf}$ does not converge from s_0. Yet, this MDP is inf-decisive w.r.t. $\smiley$ from s_0.

Despite the differences between the two approximation schemes, we have that for finitely branching MDPs, the convergence of $\mathtt{Approx\_Scheme}_2^{\inf}$ implies the convergence of $\mathtt{Approx\_Scheme}_1^{\inf}$.

Theorem 5. *Let* $\mathcal{M} = (S, \mathsf{Act}, \mathrm{P})$ *be a finitely branching MDP,* $s_0 \in S$ *be an initial state, and* $\smiley$ *be a target state. If* $\mathtt{Approx\_Scheme}_2^{\inf}$ *converges on* $\mathcal{M}$ *from* s_0, *then* $\mathtt{Approx\_Scheme}_1^{\inf}$ *converges on* $\mathcal{M}$ *from* s_0.

5.2   Convergence of $\mathtt{Approx\_Scheme}_2^{\mathrm{opt}}$

By applying the result and the discussion of Sect. 5.1, we already know that $\mathtt{Approx\_Scheme}_2^{\mathrm{opt}}$ converges under the conditions of Theorem 4, that is, when the MDP is finitely action-branching and opt-decisive. The sup-decisiveness property is rather restrictive and is not satisfied by finite MDPs, while $\mathtt{Approx\_Scheme}_2^{\sup}$ obviously converges on finite MDPs. We therefore propose alternative conditions that ensure the convergence of $\mathtt{Approx\_Scheme}_2^{\sup}$.

Definition 7. *Let* $\mathcal{M} = (S, \mathsf{Act}, \mathrm{P})$ *be an MDP,* $\smiley \in S$ *be a target state, and* $s \in S$ *be an initial state. The MDP* $\mathcal{M}$ *is* non-fleeing *w.r.t.* $\smiley$ *whenever for* *every* $\sigma \in \mathsf{Sched}_{\mathsf{pp}}(\mathcal{M})$,

$$\mathbb{P}^{\sigma}_{\mathcal{M}^{\mathrm{sup}}, s_0} \left(\mathrm{div} \cap \mathbf{F} \, \mathsf{Avoid}^{\sigma}_{\mathcal{M}^{\mathrm{sup}}} \, (\odot) \right) = 0$$

where div *is the event* $\bigcap_{n \in \mathbb{N}} \left(\mathbf{F} \left(S^{\mathrm{sup}}_{n+1} \backslash S^{\mathrm{sup}}_n \right) \right)$.

We explain the intuition of that notion through its negation: being fleeing corresponds to the possibility (in a probabilistic sense) to fly away from the origin of the MDP (and in particular never reach the target)—and even reach the avoid set of the current scheduler—in such a way that at any point, there exists a deviating scheduler that would reach the target (otherwise it would hit the avoid set summarized as $\odot^{\mathrm{sup}}$ in $\mathcal{M}^{\mathrm{sup}}$).

Theorem 6. *Let* $\mathcal{M} = (S, \mathsf{Act}, \mathsf{P})$ *be an MDP,* $s_0 \in S$ *be an initial state, and* $\odot$ *be a target state. Assume that* $\mathcal{M}$ *is finitely branching, univ-decisive w.r.t.* $\odot$ *from* s_0, *and non-fleeing. Then,* $\mathtt{Approx\_Scheme}^{\mathrm{opt}}_2$ *converges on* $\mathcal{M}$ *from* s_0.

The convergence condition in this theorem is incomparable to the one in Theorem 4: on the one hand, sup-decisiveness implies univ-decisiveness and non-fleeingness, so this condition is less restrictive; on the other hand, this theorem deals with finitely branching MDPs, as opposed to the more general finitely action-branching MDPs of Theorem 4. To prove this theorem, thanks to Theorem 3, it suffices to show the following (proof in [9, Appendix D]).

Lemma 7. *If* $\mathcal{M}$ *is finitely branching, univ-decisive, and non-fleeing, then*

$$\lim_{n \to \infty} \mathbb{P}^{\mathrm{sup}}_{\mathcal{M}^{\mathrm{sup}}_n, s_0} \left(\mathbf{F} \, s^n_{\perp} \right) = 0 \ .$$

6 Applications

We discuss the instantiation of the above approximation schemes into approximation *algorithms* for two concrete classes of systems: *non-deterministic and probabilistic lossy channel systems* (NPLCSs) and *partially observable MDPs* (POMDPs). Although theses models have distinct sources of randomness and infiniteness, they both induce countably infinite MDPs, where states are "configurations" of the system (control states and channel contents for NPLCSs, rational beliefs for POMDPs). In each case, we show that the induced MDPs (or small modifications thereof) satisfy a kind of decisiveness, which allows to use approximation schemes. We then show how to effectively compute approximations of the infimum reachability probabilities.

6.1 Lossy Channel Systems

In our first application, we consider the case where MDPs are induced by a probabilistic variant of *lossy channel systems*. *Non-deterministic and probabilistic lossy channel systems* build on channel systems, incorporating probabilistic message losses and allowing non-deterministic choices between possible read/write actions [6, 10].

Lossy Channel Systems and Induced MDP Semantics. A *channel system* is a tuple $\mathcal{S} = (Q, \mathcal{C}, \mathsf{M}, \mathsf{L}, \Delta)$ consisting of a finite set Q of *control states*, a finite set $\mathcal{C}$ of *channels*, a finite *message alphabet* M, a finite set L of *silent action labels*, and a finite set Δ of *transition rules*. Each transition rule has the form $q \xrightarrow{\mathsf{op}} p$ where op is an *operation* of the form

- $c!m$ for $c \in \mathcal{C}$ and $m \in \mathsf{M}$, representing the sending of message m along channel c;
- $c?m$ for $c \in \mathcal{C}$ and $m \in \mathsf{M}$, representing the reception of message m from channel c;
- $\ell \in \mathsf{L}$, representing an internal action labeled with ℓ with no corresponding sending/reception.

Messages are stored in FIFO queues, and the contents of the queues are naturally represented by finite words over M. A *configuration* of a channel system $\mathcal{S}$ is a pair $(q, \mathsf{w}) \in Q \times (\mathsf{M}^*)^{\mathcal{C}}$ consisting of a control state and of words describing each channel's contents. A transition rule $\delta = (q, \mathsf{op}, q')$ is *enabled* in a configuration (p, w) if $p = q$ and one of the following conditions applies: $\mathsf{op} = c?m$ and $\mathsf{w}(c) = mv$ with $v \in \mathsf{M}^*$, or $\mathsf{op} = c!m$, or $\mathsf{op} \in \mathsf{L}$. If so, firing δ from (q, w) yields the configuration $\delta(q, \mathsf{w}) = (q', \mathsf{w}')$ where, if $\mathsf{op} = c?m$ then $\mathsf{w}'(c) = v$ and for every $c' \neq c$, $\mathsf{w}'(c') = \mathsf{w}(c')$ (message m is read from channel c), if $\mathsf{op} = c!m$ then $\mathsf{w}'(c) = \mathsf{w}(c)m$ and for every $c' \neq c$, $\mathsf{w}'(c') = \mathsf{w}(c')$ (message m is written to channel c), and if $\mathsf{op} = \ell \in \mathsf{L}$ then for every $c \in \mathcal{C}$, $\mathsf{w}'(c) = \mathsf{w}(c)$ (no operation is performed on the channels contents).

A *non-deterministic and probabilistic lossy channel system* (NPLCS) is a pair $\mathcal{N} = (\mathcal{S}, \lambda)$ consisting of a channel system $\mathcal{S}$ and a loss rate $\lambda \in (0, 1)$. Its semantics is the MDP $\mathcal{M}[\mathcal{N}] = (S, \mathsf{Act}, \mathsf{P})$ where

- $S = Q \times (\mathsf{M}^*)^{\mathcal{C}}$: states are configurations of $\mathcal{S}$;
- $\mathsf{Act} = \Delta$: actions are transition rules of $\mathcal{S}$;
- the probabilistic transition function P is defined as follows

$$\mathsf{P}((q, \mathsf{w}), \delta, (q', \mathsf{w}')) = \begin{cases} \lambda^{|v| - |\mathsf{w}'|}(1 - \lambda)^{|\mathsf{w}'|}\binom{v}{\mathsf{w}'} & \text{if } \delta(q, \mathsf{w}) = (q', v) \\ 0 & \text{in all other cases.} \end{cases}$$

where the combinatorial coefficient $\binom{v}{\mathsf{w}'}$ is the number of different embeddings of w' in v. When writing $\delta(q, \mathsf{w}) = (q', v)$, we implicitly assume that δ is enabled in (q, w).

So defined, actions available from a configuration of an NPLCS correspond to transition rules that are enabled in the underlying channel system, and the successor configuration is obtained in two steps: first the rule is applied (possibly modifying the channels contents from w to v), and second each message is lost independently with probability λ (and kept with probability $(1 - \lambda)$) so that the resulting channels contents is w'.

In the sequel, when $\mathcal{S}$ and λ are clear from the context, we may simply write $\mathcal{M}$ for the MDP $\mathcal{M}[\mathcal{S}, \lambda]$.

Figure 5 represents a simple example of a channel system with a single channel (thus omitted in the action labels). Figure 6 shows an excerpt of the MDP induced by this channel system with a loss rate $\lambda = .2$. Because of the FIFO policy, the control state $\odot$ can only be reached from the initial configuration (q, ε) if messages are lost, for instance along this execution where a message is lost in the first step:

$$(p, \varepsilon) \xrightarrow{!b} (q, \varepsilon) \xrightarrow{!a} (q, a) \xrightarrow{!a} (q, aa) \xrightarrow{?a} (p, a) \xrightarrow{!b} (q, ab) \xrightarrow{?a} (p, b) \xrightarrow{?b} (\odot, \varepsilon) \ .$$

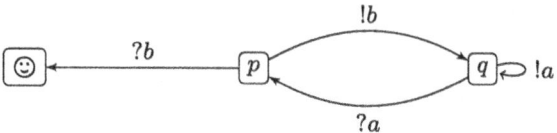

Fig. 5. A simple example of a channel system (with a single FIFO channel).

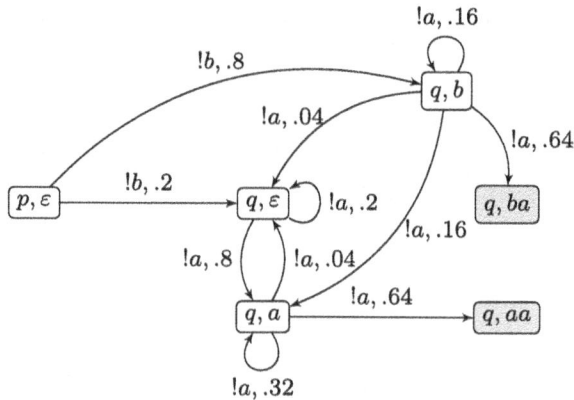

Fig. 6. An excerpt of MDP $\mathcal{M}[\mathcal{N}]$ induced by the NPLCS $\mathcal{N}$ from Fig. 5 with $\lambda = .2$: actions and states beyond the gray configurations are omitted.

We first state (un)decidability of comparing optimum reachability probabilities to qualitative threshold (0 or 1), and then use the inf-decisiveness property to show infimum reachability probability can be approximated in NPLCSs.

Qualitative Problems. We start with qualitative reachability in NPLCSs. Missing proofs are provided in [9, Appendix E].

Theorem 7. *When $\odot \subseteq Q$ is a set of control states, the following problems are decidable for NPLCSs:*

1. $\mathbb{P}^{\inf}(\mathbf{F} \; ☺) = 1$;
2. $\mathbb{P}^{\sup}(\mathbf{F} \; ☺) = 0$;
3. $\mathbb{P}^{\inf}(\mathbf{F} \; ☺) = 0$.

Yet, we establish the undecidability of the *value*-1 problem for NPLCSs, which also contrasts with the fact that the existence of a scheduler ensuring almost surely a reachability objective is decidable for NPLCSs [6].

Theorem 8. *The problem whether* $\mathbb{P}^{\sup}(\mathbf{F} \; ☺) = 1$ *is undecidable for NPLCSs.*

Approximation of the Infimum Reachability Probability. Iyer and Narasihma provided an approximation scheme for reachability probabilities in probabilistic channel systems, whose semantics is given by a countable Markov chain [25]. This result was then generalized to all decisive Markov chains by Abdulla, Ben Henda, and Mayr [1]. Here we show that, as far as infimum reachability probabilities are concerned, our approximation schemes can be used for MDPs induced by NPLCSs, thus lifting the early result of [25] from Markov chains to MDPs.

The key to prove the feasibility of approximating infimum reachability probabilities for NPLCSs is their finite attractor property:

Lemma 8 ([6], Proposition 4.2). *Let* $\mathcal{N}$ *be an NPLCS. Then* $\mathcal{M}[\mathcal{N}]$ *has a finite attractor.*

More precisely, the set of configurations with empty channels is a finite attractor for $\mathcal{M}[\mathcal{N}]$. We deduce by Proposition 1 that $\mathcal{M}[\mathcal{N}]$ is univ-decisive, hence inf-decisive, from (q_0, ε) w.r.t. any set F. The two approximation schemes `Approx_Scheme`$_1^{\inf}$ and `Approx_Scheme`$_2^{\inf}$ thus converge and are correct by Theorems 4 and 2.

Theorem 9. *There exists an algorithm that, given an NPLCS* $\mathcal{N}$ *with initial state* q_0, *goal set* $☺ \subseteq Q$ *and a rational number* $\varepsilon > 0$, *returns a value* ε-*close to* $\mathbb{P}^{\inf}_{\mathcal{M}[\mathcal{N}], q_0}(\mathbf{F} \; ☺)$.

It remains to discuss the effectiveness of the schemes. Assuming finite action-branching, thanks to Lemma 2 (item 1), computing $\mathsf{Avoid}^{\inf}_{\mathcal{M}}(☺)$ amounts to computing states from which one can almost-surely avoid $☺$ under a pure and positional scheduler, which amounts to computing states from which one can (surely) avoid $☺$. The latter set can be effectively computed as a fixed point [6].

6.2 Partially Observable MDPs

We focus in this section on *partially observable Markov decision processes*, abbreviated *POMDPs* [16,26]. Like MDPs, they exhibit both nondeterminism and probabilistic transitions; they are more general in that the scheduler making decisions does not know the current state of the system in general, but only receives a *signal* at each step that gives *partial information* about the current state. All decisions must be based on the sequence of signals received (and not the states visited) up to some point. Given such a sequence, a common way to represent the most accurate information about our current knowledge of the state of the system is through the probability distribution on the possible states, called a *belief*. Even though we consider POMDPs with finitely many states, actions, and signals, POMDPs are relevant in our framework as they each induce naturally an infinite MDP on the state space of beliefs.

Most natural quantitative problems in POMDPs are undecidable, already for simple reachability and safety (i.e., sup and inf reachability) objectives. This undecidability stems from results on the less general model of *probabilistic automata* [21,23,32,34]. Here are some examples of undecidable problems for probabilistic automata (and thus for POMDPs):

- Given a probabilistic automaton and a threshold $\lambda \in (0,1)$, decide whether there is a scheduler that ensures that a goal state is reached with probability at least λ [34]. The same holds replacing "reached" by "avoided".
- Given a probabilistic automaton, decide whether the supremum probability of reaching a goal state is 1 [23].
- Given $\varepsilon > 0$ and a probabilistic automaton such that either (i) there is a word accepted with probability at least $1 - \varepsilon$ or (ii) all words are accepted with probability at most ε, decide which case holds [32].

The latter problem is especially relevant to our setting, as it implies that there is no approximation algorithm for the *supremum* probability of reachability in POMDPs. Hence, we will not be able to make use of $\texttt{Approx\_Scheme}_1^{\text{sup}}$ or $\texttt{Approx\_Scheme}_2^{\text{sup}}$ on general POMDPs.

Yet, none of these results imply that the *infimum* reachability probability (i.e., the supremum value of safety) cannot be approximated in POMDPs. Using the inf-decisiveness property and $\texttt{Approx\_Scheme}_1^{\text{inf}}$, we show that there exists such an algorithm.

POMDPs and Induced MDP Semantics. We first recall basic notions on partially observable MDPs (POMDPs).

Definition 8. *A partially observable MDP is a tuple* $\mathcal{P} = (Q, \mathsf{Act}, \mathsf{Sig}, \mathrm{P})$ *where* Q *is a finite set of states,* Act *is a finite set of actions,* Sig *is a finite set of signals, and* $\mathrm{P}\colon Q \times \mathsf{Act} \times \mathsf{Sig} \times Q \to [0,1] \cap \mathbb{Q}$ *is a transition function such that for all* $q \in Q$ *and* $a \in \mathsf{Act}$, $\sum_{\mathfrak{s}\in\mathsf{Sig}} \sum_{q'\in Q} \mathrm{P}(q,a,\mathfrak{s},q') = 1$.

The main difference with the semantics of an MDP is that, in the case of POMDPs, the information of the current state is not known by schedulers in general; schedulers must base their decisions on the signals they receive (as well as the actions they have already selected). To keep this section concise, we will only express the semantics of POMDPs through the equivalent formulation of *belief MDPs* [26] below.

Let $\mathcal{P} = (Q, \mathsf{Act}, \mathsf{Sig}, \mathrm{P})$ be a POMDP. We assume without loss of generality that there is a distinguished state $q_{\odot} \in Q$ and a distinguished signal $\mathsf{done} \in \mathsf{Sig}$ such that for all $q \in Q$ and $a \in \mathsf{Act}$, $\mathrm{P}(q,a,\mathfrak{s},q_{\odot}) > 0$ implies $\mathfrak{s} = \mathsf{done}$, and for all $a \in \mathsf{Act}$, $\mathrm{P}(q_\odot, a, \mathsf{done}, q_\odot) = 1$. In other words, when the state $q_\odot$ is reached, the scheduler is aware of it (through the observation of signal done) and cannot escape it. We recall that we focus in this section on the *infimum* probability of reachability, which means we try as much as possible *not to* reach state $q_\odot$.

We write $\mathsf{Dist}_\mathbb{Q}(Q) = \{\mathfrak{b} \colon Q \to [0,1] \cap \mathbb{Q} \mid \sum_{q \in Q} \mathfrak{b}(q) = 1\}$ for the set of distributions over Q with rational values. A *belief (of $\mathcal{P}$)* is a probability distribution $\mathfrak{b} \in \mathsf{Dist}_\mathbb{Q}(Q)$. The *belief-update function* is the function $\mathcal{B} \colon \mathsf{Dist}_\mathbb{Q}(Q) \times \mathsf{Act} \times \mathsf{Sig} \to \mathsf{Dist}_\mathbb{Q}(Q)$ such that for all $(\mathfrak{b}, a, \mathfrak{s}) \in \mathsf{Dist}_\mathbb{Q}(Q) \times \mathsf{Act} \times \mathsf{Sig}$,

$$\mathcal{B}(\mathfrak{b}, a, \mathfrak{s})(q') = \frac{\sum_{q \in Q} \mathfrak{b}(q) \cdot \mathrm{P}(q, a, \mathfrak{s}, q')}{\sum_{q \in Q}(\mathfrak{b}(q) \cdot \sum_{q'' \in Q} \mathrm{P}(q, a, \mathfrak{s}, q''))} .$$

The belief $\mathcal{B}(\mathfrak{b}, a, \mathfrak{s})$ corresponds to the new belief that the scheduler has after selecting action a and observing signal $\mathfrak{s}$ from belief $\mathfrak{b}$. The *support* $\mathsf{supp}(\mathfrak{b})$ of a belief $\mathfrak{b}$ is the set $\{q \in Q \mid \mathfrak{b}(q) > 0\}$. The set of all belief supports then corresponds to the set $2^Q \setminus \{\emptyset\}$. In what follows, we denote beliefs (i.e., distributions) with font $\mathfrak{b}$, and belief supports (i.e., sets) with font b.

The *belief MDP of $\mathcal{P}$* is the infinite MDP $\mathcal{M}[\mathcal{P}] = (\mathsf{Dist}_\mathbb{Q}(Q), \mathsf{Act}, \mathrm{P}_\mathcal{P})$ where $\mathsf{Dist}_\mathbb{Q}(Q)$ is the set of beliefs, Act is the set of actions, and $\mathrm{P}_\mathcal{P} \colon \mathsf{Dist}_\mathbb{Q}(Q) \times \mathsf{Act} \times \mathsf{Dist}_\mathbb{Q}(Q) \to [0,1] \cap \mathbb{Q}$ is

$$\mathrm{P}_\mathcal{P}(\mathfrak{b}, a, \mathfrak{b}') = \sum_{\substack{\mathfrak{s} \in \mathsf{Sig} \text{ s.t.} \\ \mathcal{B}(\mathfrak{b}, a, \mathfrak{s}) = \mathfrak{b}'}} \sum_{q, q' \in Q} \mathfrak{b}(q) \cdot \mathrm{P}(q, a, \mathfrak{s}, q') .$$

Given our assumptions about the state $q_\odot$ of $\mathcal{P}$, we have that as soon as $q_\odot$ is reached in the POMDP, we reach the corresponding belief $q_\odot \mapsto 1$ in the belief MDP. We denote this belief by $\odot$.

If $q_0 \in Q$, we abusively write $\mathbb{P}^{\inf}_{\mathcal{M}[\mathcal{P}], q_0}(\mathbf{F} \; \odot)$ for the infimum probability of reaching $\odot$ in $\mathcal{M}[\mathcal{P}]$ starting from the belief $q_0 \mapsto 1$ (i.e., we assimilate the notation q_0 to the belief $q_0 \mapsto 1$).

Approximation of the Infimum Reachability Probability. Our plan is to apply Theorem 4 (and thus $\mathtt{Approx\_Scheme}^{\inf}_1$) to the infinite MDP $\mathcal{M}[\mathcal{P}]$ to approximate the infimum probability of reaching the goal state in a POMDP $\mathcal{P}$. Observe first that $\mathcal{M}[\mathcal{P}]$ is finitely action-branching as there are only finitely many actions in Act (and actually, even finitely branching as there are only finitely many signals in Sig, but this is not necessary to use Theorem 4). We would therefore need some kind of inf-decisiveness for $\mathcal{M}[\mathcal{P}]$. However, in general, $\mathcal{M}[\mathcal{P}]$ is not inf-decisive.

Example 4. Consider the POMDP $\mathcal{P}$ in Fig. 7. It has a single action α. Starting from q_0, if q_1 is reached, then only signal $\mathfrak{s}$ will ever be seen. Yet, through successive observations of $\mathfrak{s}$, the scheduler can never be sure to be in q_1; there is a decreasing but positive probability that the current state is q_2. Formally, the belief $\mathfrak{b}_n$ after seeing the sequence of signals $\mathfrak{s}^n$ (with $n \geq 1$) is defined by $\mathfrak{b}_n(q_1) = 1 - \frac{1}{2^n}$ and $\mathfrak{b}_n(q_2) = \frac{1}{2^n}$. All these beliefs still have a positive probability to reach ☺ (from q_2), so none of them are in $\mathsf{Avoid}^{\mathrm{inf}}_{\mathcal{M}[\mathcal{P}]}(☺)$. There is therefore a positive probability to stay in a region of $\mathcal{M}[\mathcal{P}]$ that is neither ☺ nor in $\mathsf{Avoid}^{\mathrm{inf}}_{\mathcal{M}[\mathcal{P}]}(☺)$, which shows that $\mathcal{M}[\mathcal{P}]$ is not inf-decisive w.r.t. ☺ from q_0.

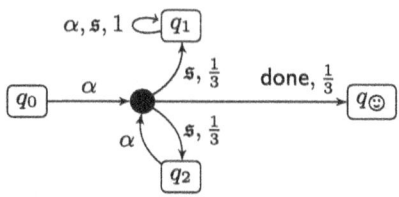

Fig. 7. A POMDP $\mathcal{P}$ such that $\mathcal{M}[\mathcal{P}]$ is not inf-decisive w.r.t. ☺ from q_0.

Our proof scheme is as follows: even though $\mathcal{M}[\mathcal{P}]$ is not inf-decisive in general, we show that from every POMDP $\mathcal{P}$ and every $\varepsilon > 0$, we can modify $\mathcal{M}[\mathcal{P}]$ slightly to obtain an infinite MDP $\mathcal{M}[\mathcal{P}]^\varepsilon$ such that:

- $\mathcal{M}[\mathcal{P}]^\varepsilon$ is inf-decisive (Lemma 9),
- the infimum probability of reaching the goal state in $\mathcal{M}[\mathcal{P}]^\varepsilon$ is within ε of the infimum probability of reaching the goal state in $\mathcal{M}[\mathcal{P}]$ (Lemma 10).

To obtain an approximation *algorithm*, we then discuss how to compute effectively the sequences $(p_n^{\mathrm{inf},-})_n$ and $(p_n^{\mathrm{inf},+})_n$ in the infinite $\mathcal{M}[\mathcal{P}]^\varepsilon$ given the finite representation of $\mathcal{P}$ (Theorem 10).

Let $\mathcal{P} = (Q, \mathsf{Act}, \mathsf{Sig}, \mathsf{P})$ be a POMDP, $q_0 \in Q$ be an initial state, and $\varepsilon > 0$. We construct $\mathcal{M}[\mathcal{P}]^\varepsilon$ from $\mathcal{M}[\mathcal{P}]$.

Observe that if a belief $\mathfrak{b}$ is such that there is a scheduler σ such that $\mathbb{P}^\sigma_{\mathcal{M}[\mathcal{P}],\mathfrak{b}}(\mathbf{F}\ ☺) = 0$, then for all beliefs $\mathfrak{b}'$ with $\mathsf{supp}(\mathfrak{b}') = \mathsf{supp}(\mathfrak{b})$, we also have $\mathbb{P}^\sigma_{\mathcal{M}[\mathcal{P}],\mathfrak{b}'}(\mathbf{F}\ ☺) = 0$. We define

$$B_{=0} = \{b \in 2^Q \setminus \{\emptyset\} \mid \exists\sigma, \forall\mathfrak{b} \text{ s.t. } \mathsf{supp}(\mathfrak{b}) = b, \mathbb{P}^\sigma_{\mathcal{M}[\mathcal{P}],\mathfrak{b}}(\mathbf{F}\ ☺) = 0\}.$$

To build $\mathcal{M}[\mathcal{P}]^\varepsilon$, we merge some specific beliefs of $\mathcal{M}[\mathcal{P}]$ into a single, new absorbing state $☺^\varepsilon$. The beliefs that are merged are the beliefs $\mathfrak{b}$ such that

$$\exists b' \subseteq \mathsf{supp}(\mathfrak{b}) \text{ s.t. } b' \in B_{=0} \text{ and } \sum_{q \in b'} \mathfrak{b}(q) \geq 1 - \varepsilon\ .$$

We call such a belief a $(1 - \varepsilon)$-*avoiding* belief. Intuitively, such a belief is one such that, if the current state is in a specific subset b' of the support (which occurs with probability $\geq 1 - \varepsilon$), a scheduler can ensure that the goal state is never reached.

Formally, we define the state space of $\mathcal{M}[\mathcal{P}]^\varepsilon$ as

$$S^\varepsilon = \{☺^\varepsilon\} \cup \{\mathfrak{b} \in \mathsf{Dist}_{\mathbb{Q}}(Q) \mid \mathfrak{b} \text{ is not } (1 - \varepsilon)\text{-avoiding}\}\ .$$

The transitions are then kept the same as in $\mathcal{M}[\mathcal{P}]$, except that the transitions to a $(1 - \varepsilon)$-avoiding belief are redirected to the absorbing state $\odot^{\varepsilon}$.

Example 5. We build the MDP $\mathcal{M}[\mathcal{P}]^{\varepsilon}$ obtained from the POMDP $\mathcal{P}$ in Example 4, with $\varepsilon = \frac{1}{8}$. It is shown in Fig. 8; we recall that state $\odot$ in $\mathcal{M}[\mathcal{P}]$ corresponds to the belief $q_{\odot} \mapsto 1$.

Observe that the only belief support in $B_{=0}$ is $\{q_1\}$. Hence, the beliefs that are $(1 - \varepsilon)$-avoiding are the beliefs $\mathfrak{b}$ such that $\mathfrak{b}(q_1) \geq \frac{7}{8}$. The MDP $\mathcal{M}[\mathcal{P}]^{\varepsilon}$ is even finite, and so is inf-decisive.

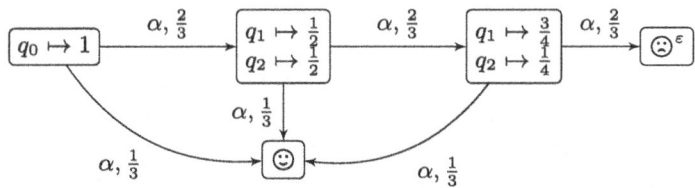

Fig. 8. The MDP $\mathcal{M}[\mathcal{P}]^{\varepsilon}$ obtained from $\mathcal{P}$ in Example 4 with $\varepsilon = \frac{1}{8}$.

We can now state the aforementioned results leading to an approximation algorithm for the infimum probability of reachability in POMDPs (proofs in [9, Appendix F]).

Lemma 9. *The infinite MDP $\mathcal{M}[\mathcal{P}]^{\varepsilon}$ is inf-decisive.*

Lemma 10. *We have that*

$$\mathbb{P}^{\inf}_{\mathcal{M}[\mathcal{P}]^{\varepsilon},q_0}(\mathbf{F}\,\odot) \leq \mathbb{P}^{\inf}_{\mathcal{M}[\mathcal{P}],q_0}(\mathbf{F}\,\odot) \leq \mathbb{P}^{\inf}_{\mathcal{M}[\mathcal{P}]^{\varepsilon},q_0}(\mathbf{F}\,\odot) + \varepsilon\ .$$

Theorem 10. *There exists an algorithm that, given any POMDP $\mathcal{P}$ and rational number $\varepsilon > 0$, returns a value ε-close to $\mathbb{P}^{\inf}_{\mathcal{M}[\mathcal{P}],q_0}(\mathbf{F}\,\odot)$.*

Proof. We describe the algorithm. Let $\mathcal{P}$ be a POMDP and $\varepsilon > 0$ be rational. We consider the MDP $\mathcal{M}[\mathcal{P}]^{\varepsilon/2}$, which is inf-decisive by Lemma 9. As $\mathcal{M}[\mathcal{P}]^{\varepsilon/2}$ is finitely action-branching, approximation scheme $\texttt{Approx\_Scheme}^{\inf}_1$ is converging (Theorem 4).

It remains to argue that the sequences $(p^{\inf,-}_n)_n$ and $(p^{\inf,+}_n)_n$ appearing in $\texttt{Approx\_Scheme}^{\inf}_1$ can be computed effectively. We first compute the set $B_{=0}$; this corresponds to multiple almost-sure safety problems on $\mathcal{P}$, which are decidable [16]. All beliefs up to a fixed depth can be computed exactly (they are all arrays of rational numbers), and since $B_{=0}$ was precomputed, we can decide whether a belief is $(1 - \frac{\varepsilon}{2})$-avoiding (and thus whether we have reached $\odot^{\varepsilon/2}$ in $\mathcal{M}[\mathcal{P}]^{\varepsilon/2}$).

Hence, we have an effective algorithm that returns a value v such that $\mathbb{P}^{\inf}_{\mathcal{M}[\mathcal{P}]^{\varepsilon/2},q_0}(\mathbf{F}\ \smiley) - \frac{\varepsilon}{2} \leq v \leq \mathbb{P}^{\inf}_{\mathcal{M}[\mathcal{P}]^{\varepsilon/2},q_0}(\mathbf{F}\ \smiley) + \frac{\varepsilon}{2}$. By Lemma 10, we have that $\mathbb{P}^{\inf}_{\mathcal{M}[\mathcal{P}]^{\varepsilon/2},q_0}(\mathbf{F}\ \smiley) \leq \mathbb{P}^{\inf}_{\mathcal{M}[\mathcal{P}],q_0}(\mathbf{F}\ \smiley) \leq \mathbb{P}^{\inf}_{\mathcal{M}[\mathcal{P}]^{\varepsilon/2},q_0}(\mathbf{F}\ \smiley) + \frac{\varepsilon}{2}$. Hence,

$$\mathbb{P}^{\inf}_{\mathcal{M}[\mathcal{P}],q_0}(\mathbf{F}\ \smiley) - \varepsilon \leq v \leq \mathbb{P}^{\inf}_{\mathcal{M}[\mathcal{P}],q_0}(\mathbf{F}\ \smiley) + \frac{\varepsilon}{2}\ ,$$

which suffices for an approximation algorithm with precision ε.　　□

7　Conclusion

We extended the decisiveness notion from Markov chains to Markov decision processes (MDPs) and demonstrated how to leverage this property to derive approximation schemes for optimum reachability probabilities (corresponding to maximizing the probability of reachability or safety objectives). The notion of inf-decisiveness appears to be of practical relevance, as we showed that it enables the approximation of the infimum reachability probability in two important classes of models: nondeterministic and probabilistic lossy channel systems and partially observable MDPs. The stronger notion of sup-decisiveness, while not yielding here new decidability results for specific MDP classes, provides valuable insights through its connection to the *stopping* notion and the convergence of value iteration for finite MDPs.

Natural directions for future research include extending our framework to richer objectives (e.g., repeated reachability) and exploring its applicability to broader classes of models (e.g., probabilistic vector addition systems with states).

Acknowledgments. Thomas Brihaye is partly supported by the Fonds de la Recherche Scientifique - FNRS under grant n° T.0027.21 and by the Belgian National Lottery.

References

1. Abdulla, P.A., Ben Henda, N., Mayr, R.: Decisive Markov chains. Logical Methods Comput. Sci. **3**(4:7), 1–32 (2007). https://doi.org/10.2168/LMCS-3(4:7)2007
2. Abdulla, P.A., Bertrand, N., Rabinovich, A., Schnoebelen, P.: Verification of probabilistic systems with faulty communication. Inf. Comput. **202**(2), 141–165 (2005). https://doi.org/10.1016/J.IC.2005.05.008
3. Abdulla, P.A., Clemente, L., Mayr, R., Sandberg, S.: Stochastic parity games on lossy channel systems. Logical Methods Comput. Sci. **10**(4) (2014). https://doi.org/10.2168/LMCS-10(4:21)2014
4. de Alfaro, L.: Formal Verification of Probabilistic Systems. Ph.D. thesis, Stanford University, Stanford, CA (1997). http://i.stanford.edu/pub/cstr/reports/cs/tr/98/1601/CS-TR-98-1601.pdf
5. Baier, C., Bertrand, N., Schnoebelen, P.: A note on the attractor-property of infinite-state Markov chains. Inf. Process. Lett. **97**(2), 58–63 (2006). https://doi.org/10.1016/j.ipl.2005.09.011

6. Baier, C., Bertrand, N., Schnoebelen, Ph.: Verifying nondeterministic probabilistic channel systems against ω-regular linear-time properties. ACM Trans. Comput. Logic **9**(1) (2007). https://doi.org/10.1145/1297658.1297663
7. Baier, C., Katoen, J.P.: Principles of Model Checking. MIT Press (2008)
8. Barbot, B., Bouyer, P., Haddad, S.: Beyond decisiveness of infinite Markov chains. In: Proceedings of 44th IARCS Annual Conference on Foundations of Software Technology and Theoretical Computer Science (FSTTCS 2024). LIPIcs, vol. 323, pp. 8:1–8:22. Leibniz-Zentrum für Informatik (2024). https://doi.org/10.4230/LIPIcs.FSTTCS.2024.8
9. Bertrand, N., Bouyer, P., Brihaye, T., Fournier, P., Vandenhove, P.: Decisiveness for countable MDPs and insights for NPLCSs and POMDPs. CoRR **abs/2008.10426** (2020). https://arxiv.org/abs/2008.10426
10. Bertrand, N., Schnoebelen, P.: Model checking lossy channels systems is probably decidable. In: Gordon, A.D. (ed.) FoSSaCS 2003. LNCS, vol. 2620, pp. 120–135. Springer, Heidelberg (2003). https://doi.org/10.1007/3-540-36576-1_8
11. Bertrand, N., Schnoebelen, Ph.: Solving stochastic Büchi games on infinite arenas with a finite attractor. In: Proceedings of 11th International Workshop on Quantitative Aspects of Programming Languages and Systems (QAPL 2013). EPTCS, vol. 117, pp. 116–131 (2013). https://doi.org/10.4204/EPTCS.117.8
12. Brázdil, T., Brozek, V., Etessami, K.: One-counter stochastic games. In: Proceedings IARCS Annual Conference on Foundations of Software Technology and Theoretical Computer Science (FSTTCS'10). LIPIcs, vol. 8, pp. 108–119. Leibniz-Zentrum für Informatik (2010). https://doi.org/10.4230/LIPIcs.FSTTCS.2010.108
13. Brázdil, T., Brožek, V., Etessami, K., Kučera, A.: Approximating the termination value of one-counter MDPs and stochastic games. In: Aceto, L., Henzinger, M., Sgall, J. (eds.) ICALP 2011. LNCS, vol. 6756, pp. 332–343. Springer, Heidelberg (2011). https://doi.org/10.1007/978-3-642-22012-8_26
14. Brázdil, T., Brozek, V., Kucera, A., Obdrzálek, J.: Qualitative reachability in stochastic BPA games. Inf. Comput. **209**(8), 1160–1183 (2011). https://doi.org/10.1016/J.IC.2011.02.002
15. Brázdil, T., et al.: Verification of Markov decision processes using learning algorithms. In: Cassez, F., Raskin, J.-F. (eds.) ATVA 2014. LNCS, vol. 8837, pp. 98–114. Springer, Cham (2014). https://doi.org/10.1007/978-3-319-11936-6_8
16. Chatterjee, K., Chmelik, M., Tracol, M.: What is decidable about partially observable Markov decision processes with ω-regular objectives. J. Comput. Syst. Sci. **82**(5), 878–911 (2016). https://doi.org/10.1016/J.JCSS.2016.02.009
17. Condon, A.: The complexity of stochastic games. Inf. Comput. **96**(2), 203–224 (1992). https://doi.org/10.1016/0890-5401(92)90048-K
18. Dehnert, C., Junges, S., Katoen, J.-P., Volk, M.: A Storm is coming: a modern probabilistic model checker. In: Majumdar, R., Kunčak, V. (eds.) CAV 2017. LNCS, vol. 10427, pp. 592–600. Springer, Cham (2017). https://doi.org/10.1007/978-3-319-63390-9_31
19. Etessami, K., Wojtczak, D., Yannakakis, M.: Quasi-birth-death processes, tree-like QBDs, probabilistic 1-counter automata, and pushdown systems. Perform. Eval. **67**(9), 837–857 (2010). https://doi.org/10.1016/j.peva.2009.12.009
20. Etessami, K., Yannakakis, M.: Recursive concurrent stochastic games. Logical Methods Comput. Sci. **4**(4) (2008). https://doi.org/10.2168/LMCS-4(4:7)2008
21. Fijalkow, N.: Undecidability results for probabilistic automata. ACM SIGLOG News **4**(4), 10–17 (2017). https://doi.org/10.1145/3157831.3157833

22. Finkel, A., Haddad, S., Ye, L.: About decisiveness of dynamic probabilistic models. In: Proceedings of 34th International Conference on Concurrency Theory (CONCUR'23). LIPIcs, vol. 279, pp. 14:1–14:17. Leibniz-Zentrum für Informatik (2023). https://doi.org/10.4230/LIPICS.CONCUR.2023.14
23. Gimbert, H., Oualhadj, Y.: Probabilistic automata on finite words: decidable and undecidable problems. In: Abramsky, S., Gavoille, C., Kirchner, C., Meyer auf der Heide, F., Spirakis, P.G. (eds.) ICALP 2010. LNCS, vol. 6199, pp. 527–538. Springer, Heidelberg (2010). https://doi.org/10.1007/978-3-642-14162-1_44
24. Haddad, S., Monmege, B.: Interval iteration algorithm for MDPs and IMDPs. Theoret. Comput. Sci. **735**, 111–131 (2018). https://doi.org/10.1016/j.tcs.2016.12.003
25. Iyer, S., Narasimha, M.: Probabilistic lossy channel systems. In: Proceedings 7th International Conference on Theory and Practice of Software Development (TAPSOFT'97). Lecture Notes in Computer Science, vol. 1214, pp. 667–681. Springer, Cham (1997). https://doi.org/10.1007/BFB0030633
26. Kaelbling, L., Littman, M.L., Cassandra, A.R.: Planning and acting in partially observable stochastic domains. Artif. Intell. **101**(1–2), 99–134 (1998). https://doi.org/10.1016/S0004-3702(98)00023-X
27. Kiefer, S., Mayr, R., Shirmohammadi, M., Totzke, P., Wojtczak, D.: How to play in infinite MDPs (invited talk). In: Czumaj, A., Dawar, A., Merelli, E. (eds.) Proc. 47th International Colloquium on Automata, Languages, and Programming (ICALP 2020). LIPIcs, vol. 168, pp. 3:1–3:18. Leibniz-Zentrum für Informatik (2020). https://doi.org/10.4230/LIPICS.ICALP.2020.3
28. Kiefer, S., Mayr, R., Shirmohammadi, M., Wojtczak, D.: Parity objectives in countable MDPs. In: Proceedings of 32th Annual Symposium on Logic in Computer Science (LICS 2017), pp. 1–11. IEEE Computer Society Press (2017). https://doi.org/10.1109/LICS.2017.8005100
29. Kucera, A.: Lectures in Game Theory for Computer Scientists. Cambridge University Press, Turn-Based Stochastic Games (2011). chap
30. Kucera, A., Esparza, J., Mayr, R.: Model checking probabilistic pushdown automata. Logical Methods Comput. Sci. **2**(1) (2006). https://doi.org/10.2168/LMCS-2(1:2)2006
31. Kwiatkowska, M., Norman, G., Parker, D.: PRISM 4.0: verification of probabilistic real-time systems. In: Gopalakrishnan, G., Qadeer, S. (eds.) CAV 2011. LNCS, vol. 6806, pp. 585–591. Springer, Heidelberg (2011). https://doi.org/10.1007/978-3-642-22110-1_47
32. Madani, O., Hanks, S., Condon, A.: On the undecidability of probabilistic planning and infinite-horizon partially observable Markov decision problems. In: Hendler, J., Subramanian, D. (eds.) Proc. 16th National Conference on Artificial Intelligence and Eleventh Conference on Innovative Applications of Artificial Intelligence, pp. 541–548. AAAI Press / The MIT Press (1999)
33. Ornstein, D.: On the existence of stationary optimal strategies. Proc. Am. Math. Soc. **20**(2), 563–569 (1969). http://www.jstor.org/stable/2035700
34. Paz, A.: Introduction to Probabilistic Automata. Academic Press (1971)
35. Puterman, M.L.: Markov Decision Processes: Discrete Stochastic Dynamic Programming. Wiley Series in Probability and Statistics, Wiley, 1st edn. (1994). https://doi.org/10.1002/9780470316887
36. Sericola, B.: Markov Chains: Theory, Algorithms and Applications. Wiley, Hoboken (2013)
37. Shapley, L.S.: Stochastic games. Proc. Natl. Acad. Sci. **39**(10), 1095–1100 (1953). https://doi.org/10.1073/pnas.39.10.1095

Polytopal Stochastic Games

Pablo F. Castro[1,3]($\boxtimes$) and Pedro R. D'Argenio[2,3]

[1] Departamento de Computación, Río Cuarto, Universidad Nacional de Río Cuarto, FCEFQyN, Córdoba, Argentina
castro@dc.exa.unrc.edu.ar
[2] Universidad Nacional de Córdoba, FAMAF, Córdoba, Argentina
pedro.dargenio@unc.edu.ar
[3] Consejo Nacional de Investigaciones Científicas y Técnicas (CONICET), Godoy Cruz , Argentina

Abstract. In this paper we introduce *polytopal stochastic games*, an extension of two-player, zero-sum, turn-based stochastic games, in which we may have uncertainty over the transition probabilities. In these games the uncertainty over the probability distributions is captured via linear (in)equalities whose space of solutions forms a polytope. We give a formal definition of these games and prove their basic properties: determinacy and existence of optimal memoryless and deterministic strategies. We do this for reachability and different types of reward objectives and show that the solution exists in a finite representation of the game. We also state that the corresponding decision problems are in NP∩coNP. We motivate the use of polytopal stochastic games via a simple example.

1 Introduction

In the last decades, stochastic systems have become ubiquitous in computer science: communication and security protocols, fault analysis in critical systems, autonomous devices, to name a few examples, typically use techniques coming from probability theory. Furthermore, well-known techniques in artificial intelligence, such as reinforcement learning [28], are based on stochastic models. In view of this, the verification and formal analysis of stochastic systems is one of the most active areas of research in software verification. Christel Baier and Joost-Pieter Katoen's book [4] is considered a standard reference in the area, it introduces common concepts and techniques for model checking probabilistic systems, this includes algorithms for verifying temporal assertions over Markov chains (MCs) and Markov Decision processes (MDPs). The latter can be considered as one player stochastic games, in which the system has to select strategies to solve non-determinism in stochastic settings. In general, game theory offers a powerful mathematical framework for specifying and verifying computing systems. The idea is appealing, a computing system can be thought of as a player

This work was supported by Agencia I+D+i PICT 2019-03134, SeCyT-UNC 33620230100384CB (MECANO), and EU Grant agreement ID: 101008233 (MISSION).

N. Bertrand et al. (Eds.): Christel Baier Festschrift, LNCS 15760, pp. 99–117, 2026.
https://doi.org/10.1007/978-3-031-97439-7_4

playing against an environment, or another system, while trying to achieve certain goals. For instance, a security system can be seen as a player that selects different countermeasures to possibly different types of maneuvers executed by an attacker (a second player) each of which may succeed with certain probabilities. The objective of the defense system is to minimize the probability that the attack succeeds while the attacker wants to maximize it. This scenario can be modeled as a stochastic game, and then analysed using techniques coming from game theory. Examples of applications of game theory to the analysis of systems can be found almost everywhere in the last years: self-driving cars [31], robotics [18], UAVs [14], security [3], etc. Furthermore, in recent years, some model checkers have been extended to provide support for stochastic games, e.g., this is the case of PRISM-Games [11], which offers support for several versions of stochastic games.

In this paper we focus on two-player, zero-sum, turn-based perfect-information stochastic games. Intuitively, they are non-deterministic probabilistic transition systems in which the vertices are partitioned into two sets: vertices belonging to player □ and vertices belonging to player ◇. When the current state belongs to a given player, say □, she performs an action by selecting one of the non-deterministic outgoing transitions which would lead to different states with some given probabilities. Typically, the players want to fulfill or maximize/minimize some objectives. Standard quantitative objectives are discounted sum (the players collect an amount of rewards during the play which are multiplied by a discount factor in each step), total sum (the players want to maximize/minimize the cumulative sum of the rewards collected during a play), mean-payoff (the objective is to maximize or minimize the long-run average reward), or simply a reachability objective, that is, they aim to maximize/minimize the probability of reaching certain subset of states. These kinds of objectives can be used and combined to model different kinds of systems, e.g., the case of a self-driving car intending to maximize the probability of reaching some zone in a city can be seen as a multiobjective game [12].

Most of the time, when modeling stochastic systems, one assumes that the probability distributions are exactly known, which may not always be the case due to measurement inaccuracies, lack of data, or other issues. In this paper we propose an extension of stochastic games that adds the possibility of having uncertainty over the probabilities. Games with some kinds of uncertainty have been considered for $1\frac{1}{2}$-player games, i.e., Markov Decision Processes (MDPs). For instance, Interval-valued Discrete-Time Markov Chains (IDTMCs) [17,20,29], Interval Markov Decision Process (IMDP) [29], and Convex MDPs [26]. To the best of our knowledge, these approaches have not been extended to stochastic games (i.e., $2\frac{1}{2}$-player games). A key challenge for doing so is that in multiplayer games one needs to prove determinacy results, this ensures that the games possess a well-defined value, which does not depend on the players' knowledge. In the aforementioned approaches the notion of uncertainty is usually adversarially resolved, that is, each time a state is visited, the adversary picks a transition distribution that respects the constraints, and takes a probabilistic step according to the chosen distribution. However, it is interesting to note that, in two-player games we may adopt two ways of resolving uncertainty: a controllable one, in which the actual player resolves the uncertainty following her goals; and an adversarial one in which the adversary resolves the uncertainty in her favor. The former approach is useful in those scenarios in which the uncertainty affects the adversary as she does not precisely know the possible movements of our player; while the latter is helpful to reason in worst-case scenarios.

We therefore introduce *polytopal stochastic games (PSG)*. PSGs, as defined in Sect. 4, allow one to model uncertainties over probability distributions using linear (closed) inequalities. Geometrically, these linear inequalities correspond to polytopes, i.e., bounded polyhedra. As PSGs are two-player games, both ways of resolving uncertainty are possible: the adversarial approach and the controllable one. Furthermore, we show that in all the cases these kinds of games preserve some good properties of standard stochastic games for several objectives: reachability, total rewards, average sum, and mean payoff. In particular we show that these games are determined and admit optimal memoryless and deterministic strategies. We also show that these inherently infinite games can be reduced to equivalent finite stochastic games that traverse exclusively through the vertices of the original polytopes. As such, they are amenable to standard algorithmic solutions. Finally, we prove that the complexity of these games for the aforementioned objectives remain in NP ∩ coNP, that is, they stay in the standard complexity class of simple stochastic games, even when polytopal games support for an uncountable number of actions for the players and the discretization may grow exponentially.

Related Work. Definitions of infinite stochastic games do exist (see, for instance, [21]) though they are of discrete nature, contrary to the type of games presented here. In fact, PSGs are related to IDTMCs [17, 20, 29], IMDPs [29], and Convex MDPs [26], but they are variants of MDPs and hence they are $1\frac{1}{2}$-games. In particular, PSGs adopt a semantics similar to IMDPs and Convex MDPs [26] in which the uncertainty introduced by the polytope is interpreted as an uncountable non-deterministic branching. While in [26] interior-point algorithms are used to solve Convex MDPs, we use a discretization through the vertices of polytopes to solve PSGs. Though this has an exponential impact, this is very mild in practice as we will show later. A much simpler variant of PSG was used in [8] to provide an algorithmic solution for a fault tolerant measure. This incipient idea served as the starting point for the generalization presented here.

Somewhat related are the stochastic timed games (STGs) [1, 7]. However, the continuous non-determinism introduced by the time in STGs is resolved by uniform and exponential distributions and the remnant non-determinism (resolved by the strategies) is still discrete. This does not make these models simpler since undecidability has been shown for games with at least 3 clocks [7].

Outline of the Paper. Section 2 presents a motivating example. Section 3 introduces the background needed for tackling the rest of the paper. The definition of PSGs, their semantics and basic properties are given in Sect. 4. The main results are presented in Sect. 5. Full proofs are gathered in the Appendix.

2   Roborta vs. Rigoborto in the Land of Uncertainties

We illustrate our approach by means of a simple example. Consider a field represented as a bidimensional (finite) grid of size $n \times n$, for $n \in \mathbb{N}$; and two robots –which we call Roborta and Rigoborto– that navigate it. Roborta can move sideways and forward, Rigoborto can move sideways and backward. The robots start at a certain initial position. Roborta intends to reach the end of the grid, i.e., she wants to reach position $(i, n+1)$ for any i, whereas Rigoborto wants to stop Roborta. He can achieve this by reaching Roborta's location. The robots play in turns. The objective of Roborta is to maximize the probability of reaching the exit, while the objective of Rigob-

Fig. 1. An example of a grid for the Roborta vs Rigoborto game.

orto is to minimize this value. We spice up this example by considering the terrain quality (which depends on factors like, e.g., stones, mud or grass) and slope, which may cause imprecisions and uncertainties in the robots mobility, probably making them slide towards some undesired direction. The terrain quality and slope may vary in each grid position. In Fig. 1, we show an example of such a scenario. Therein, the robots start at the corners, the arrows indicate the slopes in the terrain, and the colors in the cells indicate the terrain quality. Darker arrows correspond to sharper slopes. Similarly, cells with lower quality are colored with stronger red colors.

More precisely, for each (x, y)-cell, the terrain quality $q_{xy} \in [0, 0.5]$ gives the uncertainty factor, where $q_{xy} = 0$ means that probabilities are completely determined, and, as q_{xy} grows, the probability values become increasingly fuzzier. In addition, we consider two factors associated with the terrain slopes: $l_{xy}, f_{xy} \in [-1, 1]$, representing the inclination of the lateral and frontal slopes respectively. Thus, as l_{xy} get closer to 1 (-1), the likelihood of shifting to the right (left) increases, with $l_{xy} = 0$ not favouring any particular side. Similarly $f_{xy} > 1$ ($f_{xy} < -1$) biases the robot towards the front (back). Let p_c be the probability that the robot command is successful (that is, that it moves in the intended direction), and let p_l, p_r, p_f, and p_b be the probabilities that the command is unsuccessful and the robot uncontrollably slides respectively to the left, right, front, and back. Then, the space of all probability values can be defined by the following set of inequalities:

$$
\begin{aligned}
1 &= p_c + p_l + p_r + p_f + p_b \\
p_c &\geq 0, \quad p_l \geq 0, \quad p_r \geq 0, \quad p_f \geq 0, \quad p_b \geq 0 \\
p_c &\leq 1 - (q_{xy} + \tfrac{1}{2} \cdot (1 - (1 - |l_{xy}|) \cdot (1 - |f_{xy}|))) \\
0 &\leq (1 - \max(0, -l_{xy})) \cdot p_l - (1 - q_{xy}) \cdot (1 - \max(0, l_{xy})) \cdot p_r \\
0 &\leq (1 - \max(0, l_{xy})) \cdot p_r - (1 - q_{xy}) \cdot (1 - \max(0, -l_{xy})) \cdot p_l
\end{aligned}
$$

```
// action specification for Roborta moving to the left
[robl] (turn = 0) & (roby<L) & !Collision -> (rob_mov'=1) & (turn'=1)

[robl-cont] (turn = 1) & (rob_mov = 1) ->
  //The first four probabilistic options  correspond to environments setbacks
  pl : (robx'=max(0,robx-1)) & (rob_mov'=0) + pr : (robx'=min(W-1,robx+1)) & (rob_mov'=0)
  + pf : (roby'=roby+1) & (rob_mov'=0) + pb : (roby'=max(0,roby+1)) & (rob_mov'=0)
  + pc : (robx'=max(0,robx-1)) & (rob_mov'=0)
  {// inequations for uncertainty
    1-(Q[robx,roby]+(1-(1-abs(L[robx,roby]))*(1-abs(F[robx,roby])))/2) >= pc,
    (1-max(0,-L[robx,roby]))*pl - (1-Q[robx,roby])*(1-max(0,L[robx,roby]))*pr >= 0,
    (1-max(0,L[robx,roby]))*pr - (1-Q[robx,roby])*(1-max(0,-L[robx,roby]))*pl >= 0,
    (1-max(0,F[robx,roby]))*pf - (1-Q[robx,roby])*(1-max(0,-F[robx,roby]))*pb >= 0,
    (1-max(0,-F[robx,roby]))*pb - (1-Q[robx,roby])*(1-max(0,F[robx,roby]))*pf >= 0
  };
```

(a) Roborta moves left

```
// action specification for Rigoborto moving to the left
[rigl] (turn = 1) & (rob_mov = 0) & (rigy<L) & !Collision ->
  //The first four probabilistic options  correspond to environments setbacks
  pl : (rigx'=max(0,rigx-1)) & (turn'=0) & (Collision'=(robx=rigx && roby=rigy))
  + pr : (rigx'=min(W-1,rigx+1)) & (turn'=0) & (Collision'=(robx=rigx && roby=rigy))
  + pf : (rigy'=rigy+1) & (turn'=0) & (Collision'=(robx=rigx && roby=rigy))
  + pb : (rigy'=max(0,rigy+1)) & (turn'=0) & (Collision'=(robx=rigx && roby=rigy))
  + pc : (rigx'=max(0,rigx-1)) & (turn'=0) & (Collision'=(robx=rigx && roby=rigy))
  {// inequations for uncertainty
    1-(Q[rigx,rigy]+(1-(1-abs(L[rigx,rigy]))*(1-abs(F[rigx,rigy])))/2) >= pc,
    (1-max(0,-L[rigx,rigy]))*pl - (1-Q[rigx,rigy])*(1-max(0,L[rigx,rigy]))*pr >= 0,
    (1-max(0,L[rigx,rigy]))*pr - (1-Q[rigx,rigy])*(1-max(0,-L[rigx,rigy]))*pl >= 0,
    (1-max(0,F[rigx,rigy]))*pf - (1-Q[rigx,rigy])*(1-max(0,-F[rigx,rigy]))*pb >= 0,
    (1-max(0,-F[rigx,rigy]))*pb - (1-Q[rigx,rigy])*(1-max(0,F[rigx,rigy]))*pf >= 0
  };
```

(b) Rigoborto moves left

Fig. 2. Fragment of code for Roborta vs Rigoborto

$$0 \leq (1 - \max(0, f_{xy})) \cdot p_f - (1 - q_{xy}) \cdot (1 - \max(0, -f_{xy})) \cdot p_b$$
$$0 \leq (1 - \max(0, -f_{xy})) \cdot p_b - (1 - q_{xy}) \cdot (1 - \max(0, f_{xy})) \cdot p_f$$

Note that if $q_{xy} = 0$, the system has a unique solution. If, in addition, $l_{xy} > 0$, $1/(1 - l_{xy}) = p_r/p_l$ giving the likelihood ratio of sliding towards the right.

Our aim is to find the best strategy for Roborta to win against all odds. This implies that the terrain uncertainty behaves adversarially to Roborta but favourably to Rigoborto. Thus, in our model, Rigoborto controls the non-determinism introduced by the terrain uncertainty. Assuming an extension of the PRISM-Games language, the code could look like in Fig. 2, where subfigures 2a and 2b show the decisions to move left by Roborta and Rigoborto respectively.

Variable **turn** indicates who is the next player to move (with 0 for Roborta and 1 for Rigoborto). If it is Roborta's turn (see first line in Fig. 2a) and she decides to move left, she indicates it by setting **rob_mov'=1** (1 indicates a left move while 2, 3, and 4 are used for the other directions, and 0 to indicate that Roborta is not moving). At the same time, she yields her turn by setting **turn'=1**. Notice that the action is not yet complete: the reaction of the terrain to the move is encoded in the next line (action **robl-cont** in Fig. 2a). Notice that this action happens in a state in which **turn=1**, making the terrain uncertainty –defined by the polytope– adversarial to Roborta. Here, variables `robx` and `roby` correspond to Roborta's coordinates x and y and constant matrices `Q`, `L`, and `F` contain the respective values for q_{xy}, l_{xy}, and f_{xy}. The rest of the variables are as expected. Once this step is taken, variable `rob_mov` is set to 0, thus enabling Rigoborto's move. Rigoborto's decision to move left is given in Fig 2b. Notice that this is performed in a single action since we assume that the terrain uncertainty plays in favour of him. Something particular to this transition is the setting of variable `Collision` to indicate whether Rigoborto has caught Roborta.

3   Preliminaries

In this section we introduce notation and basic concepts of polytopes and games. Interested readers are referred to [21,32].

In the following $\mathscr{P}(S)$ denotes the powerset of set S, and $\mathscr{P}_f(S)$ denotes the set of finite subsets of set S. A *convex polytope* in $\mathbb{R}^n$ is a bounded set $K = \{x \in \mathbb{R}^n \mid Ax \le b\}$, with $A \in \mathbb{R}^{m \times n}$ and $b \in \mathbb{R}^m$, for some $m \in \mathbb{N}$. By *bounded* we mean (in our case) that there exists $M \in \mathbb{R}_{\ge 0}$ such that $\sum_{i=1}^{n} |x_i| \le M$ for all $x \in K$ (x_i denotes the ith element of x). Let S be a finite set. As functions in $\mathbb{R}^S$ can be equivalently seen as vectors in $\mathbb{R}^{|S|}$, we will in general refer to polytopes in $\mathbb{R}^S$. Let $\mathsf{Poly}(S)$ be the set of all convex polytopes in $\mathbb{R}^S$. Notice that the set of all probability functions on S form the convex polytope $\mathsf{Dist}(S) = \{\mu \in \mathbb{R}^S \mid \sum_{s \in S} \mu(s) = 1 \text{ and } \forall s \in S: \mu(s) \ge 0\}$. Let $\mathsf{DPoly}(S) = \{K \cap \mathsf{Dist}(S) \mid K \in \mathsf{Poly}(S)\}$. Thus, $K \in \mathsf{DPoly}(S)$ is a convex polytope whose elements are also probability functions on S and therefore its defining set of inequality $Ax \le b$ already encodes the inequalities $\sum_{s \in S} x_s = 1$ and $x_s \ge 0$ for $s \in S$.

Any convex polytope $K \in \mathsf{Poly}(S)$ can alternatively be characterized as the convex hull of its finite set of vertices. Let $\mathbb{V}(K)$ denote the set of all vertices of polytope K. If $\mathbb{V}(K) = \{v^1, \dots, v^k\}$, then every $x \in K$ is a convex combination of $\{v^1, \dots, v^k\}$, that is, $x = \sum_{i=1}^{k} \lambda_i v^k$ with $\lambda_i \ge 0$, for $i \in [1..k]$, and $\sum_{i=1}^{k} \lambda_i = 1$. A *simplex* is any convex polytope $K \in \mathsf{Poly}(S)$ whose set of vertices $\mathbb{V}(K)$ is affinely independent, that is, for any family $\{\lambda_v \in \mathbb{R}\}_{v \in \mathbb{V}(K)}$ such that $\sum_{v \in \mathbb{V}(K)} \lambda_v = 0$, $\sum_{v \in \mathbb{V}(K)} \lambda_v v = 0$ implies that $\lambda_v = 0$ for all $v \in \mathbb{V}(K)$. This implies that for every $x \in K$, with K being a simplex, the convex combination $x = \sum_{i=1}^{k} \lambda_i v^k$ is unique. We also remark that any convex polytope K can be expressed as the union of a (finite) set of simplices $\{K_i\}_{i \in I}$ so that $\mathbb{V}(K) = \bigcup_{i \in I} \mathbb{V}(K_i)$ (this is a consequence of Charathéodory's Theorem [24,32]). We will call such decomposition a *vertex-preserving triangulation*. Let $\mathsf{Simp}(S)$ denote the set of all simplices in $\mathbb{R}^S$ and $\mathsf{DSimp}(S) = \mathsf{Simp}(S) \cap \mathsf{DPoly}(S)$.

A *stochastic game* [13,15,30] is a tuple $\mathcal{G} = (\mathcal{S}, (\mathcal{S}_\square, \mathcal{S}_\diamond), \mathcal{A}, \theta)$, where $\mathcal{S}$ is a finite set of *states* with $\mathcal{S}_\square, \mathcal{S}_\diamond \subseteq \mathcal{S}$ being a partition of $\mathcal{S}$, $\mathcal{A}$ is a (finite) set of *actions*, and $\theta : \mathcal{S} \times \mathcal{A} \times \mathcal{S} \to [0,1]$ is a *probabilistic transition function* such that for every $s \in \mathcal{S}$ and $a \in \mathcal{A}$, $\theta(s, a, \cdot) \in \mathsf{Dist}(\mathcal{S})$ or $\theta(s, a, \mathcal{S}) = 0$. Let $\mathcal{A}(s) = \{a \in \mathcal{A} \mid \theta(s, a, \mathcal{S}) = 1\}$ be the set of actions enabled at state s. If $\mathcal{S}_\square = \emptyset$ or $\mathcal{S}_\diamond = \emptyset$, then $\mathcal{G}$ is a *Markov decision process* (or MDP). If, in addition, $|\mathcal{A}(s)| = 1$ for all $s \in \mathcal{S}$, $\mathcal{G}$ is a *Markov chain* (or MC). A *path* in the game $\mathcal{G}$ is an infinite sequence of states $\rho = s_0 s_1 \ldots$ such that, for every $k \in \mathbb{N}$, there is an $a \in \mathcal{A}$ with $\theta(s_k, a, s_{k+1}) > 0$. For $i \geq 0$, ρ_i indicates the ith state in the path ρ (notice that ρ_0 is the first state in ρ). Paths$_\mathcal{G}$ denotes the set of all paths, and FPaths$_\mathcal{G}$ denotes the set of finite prefixes of paths. Similarly, Paths$_{\mathcal{G},s}$ and FPaths$_{\mathcal{G},s}$ denote the set of paths and the set of finite paths starting at state s. A *strategy* for the i-player (for $i \in \{\Box, \Diamond\}$) in a game $\mathcal{G}$ is a function $\pi_i : \mathcal{S}^* \mathcal{S}_i \to \text{Dist}(\mathcal{A})$ that assigns a probabilistic distribution to each finite sequence of states such that $\pi_i(\hat{\rho}s)(a) > 0$ only if $a \in \mathcal{A}(s)$. The set of all strategies for the i-player is named Π_i. Whenever convenient, we indicate that the set of strategies Π_i belongs to the game $\mathcal{G}$ by writing by $\Pi_{\mathcal{G},i}$ A strategy π_i is said to be *pure* or *deterministic* if, for every $\hat{\rho}s \in \mathcal{S}^* \mathcal{S}_i$, $\pi_i(\hat{\rho}s)$ is a Dirac distribution (that is a distribution δ_a s.t., $\delta_a(a) = 1$ and $\delta_a(b) = 0$ for all $b \neq a$), and it is called *memoryless* if $\pi_i(\hat{\rho}s) = \pi_i(s)$, for every $\hat{\rho} \in \mathcal{S}^*$. Let Π_i^M be the set of all memoryless strategies for the i-player and Π_i^{MD} be the set of all its deterministic and memoryless strategies. Note that the definition of strategy given above works for set of actions that are finite, in Sect. 4 we define strategies for uncountable sets of actions.

Given strategies $\pi_\Box \in \Pi_\Box$ and $\pi_\Diamond \in \Pi_\Diamond$, and an initial state s, the *result* of the game is a Markov chain [10], denoted $\mathcal{G}_s^{\pi_\Box, \pi_\Diamond}$. The Markov chain $\mathcal{G}_s^{\pi_\Box, \pi_\Diamond}$ defines a probability measure $\mathbb{P}_{\mathcal{G},s}^{\pi_\Box, \pi_\Diamond}$ on the Borel σ-algebra generated by the cylinders of Paths$_{\mathcal{G},s}$. If ξ is a measurable set in such a Borel σ-algebra, $\mathbb{P}_{\mathcal{G},s}^{\pi_\Box, \pi_\Diamond}(\xi)$ is the probability that strategies $\pi_\Box$ and $\pi_\Diamond$ follow a path in ξ starting from state s. We use LTL notation to represent specific set of paths, in particular, $D \cup^n C = \{\rho \in \mathcal{S}^\omega \mid \rho_n \in C \wedge \forall j < n \colon \rho_j \in D\} = D^n \times C \times \mathcal{S}^\omega$ is the set of paths that reach $C \subseteq \mathcal{S}$ in exactly $n \geq 0$ steps traversing before only states in $D \subseteq \mathcal{S}$; $\Diamond^n C = \mathcal{S} \cup^n C$ is the set of all paths reaching states in C in exactly n steps; and $\Diamond C = \bigcup_{n \geq 0}(\mathcal{S} \backslash C) \cup^n C$ is the set of all paths that reach a state in C.

A stochastic game is said to be *almost surely stopping* [13,15] if for all pair of strategies $\pi_\Box$, $\pi_\Diamond$ the probability of reaching a terminal state is 1. A state s is *terminal* if $\theta(s, a, s) = 1$, for all $a \in \mathcal{A}(s)$. In other words, a game is stopping if $\inf_{\pi_\Diamond \in \Pi_\Diamond} \inf_{\pi_\Box \in \Pi_\Box} \mathbb{P}_s^{\pi_\Box, \pi_\Diamond}(\Diamond T) = 1$, where $T \subseteq \mathcal{S}$ is the set of terminal states. A stochastic game is *irreducible* [15] if for all pair of strategies, the probability of reaching a state from any other state is positive, that is, if $\inf_{\pi_\Diamond \in \Pi_\Diamond} \inf_{\pi_\Box \in \Pi_\Box} \mathbb{P}_s^{\pi_\Box, \pi_\Diamond}(\Diamond s') > 0$ for all pair of states $s, s' \in \mathcal{S}$.

A *quantitative objective* or *payoff function* is a measurable function $f : \mathcal{S}^\omega \to \mathbb{R}$. Let $\mathbb{E}_{\mathcal{G},s}^{\pi_\Box, \pi_\Diamond}[f]$ be the expectation of measurable function f under probability $\mathbb{P}_{\mathcal{G},s}^{\pi_\Box, \pi_\Diamond}$. The goal of the $\Box$-player is to maximize this value whereas the goal of the $\Diamond$-player is to minimize it. Sometimes quantitative objective functions can be defined via *rewards*. These are assigned by a *reward function* $r : S \to \mathbb{R}^+$. We usually consider stochastic games augmented with a reward function. Moreover, we assume that for every terminal state s, $r(s) = 0$. The value of the game for the $\Box$-player at state s under strategy $\pi_\Box$ is defined as the infimum over all the values resulting from the $\Diamond$-player strategies in that state, i.e., $\inf_{\pi_\Diamond \in \Pi_\Diamond} \mathbb{E}_{\mathcal{G},s}^{\pi_\Box, \pi_\Diamond}[f]$. The *value of the game* for the $\square$-player is defined as the supremum of the values of all the $\square$-player strategies, i.e., $\sup_{\pi_\square \in \Pi_\square} \inf_{\pi_\diamond \in \Pi_\diamond} \mathbb{E}_{\mathcal{G},s}^{\pi_\square,\pi_\diamond}[f]$. Similarly, the value of the game for the $\diamond$-player under strategy $\pi_\diamond$ and the value of the game for the $\diamond$-player are defined as $\sup_{\pi_\square \in \Pi_\square} \mathbb{E}_{\mathcal{G},s}^{\pi_\square,\pi_\diamond}[f]$ and $\inf_{\pi_\diamond \in \Pi_\diamond} \sup_{\pi_\square \in \Pi_\square} \mathbb{E}_{\mathcal{G},s}^{\pi_\square,\pi_\diamond}[f]$, respectively. We say that a game is *determined* if both values are the same, that is, $\sup_{\pi_\square \in \Pi_\square} \inf_{\pi_\diamond \in \Pi_\diamond} \mathbb{E}_{\mathcal{G},s}^{\pi_\square,\pi_\diamond}[f] = \inf_{\pi_\diamond \in \Pi_\diamond} \sup_{\pi_\square \in \Pi_\square} \mathbb{E}_{\mathcal{G},s}^{\pi_\square,\pi_\diamond}[f]$.

In this paper we focus on *total accumulated reward*, where the payoff function is defined by $\mathsf{rew}_t(\rho) = \lim_{n \to \infty} \sum_{i=0}^n r(\rho_i)$, *total discounted reward*, defined by $\mathsf{rew}_\gamma(\rho) = \lim_{n \to \infty} \sum_{i=0}^n \gamma^i r(\rho_i)$, where $\gamma \in (0,1)$ is the discount factor, and *average reward*, defined by $\mathsf{rew}_a(\rho) = \lim_{n \to \infty} \frac{1}{n+1} \sum_{i=0}^n r(\rho_i)$. By taking, respectively, $\mathsf{f}(i,n) = 1$, $\mathsf{f}(i,n) = \gamma^i$, or $\mathsf{f}(i,n) = \frac{1}{n+1}$, we refer simultaneously to the above payoff functions with the single function $\mathsf{rew}_\mathsf{f}(\rho) = \lim_{n \to \infty} \sum_{i=0}^n \mathsf{f}(i,n) r(\rho_i)$.

We also focus on *reachability objective*. In this case, the goal of the $\square$-player is to maximize the probability of reaching a state on a goal set $G \subseteq S$ whereas the goal of the $\diamond$-player is to minimize it. Therefore, similar to quantitative objectives, the *value of the reachability game for the $\square$-player* is defined by $\sup_{\pi_\square \in \Pi_\square} \inf_{\pi_\diamond \in \Pi_\diamond} \mathbb{P}_{\mathcal{G},s}^{\pi_\square,\pi_\diamond}(\diamond G)$ and the *value of the reachability game for the $\diamond$-player* is defined by $\inf_{\pi_\diamond \in \Pi_\diamond} \sup_{\pi_\square \in \Pi_\square} \mathbb{P}_{\mathcal{G},s}^{\pi_\square,\pi_\diamond}(\diamond G)$, and the game is *determined* if both values are the same.

4 Polytopal Stochastic Games

A polytopal stochastic game is characterized through a structure that contains a finite set of states divided into two sets, each owned by a different player. In addition, each state has assigned a finite set of convex polytopes of probability distributions over states. The formal definition is as follows.

Definition 1. *A polytopal stochastic game (PSG, for short) is a structure* $\mathcal{K} = (\mathcal{S}, (\mathcal{S}_\square, \mathcal{S}_\diamond), \Theta)$ *such that $\mathcal{S}$ is a finite set of states partitioned into $\mathcal{S} = \mathcal{S}_\square \uplus \mathcal{S}_\diamond$ and $\Theta : \mathcal{S} \to \mathscr{P}_\mathsf{f}(\mathsf{DPoly}(\mathcal{S}))$. If, in particular, $\Theta : \mathcal{S} \to \mathscr{P}_\mathsf{f}(\mathsf{DSimp}(\mathcal{S}))$, we call $\mathcal{K}$ a simplicial stochastic game (SSG for short).*

The idea of a PSG is as expected: in a state $s \in \mathcal{S}_i$ ($i \in \{\square, \diamond\}$), player i chooses to play a polytope $K \in \Theta(s)$ and a distribution $\mu \in K$. The next state s' is sampled according to distribution μ and the game continues from s' repeating the same process.

As a particular example, one can devise a stochastic game variant of Interval Markov Decision Processes (IMDPs) [17,20]. This type of games can be interpreted as a PSG where every polytope $K \in \Theta(s)$, for all $s \in \mathcal{S}$, is defined by $\mu \in K$ iff $\sum_{s' \in \mathcal{S}} \mu(s') = 1$ and, for all $s' \in \mathcal{S}$ and some fixed $0 \leq l_{s'} \leq u_{s'} \leq 1$, $l_{s'} \leq \mu(s') \leq u_{s'}$ (note that the intervals need to be closed).

The behaviour of a polytopal stochastic game is formally interpreted in terms of a stochastic game where the number of transitions outgoing the players' states may be uncountably large. We choose a controllable view on the uncertainty introduced by the polytope since the adversarial alternative can be encoded as was shown in Sect. 2. Formally, the interpretation of a PSG is as follows.

Definition 2. *The* interpretation *of the polytopal stochastic game* $\mathcal{K}$ *is defined by the stochastic game* $\mathcal{G}_{\mathcal{K}} = (\mathcal{S}, (\mathcal{S}_{\square}, \mathcal{S}_{\diamond}), \mathcal{A}, \theta)$, *where* $\mathcal{A} = \left(\bigcup_{s \in \mathcal{S}} \Theta(s)\right) \times \mathsf{Dist}(\mathcal{S})$ *and*

$$\theta(s, (K, \mu), s') = \begin{cases} \mu(s') & \text{if } K \in \Theta(s) \text{ and } \mu \in K \\ 0 & \text{otherwise} \end{cases}$$

Notice that the set of actions $\mathcal{A}$ can be uncountably large, as well as each set $\mathcal{A}(s) = \bigcup_{K \in \Theta(s)} \{K\} \times K$. Therefore we need to extend the strategies to this uncountable domain which should be properly endowed with a σ-algebra. For this we make use of a standard construction to provide a σ-algebra to $\mathsf{Dist}(\mathcal{S})$ [16]: $\Sigma_{\mathsf{Dist}(\mathcal{S})}$ is defined as the smallest σ-algebra containing the sets $\{\mu \in \mathsf{Dist}(\mathcal{S}) \mid \mu(S) \geq p\}$ for all $S \subseteq \mathcal{S}$ and $p \in [0,1]$. Now, we endow $\mathcal{A}$ with the product σ-algebra $\Sigma_{\mathcal{A}} = \mathscr{P}\left(\bigcup_{s \in \mathcal{S}} \Theta(s)\right) \otimes \Sigma_{\mathsf{Dist}(\mathcal{S})}$ (i.e., the smallest σ-algebra containing all rectangles $\boldsymbol{K} \times \boldsymbol{M}$ with $\boldsymbol{K} \subseteq \bigcup_{s \in \mathcal{S}} \Theta(s)$ and $\boldsymbol{M} \in \Sigma_{\mathsf{Dist}(\mathcal{S})}$) and let $\mathsf{PMeas}(\mathcal{A})$ be the set of all probability measures on $\Sigma_{\mathcal{A}}$. It is not difficult to check that each set of enabled actions $\mathcal{A}(s)$ is measurable (i.e., $\mathcal{A}(s) \in \Sigma_{\mathcal{A}}$) and that function $\theta(s, \cdot, s')$ is measurable (i.e., $\{a \in \mathcal{A} \mid \theta(s, a, s') \leq p\} \in \Sigma_{\mathcal{A}}$ for all $p \in [0,1]$).

We extend the definition of *strategy* for the i-player ($i \in \{\square, \diamond\}$) in $\mathcal{G}_{\mathcal{K}}$ to be a function $\pi_i : \mathcal{S}^* \mathcal{S}_i \to \mathsf{PMeas}(\mathcal{A})$ that assigns a probability measure to each finite sequence of states such that $\pi_i(\hat{\rho}s)(\mathcal{A}(s)) = 1$. All other concepts on strategies defined in Sect. 3 apply to this new definition as well.

In the following we present the formal definition of $\mathbb{P}_{\mathcal{G}_{\mathcal{K}}, s}^{\pi_{\square}, \pi_{\diamond}}$. First, for each $n \geq 0$ and $s \in \mathcal{S}$, define $\mathbb{P}_{\mathcal{G}_{\mathcal{K}}, s}^{\pi_{\square}, \pi_{\diamond}, n} : \mathcal{S}^{n+1} \to [0,1]$ for all $s' \in \mathcal{S}$ and $\hat{\rho} \in \mathcal{S}^{n+1}$ inductively as follows:

$$\mathbb{P}_{\mathcal{G}_{\mathcal{K}}, s}^{\pi_{\square}, \pi_{\diamond}, 0}(s') = \delta_s(s')$$

$$\mathbb{P}_{\mathcal{G}_{\mathcal{K}}, s}^{\pi_{\square}, \pi_{\diamond}, n+1}(\hat{\rho}s') = \begin{cases} \mathbb{P}_{\mathcal{G}_{\mathcal{K}}, s}^{\pi_{\square}, \pi_{\diamond}, n}(\hat{\rho}) \int_{\mathcal{A}} \theta(last(\hat{\rho}), \cdot, s') \, \mathbf{d}(\pi_{\square}(\hat{\rho})(\cdot)) & \text{if } last(\hat{\rho}) \in \mathcal{S}_{\square} \\ \mathbb{P}_{\mathcal{G}_{\mathcal{K}}, s}^{\pi_{\square}, \pi_{\diamond}, n}(\hat{\rho}) \int_{\mathcal{A}} \theta(last(\hat{\rho}), \cdot, s') \, \mathbf{d}(\pi_{\diamond}(\hat{\rho})(\cdot)) & \text{if } last(\hat{\rho}) \in \mathcal{S}_{\diamond} \end{cases}$$

and extend $\mathbb{P}_{\mathcal{G}_{\mathcal{K}}, s}^{\pi_{\square}, \pi_{\diamond}, n} : \mathscr{P}(\mathcal{S}^{n+1}) \to [0,1]$ to sets as the sum of the points.

Let $\Sigma_{\mathcal{S}}$ denote the discrete σ-algebra on $\mathcal{S}$ and $\Sigma_{\mathcal{S}^{\omega}}$ the usual product σ-algebra on $\mathcal{S}^{\omega}$. By Carathéodory extension theorem [2], $\mathbb{P}_{\mathcal{G}_{\mathcal{K}}, s}^{\pi_{\square}, \pi_{\diamond}} : \Sigma_{\mathcal{S}^{\omega}} \to [0,1]$ is defined as the unique probability measure such that for all $n \geq 0$, and $S_i \in \Sigma_{\mathcal{S}}$, $0 \leq i \leq n$,

$$\mathbb{P}_{\mathcal{G}_{\mathcal{K}}, s}^{\pi_{\square}, \pi_{\diamond}}(S_0 \times \cdots \times S_n \times \mathcal{S}^{\omega}) = \mathbb{P}_{\mathcal{G}_{\mathcal{K}}, s}^{\pi_{\square}, \pi_{\diamond}, n}(S_0 \times \cdots \times S_n)$$

The notions of *deterministic* and *memoryless* extends directly to this type of strategies. In addition, a strategy π_i, $i \in \{\square, \diamond\}$, is *semi-Markov* if for every $\hat{\rho}, \hat{\rho}' \in \mathcal{S}^*$ and $s \in \mathcal{S}_i$, $|\hat{\rho}| = |\hat{\rho}'|$ implies $\pi_i(\hat{\rho}s) = \pi_i(\hat{\rho}'s)$, that is, the decisions of π_i depend only on the length of the run and its last state. Thus, we write $\pi_i(n, s)$ instead of $\pi_i(\hat{\rho}s)$ whenever $|\hat{\rho}| = n$. Let Π_i^S denote the set of all semi-Markov strategies for the i-player. Also, we say that a strategy $\pi_i \in \Pi_i$ is *extreme* if for all $\hat{\rho} \in \mathcal{S}^*$, $\pi_i(\hat{\rho}s)(\{(K, \mu) \in \mathcal{A}(s) \mid \mu \in \mathbb{V}(K)\}) = 1$. Notice that extreme strategies only selects transitions on vertices of polytopes. Let Π_i^{XS} and Π_i^{XMD}

be, respectively, the set of all extreme semi-Markov strategies and the set of all extreme deterministic and memoryless strategies for the i-player.

Polytopal stochastic games can be translated into simplicial stochastic games preserving all the stochastic behaviour. More precisely, for every PSG $\mathcal{K}$ there is a SSG $\mathcal{K}'$ such that for every pair of strategies for $\mathcal{K}$ in a particular class (i.e., memoryless, semi-Markov, etc.), there is a pair of strategies for $\mathcal{K}'$ in the same class that yields the same probability measure and vice versa. Let $\mathsf{Triang}\colon \mathsf{DPoly} \to \mathscr{P}(\mathsf{DSimp})$ be a function that assigns a vertex-preserving triangulation $\mathsf{Triang}(K)$ to each polytope K. Then:

Proposition 1. *Let $\mathcal{K} = (\mathcal{S}, (\mathcal{S}_\square, \mathcal{S}_\diamond), \Theta)$ be a PSG and define the SSG $\mathcal{K}' = (\mathcal{S}, (\mathcal{S}_\square, \mathcal{S}_\diamond), \Theta')$ such that $\Theta'(s) = \bigcup_{K \in \Theta(s)} \mathsf{Triang}(K)$. Let $\mathcal{G}_\mathcal{K}$ and $\mathcal{G}_{\mathcal{K}'}$ be their respective interpretations. Then,*

1. *for all pair of strategies $\pi_\square$ and $\pi_\diamond$ for $\mathcal{G}_\mathcal{K}$ there is a pair of strategies $\pi'_\square$ and $\pi'_\diamond$ for $\mathcal{G}_{\mathcal{K}'}$ such that (a) $\mathbb{P}^{\pi_\square,\pi_\diamond}_{\mathcal{G}_\mathcal{K},s} = \mathbb{P}^{\pi'_\square,\pi'_\diamond}_{\mathcal{G}_{\mathcal{K}'},s}$ for all $s \in \mathcal{S}$, and (b) if π_i, $i \in \{\square, \diamond\}$, is memoryless (resp. deterministic, semi-Markov or extreme) then so is π'_i; and*
2. *the same holds with the roles of $\mathcal{G}_\mathcal{K}$ and $\mathcal{G}_{\mathcal{K}'}$ exchanged.*

Proof (Sketch). Let $\mathcal{G}_\mathcal{K} = (\mathcal{S}, (\mathcal{S}_\square, \mathcal{S}_\diamond), \mathcal{A}, \theta)$ and $\mathcal{G}_{\mathcal{K}'} = (\mathcal{S}, (\mathcal{S}_\square, \mathcal{S}_\diamond), \mathcal{A}', \theta')$. To prove item 1, the new strategies are defined so that they preserve the same measure on the probability part of the labels in $\mathcal{A}'$ as the one the old strategies measure on the probability part of $\mathcal{A}$ while properly distributing the probabilities on the simplices of the triangulation of the original polytopes. For this, first fix a function $f_K \colon \mathsf{Triang}(K) \to \mathscr{P}(K)$ for each polytope $K \in \mathsf{DPoly}(\mathcal{S})$ satisfying (i) $\forall K' \in \mathsf{Triang}(K)\colon f_K(K') \subseteq K'$, (ii) $\bigcup_{K' \in \mathsf{Triang}(K)} f_K(K') = K$, and (iii) $\forall K'_1, K'_2 \in \mathsf{Triang}(K)\colon f_K(K'_1) \cap f_K(K'_2) \neq \emptyset \Rightarrow K'_1 = K'_2$. Thus, $f_K(K')$ is almost the simplex K' but ensuring that distributions on the faces of K' are exactly in one of the $f_K(K'')$, $K'' \in \mathsf{Triang}(K)$.

Given strategies π_i, $i \in \{\square, \diamond\}$, for $\mathcal{G}_\mathcal{K}$ define π'_i for $\mathcal{G}_{\mathcal{K}'}$, for all $\hat{\rho} \in \mathcal{S}^*$, $s \in \mathcal{S}_i$, and $A' \in \Sigma_{\mathcal{A}'}$ by

$$\pi'_i(\hat{\rho}s)(A') = \sum_{K \in \Theta(s)} \sum_{K' \in \mathsf{Triang}(K)} \pi_i(\hat{\rho}s)(\{K\} \times (A'|_{K'} \cap f_K(K'))) \quad (1)$$

where $A'|_{K'} = \{\mu \mid (K', \mu) \in A'\}$ is the K' section of the measurable set A'. Notice that f_K ensures that the faces of each $K' \in \mathsf{Triang}(K)$ are considered in exactly one summand of the inner summation of (1).

For item 2, the new strategies preserve the same measure on the probability part of $\mathcal{A}$ as the old strategies while gathering the probability of the simplices in the original polytope. So, for each state $s \in \mathcal{S}$, fix $f_s \colon \Theta(s) \to \mathscr{P}(\mathsf{DSimp}(\mathcal{S}))$ such that (i) $\forall K \in \Theta(s)\colon f_s(K) \subseteq \mathsf{Triang}(K)$, (ii) $\bigcup_{K \in \Theta(s)} f_s(K) = \bigcup_{K \in \Theta(s)} \mathsf{Triang}(K)$, and (iii) $\forall K_1, K_2 \in \Theta(s)\colon f_s(K_1) \cap f_s(K_2) \neq \emptyset \Rightarrow K_1 = K_2$. Given strategies π'_i, $i \in \{\square, \diamond\}$, for $\mathcal{G}_{\mathcal{K}'}$ define π_i for $\mathcal{G}_\mathcal{K}$, for all $\hat{\rho} \in \mathcal{S}^*$, $s \in \mathcal{S}_i$, and $A \in \Sigma_\mathcal{A}$ by

$$\pi_i(\hat{\rho}s)(A) = \sum_{K \in \Theta(s)} \sum_{K' \in f_s(K)} \pi'_i(\hat{\rho}s)(\{K'\} \times A|_K) \quad (2)$$

Notice that, by definition, $K' \in \Theta'(s)$. Moreover, notice that f_s ensures that a simplex in a triangulation of a polytope outgoing s is considered in exactly one summand of (2).

In both cases, it requires some straightforward calculations to check that the properties of memoryless, semi-Markov, deterministic, and extreme are preserved by the new strategies. Also in both cases, to prove that $\mathbb{P}^{\pi_\square,\pi_\diamond}_{\mathcal{G}_K,s} = \mathbb{P}^{\pi'_\square,\pi'_\diamond}_{\mathcal{G}_{K'},s}$ it sufficies to state that $\mathbb{P}^{\pi_\square,\pi_\diamond,n}_{\mathcal{G}_K,s} = \mathbb{P}^{\pi'_\square,\pi'_\diamond,n}_{\mathcal{G}_{K'},s}$ for all $n \geq 0$ which is done by induction using results from measure theory. $\qquad\square$

5 Discretizing Polytopal Stochastic Games

In this section we show that a PSG can be solved by translating it into a finite stochastic game that is just like the original PSG but it only has the transitions corresponding to the vertices of the polytopes. We focus on reachability games, and the reward games introduced above: total accumulated reward, total discounted reward, and average reward.

The first lemma we introduce states that the calculation of the expected values of the different reward games only depend on the probability of reaching each state and the reward collected in each state regardless the path that lead to such states. In particular, Lemma 1.1 refers to the reward collected in a finite number of steps while Lemma 1.2 refers to the general case stated before.

For $k \geq 0$ define $\diamond^k s = \mathcal{S}^k \times \{s\} \times \mathcal{S}^\omega$ to be the set of all runs in which $s \in \mathcal{S}$ is reached in exactly k steps. Let $\widehat{\mathsf{rew}}^n_f(\hat{\rho}) = \sum_{i=0}^n f(i,n) r(\hat{\rho}_i)$ for all $\hat{\rho} \in \mathcal{S}^{n+1}$. Then $\mathsf{rew}_f(\rho) = \lim_{n\to\infty} \widehat{\mathsf{rew}}^n_f(\rho[..n+1])$ where $\rho[..n+1]$ is the $(n+1)$th prefix of ρ, i.e., $\rho[..n+1] = \rho_0\rho_1\rho_2...\rho_n$.

Lemma 1. *Let $\mathcal{G}_K$ be a stochastic game resulting from interpreting a PSG K. For all strategies $\pi_\square \in \Pi_\square$ and $\pi_\diamond \in \Pi_\diamond$,*

1. $\sum_{\hat{\rho}\in\mathcal{S}^{n+1}} \mathbb{P}^{\pi_\square,\pi_\diamond,n}_{\mathcal{G}_K,s}(\hat{\rho}) \, \widehat{\mathsf{rew}}^n_f(\hat{\rho}) = \sum_{i=0}^n \sum_{s'\in\mathcal{S}} \mathbb{P}^{\pi_\square,\pi_\diamond}_{\mathcal{G}_K,s}(\diamond^i s') \, f(i,n) \, r(s')$, *for all $n \geq 0$, and*

2. $\mathbb{E}^{\pi_\square,\pi_\diamond}_{\mathcal{G}_K,s}[\mathsf{rew}_f] = \lim_{n\to\infty} \sum_{i=0}^n \sum_{s'\in\mathcal{S}} \mathbb{P}^{\pi_\square,\pi_\diamond}_{\mathcal{G}_K,s}(\diamond^i s') \, f(i,n) \, r(s')$.

The proof of Lemma 1.1 follows by induction on n while Lemma 1.2 can be calculated using the first item.

The next lemma states that if the $\diamond$-player plays a semi-Markov strategy, the $\square$-player can achieve equal results whether she plays an arbitrary strategy or limits to playing only semi-Markov strategies.

Lemma 2. *Let $\mathcal{G}_K$ be a stochastic game resulting from interpreting a PSG K. If $\pi_\diamond \in \Pi^S_\diamond$ is a semi-Markov strategy, then, for any $\pi_\square \in \Pi_\square$, there is a semi-Markov strategy $\pi^*_\square \in \Pi^S_\square$ such that:*

1. $\mathbb{P}^{\pi_\square,\pi_\diamond}_{\mathcal{G}_K,s}(D \, \mathsf{U}^n \, s') = \mathbb{P}^{\pi^*_\square,\pi_\diamond}_{\mathcal{G}_K,s}(D \, \mathsf{U}^n \, s')$, *for all $n \geq 0$, $D \subseteq \mathcal{S}$ and $s' \in \mathcal{S}$;*

2. $\mathbb{P}^{\pi_\square,\pi_\diamond}_{\mathcal{G}_K,s}(\diamond C) = \mathbb{P}^{\pi^*_\square,\pi_\diamond}_{\mathcal{G}_K,s}(\diamond C)$, *for all $C \subseteq \mathcal{S}$; and*

3. $\mathbb{E}^{\pi_\square,\pi_\diamond}_{\mathcal{G}_K,s}[\mathsf{rew}_f] = \mathbb{E}^{\pi^*_\square,\pi_\diamond}_{\mathcal{G}_K,s}[\mathsf{rew}_f]$.

Similarly, if $\pi_\square \in \Pi_\square^S$ then, for any $\pi_\diamond \in \Pi_\diamond$, there exists $\pi_\diamond^* \in \Pi_\diamond^S$ satisfying, mutatis mutandis, the same equalities.

Proof (Sketch). To prove item 1, we define the new strategy $\pi_\square^*$ so that the probability of choosing from $A \in \Sigma_\mathcal{A}$ after a path of length n ending on a state s with the original strategy is uniformly distributed among the paths of this type in the new strategy. Thus, $\pi_\square^*$ is formally defined as follows. For $\hat\rho \in \mathcal{S}^*$, $s' \in \mathcal{S}$, and $A \in \Sigma_\mathcal{A}$, such that $\mathbb{P}_{\mathcal{G}_\mathcal{K},s}^{\pi_\square,\pi_\diamond}(D\,\mathsf{U}^n\,s') > 0$ and $|\hat\rho| = n \geq 0$, let

$$\pi_\square^*(\hat\rho s')(A) = \frac{\sum_{\hat\rho' \in D^n} \mathbb{P}_{\mathcal{G}_\mathcal{K},s}^{\pi_\square,\pi_\diamond,n}(\hat\rho's')\,\pi_\square(\hat\rho's')(A)}{\mathbb{P}_{\mathcal{G}_\mathcal{K},s}^{\pi_\square,\pi_\diamond}(D\,\mathsf{U}^n\,s')}$$

For $s' \in \mathcal{S}$ with $\mathbb{P}_{\mathcal{G}_\mathcal{K},s}^{\pi_\square,\pi_\diamond}(D\,\mathsf{U}^n\,s') = 0$ and $|\hat\rho s'| = n$, define $\pi_\square^*(\hat\rho s')$ to be $\delta_{\mathsf{f}(s')}$ for a globally fixed function f such that $\mathsf{f}(s') \in \mathcal{A}(s')$. Notice that $\pi_\square^* \in \Pi_\square^S$.

Then, the proof of item 1 follows by induction with particular care in the case of $\mathbb{P}_{\mathcal{G}_\mathcal{K},s}^{\pi_\square,\pi_\diamond}(D\,\mathsf{U}^n\,s') = 0$. Item 2 follows straightforwardly from item 1 and item 3 follows directly from item 2 using Lemma 1.2. The proof can be replicated mutatis mutandi with $\square$ and $\diamond$ exchanged yielding the last part of the lemma.
□

Since $\Theta(s)$ is finite, there can be finitely many polytopes K such that $(K,\mu) \in \mathcal{A}(s)$. Besides, the set of vertices $\mathbb{V}(K)$ of K is finite. Therefore the set $\{(K,\mu) \in \mathcal{A}(s) \mid \mu \in \mathbb{V}(K)\}$ is also finite and, as a consequence, extreme strategies only resolve with discrete (finite) probability distributions. That is, if π_i is extreme, $\pi_i(\hat\rho s)$ has finite support for all $\hat\rho \in \mathcal{S}^*$ and $s \in \mathcal{S}$.

It turns out that Lemma 2 can be strengthened to obtain *extreme* semi-Markov strategies. We first prove this new lemma for simplicial stochastic games since simplices have the particular property that any vector in a simplex can be uniquely defined as a convex combination of the simplex vertices which is crucial for the proof of the lemma.

Lemma 3. *Let $\mathcal{G}_\mathcal{K}$ be a stochastic game resulting from interpreting a SSG $\mathcal{K}$. If $\pi_\diamond \in \Pi_\diamond^S$ is a semi-Markov strategy, then, for any $\pi_\square \in \Pi_\square^S$, there is an extreme semi-Markov strategy $\pi_\square^* \in \Pi_\square^{XS}$ such that:*

1. $\mathbb{P}_{\mathcal{G}_\mathcal{K},s}^{\pi_\square,\pi_\diamond}(D\,\mathsf{U}^n\,s') = \mathbb{P}_{\mathcal{G}_\mathcal{K},s}^{\pi_\square^*,\pi_\diamond}(D\,\mathsf{U}^n\,s')$, for all $n \geq 0$, $D \subseteq \mathcal{S}$ and $s' \in \mathcal{S}$;
2. $\mathbb{P}_{\mathcal{G}_\mathcal{K},s}^{\pi_\square,\pi_\diamond}(\Diamond C) = \mathbb{P}_{\mathcal{G}_\mathcal{K},s}^{\pi_\square^*,\pi_\diamond}(\Diamond C)$, for all $C \subseteq \mathcal{S}$; and
3. $\mathbb{E}_{\mathcal{G}_\mathcal{K},s}^{\pi_\square,\pi_\diamond}[\mathsf{rew}_\mathsf{f}] = \mathbb{E}_{\mathcal{G}_\mathcal{K},s}^{\pi_\square^*,\pi_\diamond}[\mathsf{rew}_\mathsf{f}]$.

Similarly, if $\pi_\square \in \Pi_\square^S$ then, for any $\pi_\diamond \in \Pi_\diamond^S$, there exists $\pi_\diamond^* \in \Pi_\diamond^{XS}$ satisfying, mutatis mutandis, the same equalities.

Proof (Sketch). For any $K \in \mathsf{DSimp}(\mathcal{S})$, $\mu \in K$ and $\hat\mu \in \mathbb{V}(K)$ define $\mathsf{p}^K(\mu,\hat\mu) \in [0,1]$ such that $\sum_{\hat\mu \in \mathbb{V}(K)} \mathsf{p}^K(\mu,\hat\mu)\,\hat\mu = \mu$. That is, all $\mathsf{p}^K(\mu,\hat\mu)$, $\hat\mu \in \mathbb{V}(K)$, are the unique factors that define the convex combination for μ in the simplex K. In any other case, let $\mathsf{p}^K(\mu,\hat\mu) = 0$.

Let $\mathsf{p}((K,\mu),(K,\hat{\mu})) = \mathsf{p}^K(\mu,\hat{\mu})$ for all $K \in \mathsf{DSimp}(\mathcal{S})$, $\mu \in K$ and $\hat{\mu} \in \mathbb{V}(K)$, and let $\mathsf{p}(a,b) = 0$ for any other $a, b \in \mathcal{A}$. For every $(K,\mu) \in \mathcal{A}$ such that $\mu \in K$, let $\mathbb{V}(K,\mu) = \{(K,\hat{\mu}) \mid \hat{\mu} \in \mathbb{V}(K)\}$ and let $\mathbb{V}(K,\mu) = \emptyset$ otherwise. Thus, for every $s \in \mathcal{S}$ and $a \in \mathcal{A}$, $\theta(s,a,\cdot) = \sum_{b \in \mathbb{V}(a)} \mathsf{p}(a,b)\,\theta(s,b,\cdot)$.

We also extend p to measurable sets $B \in \Sigma_{\mathcal{A}}$ and $a \in \mathcal{A}$ by $\mathsf{p}(a,B) = \sum_{b \in B \cap \mathbb{V}(a)} \mathsf{p}(a,b)$.

For every $\hat{\rho} \in \mathcal{S}^*$, $s' \in \mathcal{S}$ and $B \in \Sigma_{\mathcal{A}}$, define $\pi_\square^*$ by

$$\pi_\square^*(\hat{\rho}s')(B) = \int_{\mathcal{A}} \mathsf{p}(\cdot, B)\,\mathbf{d}(\pi_\square(\hat{\rho}s')(\cdot)).$$

$\pi_\square^*(\hat{\rho}s')$ is defined so that it assigns to each vertex of a simplex the weighted contribution (according to $\pi_\square(\hat{\rho}s')$) of each distribution (in the said simplex) to such vertex.

Because $\pi_\square$ is semi-Markov, so is $\pi_\square^*$. Moreover, notice that if b is not a vertex label, then $\mathsf{p}(a,b) = 0$ (and hence $\mathsf{p}(a,B) > 0$ only if B contains vertices). This should hint that $\pi_\square^*$ is also extreme.

Item 1 proceeds by induction on n. Item 2 follows straightforwardly using 1, and item 3 follows from item 2 using Lemma 1.2. The proof can be replicated mutatis mutandi with $\square$ and $\Diamond$ exchanged which yields the last part of the lemma. $\qquad\square$

Because of Proposition 1, Lemma 3 extends immediately to PSG. Moreover, by applying Lemma 3 twice and Proposition 1, we have the next corollary.

Corollary 1. *Let $\mathcal{G}_K$ be a stochastic game resulting from interpreting a PSG K. For all semi-Markov strategies $\pi_\Diamond \in \Pi_\Diamond^S$ and $\pi_\square \in \Pi_\square^S$, there are extreme semi-Markov strategies $\pi_\Diamond^* \in \Pi_\Diamond^{XS}$ and $\pi_\square^* \in \Pi_\square^{XS}$ such that*

1. $\mathbb{P}_{\mathcal{G}_K,s}^{\pi_\square,\pi_\Diamond}(\Diamond C) = \mathbb{P}_{\mathcal{G}_K,s}^{\pi_\square^*,\pi_\Diamond^*}(\Diamond C)$, *for all $C \subseteq \mathcal{S}$; and*
2. $\mathbb{E}_{\mathcal{G}_K,s}^{\pi_\square,\pi_\Diamond}[\mathsf{rew}_\mathsf{f}] = \mathbb{E}_{\mathcal{G}_K,s}^{\pi_\square^*,\pi_\Diamond^*}[\mathsf{rew}_\mathsf{f}]$.

Given $\mathcal{G}_K$, define the *extreme interpretation of K* as the stochastic game $\mathcal{H}_K = (\mathcal{S}, (\mathcal{S}_\square, \mathcal{S}_\Diamond), \mathbb{V}(\mathcal{A}), \theta_{\mathcal{H}_K})$ where $\theta_{\mathcal{H}_K}$ is the restriction of θ to actions in $\mathbb{V}(\mathcal{A}) = \{(K,\mu) \in \mathcal{A} \mid \mu \in \mathbb{V}(K)\}$, that is, $\theta_{\mathcal{H}_K}(s,a,s) = \theta(s,a,s')$ for all $s, s' \in \mathcal{S}$ and $a \in \mathbb{V}(\mathcal{A})$. Since $\mathbb{V}(\mathcal{A})$ is finite, $\mathcal{H}_K$ is a finite stochastic game.

Given an extreme semi-Markov strategy $\pi_i \in \Pi_{\mathcal{G}_K,i}^{XS}$ for the i-player in the stochastic game $\mathcal{G}_K$, $i \in \{\square, \Diamond\}$, define $\pi_i^\vee(\hat{\rho}s)(A) = \pi_i(\hat{\rho}s)(A)$ for all $\hat{\rho} \in \mathcal{S}^*$, $s \in \mathcal{S}$, and $A \subseteq \mathbb{V}(\mathcal{A})$ ($A \in \Sigma_{\mathcal{A}}$ since it is finite). Notice that $\pi_i(\hat{\rho}s)(\mathcal{A}_{\mathcal{H}_K}(s)) = \pi_i(\hat{\rho}s)(\mathbb{V}(\mathcal{A}(s))) = 1$. Therefore $\pi_i^\vee \in \Pi_{\mathcal{H}_K,i}^S$ is a semi-Markov strategy in $\mathcal{H}_K$. Conversely, for a semi-Markov strategy $\pi_i \in \Pi_{\mathcal{H}_K,i}^S$ for the i-player in the stochastic game $\mathcal{H}_K$, define $\pi_i^\times(\hat{\rho}s)(A) = \pi_i(\hat{\rho}s)(A \cap \mathbb{V}(\mathcal{A}))$ for all $\hat{\rho} \in \mathcal{S}^*$, $s \in \mathcal{S}$, and $A \in \Sigma_{\mathcal{A}}$. $\pi_i^\times \in \Pi_{\mathcal{G}_K,i}^{XS}$ is a well defined extreme semi-Markov strategy in $\mathcal{G}_K$ since $\pi_i^\times(\hat{\rho}s)(\mathbb{V}(\mathcal{A}(s))) = \pi_i(\hat{\rho}s)(\mathcal{A}_{\mathcal{H}_K}(s)) = 1$ and $\pi_i^\times(\hat{\rho}s)(\mathcal{A}\backslash\mathbb{V}(\mathcal{A})) = \pi_i(\hat{\rho}s)(\emptyset) = 0$. Then, it can be calculated by induction on n that $\mathbb{P}_{\mathcal{G}_K,s}^{\pi_\square,\pi_\Diamond,n} = \mathbb{P}_{\mathcal{H}_K,s}^{\pi_\square^\vee,\pi_\Diamond^\vee,n}$ and $\mathbb{P}_{\mathcal{G}_K,s}^{\pi_\square^\times,\pi_\Diamond^\times,n} = \mathbb{P}_{\mathcal{H}_K,s}^{\pi_\square,\pi_\Diamond,n}$ which yield

$$\mathbb{P}_{\mathcal{G}_K,s}^{\pi_\square,\pi_\Diamond} = \mathbb{P}_{\mathcal{H}_K,s}^{\pi_\square^\vee,\pi_\Diamond^\vee} \text{ and } \mathbb{P}_{\mathcal{G}_K,s}^{\pi_\square^\times,\pi_\Diamond^\times} = \mathbb{P}_{\mathcal{H}_K,s}^{\pi_\square,\pi_\Diamond}. \tag{3}$$

This implies that the solution of a PSG under extreme semi-Markov strategies is equivalent to the solution the game on its extreme interpretation limited to semi-Markov strategies, which is stated in the following:

Proposition 2. *Let $\mathcal{G}_\mathcal{K}$ and $\mathcal{H}_\mathcal{K}$ be respectively the interpretation and the extreme interpretation of $\mathcal{K}$. Then, the following equalities hold*

1. $\inf_{\pi_\diamond \in \Pi^{XS}_{\mathcal{G}_\mathcal{K},\diamond}} \sup_{\pi_\square \in \Pi^{XS}_{\mathcal{G}_\mathcal{K},\square}} \mathbb{P}^{\pi_\square,\pi_\diamond}_{\mathcal{G}_\mathcal{K},s}(\Diamond C) = \inf_{\pi_\diamond \in \Pi^{S}_{\mathcal{H}_\mathcal{K},\diamond}} \sup_{\pi_\square \in \Pi^{S}_{\mathcal{H}_\mathcal{K},\square}} \mathbb{P}^{\pi_\square,\pi_\diamond}_{\mathcal{H}_\mathcal{K},s}(\Diamond C)$

2. $\sup_{\pi_\square \in \Pi^{XS}_{\mathcal{G}_\mathcal{K},\square}} \inf_{\pi_\diamond \in \Pi^{XS}_{\mathcal{G}_\mathcal{K},\diamond}} \mathbb{P}^{\pi_\square,\pi_\diamond}_{\mathcal{G}_\mathcal{K},s}(\Diamond C) = \sup_{\pi_\square \in \Pi^{S}_{\mathcal{H}_\mathcal{K},\square}} \inf_{\pi_\diamond \in \Pi^{S}_{\mathcal{H}_\mathcal{K},\diamond}} \mathbb{P}^{\pi_\square,\pi_\diamond}_{\mathcal{H}_\mathcal{K},s}(\Diamond C)$

3. $\inf_{\pi_\diamond \in \Pi^{XS}_{\mathcal{G}_\mathcal{K},\diamond}} \sup_{\pi_\square \in \Pi^{XS}_{\mathcal{G}_\mathcal{K},\square}} \mathbb{E}^{\pi_\square,\pi_\diamond}_{\mathcal{G}_\mathcal{K},s}[\mathsf{rew}_\mathsf{f}] = \inf_{\pi_\diamond \in \Pi^{S}_{\mathcal{H}_\mathcal{K},\diamond}} \sup_{\pi_\square \in \Pi^{S}_{\mathcal{H}_\mathcal{K},\square}} \mathbb{E}^{\pi_\square,\pi_\diamond}_{\mathcal{H}_\mathcal{K},s}[\mathsf{rew}_\mathsf{f}]$

4. $\sup_{\pi_\square \in \Pi^{XS}_{\mathcal{G}_\mathcal{K},\square}} \inf_{\pi_\diamond \in \Pi^{XS}_{\mathcal{G}_\mathcal{K},\diamond}} \mathbb{E}^{\pi_\square,\pi_\diamond}_{\mathcal{G}_\mathcal{K},s}[\mathsf{rew}_\mathsf{f}] = \sup_{\pi_\square \in \Pi^{S}_{\mathcal{H}_\mathcal{K},\square}} \inf_{\pi_\diamond \in \Pi^{S}_{\mathcal{H}_\mathcal{K},\diamond}} \mathbb{E}^{\pi_\square,\pi_\diamond}_{\mathcal{H}_\mathcal{K},s}[\mathsf{rew}_\mathsf{f}]$

The next proposition, whose proof also uses (3), provides necessary conditions for the polytopal stochastic game to be almost surely stopping or irreducible in terms of the extreme interpretation.

Proposition 3. *Let $\mathcal{G}_\mathcal{K}$ and $\mathcal{H}_\mathcal{K}$ be respectively the interpretation and the extreme interpretation of $\mathcal{K}$. Then, (1) if $\mathcal{G}_\mathcal{K}$ is almost surely stopping, so is $\mathcal{H}_\mathcal{K}$, and (2) if $\mathcal{G}_\mathcal{K}$ is irreducible, so is $\mathcal{H}_\mathcal{K}$.*

Notice that by fixing one strategy in $\mathcal{H}_\mathcal{K}$ to be the memoryless, the remaining structure is a Markov decision process. Then the statements in the following proposition are consequences of standard results in MDP [27].

Proposition 4. *For all $\pi^*_\square \in \Pi^{MD}_{\mathcal{H}_\mathcal{K},\square}$ and $\pi^*_\diamond \in \Pi^{MD}_{\mathcal{H}_\mathcal{K},\diamond}$,*

1. $\sup_{\pi_\square \in \Pi^{S}_{\mathcal{H}_\mathcal{K},\square}} \mathbb{P}^{\pi_\square,\pi^*_\diamond}_{\mathcal{H}_\mathcal{K},s}(\Diamond C) = \sup_{\pi_\square \in \Pi^{MD}_{\mathcal{H}_\mathcal{K},\square}} \mathbb{P}^{\pi_\square,\pi^*_\diamond}_{\mathcal{H}_\mathcal{K},s}(\Diamond C);$

2. $\inf_{\pi_\diamond \in \Pi^{S}_{\mathcal{H}_\mathcal{K},\diamond}} \mathbb{P}^{\pi^*_\square,\pi_\diamond}_{\mathcal{H}_\mathcal{K},s}(\Diamond C) = \inf_{\pi_\diamond \in \Pi^{MD}_{\mathcal{H}_\mathcal{K},\diamond}} \mathbb{P}^{\pi^*_\square,\pi_\diamond}_{\mathcal{H}_\mathcal{K},s}(\Diamond C);$

3. $\sup_{\pi_\square \in \Pi^{S}_{\mathcal{H}_\mathcal{K},\square}} \mathbb{E}^{\pi_\square,\pi^*_\diamond}_{\mathcal{H}_\mathcal{K},s}(\mathsf{rew}_\mathsf{f}) = \sup_{\pi_\square \in \Pi^{MD}_{\mathcal{H}_\mathcal{K},\square}} \mathbb{E}^{\pi_\square,\pi^*_\diamond}_{\mathcal{H}_\mathcal{K},s}(\mathsf{rew}_\mathsf{f})$, *provided* $\mathbb{E}^{\pi_\square,\pi^*_\diamond}_{\mathcal{H}_\mathcal{K},s}(\mathsf{rew}_\mathsf{f})$ *is defined for all* $\pi_\square \in \Pi^{S}_{\mathcal{H}_\mathcal{K},\square}$; *and*

4. $\inf_{\pi_\diamond \in \Pi^{S}_{\mathcal{H}_\mathcal{K},\diamond}} \mathbb{E}^{\pi^*_\square,\pi_\diamond}_{\mathcal{H}_\mathcal{K},s}(\mathsf{rew}_\mathsf{f}) = \inf_{\pi_\diamond \in \Pi^{MD}_{\mathcal{H}_\mathcal{K},\diamond}} \mathbb{E}^{\pi^*_\square,\pi_\diamond}_{\mathcal{H}_\mathcal{K},s}(\mathsf{rew}_\mathsf{f})$, *provided* $\mathbb{E}^{\pi^*_\square,\pi_\diamond}_{\mathcal{H}_\mathcal{K},s}(\mathsf{rew}_\mathsf{f})$ *is defined for all* $\pi_\diamond \in \Pi^{S}_{\mathcal{H}_\mathcal{K},\diamond}$.

We are now in conditions to present our main result. The following theorem is two folded. On the one hand, it states that the polytopal stochastic games of all quantitative objectives of interest in this paper –namely, quantitative reachability, expected total accumulated reward, expected discounted accumulated rewards, and expected average rewards– are determined. On the other hand, it states that these objectives for PSG can be equivalently solved in its extreme interpretation.

Theorem 1. *Let $\mathcal{G}_\mathcal{K}$ and $\mathcal{H}_\mathcal{K}$ be respectively the interpretation and the extreme interpretation of $\mathcal{K}$. Then,*

1. $\displaystyle\inf_{\pi_\diamond\in\Pi_{\mathcal{G}_\mathcal{K},\diamond}}\sup_{\pi_\square\in\Pi_{\mathcal{G}_\mathcal{K},\square}}\mathbb{P}^{\pi_\square,\pi_\diamond}_{\mathcal{G}_\mathcal{K},s}(\Diamond C)=\inf_{\pi_\diamond\in\Pi^{MD}_{\mathcal{H}_\mathcal{K},\diamond}}\sup_{\pi_\square\in\Pi^{MD}_{\mathcal{H}_\mathcal{K},\square}}\mathbb{P}^{\pi_\square,\pi_\diamond}_{\mathcal{H}_\mathcal{K},s}(\Diamond C)=$

$\displaystyle=\sup_{\pi_\square\in\Pi^{MD}_{\mathcal{H}_\mathcal{K},\square}}\inf_{\pi_\diamond\in\Pi^{MD}_{\mathcal{H}_\mathcal{K},\diamond}}\mathbb{P}^{\pi_\square,\pi_\diamond}_{\mathcal{H}_\mathcal{K},s}(\Diamond C)=\sup_{\pi_\square\in\Pi_{\mathcal{G}_\mathcal{K},\square}}\inf_{\pi_\diamond\in\Pi_{\mathcal{G}_\mathcal{K},\diamond}}\mathbb{P}^{\pi_\square,\pi_\diamond}_{\mathcal{G}_\mathcal{K},s}(\Diamond C)$

for all $C\subseteq\mathcal{S}$; and

2. $\displaystyle\inf_{\pi_\diamond\in\Pi_{\mathcal{G}_\mathcal{K},\diamond}}\sup_{\pi_\square\in\Pi_{\mathcal{G}_\mathcal{K},\square}}\mathbb{E}^{\pi_\square,\pi_\diamond}_{\mathcal{G}_\mathcal{K},s}(\mathsf{rew_f})=\inf_{\pi_\diamond\in\Pi^{MD}_{\mathcal{H}_\mathcal{K},\diamond}}\sup_{\pi_\square\in\Pi^{MD}_{\mathcal{H}_\mathcal{K},\square}}\mathbb{E}^{\pi_\square,\pi_\diamond}_{\mathcal{H}_\mathcal{K},s}(\mathsf{rew_f})=$

$\displaystyle=\sup_{\pi_\square\in\Pi^{MD}_{\mathcal{H}_\mathcal{K},\square}}\inf_{\pi_\diamond\in\Pi^{MD}_{\mathcal{H}_\mathcal{K},\diamond}}\mathbb{E}^{\pi_\square,\pi_\diamond}_{\mathcal{H}_\mathcal{K},s}(\mathsf{rew_f})=\sup_{\pi_\square\in\Pi_{\mathcal{G}_\mathcal{K},\square}}\inf_{\pi_\diamond\in\Pi_{\mathcal{G}_\mathcal{K},\diamond}}\mathbb{E}^{\pi_\square,\pi_\diamond}_{\mathcal{G}_\mathcal{K},s}(\mathsf{rew_f}),$

provided: (a) $\mathcal{G}_\mathcal{K}$ is almost surely stopping whenever $\mathsf{rew_f}=\mathsf{rew_t}$, and (b) $\mathcal{G}_\mathcal{K}$ is irreducble whenever $\mathsf{rew_f}=\mathsf{rew_a}$.

Proof. For item 2 we calculate as follows:

$\inf_{\pi_\diamond\in\Pi_{\mathcal{G}_\mathcal{K},\diamond}}\sup_{\pi_\square\in\Pi_{\mathcal{G}_\mathcal{K},\square}}\mathbb{E}^{\pi_\square,\pi_\diamond}_{\mathcal{G}_\mathcal{K},s}(\mathsf{rew_f})$

$\leq\inf_{\pi_\diamond\in\Pi^S_{\mathcal{G}_\mathcal{K},\diamond}}\sup_{\pi_\square\in\Pi_{\mathcal{G}_\mathcal{K},\square}}\mathbb{E}^{\pi_\square,\pi_\diamond}_{\mathcal{G}_\mathcal{K},s}(\mathsf{rew_f})$ $\hfill(\Pi^S_{\mathcal{G}_\mathcal{K},\diamond}\subseteq\Pi_{\mathcal{G}_\mathcal{K},\diamond})$

$=\inf_{\pi_\diamond\in\Pi^S_{\mathcal{G}_\mathcal{K},\diamond}}\sup_{\pi_\square\in\Pi^S_{\mathcal{G}_\mathcal{K},\square}}\mathbb{E}^{\pi_\square,\pi_\diamond}_{\mathcal{G}_\mathcal{K},s}(\mathsf{rew_f})$ $\hfill\text{(by Lemma 2.3)}$

$=\inf_{\pi_\diamond\in\Pi^{XS}_{\mathcal{G}_\mathcal{K},\diamond}}\sup_{\pi_\square\in\Pi^{XS}_{\mathcal{G}_\mathcal{K},\square}}\mathbb{E}^{\pi_\square,\pi_\diamond}_{\mathcal{G}_\mathcal{K},s}(\mathsf{rew_f})$ $\hfill\text{(by Corollary 1.2)}$

$=\inf_{\pi_\diamond\in\Pi^S_{\mathcal{H}_\mathcal{K},\diamond}}\sup_{\pi_\square\in\Pi^S_{\mathcal{H}_\mathcal{K},\square}}\mathbb{E}^{\pi_\square,\pi_\diamond}_{\mathcal{H}_\mathcal{K},s}(\mathsf{rew_f})$ $\hfill\text{(by Prop. 2.3)}$

$\leq\inf_{\pi_\diamond\in\Pi^{MD}_{\mathcal{H}_\mathcal{K},\diamond}}\sup_{\pi_\square\in\Pi^S_{\mathcal{H}_\mathcal{K},\square}}\mathbb{E}^{\pi_\square,\pi_\diamond}_{\mathcal{H}_\mathcal{K},s}(\mathsf{rew_f})$ $\hfill(\Pi^{MD}_{\mathcal{H}_\mathcal{K},\diamond}\subseteq\Pi^S_{\mathcal{H}_\mathcal{K},\diamond})$

$=\inf_{\pi_\diamond\in\Pi^{MD}_{\mathcal{H}_\mathcal{K},\diamond}}\sup_{\pi_\square\in\Pi^{MD}_{\mathcal{H}_\mathcal{K},\square}}\mathbb{E}^{\pi_\square,\pi_\diamond}_{\mathcal{H}_\mathcal{K},s}(\mathsf{rew_f})$ $\hfill\text{(by Prop. 4.3)}$

$=\sup_{\pi_\square\in\Pi^{MD}_{\mathcal{H}_\mathcal{K},\square}}\inf_{\pi_\diamond\in\Pi^{MD}_{\mathcal{H}_\mathcal{K},\diamond}}\mathbb{E}^{\pi_\square,\pi_\diamond}_{\mathcal{H}_\mathcal{K},s}(\mathsf{rew_f})$ $\hfill(*)$

$=\sup_{\pi_\square\in\Pi^{MD}_{\mathcal{H}_\mathcal{K},\square}}\inf_{\pi_\diamond\in\Pi^S_{\mathcal{H}_\mathcal{K},\diamond}}\mathbb{E}^{\pi_\square,\pi_\diamond}_{\mathcal{H}_\mathcal{K},s}(\mathsf{rew_f})$ $\hfill\text{(by Prop. 4.4)}$

$\leq\sup_{\pi_\square\in\Pi^S_{\mathcal{H}_\mathcal{K},\square}}\inf_{\pi_\diamond\in\Pi^S_{\mathcal{H}_\mathcal{K},\diamond}}\mathbb{E}^{\pi_\square,\pi_\diamond}_{\mathcal{H}_\mathcal{K},s}(\mathsf{rew_f})$ $\hfill(\Pi^{MD}_{\mathcal{H}_\mathcal{K},\square}\subseteq\Pi^S_{\mathcal{H}_\mathcal{K},\square})$

$=\sup_{\pi_\square\in\Pi^{XS}_{\mathcal{G}_\mathcal{K},\square}}\inf_{\pi_\diamond\in\Pi^{XS}_{\mathcal{G}_\mathcal{K},\diamond}}\mathbb{E}^{\pi_\square,\pi_\diamond}_{\mathcal{G}_\mathcal{K},s}(\mathsf{rew_f})$ $\hfill\text{(by Prop. 2.4)}$

$=\sup_{\pi_\square\in\Pi^S_{\mathcal{G}_\mathcal{K},\square}}\inf_{\pi_\diamond\in\Pi^S_{\mathcal{G}_\mathcal{K},\diamond}}\mathbb{E}^{\pi_\square,\pi_\diamond}_{\mathcal{G}_\mathcal{K},s}(\mathsf{rew_f})$ $\hfill\text{(by Corollary 1.2)}$

$=\sup_{\pi_\square\in\Pi^S_{\mathcal{G}_\mathcal{K},\square}}\inf_{\pi_\diamond\in\Pi_{\mathcal{G}_\mathcal{K},\diamond}}\mathbb{E}^{\pi_\square,\pi_\diamond}_{\mathcal{G}_\mathcal{K},s}(\mathsf{rew_f})$ $\hfill\text{(by Lemma 2.3)}$

$\leq\sup_{\pi_\square\in\Pi_{\mathcal{G}_\mathcal{K},\square}}\inf_{\pi_\diamond\in\Pi_{\mathcal{G}_\mathcal{K},\diamond}}\mathbb{E}^{\pi_\square,\pi_\diamond}_{\mathcal{G}_\mathcal{K},s}(\mathsf{rew_f})$ $\hfill(\Pi^S_{\mathcal{G}_\mathcal{K},\square}\subseteq\Pi_{\mathcal{G}_\mathcal{K},\square})$

$\leq\inf_{\pi_\diamond\in\Pi_{\mathcal{G}_\mathcal{K},\diamond}}\sup_{\pi_\square\in\Pi_{\mathcal{G}_\mathcal{K},\square}}\mathbb{E}^{\pi_\square,\pi_\diamond}_{\mathcal{G}_\mathcal{K},s}(\mathsf{rew_f})$ $\hfill\text{(by prop. of sup and inf)}$

Since the last term is equal to the first term in the calculation, item 2 is concluded. In particular, step $(*)$ is justified as follows, depending on $\mathsf{rew_f}$: For $\mathsf{rew_f}=\mathsf{rew_t}$, $(*)$ follows by [15, Theorem 4.2.6] since, by Proposition 3.(1), the game $\mathcal{H}_\mathcal{K}$ is also almost surely stopping. For $\mathsf{rew_f}=\mathsf{rew_\gamma}$ $(*)$ follows by [15, Theorem 4.3.2]. For $\mathsf{rew_f}=\mathsf{rew_a}$ $(*)$ follows by [15, Theorem 5.1.5] since, by Proposition 3.(2), the game $\mathcal{H}_\mathcal{K}$ is also irreducible.

Item 1 of the theorem follows similarly. In each step, propositions, lemmas and corollaries are the same only differing on the item, while step $(*)$ follows from [13, Lemma 6]. $\hfill\square$

Since extreme interpretations are finite, the values of the different games can be calculated following known algorithms [13,15]. Thus, Theorem 1 immediately provides an algorithmic solution for PSGs: one first solves the linear equation systems defining the polytopes, obtaining the extreme interpretation of the game; afterwards, value iteration, or similar methods, can be used to compute the solution of the game.

The number of vertices of a polytope grows exponentially in the dimension of the polytope [19]. More precisely if d is the dimension of a polytope K and m is the number of inequalities that defines it, $\mathbb{V}(K) \sim \Omega(m^{\lfloor d/2 \rfloor})$. This implies that the extreme interpretation $\mathcal{H}_\mathcal{K}$ grows exponentially on the largest size of the support sets of the distributions involved in the original PSG $\mathcal{K}$ which we expect not to be too large. (In our example of Sect. 2, $\lfloor d/2 \rfloor = 2$)

Condon [13] showed that deciding reachability in stochastic games is in NP ∩ coNP. Despite the exponential grow, this is still our case as we show in the following. Let $Val_s(\mathcal{K})$ denote the value of the game at state s, that is, it is equal to $\sup_{\pi_\Box \in \Pi_\Box} \inf_{\pi_\Diamond \in \Pi_\Diamond} \mathbb{P}^{\pi_\Box, \pi_\Diamond}_{\mathcal{G}_\mathcal{K}, s}(\Diamond G)$, or $\sup_{\pi_\Box \in \Pi_\Box} \inf_{\pi_\Diamond \in \Pi_\Diamond} \mathbb{E}^{\pi_\Box, \pi_\Diamond}_{\mathcal{G}_\mathcal{K}, s}[\text{rew}_f]$. The problem is then to decide whether $Val_s(\mathcal{K}) \geq q$, for a given $q \in \mathbb{Q}$ and $s \in \mathcal{S}$ Since for all the cases (total reward, discounted reward, average reward and reachability objectives under the respective conditions) the value $Val_s(\mathcal{K})$ of the game can be achieved with an extreme memoryless and deterministic strategies, we can reason as follows: (i) guess a memoryless and deterministic strategy for each player, (ii) on the resulting Markov chain compute the corresponding measure (i.e. total reward, discounted reward, average reward or reachability) on the respective set of linear equations, which can be done in polynomial time (for rew_a an extra linear summation is needed) [22], (iii) verify if it is a fixpoint of Bellman equations (for reachability, discounted, or total reward), or a fixpoint of the Alg. 5.1.1 of [15], in the case of average reward, and (iv) check whether $Val_s(\mathcal{K}) \geq q$. This puts our problem in NP. With the same process we can check whether $Val_s(\mathcal{K}) < q$ which puts the problem also in coNP. Hence we have the next theorem.

Theorem 2. *For any PSG $\mathcal{K}$, $q \in \mathbb{Q}$, and $s \in \mathcal{S}$, the problem of deciding whether $Val_s(\mathcal{K}) \geq q$ is in* NP ∩ coNP. *For* $\text{rew}_f \in \{\text{rew}_t, \text{rew}_a\}$ *the decision problem is restricted to $\mathcal{G}_\mathcal{K}$ being almost surely stopping and irreducible, respectively.*

6   Concluding Remarks

We believe that polytopal games may have several applications in practice, particularly, in scenarios where the probabilities are not exact but can be characterized with linear equations. We observe that one may expect that the number of vertices of the polytopes keep small in practical examples, hence the game discretization may have no impact on the runtime of a tool implementing the approach described in the paper. However, we leave as further work the implementation of such a tool and an in-depth evaluation of it.

In addition, it would be also be of interest to explore other types of objectives, including ω-regular objectives as already study for standard stochastic games in [9] or even solving stochastic games for conditional probabilities of temporal properties or conditional expectations of rewards models as widely studied by Christel Baier and her team in the context of Markov decision processes [5,6,23, 25].

References

1. Akshay, S., Bouyer, P., Krishna, S.N., Manasa, L., Trivedi, A.: Stochastic timed games revisited. In: Faliszewski, P., Muscholl, A., Niedermeier, R. (eds.) 41st International Symposium on Mathematical Foundations of Computer Science, MFCS 2016. LIPIcs, vol. 58, pp. 8:1–8:14. Schloss Dagstuhl - Leibniz-Zentrum für Informatik (2016). https://doi.org/10.4230/LIPICS.MFCS.2016.8
2. Ash, R.B., Doléans-Dade, C.A.: Probability and Measure Theory. Harcourt/Academic Press, 2nd edn. (1999)
3. Aslanyan, Z., Nielson, F., Parker, D.: Quantitative verification and synthesis of attack-defence scenarios. In: IEEE 29th Computer Security Foundations Symposium, CSF 2016, pp. 105–119. IEEE Computer Society (2016). https://doi.org/10.1109/CSF.2016.15
4. Baier, C., Katoen, J.: Principles of Model Checking. MIT Press (2008)
5. Baier, C., Klein, J., Klüppelholz, S., Märcker, S.: Computing conditional probabilities in Markovian models efficiently. In: Ábrahám, E., Havelund, K. (eds.) TACAS 2014. LNCS, vol. 8413, pp. 515–530. Springer, Heidelberg (2014). https://doi.org/10.1007/978-3-642-54862-8_43
6. Baier, C., Klein, J., Klüppelholz, S., Wunderlich, S.: Maximizing the conditional expected reward for reaching the goal. In: Legay, A., Margaria, T. (eds.) TACAS 2017. LNCS, vol. 10206, pp. 269–285. Springer, Heidelberg (2017). https://doi.org/10.1007/978-3-662-54580-5_16
7. Bouyer, P., Forejt, V.: Reachability in stochastic timed games. In: Albers, S., Marchetti-Spaccamela, A., Matias, Y., Nikoletseas, S., Thomas, W. (eds.) ICALP 2009. LNCS, vol. 5556, pp. 103–114. Springer, Heidelberg (2009). https://doi.org/10.1007/978-3-642-02930-1_9
8. Castro, P.F., D'Argenio, P.R., Demasi, R., Putruele, L.: Quantifying masking fault-tolerance via fair stochastic games. In: Mezzina, C.A., Caltais, G. (eds.) Proceedings Combined 30th International Workshop on Expressiveness in Concurrency and 20th Workshop on Structural Operational Semantics, EXPRESS/SOS 2023, and 20th Workshop on Structural Operational Semantics. EPTCS, vol. 387, pp. 132–148 (2023). https://doi.org/10.4204/EPTCS.387.10
9. Chatterjee, K., de Alfaro, L., Henzinger, T.A.: The complexity of stochastic rabin and streett games. In: Caires, L., Italiano, G.F., Monteiro, L., Palamidessi, C., Yung, M. (eds.) ICALP 2005. LNCS, vol. 3580, pp. 878–890. Springer, Heidelberg (2005). https://doi.org/10.1007/11523468_71
10. Chatterjee, K., Henzinger, T.A.: A survey of stochastic ω-regular games. J. Comput. Syst. Sci. **78**(2), 394–413 (2012). https://doi.org/10.1016/j.jcss.2011.05.002
11. Chen, T., Forejt, V., Kwiatkowska, M., Parker, D., Simaitis, A.: PRISM-games: a model checker for stochastic multi-player games. In: Piterman, N., Smolka, S.A. (eds.) TACAS 2013. LNCS, vol. 7795, pp. 185–191. Springer, Heidelberg (2013). https://doi.org/10.1007/978-3-642-36742-7_13

12. Chen, T., Kwiatkowska, M., Simaitis, A., Wiltsche, C.: Synthesis for multi-objective stochastic games: an application to autonomous urban driving. In: Joshi, K., Siegle, M., Stoelinga, M., D'Argenio, P.R. (eds.) QEST 2013. LNCS, vol. 8054, pp. 322–337. Springer, Heidelberg (2013). https://doi.org/10.1007/978-3-642-40196-1_28
13. Condon, A.: The complexity of stochastic games. Inf. Comput. **96**(2), 203–224 (1992). https://doi.org/10.1016/0890-5401(92)90048-K
14. Feng, L., Wiltsche, C., Humphrey, L.R., Topcu, U.: Synthesis of human-in-the-loop control protocols for autonomous systems. IEEE Trans. Autom. Sci. Eng. **13**(2), 450–462 (2016). https://doi.org/10.1109/TASE.2016.2530623
15. Filar, J., Vrieze, K.: Competitive Markov Decision Processes. Springer, Heidelberg (1996)
16. Giry, M.: A categorical approach to probability theory. In: Banaschewski, B. (ed.) Categorical Aspects of Topology and Analysis. LNM, vol. 915, pp. 68–85. Springer, Heidelberg (1982). https://doi.org/10.1007/BFb0092872
17. Jonsson, B., Larsen, K.G.: Specification and refinement of probabilistic processes. In: Proceedings of the 6th Annual Symposium on Logic in Computer Science (LICS 1991), pp. 266–277. IEEE Computer Society (1991). https://doi.org/10.1109/LICS.1991.151651
18. Junges, S., Jansen, N., Katoen, J.-P., Topcu, U., Zhang, R., Hayhoe, M.: Model checking for safe navigation among humans. In: McIver, A., Horvath, A. (eds.) QEST 2018. LNCS, vol. 11024, pp. 207–222. Springer, Cham (2018). https://doi.org/10.1007/978-3-319-99154-2_13
19. Kaibel, V., Pfetsch, M.E.: Some algorithmic problems in polytope theory. In: Joswig, M., Takayama, N. (eds.) Algebra, Geometry, and Software Systems [outcome of a Dagstuhl seminar], pp. 23–47. Springer, Cham (2003). https://doi.org/10.1007/978-3-662-05148-1_2
20. Kozine, I., Utkin, L.V.: Interval-valued finite Markov chains. Reliab. Comput. **8**(2), 97–113 (2002). https://doi.org/10.1023/A:1014745904458
21. Kucera, A.: Turn-based stochastic games. In: Apt, K.R., Grädel, E. (eds.) Lectures in Game Theory for Computer Scientists, pp. 146–184. Cambridge University Press (2011)
22. Kulkarni, V.G.: Modeling and Analysis of Stochastic Systems, 3rd edn. Texts in Statistical Science, CRC Press (2017)
23. Märcker, S., Baier, C., Klein, J., Klüppelholz, S.: Computing conditional probabilities: implementation and evaluation. In: Cimatti, A., Sirjani, M. (eds.) SEFM 2017. LNCS, vol. 10469, pp. 349–366. Springer, Cham (2017). https://doi.org/10.1007/978-3-319-66197-1_22
24. McMullen, P.: Geometric Regular Polytopes. Encyclopedia of Mathematics and its Applications. Cambridge University Press, Cambridge (2020)
25. Piribauer, J., Baier, C.: Partial and conditional expectations in Markov decision processes with integer weights. In: Bojańczyk, M., Simpson, A. (eds.) FoSSaCS 2019. LNCS, vol. 11425, pp. 436–452. Springer, Cham (2019). https://doi.org/10.1007/978-3-030-17127-8_25
26. Puggelli, A., Li, W., Sangiovanni-Vincentelli, A.L., Seshia, S.A.: Polynomial-time verification of PCTL properties of MDPs with convex uncertainties. In: Sharygina, N., Veith, H. (eds.) CAV 2013. LNCS, vol. 8044, pp. 527–542. Springer, Heidelberg (2013). https://doi.org/10.1007/978-3-642-39799-8_35
27. Puterman, M.L.: Markov Decision Processes: Discrete Stochastic Dynamic Programming. Wiley Series in Probability and Statistics, Wiley (1994). https://doi.org/10.1002/9780470316887

28. R. S. Sutton, A.G.B..: Reinforcement Learning: An Introduction. Bradford Books (2018)
29. Sen, K., Viswanathan, M., Agha, G.: Model-checking Markov chains in the presence of uncertainties. In: Hermanns, H., Palsberg, J. (eds.) TACAS 2006. LNCS, vol. 3920, pp. 394–410. Springer, Heidelberg (2006). https://doi.org/10.1007/11691372_26
30. Shapley, L.S.: Stochastic games. Proc. Natl. Acad. Sci. U.S.A. **39**(10), 1095–100 (1953). https://doi.org/10.1073/pnas.39.10.1095
31. Wang, M., Wang, Z., Talbot, J., Gerdes, J.C., Schwager, M.: Game-theoretic planning for self-driving cars in multivehicle competitive scenarios. IEEE Trans. Robot. **37**(4), 1313–1325 (2021). https://doi.org/10.1109/TRO.2020.3047521
32. Ziegler, G.M.: Lectures on Polytopes, Graduate Texts in Mathematics, vol. 152. Springer, New York (1995)

Algorithmic Applications of Schanuel's Conjecture

Toghrul Karimov[1], Joris Nieuwveld[1], Joël Ouaknine[1(✉)], Mihir Vahanwala[1], and James Worrell[2]

[1] Max Planck Institute for Software Systems, Saarland Informatics Campus, Saarbrücken, Germany
{toghs,jnieuwve,joel,mvahanwa}@mpi-sws.org
[2] Department of Computer Science, University of Oxford, Oxford, UK
jbw@cs.ox.ac.uk

Abstract. We present a survey of algorithms, mostly drawn from the broad field of logic in computer science, that rely on Schanuel's conjecture for termination and/or correctness. Schanuel's conjecture is a central hypothesis in transcendental number theory that generalises many existing classical results such as the Lindemann-Weierstrass theorem and Baker's theorem on linear independence of logarithms of algebraic numbers. The algorithmic use of Schanuel's conjecture was spearheaded by computer algebraists in the 1970s, as well as by Macintyre and Wilkie in the 1990s, most notably to establish the decidability of real arithmetic expanded with the exponential function. Since then, many further applications have been recorded in the literature. We present and discuss several of these algorithms, with a particular focus on the precise role played by Schanuel's conjecture.

1 Introduction

It is not uncommon for correctness properties of algorithms to be conditional upon hypotheses that are unproven, but plausibly true. The most emblematic examples come from cryptography, where the security of protocols, i.e., the computational infeasibility of breaking a scheme, is predicated on conjectures such as the impossibility to factor large numbers efficiently, or more generally $\mathsf{P} \neq \mathsf{NP}$.

In the fields of computer algebra, automata theory, and dynamical systems, various decision problems hinge on variants of the following subroutine:

Problem 1. Given a polynomial $p \in \mathbb{Q}[x_1, \ldots, x_k]$, and $q_1, \ldots, q_k \in \mathbb{Q}_{>0}$, decide whether $p(\log q_1, \ldots, \log q_k) > 0$.

The number-theoretic hurdle in following the obvious approach of using increasingly precise approximations of $\log q_1, \ldots, \log q_k$ is that these numbers might "unexpectedly" be *algebraically dependent*[1] with p as a witness, i.e.,

---

[1] Throughout this paper, algebraic (in)dependence and transcendence are meant to be over the field $\mathbb{Q}$ of rational numbers, unless otherwise specified.

© The Author(s), under exclusive license to Springer Nature Switzerland AG 2026
N. Bertrand et al. (Eds.): Christel Baier Festschrift, LNCS 15760, pp. 118–138, 2026.
https://doi.org/10.1007/978-3-031-97439-7_5

$p(\log q_1, \ldots, \log q_k) = 0$. The purported decision procedure could not effectively detect this case because it would never terminate if it were to arise.

However, if we are promised that $\log q_1, \ldots, \log q_k$ are *algebraically independent*, i.e., $f(\log q_1, \ldots, \log q_k) \neq 0$ for all nonzero $f \in \mathbb{Q}[x_1, \ldots, x_k]$, then instead of the usual trichotomy, we are only faced with a dichotomy between strict inequalities, which the strategy above is guaranteed to resolve. Crucially, only *termination* is conditional upon the promise: *correctness* of the output is unconditional, since there are well-known techniques for approximating logarithms within any required error bound.

Unfortunately, unconditional promises of algebraic independence often stretch beyond the capabilities of contemporary number-theoretic techniques. A classical example is the widely expected algebraic independence of e and π, which has yet to be established. This is where *Schanuel's conjecture* [19, Pages 30–31], a central hypothesis in transcendental number theory going back to the 1960s, enters the picture.

Conjecture 1 (Schanuel's conjecture). If the complex numbers $\beta_1, \ldots, \beta_k$ are linearly independent over $\mathbb{Q}$, then the set $\{\beta_1, \ldots, \beta_k, \exp(\beta_1), \ldots, \exp(\beta_k)\}$ contains a subset of k algebraically independent numbers.

As immediate examples, consider:

- The numbers 1 and $i\pi$, where i is the imaginary unit, are linearly independent over $\mathbb{Q}$. If Schanuel's conjecture holds, the set $\{1, i\pi, e, -1\}$ contains a subset of two algebraically independent numbers: by inspection, these must be $i\pi$ and e. From this assertion we can then prove the algebraic independence of π and e through elementary algebra.
- The transcendental numbers $\log 2, \log 3, \log 5$ are linearly independent over $\mathbb{Q}$, thanks to the fundamental theorem of arithmetic. Indeed, if they weren't, there would be an integer linear relationship between $\log 2, \log 3, \log 5$, e.g., $a \log 2 = b \log 3 + c \log 5$ with a, b, c positive (the other possibilities are analogous). Upon exponentiating, it would imply that 2^a can also be factored as $3^b 5^c$, a contradiction. If Schanuel's conjecture holds, these linearly independent numbers must also be the three *algebraically* independent numbers in the set $\{\log 2, \log 3, \log 5, 2, 3, 5\}$. There being no nonzero polynomial $p \in \mathbb{Q}[x_1, x_2, x_3]$ such that $p(\log 2, \log 3, \log 5) = 0$ guarantees that the obvious approach discussed earlier would resolve this specific class of instances of Problem 1.

Schanuel's conjecture is in fact a powerful and far-reaching generalisation of a number of classical results in transcendental number theory, three of which we state below. Recall that a complex number $\alpha \in \mathbb{C}$ is *algebraic* if $p(\alpha) = 0$ for some polynomial $p \in \mathbb{Q}[x]$, and is *transcendental* otherwise. The collection of algebraic numbers forms an algebraically closed field, which we denote by $\overline{\mathbb{Q}}$.

Theorem 1 (Lindemann-Weierstrass, 1885). *If $\alpha_1, \ldots, \alpha_k$ are algebraic numbers that are linearly independent over $\mathbb{Q}$, then $\exp(\alpha_1), \ldots, \exp(\alpha_k)$ are algebraically independent over $\overline{\mathbb{Q}}$.*

Theorem 2 (Gelfond-Schneider, 1934). *If α and β are algebraic numbers such that $\alpha \neq 0, 1$ and β is irrational, then α^β is transcendental.*

Theorem 3 (Baker, 1966). *If $\alpha_1, \ldots, \alpha_k$ are algebraic numbers such that $\log(\alpha_1), \ldots, \log(\alpha_k)$ are linearly independent over $\mathbb{Q}$, then $1, \log(\alpha_1), \ldots, \log(\alpha_k)$ are linearly independent over $\overline{\mathbb{Q}}$.*

As we observed in the example of $\log 2, \log 3, \log 5$ above, Schanuel's conjecture would strengthen the consequence of Baker's theorem to assert that $\log(\alpha_1), \ldots, \log(\alpha_k)$ are in fact *algebraically* independent.

In this paper, we survey the algorithmic applications of Schanuel's conjecture. We start with the role it plays in the foundations of computer algebra, and consequently, the logical reasoning about real numbers. The latter proves to be an especially convenient interface for further applications in logic, as well as in both discrete and continuous dynamical systems, which include recurrence sequences, automata, and stochastic processes.

Schanuel's conjecture was naturally of immediate interest to computer algebraists, who sought to expand their domain of operations to encompass elementary functions such as $\exp x, \log x, \sin x, \arccos x$, without compromising on the critical ability to recognise when an expression evaluates to 0. There were significant strides in this endeavour as early as 1970 [7]. Throughout the last three decades of the 20th century, Richardson developed an influential line of work culminating in [28], which achieved the above goal: this body of research showed, through both theory and practice, that assuming Schanuel's conjecture, the exponential field of *elementary* numbers (Sect. 2.2), which is countable, algebraically closed, and also closed under the elementary functions, is computable, i.e., there is an effective representation scheme that is amenable to arithmetic operations, elementary functions such as $\exp, \log, \sin, \arccos$, etc., and zero testing. The specific role of Schanuel's conjecture is to guarantee the termination of the fundamental zero-testing algorithm for elementary numbers.

As [28, Introduction] notes, the consensus [2,3,30] among the eminent number theorists of the 1970s was that Schanuel's conjecture is very likely correct, but would be extremely hard to prove. Little has changed since. In order to approach the resolution of Schanuel's conjecture, Zilber [38] showed in 2005 that there is a unique exponential field $\mathbb{B}$ with cardinality $|\mathbb{C}|$ that axiomatically "imitates" the complex numbers with exponentiation, and satisfies Schanuel's conjecture along with a property called *strong exponential-algebraic closure*. If $\mathbb{B}$ and $\mathbb{C}$ are isomorphic, then Schanuel's conjecture for $\mathbb{C}$ follows. Conversely, if $\mathbb{B}$ and $\mathbb{C}$ are not isomorphic, then at least one among Schanuel's conjecture and strong exponential-algebraic closure fails to hold for the complex numbers. Unfortunately, neither proof nor refutation of the isomorphism seems accessible.

The connection between model theory and transcendence theory, however, had come to the fore nearly a decade prior to Zilber's work when in 1996, Macintyre and Wilkie [22] used Schanuel's conjecture to prove the termination of their now celebrated algorithm to decide the first-order theory $T_{\exp}$ of the real numbers with the exponential function. As we survey decision procedures that rely upon Schanuel's conjecture, we observe that several do so solely by virtue of queries to $T_{\exp}$: such is the influence of [22] on making the algorithmic consequences of Schanuel's conjecture accessible. Although Schanuel's conjecture is only used to prove termination of the algorithm deciding $T_{\exp}$, [22, Sec. 5] nevertheless hedges against its failure by identifying that the decidability of $T_{\exp}$ is equivalent to an ostensibly weaker hypothesis (Conjecture 2), whose resolution was still expected to be inaccessible.

The synergy between transcendence theory, computer algebra, and model theory continues to be explored to this day. In 2016, Macintyre [21] exhibited remarkable evidence to suggest that Schanuel's conjecture is fundamental to computability: more precisely, he showed that any countable exponential field that obeys Zilber's axioms (which, in particular, entail Schanuel's conjecture) is computable.

In this survey, we begin by explaining the foundational computer-algebraic applications in Sect. 2. In Sect. 3.1 and 3.2, we then explain how these advances, when combined with model theory, established the decidability of first-order logical theories. Queries formulated in first-order logic are typically (but not exclusively) the means through which Schanuel's conjecture is invoked by algorithms that decide problems in monadic second-order logic (Sect. 3.3), discrete dynamical systems (Sect. 4), quantitative verification (Sect. 5), and continuous dynamical systems and MDPs (Sect. 6).

2   Computer Algebra

As discussed earlier, Schanuel's conjecture is useful for zero-testing in computer algebra because it can be intuited as a promise of algebraic dependence among numbers being formally certifiable. We shall next make this intuition concrete by summarising [29], and subsequently survey the more general case of recognising zero among the elementary numbers [27,28]. These procedures are unconditionally correct, and require Schanuel's conjecture only to guarantee termination.

2.1   Zero-Testing Elementary Expressions

The problem considered in [29] is that of testing whether expressions involving addition, subtraction, multiplication, division, taking n-th roots for a positive integer n, division, exp, and log[2] evaluate to zero. Formally, an expression E is given as a parse tree whose leaves represent rational numbers, and internal nodes carry elementary operations, and we have to decide whether it is the case that the value $V(E) = 0$. Observe that this entails recursively solving subproblems:

---

[2] We take the principal branch of the logarithm. A complex number $z = p + iq$ can also be expressed as $z = e^{x+iy}$, choosing $y \in (\pi, \pi]$. This choice defines $\log z = x + iy$. Similarly, we assume that n-th roots of numbers except 0 are interpreted as $\exp((\log z)/n)$. Note that with this set of operations, we can also implement trigonometric, inverse trigonometric functions, and their hyperbolic analogues.

in order for $V(E)$ to be well defined, for every subexpression[3] E_i that evaluates a divisor, or the argument of log, it is necessary that $V(E_i) \neq 0$.

For example, i can be expressed as either $(-1)^{1/2}$ or $\exp(\log(-1)/2)$, and thus the difference of these two expressions is 0. Similarly, π can be expressed as $\log(-1)/(-1)^{1/2}$. We shall adopt $E = (-1)^{1/2} - \exp(\log(-1)/2)$ as our running (toy) example.

We begin by classifying subexpressions. We shall use the letter A to denote subexpressions that are either arguments to exp or outputs of log, and the letter B to denote subexpressions that are either outputs of exp or arguments to log. In our example, we have $A_1 = \log(-1)$, $A_2 = \log(-1)/2$, $B_1 = -1$, $B_2 = \exp(\log(-1)/2)$. In this way, we shall maintain that either $A_j = \log(B_j)$, or $B_j = \exp(A_j)$. We note that an expression can be both an A_j and a B_k, e.g., if $E' = \exp(\exp(1))$, then $B_1' = A_2' = \exp(1)$.

We shall use α_j, β_j to respectively denote $V(A_j), V(B_j)$. In our running example, $A_1 = i\pi, A_2 = i\pi/2, B_1 = -1, B_2 = i$. By definition, we always have that $\beta_j = \exp(\alpha_j)$, and for every pair (α_j, β_j), there is an input to the operand, and an output to the operand, e.g., if $A_j = \log(B_j)$ then α_j is the output number. We use γ_j to denote the output component of the pair. By the structure of E, we have that $V(E)$, and in fact each $\alpha_1, \beta_1, \ldots, \alpha_k, \beta_k$ is an algebraic expression in $\gamma_1, \ldots, \gamma_k$ (where k is the number of subexpressions rooted at exp or log).

Now, if we assume that $\alpha_1, \ldots, \alpha_k$ are linearly independent over $\mathbb{Q}$, then Schanuel's conjecture promises us that there are k algebraically independent numbers among $\alpha_1, \ldots, \alpha_k, \beta_1, \ldots, \beta_k$. By the preceding observation, the only way to ensure this is for $\gamma_1, \ldots, \gamma_k$ to be algebraically independent. Recall that we have $V(E) = f(\gamma_1, \ldots, \gamma_k)$ for some algebraic function f. If $f \equiv 0$, then we have proof that $V(E) = 0$ regardless of our assumption of linear independence above. If not, then we are promised in particular that $V(E) \neq 0$. We thus run a semi-algorithm to compute $V(E)$ to arbitrary precision to cover this case, Schanuel's conjecture assuring us that this will eventually certify that $V(E) \neq 0$.

If the linear independence assumption fails, then there exist $c_1, \ldots, c_j \in \mathbb{Z}$ with $c_j \neq 0$ such that $c_1\alpha_1 + \cdots + c_j\alpha_j = 0$. If we find these integers, we can use them to obtain an equivalent expression E' of a "reduced order", i.e., E' has fewer subexpressions using exp, log, and $V(E') = V(E)$. In our example, we have that $2\alpha_2 = \alpha_1$, and $B_2 = \exp(A_2)$. We can replace B_2 with $(B_1)^{1/2}$, to get the reduced order $E' = (-1)^{1/2} - (-1)^{1/2}$. More generally, if $B_j = \exp(A_j)$, then we replace B_j with $(B_1^{-c_1} \cdots B_{j-1}^{-c_{j-1}})^{1/c_j}$, and if $A_j = \log(B_j)$, then we replace A_j with $-(c_1A_1 + \cdots + c_{j-1}A_{j-1})/c_j$. Clearly, the order can be reduced only finitely often.

However, in order to soundly obtain such an equivalent E', one needs to *prove* that a purported[4] linear dependency holds. Towards such a formal proof,

---

[3] Technically, we take the parse tree to be a directed acyclic graph in order to avoid duplicate subexpressions. We assume an ordering such that if $i < j$, then E_i cannot have E_j as a subexpression.

[4] Algorithms such as LLL (Lenstra-Lenstra-Lovász lattice basis reduction) or PSLQ [13] can greatly optimise the enumeration of potential dependencies.

we shall recursively associate with each subexpression E_i of E, an algebraic function $\eta(E_i)$ as follows. If E_i is $\exp(A_j)$ for some j, then $\eta(E_i) = x_j$, likewise if E_i is $\log(B_j)$ for some j, then $\eta(E_i) = y_j$. Otherwise, if $E_i = \mathsf{op}(E_{i_1}, \ldots)$ for some algebraic operation op, then we set $\eta(E_i) = \mathsf{op}(\eta(E_{i_1}), \ldots)$. Finally, if E_i is a rational constant, then $\eta(E_i)$ returns the same constant, e.g., $\eta(-1) = -1$.

Returning to our running example, we have that $\eta(A_1) = x_1, \eta(B_1) = -1, \eta(A_2) = x_1/2, \eta(B_2) = y_2$. By easy symbolic reasoning, we see that if $c_1\eta(A_1) + \cdots + c_j\eta(A_j) \equiv 0$ (is identically 0), then $c_1\alpha_1 + \cdots + c_j\alpha_j = 0$. Similarly, if $\eta(B_1)^{c_1} \cdots \eta(B_j)^{c_j} - 1 \equiv 0$ and $|c_1\alpha_1 + \cdots + c_j\alpha_j| < 1$, then $c_1\alpha_1 + \cdots + c_j\alpha_j = 0$ because it must be an integer multiple of $2\pi i$. In this manner, associated algebraic functions serve as certificates; however, if a linear dependency exists, is it guaranteed there to be a certificate?

We make a thematic invocation of Schanuel's conjecture to answer affirmatively. Suppose that we have $c_1\alpha_1 + \cdots + c_j\alpha_j = 0$ with j minimal, i.e., $c_j \neq 0$ and $\alpha_1, \ldots, \alpha_{j-1}$ are linearly independent. Let λ_j be the input component among α_j, β_j. We have by structure that $\alpha_1, \beta_1, \ldots, \alpha_{j-1}, \beta_{j-1}, \lambda_j$ are algebraic in $\gamma_1, \ldots, \gamma_{j-1}$. In particular, either $\eta(c_1A_1 + \cdots + c_jA_j)$, or $\eta(B_1^{c_1} \cdots B_j^{c_j} - 1)$ is of the form $f(z_1, \ldots, z_{j-1})$, where $f(\gamma_1, \ldots, \gamma_{k-1})$ would return $c_1\alpha_1 + \cdots + c_j\alpha_j$ or $\beta_1^{c_1} \cdots \beta_j^{c_j} - 1$ respectively. The arguments $\gamma_1, \ldots, \gamma_{k-1}$, by Schanuel's conjecture, are algebraically independent. The only way to realise $f(\gamma_1, \ldots, \gamma_{k-1}) = 0$ therefore, would be $f \equiv 0$, implying that the minimal dependency, when enumerated, would constitute a desired symbolic proof.

Note that in the worst case, if Schanuel's conjecture is false, then no proof would be observed, and we would continue our search for a dependency, but never make unsound progress towards a decision in the algorithm.

In summary, we run a semi-algorithm that seeks to prove that $V(E) \neq 0$ by brute approximation, and a semi-algorithm that seeks to reduce the (finite) order of E. If Schanuel's conjecture holds, at least one of them will terminate. Having terminated, the algorithm gives a certificate (an approximate, or an identically zero function) of whether $V(E) = 0$.

2.2 Zero-Testing Elementary Numbers

Recall that elementary numbers constitute an algebraically closed subfield of the complex numbers, which is closed under applications of elementary functions and computable (i.e., there is an effective representation scheme that is amenable to arithmetic operations, elementary functions such as $\exp, \log, \sin, \arccos$, etc., and zero testing) assuming Schanuel's conjecture. We now follow [28, Sec. 2] and record the concepts required to define elementary numbers.

The first building block is an *exponential system*, which consists of polynomials $p_1, \ldots, p_r \in \mathbb{Q}[x_1, \ldots, x_n]$ and expressions $w_1 - \exp(z_1), \ldots, w_k - \exp(z_k)$ where $\{w_1, \ldots, w_k, z_1, \ldots, z_k\} \subseteq \{x_1, \ldots, x_n\}$. We use $F = \langle F_1, \ldots, F_{r+k} \rangle$ to collectively denote the entire exponential system.

An *elementary point* $\overline{\gamma} = (\gamma_1, \ldots, \gamma_n) \in \mathbb{C}^n$ is a nonsingular root of an exponential system $F = \langle F_1, \ldots, F_n \rangle$, i.e., $\overline{\gamma}$ satisfies $F(\overline{\gamma}) = 0$, and the Jacobian determinant $J_F(\overline{\gamma}) = \det\left(\frac{\partial F_i}{\partial x_j}\right)_{1 \le i,j \le n} (\overline{\gamma}) \ne 0$. A number $\zeta \in \mathbb{C}$ is *elementary* if there is an elementary point $\overline{\gamma} \in \mathbb{C}^n$ and a polynomial $p \in \mathbb{Q}[x_1, \ldots, x_n]$ such that $\zeta = p(\overline{\gamma})$. Naturally, an elementary number ζ is represented by $(\overline{\gamma}, p)$. It is clear how to represent p; we discuss how we represent an elementary point $\overline{\gamma}$.

1. We are given an exponential system $F = \langle F_1, \ldots, F_n \rangle$, which is straightforward to represent.
2. We are given an approximate $\overline{\alpha} \in (\mathbb{Q}[i])^n$, i.e., all coordinates of $\overline{\alpha}$ have rational real and imaginary parts.
3. We are given a rational precision $\varepsilon > 0$, which defines an ε-neighbourhood $B(\overline{\alpha}, \varepsilon)$ around $\overline{\alpha}$ such that:
   - The infimum over $\overline{\beta} \in B(\overline{\alpha}, \varepsilon)$ of $|J_F(\overline{\beta})|$ is greater than 0, and is effectively computable.
   - The neighbourhood passes a fixed standard test (e.g., [27, Sec. 2], see also [28, Sec. 2.1] for a survey of alternatives) that Newton's iteration[5] converges to the unique root of F in the neighbourhood, thereby proving that $\overline{\alpha}$ approximates $\overline{\gamma}$ to precision ε.

The algorithm of [28] decides whether a representation as described above encodes 0, and relies on Schanuel's conjecture to guarantee termination.

3 Logic

In this section, we discuss how decidability results for expansions of classical structures follow when the computability enabled by Schanuel's conjecture is combined with model theory in the case of first-order logic, and with automata theory, dynamical systems, and infinite-word combinatorics in the case of monadic second-order logic.

3.1 First-Order Theory of the Reals with Exponentiation

Tarski [31] famously showed that the first-order theory T_0 of the structure $\langle \mathbb{R}; +, -, \cdot, <, 0, 1 \rangle$ is decidable by virtue of admitting quantifier elimination. In other words, there is an algorithm to translate any first-order formula that is interpreted over the real numbers, and uses the functions $+, -, \cdot$, the predicate $<$, and the constants $0, 1$, into an equivalent formula (over the same signature) without quantified variables. In particular, sentences (formulas without free variables) are translated into Boolean combinations of inequalities involving integers. The question naturally arose: could we expand the signature with, e.g., the exponentiation function and yet obtain a structure with a decidable theory?

Macintyre and Wilkie [22] proved that the first-order theory $T_{\exp}$ of the structure $\langle \mathbb{R}; +, -, \cdot, <, 0, 1, \exp \rangle$, i.e., the real numbers expanded with the exponential function is decidable subject to Schanuel's conjecture (for $\mathbb{R}$). To be specific, the

---

[5] This approximates a root as the limit of the recurrence $\overline{\alpha_{i+1}} = \overline{\alpha_i} - J_F^{-1}(\overline{\alpha_i})F(\overline{\alpha_i})$.

decidability of $T_{\exp}$ is *equivalent* to a weaker version of Schanuel's conjecture [22, Sec. 5], which is stated in computational terms (Conjecture 2). In view of the fact that this work built an accessible interface for the algorithmic applications of Schanuel's conjecture, it is worth insisting on the fact that results which are conditional because they invoke the decidability of the first-order theory of the reals with exponentiation are technically reliant upon a weaker hypothesis.

The decision procedure is underpinned by the proposition that the complete theory $T_{\exp}$ can be axiomatised by $T \cup \mathcal{E}_{\exp}$, where T is a recursive set of sentences, and $\mathcal{E}_{\exp}$ is the existential fragment of $T_{\exp}$ [22, Thm. 2.6]. In order to decide whether a given sentence φ is in $T_{\exp}$, we enumerate formulas in $T_{\exp}$ until arriving at either φ or $\neg\varphi$. This enumeration performs the following tasks in parallel: (i) enumerate sentences from the recursive set T; (ii) enumerate sentences that can be deduced from the ones enumerated thus far; (iii) iterate in parallel over existential sentences φ_e, running the subroutine described below on each φ_e until either: (a) it certifies that $\varphi_e \in \mathcal{E}_{\exp}$, (b) $\neg\varphi_e$ has been enumerated as a member of $T_{\exp}$. Schanuel's conjecture is needed to guarantee that each $\varphi_e \in \mathcal{E}_{\exp}$ will indeed be certified.

We now explain the ingredients to devise a subroutine whose termination certifies that $\varphi_e \in \mathcal{E}_{\exp}$. An arbitrary existential sentence can effectively be translated[6]mark into the form

$$\exists x_1 \cdots \exists x_n.\, p(x_1, \ldots, x_n, \exp(x_1), \ldots, \exp(x_n)) = 0\,, \tag{1}$$

where $p \in \mathbb{Z}[x_1, \ldots, x_{2n}]$. We shall use $F_p : \mathbb{R}^n \to \mathbb{R}$ to denote the function that maps $\overline{x} = (x_1, \ldots, x_n)$ to $p(x_1, \ldots, x_n, \exp(x_1), \ldots, \exp(x_n))$. The task is thus to certify that F_p has a root. Due to the following lemma [36, Lem. 6], the existence of a root of F_p implies the existence of one that can be "isolated".

Lemma 1. *Suppose F_p has a root. Then there exist $q_1, \ldots, q_n \in \mathbb{Z}[x_1, \ldots, x_{2n}]$ and $\overline{\gamma} = (\gamma_1, \ldots, \gamma_n) \in \mathbb{R}^n$ such that:*

1. *$\overline{\gamma}$ is a root of F_p, i.e., $F_p(\gamma_1, \ldots, \gamma_n) = 0$;*
2. *$\overline{\gamma}$ is a nonsingular root of the function $\langle F_{q_1}, \ldots, F_{q_n} \rangle : \mathbb{R}^n \to \mathbb{R}^n$.*

For example, $(\log 2, \log 3, \log 5)$ is a nonsingular root of $\langle \exp(x) - 2, \exp(y) - 3, \exp(z) - 5 \rangle$. Nonsingularity ensures that such roots can be approximated via (an appropriate version of) Newton's method [36, Lem. 5], and as discussed in Sect. 2.2, given a sufficiently precise neighbourhood of the root, we can use standard tests to prove that Newton's method will converge.

Our task is thus to enumerate $(q_1, \ldots, q_n, \overline{\alpha}, \varepsilon)$ until we find a tuple such that $\alpha \in \mathbb{Q}^n$ approximates to precision $\varepsilon \in \mathbb{Q}$ a point $\overline{\gamma}$ which is a nonsingular root of $\langle F_{q_1}, \ldots F_{q_n} \rangle$, and moreover satisfies $F_p(\overline{\gamma}) = 0$. The verification at each iteration is reminiscent (albeit with slight technical differences) of the zero-testing in Sect. 2.2, and requires Schanuel's conjecture for termination.

---

[6] We translate $\exists \overline{x}.\ F_1(\overline{x}) < 0 \wedge \cdots \wedge F_k(\overline{x}) < 0 \wedge F_{k+1}(\overline{x}) = 0 \wedge \cdots \wedge F_m(\overline{x}) = 0$ to
$\exists \overline{x}. \exists \overline{t}.\ (F_1(\overline{x}) + \exp(t_1))^2 + \cdots + (F_k(\overline{x}) + \exp(t_k))^2 + F_{k+1}(\overline{x})^2 + \cdots + F_m(\overline{x})^2 = 0.$

Intuitively, we use Schanuel's conjecture to argue that if some $\overline{\gamma}$ satisfies algebraic and transcendental dependencies from $\langle F_{q_1}, \ldots F_{q_n} \rangle$ as well as F_p, then it is no coincidence; rather, $p, q_1, \ldots, q_n$ are related in a manner that we can elicit, and furthermore use as a formal proof of F_p having a root [36, Cor. of SC]. In fact, [22, Sec. 5] identifies that the validity of the following weaker hypothesis is sufficient as well as *necessary* for the decidability of $T_{\exp}$:

Conjecture 2 (Weak Schanuel's conjecture). Given $p, q_1, \ldots, q_n \in \mathbb{Z}[x_1, \ldots, x_{2n}]$, we can compute a positive $\eta \in \mathbb{N}$ such that if $\overline{\gamma} \in \mathbb{R}^n$ is a nonsingular root of $\langle F_{q_1}, \ldots F_{q_n} \rangle$ and $|F_p(\overline{\gamma})| \leq 1/\eta$, then $F_p(\overline{\gamma}) = 0$.

We note that this η certainly exists: [22] argues this via Khovanskii's theorem, which implies that $\langle F_{q_1}, \ldots, F_{q_n} \rangle$ only has finitely many nonsingular roots. It is only the effectiveness of η that is conditional.

Before proceeding, we illustrate the difference between Schanuel's conjecture and the above weak variant with the example problem of determining, given nonzero $p \in \mathbb{Z}[x, y, z]$, whether $p_0 = p(\log 2, \log 3, \log 5) = 0$. Schanuel's conjecture immediately declares that this cannot be the case. The weak variant is more circumspect: it computes $\eta(e^x - 2, e^y - 3, e^z - 5, p(x, y, z))$ such that if $p_0 \neq 0$, then $|p_0| > 1/\eta$. It remains to approximate p_0 to precision $1/3\eta$.

We now return our attention to showing how to verify a purported certificate $(q_1, \ldots, q_n, \overline{\alpha}, \varepsilon)$ that F_p has a root. We first check (by a standard Newton's method-based test) that it is well-formed, i.e., $\overline{\alpha}$ indeed approximates to precision ε a nonsingular root $\overline{\gamma}$ of $\langle F_{q_1}, \ldots F_{q_n} \rangle$. We then evaluate $F_p(\overline{\alpha})$ and use continuity to check that this value implies $F_p(\overline{\gamma}) < 1/\eta$.

Conversely, it is elementary to argue by continuity of F_p that if $F_p(\overline{\gamma}) = 0$, then there is a well-formed certificate with sufficiently high precision which is guaranteed to be accepted.

3.2 Extensions and Related First-Order Decidability Results

The result above has an analogue of a more technical number-theoretic flavour: the 2013 PhD thesis of Mariaule [24] shows the decidability of the first-order theory of the ring $\mathbb{Z}_p$ of p-adic integers with the (p-adic) exponential function.

For the reals, however, Macintyre and Wilkie have actually adapted the above techniques to prove a stronger (unpublished) result (see [21, Thm. 3.1(4)]): assuming Schanuel's conjecture, the first-order theory T_{el} of the reals with exponentiation and *restricted* trigonometric functions, i.e., the theory of the structure $\langle \mathbb{R}; +, \cdot, <, \exp, \sin \upharpoonright [0, n], \cos \upharpoonright [0, n] \rangle$ is decidable. Here, a restricted function $f \upharpoonright [0, n]$ (where $n \in \mathbb{N}$ returns $f(x)$ for $x \in [0, n]$, and 0 otherwise. We believe that the proof of decidability of T_{el} requires Schanuel's conjecture for $\mathbb{C}$ only for termination of the algorithm, and proceeds by adapting the model-theoretic machinery of [35] to prove a "combined" (and effective) version of the two main results therein,[7] and hence deduce that T_{el} is effectively model-complete and

---

[7] In fact, for the applications we survey in Sect. 6, it suffices to consider the restriction of *all* three functions to bounded intervals. In this case, one only needs carefully assess the proof of the first main result of of [35] for effectiveness.

axiomatised analogously to $T_{\exp}$. Furthermore, [35, Thm. 5.1] in particular can be seen as the required analogue of Lemma 1 to enable the detection of roots of the ensuing elementary functions by enumerating guesses for an elementary-point root, and using the techniques surveyed in Sect. 2.2 to verify the guesses.

We observe that the introduction of trigonometric functions takes us to the frontiers of decidability: if we were to allow *unrestricted* trigonometric functions, we would get undecidable theories: indeed the expanded first-order theory of $\langle \mathbb{R}; +, \cdot, <, \sin \rangle$ is undecidable because one can express the predicate "x is an integer" as $\forall w.\ \sin(w) = 0 \Rightarrow \sin(xw) = 0$. With access to both addition and multiplication, we can then encode any given instance of Hilbert's 10th problem[8] as a sentence. However, decidability subject to Schanuel's conjecture is recovered for the theory $T_{+,\sin}$ of $\langle \mathbb{R}; +, <, \sin \rangle$ [6, Thm. 2.10].

The question of whether we can get decidable theories if we relinquish multiplication in favour of direct access to the integers was very recently considered in [6], which studied Presburger arithmetic with the sine function. In this setting, we only have access to addition and the sine function, the variables are interpreted over the integers, and the terms are interpreted over the reals. The first main result of [6] is that the resulting theory is undecidable, already with four alternating blocks of quantifiers. The second main result, however, is that the existential fragment is decidable subject to Schanuel's conjecture. Rather atypically, the decision procedure relies on Schanuel's conjecture for correctness (of the step [6, Thm. 4.5]) as well as termination (deciding whether a sentence obtained by [6, Thm. 4.18] is in $T_{+,\sin}$). In particular, if Schanuel's conjecture does not hold, the step of [6, Thm. 4.5] fails and the algorithm might incorrectly decide sentences of the form $\exists x_1, \ldots x_n.\ \bigwedge_{i=1}^{k} t_i(x_1, \ldots, x_n) = 0$, i.e., a conjunction of equalities.

3.3 Monadic Second-Order (MSO) Theories of the Natural Numbers with Integer-Power Predicates

In 1966, Elgot and Rabin [12] showed that monadic second-order (MSO) theories of structures of the form $\langle \mathbb{N}; <, a^{\mathbb{N}} \rangle$ are decidable, where $a \geq 2$ is a natural number and $a^{\mathbb{N}}$ denotes the predicate $\{a^n : n \in \mathbb{N}\}$. It remained open until recently whether expanding the structure with multiple such predicates results in decidable theories, when [4] provided a positive resolution: for any a, b, the MSO theory of $\langle \mathbb{N}; <, a^{\mathbb{N}}, b^{\mathbb{N}} \rangle$ is decidable. However, the decidability of the MSO theory of $\langle \mathbb{N}; <, a_1^{\mathbb{N}}, \ldots, a_d^{\mathbb{N}} \rangle$ (e.g., the MSO theory of $\langle \mathbb{N}; <, 2^{\mathbb{N}}, 3^{\mathbb{N}}, 5^{\mathbb{N}} \rangle$) was established subject to Schanuel's conjecture. In this subsection, we explain the role of Schanuel's conjecture, and present a slightly improved result.

---

[8] Hilbert's 10th problem takes as input a polynomial equation and asks whether it has an integer solution. It was famously shown undecidable in 1970 by Matiyasevich.

Theorem 4. *Let $2 \leq a_1 < \cdots < a_d$ be natural numbers. The MSO theory of $\langle \mathbb{N}; <, a_1^{\mathbb{N}}, \ldots, a_d^{\mathbb{N}} \rangle$ is decidable subject to the decidability of $T_{\exp}$ (the first-order theory of the reals with exponentiation).*[9]

We can assume, without loss of generality, that $a_1, \ldots, a_d$ are pairwise multiplicatively independent, i.e., for all $i, j \in \{1, \ldots, d\}$ and integers $p, q \geq 1$, $a_i^p \neq a_j^q$. We then define the *order word* $\alpha \in \{a_1, \ldots, a_d\}^\omega$, which records the order in which the powers of $a_1, \ldots, a_d$ (excluding 1) appear. For example, the order word of the powers $2, 3, 4, 5, 8, 9, 16, 25, 27, 32, 64, 81, 125, 128, \ldots$ of $2, 3, 5$ is $23252325322352 \cdots$. In [4], it was shown that deciding whether a sentence is in the MSO theory of $\langle \mathbb{N}; <, a_1^{\mathbb{N}}, \ldots, a_d^{\mathbb{N}} \rangle$ is equivalent to deciding whether a given automaton $\mathcal{A}$ accepts the order word α.

The key insight is that the order word enjoys a *toric* structure. Consider the example of $2, 3, 5$, and the factorisation of α based on occurrences of powers of $a_1 = 2$. We get $2 \cdot 32 \cdot 52 \cdot 32 \cdot 532 \cdot 2 \cdot 352 \cdots$. The k-th factor conveys whether there were powers of 3 and 5 between 2^{k-1} and 2^k, and in what order they occurred. Equivalently, it conveys whether there were integer multiples of $\log 3, \log 5$ (there can be at most one of each kind) between $(k-1)\log 2$ and $k \cdot \log 2$, and in what order. We capture the above through the torus $\mathbb{T}$ in Fig. 1. Due to [25, Thm. 3] (stated below), it would suffice to elicit *effective almost-periodicity* from the toric order word in order to prove our decidability result.

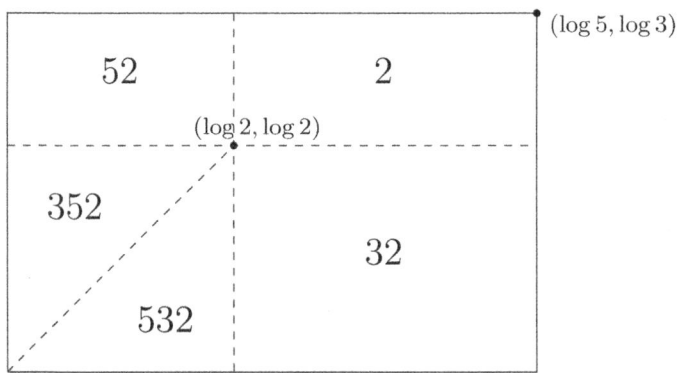

Fig. 1. The above torus $\mathbb{T}$ helps construct the order word through the orbit of a point that starts at $(\log 2, \log 2)$ and travels in discrete steps of $(\log 2, \log 2)$. The label of the (open) region it lands in determines the next letters of the order word; the starting point prints 2 and is the only point that lands on a boundary.

---

[9] Technically, we only work with the structure $\langle \mathbb{R}; <, +, \cdot, \log a_1, \ldots, \log a_d \rangle$ (the usual first-order structure of the reals, expanded with constants), which, due to Tarski [32], admits quantifier elimination. Consequently, decidability hinges on the ability to decide polynomial (in)equalities in $\log a_1, \ldots, \log a_d$.

Theorem 5. *Define a word $\alpha \in \Sigma^\omega$ to be effectively almost-periodic if, given any $u = u(0) \cdots u(\ell - 1) \in \Sigma^+$, we can compute a return time $R \in \mathbb{N}$ such that either:*

- *For all $n \geq R$, $\alpha(n) \cdots \alpha(n + \ell - 1) \neq u$, or*
- *For all n, there exists an $m \in [n, n + R)$ such that $\alpha(m) \cdots \alpha(m + \ell - 1) = u$.*

If α is effectively almost-periodic, then given any Büchi automaton $\mathcal{A}$, we can decide whether $\mathcal{A}$ accepts α.

It remains to show how, given a word $u \in \{a_1, \ldots, a_d\}^+$, the toric structure helps us to compute a return time. We work with our running example of powers of $2, 3, 5$ for ease of exposition; it is straightforward to generalise the arguments.

Let us take the example of $u = 5232532$. As the first step, we factor[10] it based on occurrences of $a_1 = 2$, and obtain $52 \cdot 32 \cdot 532 = b_0 b_1 b_2$. In order for $u = b_0 \cdots b_\ell$ to be observed, there must be a sequence of points $(x_0, y_0), \ldots, (x_\ell, y_\ell)$ in the torus $\mathbb{T}$ such that:

- For each i, (x_{i+1}, y_{i+1}) is obtained by taking a single step of $(\log 2, \log 2)$ from (x_i, y_i), i.e., $x_{i+1} = x_i + \log 2$ or $x_i + \log 2 - \log 5$ as appropriate to ensure that $0 \leq x_i < \log 5$ (and likewise for y). Thus, the sequence is uniquely determined by (x_0, y_0).
- The point (x_0, y_0) is in a region that has b_0 as its suffix. In our example, (x_0, y_0) must be in the region of 52 or 352. Likewise (x_ℓ, y_ℓ) must be in a region that has b_ℓ as its prefix.
- All other points (x_i, y_i) must be in the region that prints b_i.

Points (x_0, y_0) that satisfy the above constraints, i.e., define a sequence of points that land in the appropriate regions, are easily (e.g., by linear programming) seen to comprise a bounded union $\mathcal{I}_u$ of open sets of the torus $\mathbb{T}$. By construction, there is a direct correspondence between occurrences of u and visits to $\mathcal{I}_u$: in particular, u has gaps of at most R between consecutive occurrences in the order word if and only if the orbit visits $\mathcal{I}_u$ at least once in every R steps.

The orbit is dense in some (not necessarily proper) subtorus of $\mathbb{T}$. More specifically, by Kronecker's theorem in Diophantine approximation [15], the orbit is dense in the subtorus[11] $\mathbb{T}_{\text{orbit}}$ given by

$$\left\{ (x, y) \in \mathbb{T} : \text{for all } b, c \in \mathbb{Z}, \text{ if } \frac{b \cdot \log 2}{\log 5} + \frac{c \cdot \log 2}{\log 3} \in \mathbb{Z}, \text{ then } \frac{bx}{\log 5} + \frac{cy}{\log 3} \in \mathbb{Z} \right\}.$$

By invoking the compactness of the (sub)torus $\mathbb{T}_{\text{orbit}}$, we can prove that if $\mathcal{J}_u = \mathcal{I}_u \cap \mathbb{T}_{\text{orbit}}$ is nonempty, then there exists an R such that the orbit of any point in $\mathbb{T}_{\text{orbit}}$ visits $\mathcal{J}_u$, and hence $\mathcal{I}_u$, within R steps. On the other hand, if $\mathcal{J}_u$ is empty, then u can never occur in the order word.

---

[10] The case where there is just a single factor (because, e.g., there is no a_1) is trivial.

[11] We continue with the example of powers of $2, 3, 5$ for ease of exposition, the general statement for $a_1, \ldots, a_d$ is completely analogous.

We have thus far proven that the order word α is not merely almost-periodic, but in fact enjoys the stronger property of *uniform recurrence*: every u occurs either with bounded gaps $R(u)$, or never at all. It remains to effectively compute a return time R.[12] The problem with the strategy of brute enumeration to compute R is that we cannot unconditionally obtain $\mathbb{T}_{\text{orbit}}$: it can so happen that we enumerate R forever because $\mathcal{I}_u$ is nonempty but $\mathcal{J}_u$ is empty as a consequence of $\mathbb{T}_{\text{orbit}} \subset \mathbb{T}$.

The original paper [4] invoked Schanuel's conjecture to argue that $\mathbb{T}_{\text{orbit}} = \mathbb{T}$, since indeed in our example, the reciprocals of $\log 2, \log 3, \log 5$ would have to be linearly independent over the rationals. This guaranteed that the enumeration of potential R would terminate. In this case, Baker's theorem (Theorem 3) allows us to determine whether $\mathcal{J}_u = \mathcal{I}_u$ is nonempty.

However, we show that relying upon the decidability of T_{exp} suffices. The key observation is that it is not always necessary to know $\mathbb{T}_{\text{orbit}}$ exactly in order to compute R. We justify this as follows. Consider a "restricted" torus $\mathbb{T}_{\text{res}}$ such that $\mathbb{T} \supseteq \mathbb{T}_{\text{res}} \supseteq \mathbb{T}_{\text{orbit}}$. For an arbitrary word v, we define $\mathcal{K}_v$ as $\mathcal{I}_v \cap \mathbb{T}_{\text{res}}$, and observe that $\mathcal{I}_v \supseteq \mathcal{K}_v \supseteq \mathcal{J}_v$. If $\mathcal{K}_u$ is empty, it implies the emptiness of $\mathcal{J}_u$, and thus the non-occurrence of u. Conversely, if for some R the orbit of every point in $\mathbb{T}_{\text{res}}$ visits $\mathcal{K}_u$ within R steps, it implies that the actual orbit visits $\mathcal{I}_u$ every R steps, thereby certifying the occurrence of u with gaps bounded by R.

Each candidate $\mathbb{T}_{\text{res}}$ thus outlines a correct semi-algorithm to compute a return time R given u. In case $\mathbb{T}_{\text{res}} = \mathbb{T}_{\text{orbit}}$, then termination is guaranteed (provided the attendant linear programming is effective) and we have an algorithm. For effectiveness of R, it suffices to show that there is an enumerable set Tori of candidate $\mathbb{T}_{\text{res}}$ such that we are guaranteed $\mathbb{T}_{\text{orbit}} \in$ Tori.

Recall that $\mathbb{T}_{\text{orbit}}$ was defined in terms of the integer linear dependencies of $\frac{1}{\log 2}, \frac{1}{\log 3}, \frac{1}{\log 5}$. More generally, the integer linear dependencies of $\frac{1}{\log a_1}, \ldots, \frac{1}{\log a_d}$ will have a basis[13] $\overline{b_1}, \ldots, \overline{b_p}$, where for all i, $\overline{b_i} \in \mathbb{Z}^d$, $\sum_j \frac{b_{ij}}{\log a_j} = 0$, and $p < d-1$. Our candidate $\mathbb{T}_{\text{res}}$ will be enumerated by linearly independent sets $\{\overline{c_1}, \ldots, \overline{c_q}\} \subset \mathbb{Z}^d$, and have the form

$$\left\{(x_2, \ldots, x_d) \in \mathbb{T} : \text{for all } i, \frac{c_{i2}x_2}{\log a_2} + \cdots + \frac{c_{id}x_d}{\log a_d} \in \mathbb{Z}\right\},$$

where for each i, we have $\frac{c_{i1}}{\log a_1} + \cdots + \frac{c_{id}}{\log a_d} = 0$.

The conditions necessitate that $\overline{c_1}, \ldots, \overline{c_q}$ lie in the free module generated by $\overline{b_1}, \ldots, \overline{b_p}$, by linear independence it suffices to consider $q < d-1$, and by

---

[12] One implementation can use the open-cover characterisation above. Alternately, we can also use the property that all words v of length $R + |u|$ for which $\mathcal{J}_v$ is nonempty must contain u as a contiguous subword.

[13] These dependencies constitute a submodule L of the free module $\mathbb{Z}^d$; hence L is also free and generated by a basis B with cardinality at most d (see the text [20, App. 2, page 880] for a proof). If $|B| = d$, then by elementary linear algebra, the reciprocals of logs are all 0, a contradiction. If $|B| = d-1$, linear algebra would imply pairwise multiplicative dependencies between $a_1, \ldots, a_d$, again a contradiction.

construction $\mathbb{T}_{\mathsf{res}} \supseteq \mathbb{T}_{\mathsf{orbit}}$. By definition, $\mathbb{T}_{\mathsf{orbit}}$ will necessarily be enumerated when we guess $\overline{b}_1, \ldots, \overline{b}_p$ as the basis.

We finally address the outstanding effectiveness concerns. In order to make the above enumeration effective, we need to resolve the last condition consisting of equalities, and hence invoke the decidability of $T_{\exp}$ (the first-order theory of the reals with exponentiation). Finally, we argue that having assumed the decidability of $T_{\exp}$, one can effectively implement the linear programming required to check that $\mathcal{K}_u \subseteq \mathbb{T}_{\mathsf{res}}$ is nonempty. This completes the proof of Theorem 4.

4   Applications to Discrete Recurrence Sequences

By the turn of the century, Schanuel's conjecture had been used to prove the termination of fundamental algorithms in computer algebra and first-order logic. While choosing an instruction set to devise algorithms, the computability of elementary numbers is an immensely powerful tool; at a higher level of abstraction, the decidability of the first-order theory $T_{\exp}$ of the real numbers with the exponential function serves as a conveniently accessible interface.

In this section, we discuss how these subroutines help in solving problems about discrete recurrence sequences. We start with the Skolem problem for linear recurrence sequences (LRS). An integer LRS of order k satisfies an integer recurrence relation $u_{n+k} = a_0 u_n + \cdots + a_{k-1} u_{n+k-1}$, and is given by the coefficients $a_0, \ldots, a_{k-1} \in \mathbb{Z}$ as well as the initial terms $u_0, \ldots, u_{n-1} \in \mathbb{Z}$. The Skolem problem asks to decide whether a given LRS has a term that is equal to 0. A practical algorithm to solve the Skolem problem for *simple* LRS was given in [5]: this algorithm is unconditionally correct, but relies on two number-theoretic conjectures for termination, one of which is a p-adic version of Schanuel's conjecture. The role of Schanuel's conjecture is to test whether expressions [5, Prop. 8] of the form $f(\log \lambda_1, \ldots, \log \lambda_k)$ are equal to 0, where $\lambda_1, \ldots, \lambda_d$ are *characteristic roots* of the LRS, and f is a polynomial whose coefficients are *p-adic integers*.

A more sophisticated application is that to hypergeometric sequences [18]. These sequences are given by an initial term $u_0 \in \mathbb{Q}$, and satisfy the recurrence $q(n)u_{n+1} = p(n)u_n$, where $p, q \in \mathbb{Q}[x]$ are polynomials without roots in $\mathbb{N}$. The membership problem asks whether some term equals a given $t \in \mathbb{Q}$. Difficulties arise when p, q are *harmonious*, implying that the sequence converges to a finite nonzero limit ℓ. This limit is expressed in terms of the gamma function, and is not known to be elementary, unless we restrict the spectra of p, q [18, Sec. 4, Property S]. Because the convergence to ℓ is effectively eventually monotone, the critical task is to decide whether $\ell = t$: the result would determine an upper bound on the iterate by which t can occur in the sequence. We thus use Schanuel's conjecture for testing the resulting elementary expression for 0 [18, Sec. 4.2].

For a setting where the above kind of "invariant synthesis" is broader in scope, we return to the realm of linear algebra, and consider linear dynamical systems (LDS). An LDS is specified by a starting point $s \in \mathbb{Q}^d$ and an (invertible) update matrix $A \in \mathbb{Q}^{d \times d}$, and defines a trajectory $(s, As, A^2 s, \ldots)$. The halting problem for LDS considered in [1] additionally takes a set $F \subseteq \mathbb{R}^d$ represented by a first-order formula over the structure $\mathfrak{R}_0 = \langle \mathbb{R}; +, \cdot, < \rangle$ or $\mathfrak{R}_{\exp}$ ($\mathfrak{R}_0$ expanded with the exponential function, as discussed in Sect. 3.1), and asks to decide whether the trajectory intersects F. A "yes" answer is certified by n such that $A^n s \in F$; a "no" answer is certified by an *invariant*, i.e., a set $\mathcal{I} \subseteq \mathbb{R}^d$ such that for all $x \in \mathcal{I}$, $Ax \in \mathcal{I}$, $s \in \mathcal{I}$, and $\mathcal{I} \cap F$ is empty. The paper [1] considers the task of synthesising invariants that consist of finitely many connected components by virtue of being defined in $\mathfrak{R}_{\exp}$, a structure that is *o-minimal*.

We now outline the techniques. It is first shown [1, Thm. 5] that there is a family $\mathcal{J}$ of $\mathfrak{R}_{\exp}$-definable sets parametrised by t_0, such that the set $\mathcal{J}(t_0)$ contains the trajectory for every $t_0 \geq 1$. Furthermore, [1, Lem. 11] asserts that an invariant of the desired form must belong to this family. Critically, the set $\mathcal{T}$ of t_0 for which the set $\mathcal{J}(t_0)$ is indeed an invariant is $\mathfrak{R}_{\exp}$-definable. Finally, using the decidability of $\mathcal{T}_{\exp}$ (whose termination is conditional on Schanuel's conjecture), and the fact that $\mathcal{T}$ consists of finitely many connected components, we can test whether $\mathcal{T}$ is non-empty, and if yes, effectively return $\mathcal{I} = \mathcal{J}(t_0)$ for some $t_0 \in \mathcal{T}$ [1, Thm. 12]. We remark that if the set F is defined in $\mathfrak{R}_0$, then the $\mathfrak{R}_{\exp}$-definable invariant synthesis is unconditional [1, Thm. 13].

5   Applications to Quantitative Verification

A fundamental object in quantitative verification is the *weighted automaton*, which can be intuited as "implementing (linear) recurrence with branching." In this section, we survey two instances of algorithms for weighted automata invoking Schanuel's conjecture through queries to $\mathcal{T}_{\exp}$.

Formally, a weighted automaton computes over a (commutative) semiring $\mathcal{R}$, i.e., a set that is equipped with distinguished elements $0 \neq 1$, a commutative and associative addition operation $+$ that has 0 as its identity, and a commutative and associative multiplication operation $\cdot$ that has 1 as its identity, distributes over $+$, and has 0 as its absorbing element (for all x, $0 \cdot x = 0$). For example, the usual finite-word automata compute over the Boolean semiring $\{0, 1\}$ where $+$ is disjunction and $\cdot$ is conjunction. In this section, we shall survey results for weighted automata that compute over the semiring of nonnegative rational numbers with the usual addition and multiplication.

In general, a weighted automaton $\mathcal{A}$ over the semiring $\mathcal{R}$ is given by the tuple $(Q, \Sigma, \Delta, I, F)$, where Q is the finite set of states, Σ is the alphabet, $\Delta \subseteq Q \times \Sigma \times \mathcal{R} \times Q$ is the finite set of transitions, $I : Q \to \mathcal{R}$ is the initial weight function, and $F : Q \to \mathcal{R}$ is the final weight function.[14] Given an input $u = u(0) \cdots u(\ell - 1) \in \Sigma^*$, a path of $\mathcal{A}$ on u is given as $w_0, q_0, u(0), w_1, q_1, \ldots, q_{\ell-1}, u(\ell-1), w_\ell, q_\ell, w_{\ell+1}$, where $w_0 = I(q_0)$, for all $j \in \{0, \ldots, \ell - 1\}$, we have $(q_j, u(j), w_{j+1}, q_{j+1}) \in \Delta$, and $w_{\ell+1} = F(q_\ell)$. A path on the empty word is simply w_0, q_0, w_1, where $w_0 = I(q_0), w_1 = F(q_0)$. The weight of the path is the product $w_0 \cdot w_1 \cdots w_{\ell+1}$. The weight of the word u is the sum of weights of all paths of $\mathcal{A}$ on u (hence the weight is 0 if there is no path). The automaton thus defines a function $[\![\mathcal{A}]\!] : \Sigma^* \to \mathcal{R}$.

---

[14] In this paper, we only survey works that consider $F : Q \to \{0, 1\}$.

For example, a conventional automaton $(Q, \Sigma, \Delta, I, F)$ (where $\Delta \subseteq Q \times \Sigma \times Q$, $I, F \subseteq Q$, as is familiar) can be interpreted as a weighted automaton over the Boolean semiring as follows: we replace each $(q, a, q') \in \Delta$ with $(q, a, 1, q')$, construct the initial weight function to assign 1 to elements of I and 0 to others, and similarly the final weight function assigns 1 to elements of F and 0 to others. The weight of a path is 1 if it starts in an initial state and ends in an accepting state, and 0 otherwise; a word is assigned weight 1 if it is accepted by the automaton, and weight 0 otherwise.

As mentioned before, we shall consider weighted automata over nonnegative rational numbers. If, in addition, we have that all weights are at most 1, the sum of all initial weights is 1, all final weights are 0 or 1, and for each state q, the sum of weights of all outgoing transitions is 1, the automaton is said to be *probabilistic*. Probabilistic automata are closely related to other stochastic models such as Markov chains, and Markov decision processes.

Weighted, and indeed even probabilistic automata constitute a powerfully expressive model of computation: it is folklore [26] (see also [14] for a modern proof) that the emptiness problem for probabilistic automata, which gives $\mathcal{A}$ and $t \in [0, 1]$ and asks whether there is some word $u \in \Sigma^*$ such that $[\![\mathcal{A}]\!](u) > t$, is undecidable. Even over a unary alphabet, which makes the automaton a Markov chain, the problem is as hard as the yet unresolved positivity problem for linear recurrence sequences [33]. In order to obtain decidability results for weighted automata, works have thus restricted either the structure of the input automaton, or the language $\mathcal{L}(\mathcal{A})$ of words that are assigned nonzero weight.

The former is done by restricting *ambiguity*: an automaton $\mathcal{A}$ is k-ambiguous if there are at most k nonzero paths for every $u \in \Sigma^*$. We say that $\mathcal{A}$ is finitely ambiguous if it is k-ambiguous for some k, and unambiguous if $k = 1$. We can further parametrise ambiguity, and say that an automaton is polynomially (respectively, linearly) ambiguous if for all u, there are at most $p(|u|)$ nonzero paths, where p is a polynomial (respectively, a polynomial of degree 1). This distinction is technically relevant, because if an automaton fails to be finitely ambiguous, then it is at least linearly ambiguous [34, Sec. 3]. Furthermore, it has been shown [11, Proof of Thm. 1] that the emptiness problem is undecidable even for linearly ambiguous probabilistic automata.

The only hope to recover decidability with this restriction, therefore, is to assume finite ambiguity. Indeed, [11, Thm. 2] shows that assuming Schanuel's conjecture, the containment problem, which gives probabilistic automata $\mathcal{A}, \mathcal{B}$ over the alphabet Σ, and asks whether for all $u \in \Sigma^*$, $[\![\mathcal{A}]\!](u) \leq [\![\mathcal{B}]\!](u)$, is decidable for the class of finitely ambiguous probabilistic automata, provided at least one of the input automata is unambiguous. In fact, the decidability is unconditional if $\mathcal{A}$ is finitely ambiguous and $\mathcal{B}$ is unambiguous [11, Prop. 5].

Otherwise, the task is equivalent to the integer program with exponentiation problem that asks whether the system $M\overline{x} < \overline{c}$, $\sum_{i=1}^{\ell} r_i s_{i,1}^{x_1} \cdots s_{i,n}^{x_n} < 1$ has a solution $\overline{x} \in \mathbb{Z}^n$, where M is an integer matrix, $\overline{c}$ is an integer vector, and $r_i, s_{i,j}$ are all positive rationals. Obviously, if there is a solution, a semi-algorithm that simply enumerates integer vectors will find it. It remains to describe a semi-algorithm that would certify the absence of integral feasible points.

The key technical observation [11, Lem. 10] is that if there is no integer solution, then the feasible region X of *real* solutions must be contained in a "tube", i.e., a set $\{\overline{y} \in \mathbb{R}^n : \overline{d}^\top \overline{y} \in [a, b]\}$, where $\overline{d} \in \mathbb{Z}^n, a, b \in \mathbb{Z}$. If we find such an encompassing tube, we know that any integer feasible point must satisfy $\overline{d}^\top \overline{x} = i$ for some integer $i \in [a, b]$, and hence can reduce the search for a certificate of absence to finitely many lower-dimensional systems [11, Lem. 11].

To make the proposed semi-algorithm effective, we need to check whether a tube purported by $\overline{d}, a, b$ indeed contains the feasible set X. This is the step that uses Schanuel's conjecture for termination, because it is implemented as a query to $T_{\exp}$. However, given the lack of explicit transcendence in the integer programming with exponentiation problem, we cannot yet rule out the circumvention of Schanuel's conjecture.

The general undecidability of the containment problem motivated the consideration of an approximate variant, namely, the big-O problem for weighted automata [8], which gives as input a weighted automaton $\mathcal{A}$, two states q, q', and asks whether there exists $c > 0$ such that for all $u \in \Sigma^*$, we have $[\![\mathcal{A}_q]\!](u) \leq c \cdot [\![\mathcal{A}_{q'}]\!](u)$, where $\mathcal{A}_q$ is the automaton obtained by modifying the initial weight function of $\mathcal{A}$ to return 1 for q and 0 for all other states. Unfortunately, as [8] establishes, the big-O problem is also undecidable in general, and relatively tractable in very special cases (in P for unambiguous automata, in coNP if the alphabet is unary).

The interesting restriction is that of requiring the languages $\mathcal{L}(\mathcal{A}_q)$ and $\mathcal{L}(\mathcal{A}_{q'})$ be *bounded*: a language $L \subseteq \Sigma^*$ is bounded if it is contained in $w_1^* \cdots w_k^*$ for some words $w_1, \ldots, w_k \in \Sigma^*$. The big-O problem for these instances is decidable subject to Schanuel's conjecture [8, Thm. 28].

The technical algorithm requires a subroutine to test whether a first-order sentence is in $T_{\exp}$ [8, Lem. 37]. The expressions therein involve the logarithms of variables, as well as the logarithms of real positive algebraic constants as coefficients of linear terms, and hence, in this case, Schanuel's conjecture appears to play a necessary role.

6   Applications to Continuous-Time Systems

In this final technical section, we survey how queries about elementary functions naturally arise when reasoning about continuous-time (linear) dynamical systems. These queries often involve trigonometric functions: recall from Sect. 3.2 that in order for the first-order theory of the attendant structure to be decidable, these functions must be restricted to bounded intervals. Consequently, the literature predominantly considers decision problems whose inputs specify a bounded time interval $[0, N]$ of interest. As discussed in Sect. 3.2, Schanuel's conjecture then assures[15] the decidability of the theory $T_{el,N}$ of the structure $\langle \mathbb{R}; <, +, \cdot, \exp \upharpoonright [0, N], \sin \upharpoonright [0, N], \cos \upharpoonright [0, N] \rangle$.

A fundamental continuous-time decision problem is the bounded continuous Skolem problem, which asks whether the solution $f : \mathbb{R} \rightarrow \mathbb{R}$ of the ordinary differential equation $f^{(n)} + a_{n-1} f^{(n-1)} + \cdots + a_0 f \equiv 0$, with the coefficients $a_0, \ldots, a_{n-1}$, and initial conditions $f(0), \ldots, f^{(n-1)}(0)$ being real algebraic numbers, has a zero in the interval $[c, d]$. We have that the function f is the first component of the vector $\exp(tA) \cdot \mu_0$, where A is the *companion matrix*[16] derived from $a_0, \ldots, a_{n-1}$, and μ_0 is the vector of initial conditions. We can therefore express $f(t) = \sum_{j=1}^{m} \exp(\rho_j t)(p_j(t) \sin(\omega_j t) + q_j(t) \cos(\omega_j t))$, where ρ_j, ω_j, and the coefficients of the polynomials p_j, q_j are real algebraic numbers. As in [9, Introduction], we can choose $N = (d + 1) \cdot \max(\rho_1, \omega_1, \ldots, \rho_m, \omega_m)$, and query $T_{el,N}$ to arrive at a decision.

However, since the decidability proof of $T_{el,N}$ is not published, [9] gives a decision procedure for the bounded continuous Skolem problem by invoking Schanuel's conjecture directly (via Prop. 5 therein) in technical algebraic arguments. The technical ingredient that is the difference between the raw Schanuel's conjecture and the off-the-shelf decidability of $T_{el,N}$ is the model-completeness of the theory. Bereft of this refinement, the direct algorithm of [9] relies on Schanuel's conjecture for correctness in Case (i) of the proof of its Thm. 7 (the main result, which states that the bounded-time continuous Skolem problem is decidable subject to Schanuel's conjecture, and is proven via a technical case distinction): the algorithm trusts Schanuel's conjecture's assertion that a function f has no root without a certificate; on the other hand, in case the claim is true, the (effectively) model-complete $T_{el,N}$ would provide a proof in the form of the claim being a formal consequence of the theory. As usual, the termination is also conditional as Case (ii) of the proof of Thm. 7 indicates. See also [10] for an alternative presentation of this argument.

The choice of whether to trust Schanuel's conjecture for correctness and forego the cumbersome queries to logical theories that give unconditional correctness has practical consequences. The work [17] is an applied example of prioritising efficiency: they give an algorithm to isolate roots of an exponential-polynomial, and rely on Schanuel's conjecture to argue its completeness, i.e., that the algorithm does not discount any roots, and termination [17, Discussion of Example 16].

Such root isolation algorithms find immediate applications to verify the evolution of the probability distributions in continuous time Markov chains. A continuous time Markov chain with k states is specified by a $k \times k$ matrix M, where the entry $M(i, j) = m$ indicates that a transition from state j to state i is possible, and its timing is exponentially distributed with rate m. Succinctly, the dynamics of the probability distribution μ are given by $\mu^{(1)}(t) = M \mu(t)$,

---

[15] We recall that this claim is based on unpublished results; however, we outlined why we believe that Schanuel's conjecture is needed only for the termination of the deciding algorithm.

[16] The matrix exponential $\exp(M)$ is defined as the limit of $I + M + M^2/2! + M^3/3! + \cdots$.

and the distribution at time t is given by $\exp(Mt) \cdot \mu_0$, where μ_0 is the initial distribution at $t = 0$.

In order to verify the evolution of the distribution, [16] considers CLL, a continuous (bounded-time) counterpart of LTL, whose atomic propositions are of the form "the probability of being in state s is at least 0.8"; an example specification $true\ \mathcal{U}^{[3,7]}\langle s, \geq 0.8\rangle$ asserts "the probability of being in state s is at least 0.8 at some moment $t \in [3,7]$." Upon observing the expansion of $\exp(Mt) \cdot \mu_0$ and the semantics of CLL, we remark that the satisfaction of a formula can, in theory, be verified with unconditional correctness by querying $T_{\mathrm{el},N}$ for sufficiently large N. However, as [16] demonstrates, trusting Schanuel's conjecture for correctness makes the problem more tractable in practice.

As in the discrete case, having techniques to reason about continuous-time Markov chains equips us to reason about certain *policies* in continuous-time Markov *decision processes* (MDPs), which augment Markov chains by determining transition dynamics at any given time by an agent's resolution of a finite choice of *actions*. A policy is a function that prescribes these choices; it is called stationary if the prescription does not change with time. We observe that following a stationary policy induces a Markov chain.

MDPs form the building blocks of reinforcement learning, and [23] considers the problem of deciding whether there is a policy that achieves the objective of visiting a designated good state within given time bound is achieved with probability above a given threshold. The fact that the optimal policy for this task is piecewise stationary allows the problem to be formulated as one about Markov chains in spirit, whence the decidability of $T_{\mathrm{el},N}$ can be called upon for a solution.

Finally, we note [37] uses a slight augmentation of MDPs to model the runs of probabilistic programs while accounting for scheduling delays, and considers the problem of quantitatively verifying termination within a time bound. The number-theoretic engine [37, Lem. 6.2] for their results is the root-isolation subroutine, which is efficient in practice provided that the implementation trusts Schanuel's conjecture for correctness, but in theory can also be implemented using queries to $T_{\mathrm{el},N}$ too.

Acknowledgements. Toghrul Karimov, Mihir Vahanwala, and Joël Ouaknine were supported by DFG grant 389792660 as part of TRR 248 (see https://perspicuous-computing.science). Joël Ouaknine is also affiliated with Keble College, Oxford as emmy network Fellow (https://emmy.network/). James Worrell was supported by EPSRC Fellowship EP/X033813/1.

References

1. Almagor, S., Chistikov, D., Ouaknine, J., Worrell, J.: O-minimal invariants for linear loops. In: 45th International Colloquium on Automata, Languages, and Programming (ICALP 2018). Schloss-Dagstuhl-Leibniz Zentrum für Informatik (2018)
2. Ax, J.: On Schanuel's conjectures. Ann. Math. **93**(2), 252–268 (1971)

3. Baker, A.: Transcendental Number Theory. Cambridge Mathematical Library. Cambridge University Press (1975)
4. Berthé, V., Karimov, T., Nieuwveld, J., Ouaknine, J., Vahanwala, M., Worrell, J.: On the decidability of monadic second-order logic with arithmetic predicates. In: Proceedings of the 39th Annual ACM/IEEE Symposium on Logic in Computer Science, pp. 1–14 (2024)
5. Bilu, Y., Luca, F., Nieuwveld, J., Ouaknine, J., Purser, D., Worrell, J.: Skolem meets schanuel. In: 47th International Symposium on Mathematical Foundations of Computer Science (2022)
6. Blanchard, E., Hieronymi, P.: Decidability bounds for Presburger arithmetic extended by sine. Ann. Pure Appl. Logic **175**, 103487 (2024)
7. Caviness, B.F.: On canonical forms and simplification. J. ACM **17**(2), 385–396 (1970)
8. Chistikov, D., Kiefer, S., Murawski, A.S., Purser, D.: The big-O problem for labelled Markov chains and weighted automata. Leibniz Int. Proc. Inform. (LIPIcs) **171**, 41 (2020)
9. Chonev, V., Ouaknine, J., Worrell, J.: On the Skolem problem for continuous linear dynamical systems. In: 43rd International Colloquium on Automata, Languages, and Programming (ICALP 2016). Schloss-Dagstuhl-Leibniz Zentrum für Informatik (2016)
10. Chonev, V., Ouaknine, J., Worrell, J.: On the zeros of exponential polynomials. J. ACM **70**(4), 26:1–26:26 (2023)
11. Daviaud, L., Jurdziński, M., Lazić, R., Mazowiecki, F., Pérez, G.A., Worrell, J.: When is containment decidable for probabilistic automata? In: 45th International Colloquium on Automata, Languages, and Programming, ICALP 2018, p. 121. Schloss Dagstuhl-Leibniz-Zentrum fur Informatik GmbH, Dagstuhl Publishing (2018)
12. Elgot, C.C., Rabin, M.O.: Decidability and undecidability of extensions of second (first) order theory of (generalized) successor. J. Symb. Logic **31**(2), 169–181 (1966)
13. Ferguson, H.R.P., Bailey, D.H., Arno, S.: Analysis of PSLQ, an integer relation finding algorithm. Math. Comput. **68**(225), 351–369 (1999)
14. Gimbert, H., Oualhadj, Y.: Probabilistic automata on finite words: decidable and undecidable problems. In: Abramsky, S., Gavoille, C., Kirchner, C., Meyer auf der Heide, F., Spirakis, P.G. (eds.) ICALP 2010. LNCS, vol. 6199, pp. 527–538. Springer, Heidelberg (2010). https://doi.org/10.1007/978-3-642-14162-1_44
15. Gonek, S.M., Montgomery, H.L.: Kronecker's approximation theorem. Indagationes Mathematicae **27**(2), 506–523 (2016). In Memoriam J.G. Van der Corput (1890–1975) Part 2
16. Zingg, S., Krstić, S., Raszyk, M., Schneider, J., Traytel, D.: Verified first-order monitoring with recursive rules. In: TACAS 2022. LNCS, vol. 13244, pp. 236–253. Springer, Cham (2022). https://doi.org/10.1007/978-3-030-99527-0_13
17. Huang, C.-C., Li, J.-C., Ming, X., Li, Z.-B.: Positive root isolation for poly-powers by exclusion and differentiation. J. Symb. Comput. **85**, 148–169 (2018)
18. Kenison, G.: The threshold problem for hypergeometric sequences with quadratic parameters. In: 51st International Colloquium on Automata, Languages, and Programming (ICALP 2024). Schloss Dagstuhl–Leibniz-Zentrum für Informatik (2024)
19. Lang, S.: Introduction to Transcendental Numbers. Addison-Wesley series in mathematics. Addison-Wesley Publishing Company (1966)
20. Lang, S.: Algebra. Graduate Texts in Mathematics, 3rd edn. Springer, New York (2002)

21. Macintyre, A.: Turing meets Schanuel. Ann. Pure Appl. Logic **167**(10), 901–938 (2016). Logic Colloquium 2012
22. Macintyre, A., Wilkie, A., Odifreddi, P.: On the decidability of the real exponential field. Kreisel's Math. **115**, 451 (1996)
23. Majumdar, R., Salamati, M., Soudjani, S.: On decidability of time-bounded reachability in CTMDPs. In: 47th International Colloquium on Automata, Languages, and Programming (ICALP 2020). Schloss-Dagstuhl-Leibniz Zentrum für Informatik (2020)
24. Mariaule, N.: On the decidability of the p-adic exponential ring. The University of Manchester (United Kingdom) (2013)
25. Muchnik, A., Semenov, A., Ushakov, M.: Almost periodic sequences. Theoret. Comput. Sci. **304**(1), 1–33 (2003)
26. Paz, A.: Introduction to probabilistic automata (Computer science and applied mathematics). Academic Press, Inc., USA (1971)
27. Richardson, D.: A simplified method of recognizing zero among elementary constants. In: Proceedings of the 1995 International Symposium on Symbolic and Algebraic Computation, pp. 104–109 (1995)
28. Richardson, D.: How to Recognize Zero. J. Symb. Comput. **24**(6), 627–645 (1997)
29. Richardson, D.: Zero tests for constants in simple scientific computation. Math. Comput. Sci. **1**, 21–37 (2007)
30. Rosenlicht, M.: On Liouville's theory of elementary functions. Pac. J. Math. **65**, 485–492 (1976)
31. Tarski, A.: A decision method for elementary algebra and geometry. In: Caviness, B.F., Johnson, J.R. (eds.) Quantifier Elimination and Cylindrical Algebraic Decomposition, Vienna, pp. 24–84. Springer Vienna (1998)
32. Alfred Tarski. A Decision Method for Elementary Algebra and Geometry. In Bob F. Caviness and Jeremy R. Johnson, editors, *Quantifier Elimination and Cylindrical Algebraic Decomposition*, pages 24–84, Vienna, 1998. Springer Vienna
33. Vahanwala, M.: Skolem and positivity completeness of ergodic Markov chains. Inf. Process. Lett. **186**, 106481 (2024)
34. Weber, A., Seidl, H.: On the degree of ambiguity of finite automata. Theoret. Comput. Sci. **88**(2), 325–349 (1991)
35. Wilkie, A.J.: Model completeness results for expansions of the ordered field of real numbers by restricted pfaffian functions and the exponential function. J. Am. Math. Soc. **9**(4), 1051–1094 (1996)
36. Wilkie, A.J.: Schanuel's Conjecture and the Decidability of the Real Exponential Field, pp. 223–230. Springer, Dordrecht (1997)
37. Ming, X., Deng, Y.: Time-bounded termination analysis for probabilistic programs with delays. Inf. Comput. **275**, 104634 (2020)
38. Zilber, B.: Pseudo-exponentiation on algebraically closed fields of characteristic zero. Ann. Pure Appl. Logic **132**(1), 67–95 (2005)

Stop Gambling! It Just Takes Too Long

Sascha Klüppelholz[1]([✉]) and Jakob Piribauer[1,2]

[1] Technische Universität Dresden, Dresden, Germany
sascha.klueppelholz@tu-dresden.de
[2] Leipzig University, Leipzig, Germany

Abstract. The classical stochastic shortest path problem (SSPP) on integer-weighted Markov decision processes asks for a scheduler that maximizes the expected accumulated weight before reaching a goal state while ensuring that the goal is reached almost surely. In this article, we introduce the positively almost-surely terminating stochastic shortest path problem (PAST-SSPP), a variant of the classical SSPP, where the additional requirement is made that the expected number of steps for reaching the goal has to be finite. We show that PAST-SSPP can be solved in polynomial time. To this end, we provide an extension of the well-known spider construction as introduced by Christel Baier et al. used for the solution of the classical SSPP.

Keywords: Stochastic shortest path problem · almost-sure termination · Markov decision processes · optional stopping theorem

1 Introduction

Shortest path problems are fundamental in computer science, as they can be found at the core of numerous optimization tasks coming from many different application fields within and outside of computer science, such as network routing, navigation, and operations research. Therefore, efficient methods, techniques and algorithms contributing to solving shortest (or longest) path problems are of particular interest. Stochastic shortest (or longest) path problems (SSPP) [1,2,5] are a variant of the shortest (or longest) path problem, formulated for weighted operational models exposing stochastic behavior. In this paper, we rely on finite-state Markov decision processes (MDPs) that allow combining nondeterministic choice with probabilistic branching among successor states. These MDPs are equipped with weights that are collected in each transition step. The SSPP – formulated as a maximization problem from now on throughout this paper – translates to finding a *proper* scheduler that maximizes the expected accumulated weight until reaching a set of *goal* states, where proper schedulers ensure that *goal* is reached almost surely (i.e., with probability 1).

The classical SSPP is solvable in polynomial time. Bertsekas and Tsitsiklis [5] specify a semantic condition for MDPs that ensure that the SSPP can be solved

The authors are supported by the DFG through the DFG grant 389792660 as part of TRR 248 (see https://perspicuous-computing.science).

N. Bertrand et al. (Eds.): Christel Baier Festschrift, LNCS 15760, pp. 139–157, 2026.
https://doi.org/10.1007/978-3-031-97439-7_6

via linear programming. Namely, this condition states that the expected accumulated weight under any scheduler not reaching the goal state almost surely is $-\infty$. De Alfaro [1] provides a pre-processing technique collapsing end-components (ECs) only containing zero-weight transitions. ECs are strongly connected sub-MDPs, in which a scheduler can choose to stay arbitrarily long before proceeding to move to *goal*. This pre-processing allows to solve the problem if all weights are non-negative or non-positive, respectively. A full solution for the general problem with positive and negative weights was provided by Baier et al. [2]. This solution solves the SSPP via a fine-grained classification of ECs. This classification results in a polynomial-time pre-processing that ensures that the semantic conditions of Bertsekas and Tsitsiklis are satisfied and consequently in a polynomial-time algorithm for the SSPP.

This classification distinguishes different kinds of limiting behavior of ECs with respect to the expected accumulated weight. Formulated for bottom strongly connected components (BSCCs), which are ECs with one action per state, meaning that the behavior inside the BSCC is purely probabilistic, the key parts of the classification are as follows: In a *positively weight-divergent BSCC*, the accumulated weight almost surely eventually exceeds any threshold. If a positively weight divergent BSCC exists in an MDP, the value in the SSPP is ∞.[1] A maximizing scheduler can move to the BSCC with positive probability, stay in the BSCC until the accumulated weight is arbitrarily high, and then move to *goal*. In this way, arbitrarily high amounts of expected accumulated weight before reaching *goal* can be achieved.

If no positively weight-divergent BSCC exists, there are two types of BSCCs that might still be present: On the one hand, there can be BSCCs with negative expected mean payoff, i.e., the expected weight that is accumulated per step is negative. If all BSCCs are of the latter type, the semantic condition of Bertsekas and Tsitsiklis [5] is met, as staying in such a BSCC ensures that the accumulated weight almost surely diverges to $-\infty$. On the other hand, there can be 0-BSCCs in which every cycle has accumulated weight 0. In such a 0-BSCC, the accumulated weight always remains within a bounded interval, but the semantic conditions of Bertsekas and Tsitsiklis are violated.

Therefore, the key ingredient for a solution to the SSPP is now a construction that removes 0-BSCCs: The so-called *spider construction* of Baier et al. presented in [2] iteratively flattens those 0-BSCCs to acyclic structures, preserving the behavior of the original MDP when only considering proper schedulers reaching the goal almost surely. Finally, in [2], poly-time algorithms for the identification of BSCCs types are presented and it is shown that the spider construction can remove all 0-BSCCs in polynomial time in case no positively weight-divergent BSCCs exist. Overall, this results in a pre-processing transforming a given MDP into a new MDP to which the polynomial-time solution of [5] via linear programming can be applied.

---

[1] We assume that all states in an MDP are reachable and that *goal* can be reached almost surely from every state.

An important special case of positively weight-divergent BSCCs that will play a major role in this paper are so-called *gambling* BSCCs. Gambling BSCCs are positively weight-divergent BSCCs that are at the same time *negatively weight-divergent* meaning that the accumulated weight almost surely also drops below any (negative) threshold. The expected mean payoff, i.e., the expected amount of weight accumulated per step in the long-run, in such BSCCs is 0 and so one can view the accumulated weight in such a BSCC as a random walk without drift. In the long-run, the accumulated weight will oscillate around 0 with larger and larger magnitude reaching arbitrarily high and arbitrarily low values almost surely.

Example 1. Consider the MDP $\mathcal{M}$ depicted in Fig. 1. Starting in s_{init}, action α can be taken to flip a coin and receive weight $+1$ or -1 with probability $1/2$ each before returning to s_{init}. Alternatively, β can be chosen in s_{init} to end the run and reach *goal*.

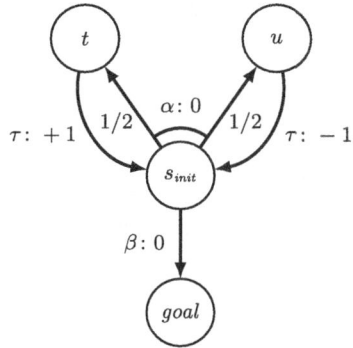

Fig. 1. MDP $\mathcal{M}$ with a gambling bottom strongly connected component.

The sub-MDP consisting of states s_{init}, s and u where only action α is enabled in s_{init} forms a gambling BSCC $\mathcal{B}$. As long as α is chosen, the execution stays in $\mathcal{B}$ and the accumulated weight follows a symmetric random walk. Following this random walk any arbitrary integer value for the accumulated weights is reached almost surely. A scheduler can hence wait until some fixed amount $k \in \mathbb{N} \setminus \{0\}$ of weight has been accumulated and then choose β. Such a scheduler $\mathfrak{S}_k$ achieves an expected accumulated weight of k and reaches *goal* almost surely.

However, there is one caveat with this scheduler: the expected number of steps it takes to reach *goal* is ∞. To see this, let E be the expected number of steps until weight $+1$ has been accumulated. Starting with weight 0, this happens within two steps (to t and back) with probability $1/2$. With the remaining probability of $1/2$, however, weight -1 is gained within two steps. This makes it necessary to accumulate weight $+1$ twice afterwards, which takes $2E$ steps in expectation. So,

$$E = \frac{1}{2} \cdot 2 + \frac{1}{2}(2 + 2E) = 2 + E.$$

Therefore, E has to be infinite. Hence, it of course also takes infinitely many steps in expectation to reach accumulated weight $k > 0$.

This observation yields the motivation for considering an alternative SSPP. As we believe, the question of how to maximize the expected weight among all schedulers that almost surely reach *goal within a finite number of steps in expectation* is highly relevant in general. This holds not only for many real-world optimization problems, but also in the context of program analysis for probabilistic programs, where termination within finite expected time states an important property (see, e.g., [12,13]). Therefore, in this paper we consider the *positively almost-surely terminating stochastic shortest path problem (PAST-SSPP)*, where the task is to maximize the expected accumulated weight among all schedulers that (1) reach *goal* with probability 1 and (2) within a finite number of steps in expectation. We call schedulers fulfilling the requirements (1) and (2) *positively almost-surely terminating schedulers (PAST-schedulers)*.

The key difference between the SSPP and the PAST-SSPP lies in the role of gambling BSCCs. As gambling BSCCs are positively weight-divergent, their existence immediately implies that the value in the SSPP is $+\infty$ as described above. For the PAST-SSPP, however, we will show in general that the gambling behavior cannot be exploited, as already illustrated in Example 1. The key technical tool to prove this result is the *optional stopping theorem* from martingale theory. This will also imply, that the only true decision PAST-schedulers can effectively make within a gambling BSCC, is the selection of the state from which to leave the BSCC. How long they stay in the BSCC does not have any influence on the total expected accumulated weight before reaching *goal*.

Example 2. Consider the MDP depicted in Fig. 2 modelling the reward gained from doing research and the price to pay in terms of stress of extra work in academia brings with it. In this MDP model, there is a gambling BSCC with states "do nice research" and "doing extra work" and only action *continue* enabled. The BSCC can be left at any point in time by choosing action *retire*.

Possibly somewhat surprisingly, we will see that any PAST-scheduler that leaves the BSCC in state "do nice research" is not a maximizing scheduler, as the expected accumulated weight before reaching "retirement" is 0 – no matter how involved the criterion on when to leave the BSCC is. Any PAST-scheduler that leaves the BSCC in state "doing extra work", on the other hand, results in an expected accumulated weight before reaching "retirement" of 4. In other words, as long as we want to reach "retirement" in a finite number of steps in expectation, there is no way to exploit the oscillating nature of the gambling BSCC to gain additional accumulated weight. It is only important to leave the BSCC in the right state. In the considered example, this corresponds to choosing action *retire* at the first visit of state "doing extra work" as choice of an optimal scheduler[2].

---

[2] Please note that this example is illustrative only and abstracts from many important real world facts. For example, we consider research to be exponentially more rewarding than the weights in our MDP might suggest.

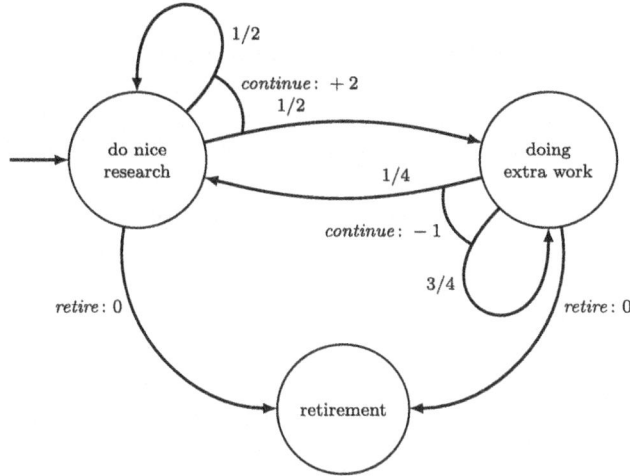

Fig. 2. An MDP capturing the reward received in academia before retirement.

Contribution

We introduce the positively almost-surely terminating stochastic shortest path problem (PAST-SSPP) and show that it can be solved in polynomial time. To this end, we exploit the optional stopping theorem to show that PAST-schedulers cannot exploit the oscillating behavior of gambling BSCCs. This allows us to provide an extension of the spider construction of [2]. Our extension can be applied to all bottom strongly connected components (BSCCs) with expected mean payoff equal 0 and not only to 0-BSCCs. It iteratively removes these BSCCs while not changing the expected accumulated weight that can be realized by PAST-schedulers.

Related Work

An early formulation of the stochastic shortest path problem (SSPP) can be found in [9] and the problem has subsequently also been studied under the name *first passage problem* [8,17]. It is known that this *classical* stochastic shortest path problem is solvable in polynomial time [1,2,5]. For MDPs with non-negative weights [4,14], the problem becomes considerably more complicated and computationally harder to solve already. The decidability status for general integer weights is still open. A decidability result would have far-reaching consequences in analytic number theory [15]. Non-classical variants of the SSPP typically weaken the requirement that *goal* has to be reached almost surely. One example are conditional expectations for MDPs [4] or works that focus on partial expectations [6,7,14]. In contrast, we strengthen this requirement by adding the condition that the expected number of steps has to be finite.

The difference between almost-sure and positive almost sure termination that provides the motivation for the PAST-SSPP also plays an important role in the

analysis of probabilistic programs. Deciding almost-sure termination of probabilistic programs is undecidable and even harder than the halting problem [11].
Deciding positive almost-sure termination is again much harder to decide and
lies outside the arithmetical hierarchy [13]. Nevertheless, sound proof methods
for almost-sure termination that are complete for restricted classes of probabilistic programs have been developed, e.g., using ranking super-martingales [10]. To
prove positive almost-sure termination a sound precondition-style calculus has
been developed in [12].

2   Preliminaries

We assume some familiarity with Markov decision processes and probability
theory. In the sequel, we briefly introduce the notions and notation used in this
paper. More details can be found, e.g., in [3,16].

Notations for Markov Decision Processes. A *Markov decision process* (MDP) is a
tuple $\mathcal{M} = (S, Act, P, s_{init}, wgt)$ where S is a finite set of states, Act a finite set of
actions, $P \colon S \times Act \times S \to [0,1] \cap \mathbb{Q}$ the transition probability function, $s_{init} \in S$
the initial state, and $wgt \colon S \times Act \to \mathbb{Z}$ the weight function. We require that
$\sum_{t \in S} P(s, \alpha, t) \in \{0,1\}$ for all $(s, \alpha) \in S \times Act$. We say that action α is *enabled*
in state s iff $\sum_{t \in S} P(s, \alpha, t) = 1$ and denote the set of all actions that are enabled
in state s by $Act(s)$. We further require that $Act(s) \neq \emptyset$ for all $s \in S$. The *size* of
$\mathcal{M}$ is the sum of the number of states plus the total sum of the encoding lengths
in binary of the non-zero probability values $P(s, \alpha, s')$ as fractions of co-prime
integers as well as the encoding length in binary of the weights.

If for a state s and all actions $\alpha \in Act(s)$, we have $P(s, \alpha, s) = 1$ and
$wgt(s, \alpha) = 0$, we say that s is *absorbing*. The paths of $\mathcal{M}$ are finite or
infinite sequences $s_0 \, \alpha_0 \, s_1 \, \alpha_1 \ldots$ where states and actions alternate such that
$P(s_i, \alpha_i, s_{i+1}) > 0$ for all $i \geq 0$. For $\pi = s_0 \, \alpha_0 \, s_1 \, \alpha_1 \ldots \alpha_{k-1} s_k$, $P(\pi) =$
$P(s_0, \alpha_0, s_1) \cdot \ldots \cdot P(s_{k-1}, \alpha_{k-1}, s_k)$ denotes the probability of π and $last(\pi) = s_k$
its last state. Further, $wgt(\pi) = wgt(s_0, \alpha_0) + \cdots + wgt(s_{k-1}, \alpha_{k-1})$ is the weight
of π. For infinite paths ρ, the weight $wgt(\rho)$ is defined analogously and hence
well-defined and finite if ρ reaches an absorbing state as from that moment on
only weight 0 is collected in each step. Given a path ρ, we denote the prefix of
the first k steps of ρ by $\rho[0.k]$. Further, given a finite path π ending in a state s
and a path ρ starting in s, we denote there concatenation by $\pi \circ \rho$.

An *end-component* of $\mathcal{M}$ is a strongly connected sub-MDP formalized by a
subset $S' \subseteq S$ of states and a non-empty subset $\mathfrak{A}(s) \subseteq Act(s)$ for each state
$s \in S'$ such that for each $s \in S'$, $t \in S$ and $\alpha \in \mathfrak{A}(s)$ with $P(s, \alpha, t) > 0$, we
have $t \in S'$ and such that in the resulting sub-MDP all states are reachable from
each other. An end-component is called a bottom strongly connected component
(BSCC) if it contains exactly one action per state. In particular, end-components
of Markov chains are BSCCs. An end-component is a 0-end-component if it only
contains state-action-pairs with weight 0; 0-BSCCs are defined analogously.

A *Markov chain* is an MDP in which the set of actions is a singleton. In
this case, we can drop the set of actions and consider a Markov chain as a tuple

$\mathcal{M} = (S, P, s_{init}, wgt)$ where P now is a function from $S \times S$ to $[0, 1]$ and wgt a function from S to $\mathbb{Z}$.

Schedulers. A *scheduler* (also called *policy*) for $\mathcal{M}$ is a function $\mathfrak{S}$ that assigns to each finite path π a probability distribution over $Act(last(\pi))$. If $\mathfrak{S}(\pi) = \mathfrak{S}(\pi')$ for all finite paths π and π' with $last(\pi) = last(\pi')$, we say that $\mathfrak{S}$ is *memoryless*. In this case, we also view schedulers as functions mapping states $s \in S$ to probability distributions over $Act(s)$. A scheduler $\mathfrak{S}$ is called deterministic if $\mathfrak{S}(\pi)$ is a Dirac distribution for each finite path π, in which case we also view the scheduler as a mapping to actions in $Act(last(\pi))$. Given an MDP $\mathcal{M}$, a scheduler $\mathfrak{S}$ for $\mathcal{M}$ induces an infinite-state Markov chain with finite paths of $\mathcal{M}$ as states, which we denote by $\mathcal{M}^{\mathfrak{S}}$.

Probability Measure. We write $\Pr_{\mathcal{M},s}^{\mathfrak{S}}$ to denote the probability measure induced by a scheduler $\mathfrak{S}$ and a state s of an MDP $\mathcal{M}$. It is defined on the σ-algebra Σ generated by the cylinder sets $Cyl(\pi)$ of all infinite extensions of a finite path $\pi = s_0 \alpha_0 s_1 \alpha_1 \ldots \alpha_{k-1} s_k$ starting in state s, i.e., $s_0 = s$, by assigning to $Cyl(\pi)$ the probability that π is realized under $\mathfrak{S}$, which is $P^{\mathfrak{S}}(\pi) \overset{\text{def}}{=} \prod_{i=0}^{k-1} \mathfrak{S}(s_0 \alpha_0 \ldots s_i)(\alpha_i) \cdot P(s_i, \alpha_0, s_{i+1})$. This can be extended to a unique probability measure on the mentioned σ-algebra. For details, see [16]. For a random variable X, i.e., a measurable function defined on infinite paths in $\mathcal{M}$, we denote the expected value of X under a scheduler $\mathfrak{S}$ and state s by $\mathbb{E}_{\mathcal{M},s}^{\mathfrak{S}}(X)$. We define $\mathbb{E}_{\mathcal{M},s}^{\min}(X) \overset{\text{def}}{=} \inf_{\mathfrak{S}} \mathbb{E}_{\mathcal{M},s}^{\mathfrak{S}}(X)$ and $\mathbb{E}_{\mathcal{M},s}^{\max}(X) \overset{\text{def}}{=} \sup_{\mathfrak{S}} \mathbb{E}_{\mathcal{M},s}^{\mathfrak{S}}(X)$. If $s = s_{init}$, we sometimes drop the subscript s in $\Pr_{\mathcal{M},s}^{\mathfrak{S}}$ and $\mathbb{E}_{\mathcal{M},s}^{\mathfrak{S}}$.

Reachability and Mean Payoff. Given an MDP $\mathcal{M}$ and a state s, we denote the event that s is reached, i.e., the set of all paths reaching s, by $\Diamond s$. We will further use the *mean payoff* measure as tool to classify end-components. For an infinite path ρ, the mean payoff is defined as

$$\mathbb{MP}(\rho) = \lim_{n \to \infty} \inf \frac{1}{n} \cdot wgt(\rho n)$$

where ρn is the prefix of length n of ρ. The maximal expected mean payoff in an MDP starting in s is now $\mathbb{E}_{\mathcal{M},s}^{\max}(\mathbb{MP})$. In a Markov chain $\mathcal{C}$, we consequently write $\mathbb{E}_{\mathcal{C},s}(\mathbb{MP})$. It is well-known that, in strongly connected Markov chains, the mean payoff of almost all paths is equal to the expected mean payoff.

Martingales in Markov Chains. Given a Markov chain $\mathcal{M} = (S, P, s_{init}, wgt)$, let Σ_n be the σ-algebra generated by all cylinder sets $Cyl(\pi)$ of paths π of $\mathcal{M}$ of length at most n. So, we obtain a sequence of σ-algebras $\Sigma_0 \subseteq \Sigma_1 \subseteq \cdots \subseteq \Sigma$ where Σ is the σ-algebra generated by all cylinder sets. Such a sequence of σ-algebras is called a *filtration*. Intuitively, the σ-algebra Σ_k captures the set of observations that are possible after k steps of the Markov chain.

A (discrete-time) *martingale* with respect to this filtration is a sequence of random variables $(X_t)_{t \in \mathbb{N}}$ such that

- X_t is Σ_t-measurable for all $t \in \mathbb{N}$ (intuitively speaking, X_t only depends on the first t steps of $\mathcal{M}$),

- $\mathbb{E}(|X_t|) < \infty$ for all $t \in \mathbb{N}$, and
- $\mathbb{E}(X_{t+1} \mid \pi) = X_t(\pi)$ for all $t \in \mathbb{N}$ and all paths π of length t in $\mathcal{M}$, where $X_t(\pi)$ denotes the unique value that X_t takes on all infinite extensions of π.

Intuitively, this can be understood as a sequence of values X_t depending on the first t steps of the Markov chain such that the next observed value is, in expectation, always as large as the current observed value. Note that, in particular, this also implies $\mathbb{E}(X_t) = \mathbb{E}(X_0)$ for all $t \in \mathbb{N}$.

Stopping Time. With $\mathcal{M}$ and the filtration $\Sigma_0 \subseteq \Sigma_1 \subseteq \cdots \subseteq \Sigma$ as above, a stopping time for a martingale $(X_t)_{t \in \mathbb{N}}$ is an $\mathbb{N}$-valued random variable τ such that $(\tau = k) \in \Sigma_k$ for all $k \in \mathbb{N}$. Intuitively, a stopping time determines whether a process $(X_t)_{t \in \mathbb{N}}$ should be stopped at time k given the observations up to time point k. Given $(X_t)_{t \in \mathbb{N}}$ and a stopping time τ, the random variable X_τ returns the current observation in the moment the martingale is stopped by τ.

We will rely on the following well-known theorem from martingale theory, which is sometimes also called Doob's optional sampling theorem.

Theorem 1 (Optional stopping theorem (see, e.g., [18])). *Given a martingale $(X_t)_{t \in \mathbb{N}}$ and a stopping time τ such that*

- $\mathbb{E}(\tau) < \infty$ *and*
- *there is a constant $c > 0$ such that $|X_{t+1} - X_t| \le c$ almost surely for all $t \in \mathbb{N}$,*

then X_τ is almost surely well-defined and $\mathbb{E}(X_\tau) = \mathbb{E}(X_0)$.

3   The Classical Stochastic Shortest Path Problem

In this section we recall the classical SSPP and summarize main results. For this let $\mathcal{M} = (S, Act, P, s_{init}, wgt)$ be an MDP with a weight function $wgt \colon S \times Act \to \mathbb{Z}$ and $goal \in S$ be an absorbing goal state. The SSPP can be formulated on the basis of the following random variable $\oplus goal$ on infinite paths ρ of $\mathcal{M}$:

$$\oplus goal(\rho) = \begin{cases} wgt(\rho) & \text{if } \rho \models \Diamond goal, \\ undefined & \text{otherwise.} \end{cases}$$

The expected accumulated weight before reaching $goal$ under a scheduler $\mathfrak{S}$ is given by the expected value $\mathbb{E}^{\mathfrak{S}}_{\mathcal{M}, s_{init}}(\oplus goal)$. It is evident that this expected value is only defined if $\Pr^{\mathfrak{S}}_{\mathcal{M}, s_{init}}(\Diamond goal) = 1$. Hence, for the classical SSPP we only range over the class of almost-surely terminating schedulers. We call a scheduler $\mathfrak{S}$ for $\mathcal{M}$ almost-surely terminating (AST-scheduler) if $\Pr^{\mathfrak{S}}_{\mathcal{M}, s_{init}}(\Diamond goal) = 1$. With this, we can now formally define the classical *stochastic shortest path problem* (SSPP) as follows.

SSPP: Given an MDP $\mathcal{M} = (S, Act, P, s_{init}, wgt)$ with an absorbing goal state *goal* as above, determine the optimal value

$$\mathbb{E}^{\max}_{\mathcal{M}, s_{init}}(\lozenge goal) = \sup_{\mathfrak{S}} \mathbb{E}^{\mathfrak{S}}_{\mathcal{M}, s_{init}}(\lozenge goal)$$

where the supremum ranges over all AST-schedulers $\mathfrak{S}$.

The analogous minimization problem can be treated by multiplying all weights with -1 and solving the resulting maximization problem.

As described also in the introduction, the classical SSPP is solvable in polynomial time [1,2,5]. In the sequel, we give a brief overview over the necessary classification of end-components reformulated in terms of BSCCs and the so-called spider construction used in the pre-processing of [2]. This pre-processing ensures that the condition that the expected accumulated weight under any scheduler not reaching the goal state almost surely is $-\infty$ introduced by [5] is met. Afterwards, the problem can be solved in polynomial time via linear programming as shown in [5].

Classification of BSCCs. Let $\mathcal{M} = (S, Act, P, s_{init}, wgt)$ be an MDP with an absorbing goal state *goal* $\in S$. We assume that all states are reachable from s_{init} and that *goal* is almost surely reachable from all states. The solution to the SSPP by Baier et al. relies on the following classification of the weight-divergent behavior in BSCCs.

Definition 1 (See [2]). *Let $\mathcal{B}$ be a BSCC of $\mathcal{M}$ viewed as a Markov chain (T, P, s, wgt) with an arbitrary initial state s in $\mathcal{B}$. The BSCC $\mathcal{B}$ is called*

- *a 0 -BSCC if all cycles in $\mathcal{B}$ have weight 0,*
- *positively weight-divergent if $\mathrm{Pr}_{\mathcal{B},s}(\lozenge(\text{accumulated weight} \geq n)) = 1$ for all $n \in \mathbb{N}$, where $\lozenge(\text{accumulated weight} \geq n)$ denotes the event that the accumulated weight of a prefix of a path is at least n,*
- *negatively weight-divergent if $\mathrm{Pr}_{\mathcal{B},s}(\lozenge(\text{accumulated weight} \leq -n)) = 1$ for all $n \in \mathbb{N}$, where $\lozenge(\text{accumulated weight} \leq -n)$ denotes the event that the accumulated weight of a prefix of a path is at most $-n$,*
- *pumping if it is positively weight-divergent and not negatively weight-divergent,*
- *gambling if it is positively and negatively weight-divergent*

The existence of positively weight-divergent BSCCs immediately implies that the value of the SSPP is $+\infty$ [2]. The reason is that a scheduler may move to the BSCC with positive probability and then stay in the BSCC until an arbitrarily high amount or weights has been accumulated before moving to goal.

Example 3. Recall the example depicted in Fig. 2. The states "do nice research" and "doing extra work" with the action *continue* form a gambling and hence positively weight-divergent BSCC. Thus, the value of the SSPP is ∞. A scheduler $\mathfrak{S}_k$ for $k \in \mathbb{N}$ can simply wait until the accumulated weight exceeds k, which will happen with probability 1 and then move to the state "retirement".

If there are no positively weight-divergent BSCCs, 0-BSCC may still cause the semantic condition of Bertsekas and Tsitsiklis to fail. So, the SSPP cannot directly be solved by linear programming. To overcome this problem, [2] introduce the so-called *spider construction* that allows to remove 0-BSCCs successively in a pre-processing procedure. We now give a brief sketch of this spider construction.

Spider Construction. Let $\mathcal{M} = (S, Act, P, s_{init}, wgt)$ be an MDP with an absorbing goal state $goal \in S$ such that $\Pr_{\mathcal{M},s_{init}}^{\max}(\Diamond goal) = 1$. Furthermore let $\mathcal{B}$ be a 0-BSCC with set of states $E \subseteq S$. The *spider construction* of [2] proceeds as follows: Pick a state $e \in E$. Then, for each state $s \in E$, all paths from s to e inside $\mathcal{B}$ have the same accumulated weight w_s. This follows from the condition that all cycles have weight 0. For each state $s \in E$, disable all actions in state s. Instead enable one action β_s in each state $s \in E \setminus \{e\}$ leading to state e with probability 1 and weight w_s. For each state $s \in E$ and each action $\alpha \in Act(s)$ not belonging to $\mathcal{B}$, enable a new action $\beta_{s,\alpha}$ in state e. Let the weight of this new action in state e be $wgt(s, \alpha) - w_s$. For each state $t \in S$, let $P(e, \beta_{s,\alpha}, t) = P(s, \alpha, t)$. In this way, taking action α in state s can be mimicked by first moving to e via β_s and then taking $\beta_{s,\alpha}$ in state e. The construction is illustrated in Fig. 3.

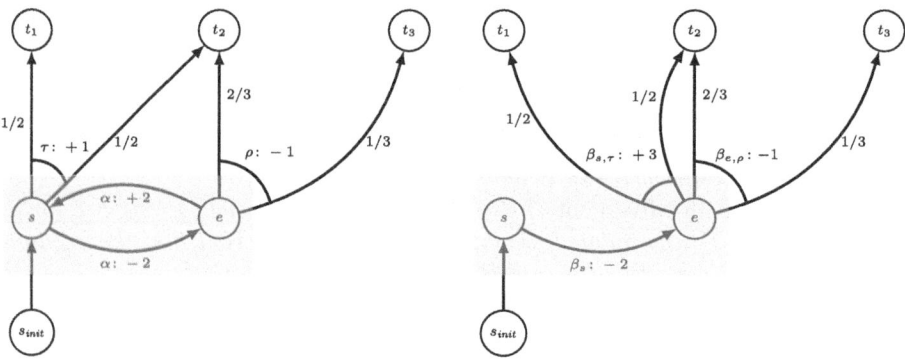

Fig. 3. Illustration of the spider construction. The gray 0-end-component in the left MDP is removed on the right. All transitions leaving the end-component are moved to start from state e after the construction. The weights are adjusted accordingly.

Solution to the Classical SSPP. With the spider construction a solution method for the classical SSPP can now proceeds as follows.

1. It can be checked in polynomial time whether there is a positively weight-divergent BSCC. If this is the case, $\mathbb{E}_{\mathcal{M},s_{init}}^{\max}(\Diamond goal) = \infty$.
2. If there is no positively weight-divergent BSCCs, there is a polynomial-time pre-processing that iteratively identifies and removes all 0-BSCCs using the spider-construction.

3. After this pre-processing, the semantic condition of Bertsekas and Tsitsiklis [5] is satisfied and the value $\mathbb{E}^{\max}_{\mathcal{M}, s_{init}}(\text{⬥}goal)$ is finite and can be computed in polynomial time, e.g., via solving a linear program.

This procedure yields the following result.

Theorem 2 (See [2,5]). *The classical SSPP is solvable in polynomial time.*

4 The Positively Almost-Surely Terminating SSPP

We now come to the main contribution of this paper and start by introducing a new variant of the classical SSPP that requires schedulers to reach the goal in a finite amount of steps in expectation. For this let again $\mathcal{M} = (S, Act, P, s_{init}, wgt)$ be an MDP with a weight function $wgt \colon S \times Act \to \mathbb{Z}$ and state $goal \in S$ be an absorbing goal such that $\Pr^{\max}_{\mathcal{M}, s_{init}}(\Diamond goal) = 1$. From now on we are interested in schedulers that represent *positively* almost-surely terminating processes.

Definition 2. *We call an AST-scheduler $\mathfrak{T}$ for $\mathcal{M}$ positively almost-surely terminating (PAST-scheduler) if $\mathbb{E}^{\mathfrak{T}}_{\mathcal{M}, s_{init}}(\oplus goal) < \infty$ where $\oplus goal$ denotes the number of steps taken before reaching goal.*

Using this notion of PAST-scheduler, the problem we formalize the *positively almost-surely terminating stochastic shortest path problem* (PAST-SSPP) as follows.

PAST-SSPP: Given an MDP $\mathcal{M} = (S, Act, P, s_{init}, wgt)$ with an absorbing goal state $goal$ as above, determine the optimal value

$$\mathbb{E}^{\max}_{\mathcal{M}, s_{init}}(\text{⬥}goal) = \sup_{\mathfrak{S}} \mathbb{E}^{\mathfrak{S}}_{\mathcal{M}, s_{init}}(\text{⬥}goal)$$

where the supremum ranges over all PAST-schedulers $\mathfrak{S}$.

In the following Sect. 4.1, we show that for PAST-schedulers there is no benefit of staying in gambling ECs with mean payoff 0 forever and that expected accumulated weights only depends on the moment a scheduler decides for leaving the EC. Motivated by these insides, Sect. 4.2 presents a variant of the original spider construction suited for PAST-SSPP. Based on this variant of the spider construction, we present a PTIME algorithm for solving PAST-SSPP in Sect. 4.3.

4.1 PAST-Schedulers and Their Gambling Habits

We now aim to apply the optional stopping theorem to gambling BSCCs of MDPs. More precisely, we observe that a PAST-scheduler has to leave any BSCC of an MDP after a finite number of steps in expectation. This induces a stopping time with finite expected value. We will show that this implies that a PAST-scheduler leaving a gambling BSCC cannot gain or lose accumulated weight inside the gambling BSCC compared to any other PAST-scheduler.

First, we consider a Markov chain in isolation and afterwards apply the result to BSCCs. Let $\mathcal{C} = (S, P, s_{init}, wgt)$ be a strongly connected Markov chain. Fix a state $s \in S$. For each $q \in S$, let $D_s^{\mathcal{C}}(q) = \mathbb{E}_{\mathcal{C},q}(\lozenge s)$. The function $D_s^{\mathcal{C}}$ measures the expected distance from q to s in terms of the weight that is expected to be accumulated on paths from q to s. Observe that $D_s^{\mathcal{C}}(s) = 0$. As $\mathcal{C}$ is strongly connected, $D_s(q)$ exists and is finite for every state $q \in S$.
We now consider the following two families of random variables:

- For each $t \in \mathbb{N}$, let X_t be the state the Markov chain is in after t time steps.
- For each $t \in \mathbb{N}$, let R_t be the weight accumulated in the first t time steps.

Lemma 1. *Let $\mathcal{C} = (S, P, s_{init}, wgt)$ be a strongly connected Markov chain and $s \in S$ a state. Let X_t and R_t for $t \in \mathbb{N}$ be the random variables as defined above. If $\mathbb{E}_{\mathcal{C},s}(\text{MP}) = 0$, then $(Z_t^{\mathcal{C}})_{t \in \mathbb{N}} \stackrel{def}{=} (R_t + D_s^{\mathcal{C}}(X_t))_{t \in \mathbb{N}}$ is a martingale.*

Proof. To show this, we first observe that $(Z_t^{\mathcal{C}})_t$ is Σ_t-measurable and has finite expectation for all $t \in \mathbb{N}$. What remains to show is, that $\mathbb{E}(Z_{t+1} \mid \pi) = Z_t(\pi)$ for all $t \in \mathbb{N}$ and all finite paths π of length t in $\mathcal{M}$, where $Z_t(\pi)$ denotes the unique value that Z_t takes on all infinite extensions of π.

Observation 1: We first consider a finite path π in $\mathcal{C}$ such that $last(\pi) = q \neq s$. Let t be the number of steps taken in π. So, after t steps in $\mathcal{C}$ following π, we have $X_t = q$, $R_t = wgt(\pi)$, and $D_s^{\mathcal{C}}(X_t) = \mathbb{E}_{\mathcal{C},q}(\lozenge s)$. Now, taking one additional step, weight $wgt(q)$ is received and so $R_{t+1} = R_t + wgt(q)$. On the other hand, we have

$$D_s^{\mathcal{C}}(X_t) = \mathbb{E}_{\mathcal{C},q}(\lozenge s) = wgt(q) + \sum_{p \in S} P(q, p) \cdot \mathbb{E}_{\mathcal{C},p}(\lozenge s)$$

and hence

$$\sum_{p \in S} P(q, p) \cdot \mathbb{E}_{\mathcal{C},p}(\lozenge s) = \mathbb{E}_{\mathcal{C},q}(\lozenge s) - wgt(q) = D_s^{\mathcal{C}}(X_t) - wgt(q).$$

The expected value of $D_s(X_{t+1})$ given X_t is exactly $\sum_{p \in S} P(q, p) \cdot D_s(p) = \sum_{p \in S} P(q, p) \cdot \mathbb{E}_{\mathcal{C},p}(\lozenge s)$. So, we have

$$\mathbb{E}(R_{t+1} + D_s^{\mathcal{C}}(X_{t+1}) \mid \pi) = R_t + wgt(q) + D_s^{\mathcal{C}}(X_t) - wgt(q) = R_t + D_s^{\mathcal{C}}(X_t).$$

Observation 2: Now, consider a path π such that $last(\pi) = s$ and again let t be the length of π. Then, $D_s(X_t) = 0$. Now, we have

$$\mathbb{E}(R_{t+1} + D_s^{\mathcal{C}}(X_{t+1}) \mid \pi) = R_t + wgt(q) + \sum_{p \in S} P(s, p) \cdot \mathbb{E}_{\mathcal{C},p}(\lozenge s).$$

The expected accumulated weight when starting in state s until reaching state s again for the first time is precisely given by

$$wgt(q) + \sum_{p \in S} P(s, p) \cdot \mathbb{E}_{\mathcal{C}, p}(\lozenge s),$$

which is 0 as the mean payoff in $\mathcal{C}$ is 0. With these two observations above we have that

$$\mathbb{E}(R_{t+1} + D_s^{\mathcal{C}}(X_{t+1}) \mid \pi) = R_t = R_t + D_s^{\mathcal{C}}(X_t)$$

and hence $(Z_t^{\mathcal{C}})_{t \in \mathbb{N}} \stackrel{\text{def}}{=} (R_t + D_s^{\mathcal{C}}(X_t))_{t \in \mathbb{N}}$ is in fact a martingale. $\qquad \square$

Stopping Time Induced by a Scheduler. Let $\mathcal{M} = (S, Act, P, s_{init}, wgt)$ be an MDP with a weight function $wgt \colon S \times Act \to \mathbb{Z}$ and an absorbing goal state $goal \in S$ that can be reached almost surely. Let $\mathcal{C}$ be a gambling BSCC of $\mathcal{M}$. As shown in [2], gambling BSCCs $\mathcal{C}$ viewed as a Markov chain satisfy $\mathbb{E}_{\mathcal{C}, s}(\mathrm{MP}) = 0$ for any state s in $\mathcal{C}$. So, $\mathcal{C}$ can be viewed as a strongly connected Markov chain with expected mean payoff 0.

Let $\mathfrak{S}$ be a PAST-scheduler for $\mathcal{M}$. Suppose there is a finite path π under $\mathfrak{S}$ leading to a state s of $\mathcal{C}$. Now, we define the following stopping time $\tau_{\mathfrak{S}}$ for the martingale $(Z_t^{\mathcal{C}})_{t \in \mathbb{N}}$ defined in Lemma 1 for the Markov chain $\mathcal{C}$. We let $\tau_{\mathfrak{S}} \stackrel{\text{def}}{=} k$ if $\mathfrak{S}$ chooses an action not belonging to the BSCC $\mathcal{C}$ for the first time after k steps (starting from s after π). As $\mathfrak{S}$ is a PAST-scheduler, $\mathfrak{S}$ has to leave $\mathcal{C}$ within finitely many steps in expectation and hence the expected value of $\tau_{\mathfrak{S}}$ is finite.

Remark 1. To make the definition of $\tau_{\mathfrak{S}}$ as a stopping time formally more precise, there is one detail that is omitted here: As schedulers may use randomization, it does not only depend on the path taken so far whether $\mathfrak{S}$ chooses an action not belonging to $\mathcal{C}$, but also on the possible outcome of the randomization. We can assume that this randomization depends on a family of independent uniform random variables U_k on $[0, 1]$, where U_k governs the randomization at time step k. Now, we can consider the product of the probability space of the Markov chain with the probability space underlying this sequence of independent uniform random variables and extend the filtration in the natural way. Then, $\tau_{\mathfrak{S}}$ is a stopping time with respect to this filtration and the martingale and all results discussed in the sequel are not affected.

Now again, let R_t be the weight that is accumulated in the first t time steps in $\mathcal{C}$ and let X_t be the state the Markov chain $\mathcal{C}$ is in after these t steps. As shown in Lemma 1, with the distance function $D_s^{\mathcal{C}}$ as defined above, $(R_t + D_s^{\mathcal{C}}(X_t))_{t \in \mathbb{N}}$ is a martingale.

Lemma 2. *Given $\mathcal{M}$, $\mathfrak{S}$, $\mathcal{C}$, $(R_t + D_s^{\mathcal{C}}(X_t))_{t \in \mathbb{N}}$, and $\tau = \tau_{\mathfrak{S}}$ as above, we have*

$$\mathbb{E}(R_\tau + D_s^{\mathcal{C}}(X_\tau)) = 0.$$

Proof. We have $R_0 = 0$ and $D_s^{\mathcal{C}}(X_0) = 0$. As τ is a stopping time with finite expected value and the changes $|R_{t+1} - R_t|$ and $|D_s^{\mathcal{C}}(X_{t+1}) - D_s^{\mathcal{C}}(X_t)|$ are bounded by a constant, the optional stopping theorem (see Theorem 1) yields the claim.   □

This means that the expected value $\mathbb{E}(R_\tau)$ at the moment the scheduler $\mathfrak{S}$ chooses an action not belonging to $\mathcal{C}$ is simply $\mathbb{E}(-D_s^{\mathcal{C}}(X_\tau))$. So, $\mathbb{E}(R_\tau)$ only depends on the distribution over states at which $\mathfrak{S}$ first chooses an action not belonging to $\mathcal{C}$. For each state q in $\mathcal{C}$, let $\delta(q)$ be the probability that $\mathfrak{S}$ first chooses such an action in q.

Recall that, for any state q, $D_s^{\mathcal{C}}(q)$ is the expected accumulated weight it takes to move from state q to state s in $\mathcal{C}$. As the mean payoff in $\mathcal{C}$ is 0, it holds that $D_q^{\mathcal{C}}(s) = -D_s^{\mathcal{C}}(q)$. Then, denoting the set of states of $\mathcal{C}$ by C,

$$\mathbb{E}(R_\tau) = \sum_{q \in C} \delta(q) \cdot D_q^{\mathcal{C}}(s) = \sum_{q \in C} \delta(q) \cdot \mathbb{E}_{\mathcal{C},s}(\lozenge q).$$

So, no matter how long the scheduler $\mathfrak{S}$ stays in $\mathcal{C}$ after entering it via π, the weight the scheduler will accumulate in expectation while staying in $\mathcal{C}$ only depends on where the BSCC is left again. The scheduler might as well leave the BSCC as soon as a desired exit state is reached. Note that this is in stark contrast to the gambling behavior of AST-schedulers. If it is not required to leave $\mathcal{C}$ within a finite number of steps in expectation, a scheduler might "gamble" in a gambling BSCC until arbitrarily high amounts of weight have been accumulated, which will happen almost surely – just not in finite expected time.

In summary, we have learned the following: A PAST-scheduler cannot exploit a gambling BSCC and might as well leave the BSCC as soon as a desired exit state is reached for the first time. This can be viewed as a translation of the optional stopping theorem to our setting.

4.2   A Spider Construction for the PAST-SSPP

The key observation from the previous section that PAST-schedulers do not profit from staying in BSCCs suggests a natural extension of the spider construction to gambling BSCCs. We call this extension the *PAST-spider construction*. The PAST-spider construction is in PTIME. When applied to 0-BSCCs, this PAST-spider construction coincides with the original spider construction. Hence, the PAST-spider construction is applicable to 0-BSCCs and gambling BSCCs. It has been shown [2] that a BSCC with mean payoff 0 is either a 0-BSCC or gambling BSCC. With this, our construction can be used within a polynomial time decision procedure for PAST-SSPP.

The PAST-spider construction generalizes the original spider construction as follows: Given a BSCC $\mathcal{B} = (B, \mathfrak{A})$ with expected mean payoff zero, we pick an arbitrary state e in $\mathcal{B}$. Then, we remove all actions belonging to $\mathcal{B}$ and instead enable a new action β_s at each state $s \neq e$ of $\mathcal{B}$ that leads to e with probability 1 and weight $D_e^{\mathcal{B}}(s)$, i.e., the expected accumulated weight when moving from s to e inside $\mathcal{B}$. If $\mathcal{B}$ is a 0-BSCC, all paths from s to e have the same accumulated weight and hence this value is of course also the expected accumulated weight from s to e. Afterwards, for each action $\alpha \neq \mathfrak{A}(s)$ enabled in $s \in B$, we enable a new action $\beta_{s,\alpha}$ in e that has the same distribution over successor states as α in s and that has weight $-D^{\mathcal{B}}_e(s) + wgt(s,\alpha) = D^{\mathcal{B}}_s(e) + wgt(s,\alpha)$, i.e., exactly the weight that is accumulated in expectation when moving from e to s inside $\mathcal{B}$ and then choosing α there (Fig. 4).

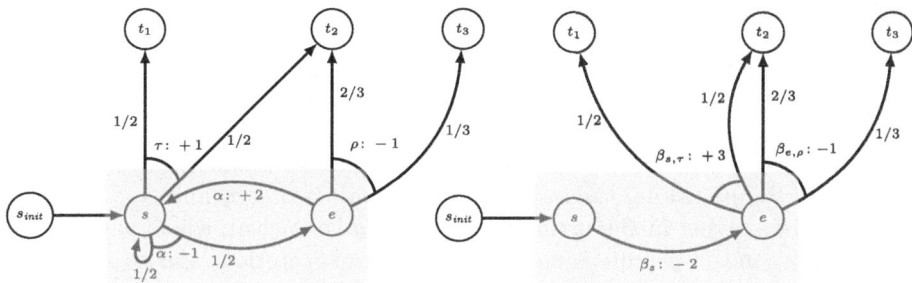

Fig. 4. Illustration of the PAST-spider construction. The gray 0-end-component in the left MDP is removed on the right. All transitions leaving the end-component are moved to start from state e after the construction. The weights are adjusted according to the expected accumulated weight encountered along paths between states of the BSCC.

It remains to prove soundness of this PAST-spider construction.

Theorem 3. *Given an MDP $\mathcal{M}$ with an absorbing goal state goal and BSCC $\mathcal{B}$ such that $\mathbb{E}_{\mathcal{B},s}(\mathbb{MP}) = 0$ for any state s of $\mathcal{B}$, let $\mathcal{N}$ be the MDP obtained by applying the PAST-spider construction to $\mathcal{B}$ in $\mathcal{M}$. Then, for each PAST-scheduler $\mathfrak{S}$ for $\mathcal{M}$, there is a PAST-scheduler $\mathfrak{T}$ for $\mathcal{N}$, and vice versa, such that*

$$\mathbb{E}^{\mathfrak{S}}_{\mathcal{M},s_{init}}(\lozenge goal) = \mathbb{E}^{\mathfrak{T}}_{\mathcal{N},s_{init}}(\lozenge goal).$$

Proof. Given a PAST-scheduler $\mathfrak{S}$ for $\mathcal{M}$, we define a PAST-scheduler $\mathfrak{T}$ for $\mathcal{N}$ in the sequel: Whenever $\mathfrak{S}$ enters the BSCC $\mathcal{B}$ in a state s, consider the probability $\delta(q,\alpha)$ that the next action not belonging to $\mathcal{B}$ chosen by $\mathfrak{S}$ is α in state q. As $\mathfrak{S}$ is a PAST-scheduler, these probabilities sum up to 1.

Now, let $\mathfrak{T}$ be the following scheduler mimicking $\mathfrak{S}$: When a run reaches s, the scheduler $\mathfrak{T}$ moves from state s to state e via action β_s (unless $s = e$) and then chooses action $\beta_{q,\alpha}$ with probability $\delta(q,\alpha)$ for each appropriate pair (q,α). Afterwards, $\mathfrak{T}$ follows the (conditional) behavior of how $\mathfrak{S}$ behaves in case $\mathfrak{S}$ leaves $\mathcal{B}$ via action α in state q, i.e., behaving like $\mathfrak{S}$ would behave conditioned on the event that indeed α in state q was the first chosen action not belonging to $\mathcal{B}$. So, $\mathfrak{S}$ and $\mathfrak{T}$ only differ in how they traverse $\mathcal{B}$ (or the remainder of $\mathcal{B}$ after the application of the PAST-spider construction, respectively). The weight accumulated in expectation during one visit to $\mathcal{B}$ only depends on the exit probabilities $\delta(q, \alpha)$ (cf. Lemma 2). Then, the weight that $\mathfrak{S}$ accumulates in expectation is

$$\sum_{(q,\alpha)\in S\times Act} \delta(q, \alpha) \cdot (\mathbb{E}_{\mathcal{B},s}(\lozenge q) + wgt(q, \alpha))\,.$$

The expected weight obtained by $\mathfrak{T}$ during the corresponding visit is

$$\mathbb{E}_{\mathcal{B},s}(\lozenge e) + \sum_{(q,\alpha)\in S\times Act} \delta(q, \alpha) \cdot (\mathbb{E}_{\mathcal{B},e}(\lozenge q) + wgt(q, \alpha))\,.$$

As the expected mean payoff in $\mathcal{B}$ is 0, we have $\mathbb{E}_{\mathcal{B},s}(\lozenge e) + \mathbb{E}_{\mathcal{B},e}(\lozenge q) = \mathbb{E}_{\mathcal{B},s}(\lozenge q)$ for all states s and q in $\mathcal{B}$ and the equality of the two expressions follows.

In the other direction, a PAST-scheduler $\mathfrak{T}$ can be mimicked in $\mathcal{M}$ with a scheduler $\mathfrak{S}$ by mimicking the sequence of chosen actions β_s and $\beta_{q,\alpha}$ for some s, q, and α by staying in $\mathcal{B}$ starting from s until q is reached, which will happen almost surely and in a finite number of steps in expectation, and the choosing α in q. The equality of the values of the two schedulers $\mathfrak{T}$ in $\mathcal{N}$ and $\mathfrak{S}$ in $\mathcal{M}$ follows analogously to the case above.   □

Example 4. Applying the PAST-spider construction to the MDP depicted in Fig. 2 when picking "doing extra work" as the center of the construction results in the MDP depicted in Fig. 5. In the light of the soundness of the PAST-spider construction we have just shown, it becomes apparent that the best way for PAST-schedulers to optimize the expected accumulated weight is to wait until state "doing extra work" is reached and then choose *retire* immediately[3].

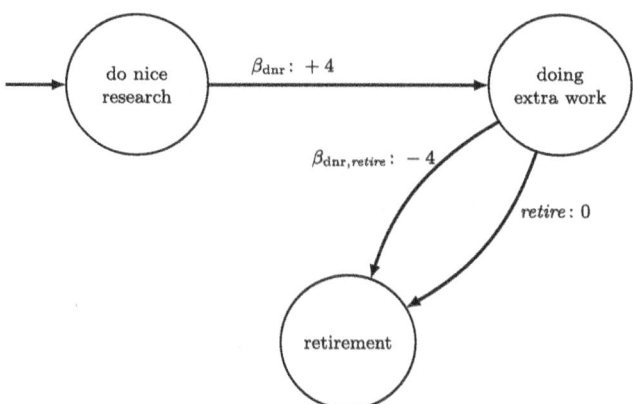

Fig. 5. The PAST-spider construction applied to the gambling BSCC in the MDP in Fig. 2. We abbreviate "do nice research" with "dnr".

---

[3] Please remember, the illustrative nature of the example.

4.3 Solving the PAST-SSPP

Using the PAST-spider construction, we can now solve the PAST-SSPP analogously to the solution of the SSPP presented in the previous chapter. Let $\mathcal{M} = (S, Act, P, s_{init}, wgt)$ be an MDP with a weight function $wgt \colon S \times Act \to \mathbb{Z}$ and an absorbing goal state $goal \in S$. We assume that all states are reachable and that $\mathrm{Pr}_{\mathcal{M}}^{\max}(\lozenge goal) = 1$.

1. If there is pumping BSCC in $\mathcal{M}$, then the optimal value in the PAST-SSPP is ∞. In pumping BSCCs, arbitrarily high amounts of accumulated weight are obtained almost surely in *finite* expected number of steps. So, a PAST-scheduler can move to a pumping BSCC, accumulate arbitrarily much weight, and then move to goal with positive probability. Pumping BSCCs are precisely the BSCCs with positive expected mean payoff and there existence can be checked in polynomial time by computing the maximal mean payoff in each maximal end-component (MEC) of $\mathcal{M}$ (see [2]).
2. If there is no pumping BSCC, we can iteratively remove BSCCs with expected mean payoff equal to 0 by applying the PAST-spider construction. Such BSCCs can be found in polynomial time by checking for each MEC whether the maximal expected mean payoff is 0 and returning a BSCC induced by a mean payoff optimal memoryless deterministic scheduler if this is the case. Analogously to the number of necessary applications of the spider construction in [2], there are no BSCCs with expected mean payoff 0 left after at most $\sum_{s \in S} |Act(s)|$ iterations and hence the procedure terminates in polynomial time.
3. After this pre-processing, we obtain an MDP $\mathcal{N}$ in which the expected mean payoff in all BSCCs is negative. So, the semantic condition of Bertsekas and Tsitsiklis [5] that every scheduler not reaching $goal$ almost surely obtains expected accumulated weight $-\infty$ holds. Hence, the SSPP on $\mathcal{N}$ can be solved in polynomial time and an optimal memoryless deterministic scheduler $\mathfrak{S}$ can be returned. As $\mathfrak{S}$ reaches $goal$ almost surely and is memoryless and deterministic, $\mathfrak{S}$ reaches $goal$ within a finite number of steps in expectation and hence is a PAST-scheduler. So, the optimal value and scheduler obtained by solving the SSPP on $\mathcal{N}$ are also the solution to the PAST-SSPP on $\mathcal{N}$. By Theorem 3, the optimal value is also the same as in the PAST-SSPP of the original MDP $\mathcal{M}$.

The results presented in this section lead to the following main theorem.

Theorem 4. *The PAST-SSPP can be solved in polynomial time.*

5 Conclusion

We introduced the PAST-SSPP that adds to the classical SSPP the requirement that a goal state has to be reached within finitely many steps in expectation. In contrast to the classical SSPP, this makes it impossible to exploit the oscillating behavior of gambling BSCCs to achieve arbitrarily high values of expected accumulated weight before reaching the goal, as we have shown using the optional stopping theorem for martingales. Based on this result, we developed an extension of the spider construction of Baier et al. that in its original form is the key ingredient for solving the classical SSPP. The extension can be applied not only to 0-BSCCs but also to gambling BSCCs and does not change the optimal value in the PAST-SSPP. Using this construction in a pre-processing step, we showed that the PAST-SSPP is solvable in polynomial time.

The application of the optional stopping theorem to the behavior of schedulers in MDPs might be useful for various optimization problems on MDPs. Adding the requirements that schedulers have to leave BSCCs within finitely many steps in expectation is a natural refinement also of other variants of the SSPP such as the optimization of conditional and partial expectations in weighted MDPs. Exploring further potential applications to other optimization problems hence constitutes an interesting direction for future research.

References

1. Alfaro, L.: Computing minimum and maximum reachability times in probabilistic systems. In: Baeten, J.C.M., Mauw, S. (eds.) CONCUR 1999. LNCS, vol. 1664, pp. 66–81. Springer, Heidelberg (1999). https://doi.org/10.1007/3-540-48320-9_7
2. Baier, C., Bertrand, N., Dubslaff, C., Gburek, D., Sankur, O.: Stochastic shortest paths and weight-bounded properties in Markov decision processes. In: Proceedings of the 33rd Annual ACM/IEEE Symposium on Logic in Computer Science (LICS), pp. 86–94. ACM (2018). https://doi.org/10.1145/3209108.32091
3. Baier, C., Katoen, J.P.: Principles of Model Checking. MIT Press (2008)
4. Baier, C., Klein, J., Klüppelholz, S., Wunderlich, S.: Maximizing the conditional expected reward for reaching the goal. In: Legay, A., Margaria, T. (eds.) TACAS 2017. LNCS, vol. 10206, pp. 269–285. Springer, Heidelberg (2017). https://doi.org/10.1007/978-3-662-54580-5_16
5. Bertsekas, D.P., Tsitsiklis, J.N.: An analysis of stochastic shortest path problems. Math. Oper. Res. **16**(3), 580–595 (1991). https://doi.org/10.1287/moor.16.3.580
6. Chen, T., Forejt, V., Kwiatkowska, M., Parker, D., Simaitis, A.: Automatic verification of competitive stochastic systems. Formal Methods Syst. Des. **43**(1), 61–92 (2013). https://doi.org/10.1007/s10703-013-0183-7
7. Chen, T., Forejt, V., Kwiatkowska, M., Parker, D., Simaitis, A.: PRISM-games: a model checker for stochastic multi-player games. In: Piterman, N., Smolka, S.A. (eds.) TACAS 2013. LNCS, vol. 7795, pp. 185–191. Springer, Heidelberg (2013). https://doi.org/10.1007/978-3-642-36742-7_13
8. Derman, C.: Finite State Markovian Decision Processes. Academic Press (1970)
9. Eaton, J., Zadeh, L.: Optimal pursuit strategies in discrete-state probabilistic systems. Trans. ASME Ser. D J. Basic Eng. **84**(1), 23–29 (1962). https://doi.org/10.1115/1.3657260
10. Fioriti, L.M.F., Hermanns, H.: Probabilistic termination: soundness, completeness, and compositionality. In: Proceedings of the 42nd Annual ACM SIGPLAN-SIGACT Symposium on Principles of Programming Languages, POPL, pp. 489–501 (2015). https://doi.org/10.1145/2676726.2677001

11. Kaminski, B.L., Katoen, J.-P.: On the hardness of almost–sure termination. In: Italiano, G.F., Pighizzini, G., Sannella, D.T. (eds.) MFCS 2015. LNCS, vol. 9234, pp. 307–318. Springer, Heidelberg (2015). https://doi.org/10.1007/978-3-662-48057-1_24

12. Kaminski, B.L., Katoen, J.-P., Matheja, C., Olmedo, F.: Weakest precondition reasoning for expected run–times of probabilistic programs. In: Thiemann, P. (ed.) ESOP 2016. LNCS, vol. 9632, pp. 364–389. Springer, Heidelberg (2016). https://doi.org/10.1007/978-3-662-49498-1_15

13. Majumdar, R., Sathiyanarayana, V.R.: Positive almost-sure termination: complexity and proof rules. Proc. ACM Program. Lang. **8**(POPL), 1089–1117 (2024). https://doi.org/10.1145/3632879

14. Piribauer, J., Baier, C.: Partial and conditional expectations in Markov decision processes with integer weights. In: Bojańczyk, M., Simpson, A. (eds.) FoSSaCS 2019. LNCS, vol. 11425, pp. 436–452. Springer, Cham (2019). https://doi.org/10.1007/978-3-030-17127-8_25

15. Piribauer, J., Baier, C.: Positivity-hardness results on Markov decision processes. TheoretiCS **3** (2024). https://doi.org/10.46298/THEORETICS.24.9

16. Puterman, M.L.: Markov Decision Processes: Discrete Stochastic Dynamic Programming. Wiley (1994)

17. Whittle, P.: Optimization Over Time, vol. 2. Wiley (1983)

18. Williams, D.: Probability with Martingales. Cambridge University Press, Cambridge (1991)

Probabilistic Model Checking: Applications and Trends

Marta Kwiatkowska[1] ⓘ, Gethin Norman[1,2] ⓘ, and David Parker[1,2(✉)] ⓘ

[1] Department of Computer Science, University of Oxford, Oxford, UK
{marta.kwiatkowska,david.parker}@cs.ox.ac.uk
[2] School of Computing Science, University of Glasgow, Glasgow, UK
gethin.norman@glasgow.ac.uk

Abstract. Probabilistic model checking is an approach to the formal modelling and analysis of stochastic systems. Over the past twenty five years, the number of different formalisms and techniques developed in this field has grown considerably, as has the range of problems to which it has been applied. In this paper, we identify the main application domains in which probabilistic model checking has proved valuable and discuss how these have evolved over time. We summarise the key strands of the underlying theory and technologies that have contributed to these advances, and highlight examples which illustrate the benefits that probabilistic model checking can bring. The aim is to inform potential users of these techniques and to guide future developments in the field.

1 Introduction

Probabilistic model checking [10] is a technique for the formal verification of stochastic systems. Properties to be verified are specified in temporal logic and then algorithmically checked against a model of the system. Some of the earliest work in the field dates from the 1980s [31,76], where algorithms were developed for computing the probability that linear temporal logic specifications are satisfied by sequential or concurrent probabilistic programs. The primary motivation was to establish the correctness of randomised algorithms, which had proved to be difficult, particularly in the context of concurrency.

Further verification techniques for these models, which are now typically referred to as discrete-time Markov chains (DTMCs) and Markov decision processes (MDPs), were soon developed. The now widely used temporal logic PCTL was proposed for DTMCs [42] and MDPs [18], expanding the range of properties that could be specified. Cited motivations included verifying the performance and reliability of, e.g., computer networks.

An extension of PCTL to the continuous-time setting was proposed in the late 1990s, called CSL [6]. This was designed for continuous-time Markov chains (CTMCs), a well established model for assessing performance and dependability properties of computing and communication systems. Efficient model checking algorithms were also developed [8]. The further integration of costs and rewards,

N. Bertrand et al. (Eds.): Christel Baier Festschrift, LNCS 15760, pp. 158–173, 2026.
https://doi.org/10.1007/978-3-031-97439-7_7

well-known notions from Markov modelling, into the probabilistic model checking framework led to additional temporal logics [44], allowing a modeller to reason about other quantitative characteristics such as power consumption. The flexibility of these formalisms meant that they applied equally well in non-traditional application domains such as biological systems.

For the model of MDPs, interest grew in synthesising "correct-by-construction" controllers or policies, using temporal logic to specify the desired behaviour of the system under control. Existing temporal logics and model checking algorithms were extended to incorporate various cost- and reward-based measures and multi-objective specifications [37]. This opened up new applications, including a variety of scheduling and planning problems. Probabilistic variants of the timed automata formalism, extending MDPs with clocks, allowed modelling of stochastic systems with real-time constraints and delays [64].

Fast forwarding to the present day, we observe a further significant expansion to the range of stochastic models for which temporal logics and model checking algorithms have been developed [57]: partially observable variants of MDPs, widely used in AI and planning, allow modelling of autonomous systems with unreliable sensors or of security protocols deploying secret keys; stochastic games provide reasoning about agents operating either competitively or collaboratively in probabilistic settings; and uncertain (or robust) Markov models formally capture epistemic uncertainty resulting, e.g., from data driven modelling.

To back up these advances in modelling formalisms and property specification languages, tool support has evolved and is now readily available. This comes in the form of both mature, general-purpose probabilistic model checkers, such as PRISM [56], Storm [47] or the Modest toolset [43], and various specialised verification tools [5]. A multitude of larger tool-chains and software frameworks have also been built, as probabilistic model checkers have become more robust and offer a variety of file formats and APIs with which to interface. The availability of this software, combined with the breadth of available formalisms, have seen probabilistic model checking used in a large and diverse range of applications, both within and outside computer science.

In this paper, we provide an overview of the broad range of problems to which probabilistic model checking has been applied, with a view to showcasing the current state of the field and giving some insight into its history. We identify a number of key areas in which the techniques have been particularly successful and consider how these have evolved over time, from the early work in the field to the present day. We also draw out the particular characteristics of the methods that make it well suited to each application area, and mention some of the key developments in theories and technologies that have been leveraged. Throughout, we highlight various case studies, illustrating the different ways in which probabilistic model checking can be deployed, the types of insight it can bring and the diversity of its usage. We hope that this paper therefore also serves as a guide to both current and potential users when selecting a formalism and analysis method for a given target application.

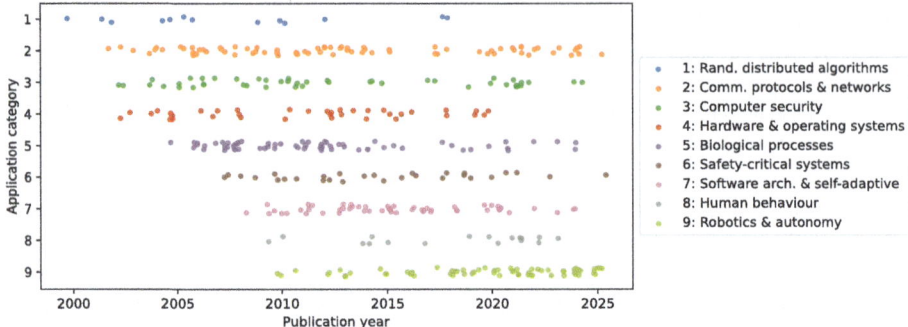

Fig. 1. Categorisation of approximately 400 application-oriented publications from the PRISM bibliography [81], illustrating how applications have evolved.

2 Applications of Probabilistic Model Checking

In the following sections we discuss, in loosely chronological order, a selection of key application areas for probabilistic model checking. For context, Fig. 1 presents a graphical illustration of how these areas have evolved over time. For this visualisation, we considered application-oriented publications from the online bibliography of the PRISM model checker [81], in which many of the examples mentioned in this paper can be found. We categorise them according to the areas discussed below and show the publications that appeared in each year.

2.1 Randomised Distributed Algorithms

Motivated by the success of existing model checking techniques in identifying bugs and unexpected behaviour in concurrent systems, a fruitful initial area of study for probabilistic verification was *randomised distributed algorithms*. In this setting, the use of randomisation is a crucial tool for breaking symmetry between concurrent processes, in order to establish correct termination or fairness between participants. However, the subtle interaction between the parallelisation of processes and their probabilistic behaviour makes it challenging to formally reason about their correctness or evaluate their efficiency.

Probabilistic model checking was used to analyse the correctness and efficiency of various randomised distributed algorithms, including for consensus, byzantine agreement, leader election and self-stabilisation [62]. MDPs (and the closely related model of probabilistic automata) are ideally suited to modelling the mixture of stochasticity and concurrency that arises here. In some cases (e.g., fully synchronous settings), DTMCs are also a useful model.

The scalability of verification techniques to these applications was enhanced significantly by the development of *symbolic* implementations of probabilistic model checking, based on (multi-terminal) binary decision diagrams [7,66]. While the size of the state spaces of the models grows rapidly, the high degree of regularity and the need for a mixture of both qualitative ("consensus is eventually established with probability 1") and quantitative ("what is the worst-case number of algorithm rounds required for termination?") properties made these applications well placed to benefit from the advantages of symbolic model checking.

In a notable example, in 2005, McIver and Morgan studied a self-stabilization algorithm due to Herman [48], establishing results on its worst-case runtime [60]. They also made a conjecture about the initial configurations of the network that would yield the worst-case expected time to achieve stabilisation, and validated this empirically through probabilistic model checking. However, a proof turned out to be elusive: a series of subsequent papers gradually established tighter bounds until the conjecture was finally proved more than 10 years later [19].

2.2 Communications and Networks

Communication Protocols. A natural evolution from early work on distributed algorithms was to verify *communication protocols*. Again, these often deploy randomisation to break symmetry, for example in combination with exponential back-off schemes to prevent collisions between data transmissions; the Ethernet protocol is a classic example. Moreover, particularly in wireless communication settings, it is desirable for a formal analysis of a protocol's reliability and efficiency to also incorporate a stochastic model of the underlying communication medium's unreliability, e.g., message delays or losses.

As above, both DTMCs and MDPs are useful for modelling communication protocols, with the latter in particular allowing an analysis of the *worst-case* expected performance or runtime over any possible concurrent scheduling. Verification efforts in this area, for example of the CSMA/CA (carrier-sense multiple access with collision avoidance), FireWire and Zigbee protocols, also benefited from the development of *probabilistic timed automata* (PTAs) [64], providing scalable modelling and verification techniques for analysing probabilistic systems with time-outs and deadlines. Models of randomised protocols with complex interactions but a degree of regularity also proved to be amenable to symbolic model checking. A good example is the verification of the Bluetooth protocol [34], which built and analysed multiple DTMC models with more than 10^{10} states in order to perform a detailed analysis of the impact of differences between variants of the protocol specification, and of the underlying model assumptions used.

Interest in this application area has continued to this day, with a wide variety of more modern wireless protocols being verified, for example, in the context of ad-hoc or vehicular networks. There has been particular interest in wireless sensor networks, see e.g., [24], where the small scale of the hardware deployed makes analysing unreliability particularly important.

Computer and Communication Networks. More broadly, probabilistic model checking has been used to formally evaluate the effectiveness of designs for communication and computer networks. PTA models, in particular, have been shown to be useful for quantifying the reliability or timeliness of transmissions across a network in the context of uncertain delays or message losses caused by individual network components. Examples include exploring the effectiveness of publish-subscribe messaging system [46], optimising the quality of networked automation systems for control problems [41] and comparing wireless token-passing schemes for networks in safety-critical systems [33].

This application domain also showcases the benefits of another important model class: *continuous-time Markov chains* (CTMCs). These provide an alternative model of stochastic systems in which the delays between each event are represented by negative exponential distributions. These can be very effective in modelling the timing of, for example, the rates with which jobs arrive or are serviced in a network queue, or with which failures occur in hardware components.

Outside formal verification, the well established fields of *performance evaluation* and *performability* use CTMCs to model and evaluate the performance and dependability of computer and communication networks (as well as others, such as manufacturing systems). The coming together of these field with probabilistic model checking [9] showed that temporal logics for CTMCs like CSL and its extensions offer a powerful means of formally specifying a wide range of important properties across both performance and reliability (e.g., "the probability of a server response taking more than 1 s is at most 0.02", "in the long-run, the availability of the network is at least 98%" or "the mean time to failure is at least 200 h"). Example applications include comparing the quality-of-service of alternative approaches to traffic shaping in wireless networks [14] and assessing the dependability of novel topologies for optical networks [72].

2.3 Computer Security

Computer security is another area where formal verification has long been an important tool, due to the need for strong guarantees on the resilience of systems to attack and the existence of unexpected weaknesses caused by subtleties in protocol designs. Probability is again a crucial modelling tool since the use of randomisation is widespread, e.g., for the generation of keys or session identifiers, or to prevent buffer overflows or DNS cache poisoning attacks.

It is natural to use MDPs for the analysis of security systems, with nondeterminism representing the potential actions of an adversary and probability used to model randomisation or other stochastic aspects. The work of [74] represents an early example of using MDP policy synthesis within probabilistic model checking. It generates optimal PIN block attacks, which are unexpected sequences of interactions with an API intended to determine the value of a PIN.

This approach can be generalised by using *stochastic games*, which provide formal modelling of the behaviour of multiple agents with differing objectives, such as an attacker and defender in a secure network. Applications of of stochastic game model checking in this area include performing threat analysis for real-world scenarios modelling using information from technical knowledge bases [75] and verifying the resilience of collective adaptive systems to attacks [40].

Other types of models are also widely used. For example, [2] uses CTMC model checking to study the well-known Kaminsky DNS cache-poisoning attack, performing a detailed analysis of the efficacy of randomisation-based fixes that have been proposed to counter the attack. Furthermore, in many security systems, there is an inherent trade-off between security and performance, in which case the expressivity and flexibility offered by probabilistic temporal logics is valuable. A good example is the work in [15], which performs a detailed analysis of a certified e-mail message delivery protocol, analysing in particular the relation between error rate and transmission cost.

2.4 Biological Processes

CTMCs have other important modelling applications beyond their usage for performance or reliability evaluation discussed above. A notable example is for modelling the stochastic dynamics of *biological processes*, from the level of cellular reactions, such as protein-protein interactions or gene expression, up to population-level models, for example of disease spreading.

This has proved to be a popular application of probabilistic model checking and has sparked adoption and collaboration with fields such as computational and systems biology. Many biological systems are relatively straightforward to represent in the PRISM modelling language and automated translations have been developed from custom languages [50] or process algebras [29].

As with other uses of CTMCs, the temporal logic CSL and its extensions are expressive enough to specify key properties of interest for biological systems (e.g., "what is the expected number of molecules of protein X after 5 min?" or "what is the probability that more than 10% of the population is infected within 2 weeks?"). However, logics such as LTL have also been used to characterise more complex temporal aspects such as oscillatory behaviour; see, e.g., [13].

From a computational perspective, verification tools implement various efficient CTMC solution methods. However, this area, in particular, has spurred the development of *statistical model checking* [79], which uses discrete-event simulation to provide approximate results to verification queries along with statistical guarantees as to their accuracy. This class of methods works particularly well when, as here, systems are modelled as Markov chains (i.e., without nondeterminism) and properties of interest are often over a finite time horizon.

In contrast to many other applications of probabilistic model checking, which feature human-engineered systems, the mechanisms underlying biological processes are often poorly understood, and these techniques provide a means to validate a hypothesised model. One example is an investigation of the impact of components of the fibroblast growth factor (FGF) signalling pathway [55], with the outputs of verification later validated against laboratory experiments. Other examples include an analysis of the efficacy of hyperthermia treatment [70], which corroborated prior beliefs as to the effectiveness of combining cancer treatment strategies, and a study of concentration-based navigation of sperm cells [53], again validating previous experimental observations. Similarly, probabilistic model checking can support *synthetic biology*, in which biological circuits are engineered; a good example here is DNA computing [58].

2.5   Safety-Critical Systems

Formal verification can be particularly beneficial for *safety-critical systems*, i.e., those where failures could result in death, serious injury or other significant damage. Probabilistic model checking can be used to rigorously quantify the likelihood with which high-risk failures or malfunctions could occur. This can be used for the assessment of *safety integrity levels* (SILs), which specify acceptable probabilities or rates of failure, and which underlie various technical standards that are either recommended or mandated for safety-critical systems.

Typically, systems are modelled as DTMCs or CTMCs, with the latter being more common if the model is being used to quantify the *rate* of failure over a specified time period, rather than the probability of a failure at any point. Safety specifications typically amount to relatively simple PCTL or CSL queries, although the analysis can still be expensive if probabilities are very low or the model comprises a large number of components. It is worth noting that probabilistic verification tools typically feature a variety of different methods for computing probabilities, trading off the accuracy required (which can be high in this setting) against computational cost [21].

An early example of the use of probabilistic model checking in this area was its integration into the *failure mode and effects* analysis of an airbag system [3]. Two contrasting designs were checked against relevant safety standards and the precise causes of safety violations were then identified. Connections have also been established between probabilistic model checking and *fault tree analysis* [77] and applied for example to reliability analysis of railway infrastructure [78] and safety analysis of vehicle guidance systems [39].

Other safety-critical applications of probabilistic model checking include those from *medical* and *aerospace* domains: verifying pacemaker designs by integrating probabilistic models of heart behaviour and quantifying the resulting probability of correct device operation [26]; analysing the reliability, availability and maintainability of satellite systems [49]; and verifying the reliability of spacecraft designs [28]. We also discuss below (in Sect. 2.7) other applications that may be safety-critical, such as autonomous driving and human-robot interaction.

2.6   Hardware and Operating Systems

Many of the instances of probabilistic verification surveyed above model systems at a relatively high level of abstraction, e.g., the interactions between participants in a protocol or components in a network. However, formal probabilistic modelling also provides value at the *hardware* or *operating system* level.

One application is to verify that hardware circuits function reliably even in the presence of failures of individual system components. This typically requires model checking of a DTMC in a fashion not dissimilar to that described above for quantifying risk in safety-critical systems. An early illustration of this was the verification of NAND multiplexing [63], a fault-tolerant design motivated by manufacturing defects found in the field of nanotechnology. Probabilistic model checking illustrated the impact of slight variations in component reliability and, in doing so, identified a flaw in an earlier analysis of the design.

A much broader range of system properties than just reliability can also be studied, including those relating to *timing, power usage* and *energy consumption*. This makes probabilistic model checking a powerful tool to explore design spaces and investigate trade-offs between competing characteristics. For example, more recently, work in [68] presented a detailed illustration of the integration of probabilistic model checking into the early design process of circuit designs using reconfigurable transistors. They computed guarantees on delay, power dissipation and energy consumption per operation, including various properties that would be difficult to analyse with simulation.

A similar spectrum of quantitative properties can be analysed in operating systems applications. For example, [71] uses MDPs for thermal modelling of multi-core systems and then analyses performance-reliability trade-offs. This is done using probabilistic model checking with *energy-utility quantiles* [11,12], which incorporate conditional probabilities and ratio constraints between cost and reward measures to investigate the interplay between multiple objectives.

2.7   Robotics and Autonomous Systems

In recent years, a clear trend within the use of probabilistic model checking has been a shift towards synthesising "correct-by-construction" controllers or policies for a system, based on a formal specification of its desired behaviour. Although this differs from the more classical approach of verifying a fixed system model to determine whether it satisfies a specification, probabilistic model checking is already well suited to this task, notably through the use of MDPs, with temporal logic used for controller specifications. In parallel, there has been a growing synergy between verification and fields of artificial intelligence such as planning and reinforcement learning, which often also use MDPs.

An application domain where these shifts have been particularly apparent is *robotics and autonomous systems*. The use of stochastic modelling is essential here since robots frequently operate in uncertain and dynamic environments, due to the presence of humans or other unknown obstacles, and imperfect or unreliable sensors and actuators. Probabilistic model checking offers a wealth of ways to formally specify desired behaviour in this setting, and the means to provide a probabilistic guarantee on the behaviour of generated policies. These can be critical for safety purposes, e.g., for robots operating in the presence of humans or high risk environments, or to guarantee performance and reliability sometimes under tight resource (e.g., battery) constraints.

The use of linear temporal logic (LTL) is common here, in order to capture more complex temporal specifications (e.g., "maximise the probability of inspecting all three sensors, in any order, whilst avoiding X"). This is often combined with multi-objective model checking [37], which allows the generation of policies that trade-off between several, competing objectives (e.g., battery life vs. mission execution time) or analysis of the corresponding Pareto front.

Robotics provides good examples of probabilistic model checking being embedded in real-world systems. For example, [45] describes a major *long-term autonomy* project deploying mobile robots in everyday environments such as offices and care homes. Multi-objective model checking on MDPs is repeatedly applied within the robots' control software to perform tasks effectively and reliably as the map of the environment is updated over time. In other recent work, probabilistic model checking is used within a planning system for *autonomous underwater vehicles* (AUVs) to retrieve data from sensor networks [20]. The resulting policies are deployed in a real-world trial and shown to outperform existing hand-designed policies.

Robotics and autonomous systems applications also highlight further challenges arising in verification today. One is the integration of components that deploy *machine learning*. In [22], a probabilistic model checking framework is proposed that incorporates deep-learning perception techniques. The uncertainty arising from the learning-driven components is factored into the verification of the overall system; this is applied to controllers for mobile robots and a driver-attentiveness management system for shared-control autonomous cars.

2.8   Software Architectures and Self-Adaptive Systems

The performance and reliability of *software architectures* has also been investigated quite extensively with probabilistic model checking. In similar fashion to other applications already discussed, such as computer and communication networks or hardware designs, formal modelling and analysis of Markov chains can be used to rigorously establish how the likelihood of failures or delays in individual components impacts the overall reliability and performance of a complex system of components. In an early illustration of this, researchers from ABB used DTMC models to evaluate the reliability of a large-scale industrial control system with more than 100 components [52].

In recent years, there has been particular interest in the modelling and verification of *cloud computing* services, where there is a need to maintain high Quality-of-Service levels, even under variable and unpredictable workloads. Moreover, it is often essential to meet precisely specified *service level agreements*, relating to response time or availability, which can be conveniently expressed and verified using probabilistic model checking; see e.g., [54]. Perhaps more importantly, these techniques can also support run-time decision making in cloud systems such as for auto-scaling (adjusting resource levels to meet demand) and load balancing. As discussed for autonomous systems above, MDP-based model checking provides an effective way to do this, particular when reasoning about trade-offs between different objectives or metrics; see e.g., [65].

More generally, *self-adaptive* systems, which monitor their environment and adapt their behaviour accordingly at run-time, have also proved to be a popular application domain for these techniques [61]. In fact, a variety of more advanced methods have been used effectively here. This includes model checking of *stochastic games*, which have been used to model worst-case assumptions about the environment of the cloud system [23]. Another example is the use of *parametric model checking* [51], which can in this context be used to apply a form of sensitivity analysis, quantifying the impact that unknown system parameters have on overall performance metrics [1]. Lastly, methods for *families* of probabilistic models [27] have been used to efficiently verify multiple configurations supported by a self-adaptive system [67].

2.9   Human Behaviour Modelling

The modelling and analysis of *human behaviour* represents another challenge for formal verification, and one where probabilistic techniques play a key role. An example is the evaluation of diabetes patients' behaviour when using insulin pumps [25]. Machine learning is used to extract representative patient behavioural patterns from a clinical dataset, and probabilistic model checking is deployed to analyse how different behavioral patterns impact an individual's glucose physiology. The results demonstrate that switching behaviour types can significantly improve a patient's glycemic control outcomes, boosting the effectiveness of diabetes patient education and peer support. The work of [80] uses DTMCs and probabilistic model checking, together with a cognitive reliability and error analysis method to transform expert estimates of relevant environmental and cognitive factors into human error rates, to assess the reliability of the procedures of a pharmacy and also analyse the effects of potential alternatives.

Another interesting application is the analysis of user interactive systems [4]. The approach is based on first inferring DTMC models of users' activity patterns by applying machine learning to logged user traces. Probabilistic model checking is then applied to the DTMCs to express hypotheses about user behaviour and relationships within and between the activity patterns. The approach has been applied to a real-world case study of a deployed app with thousands of users and the analysis performed revealed insights into real-life app usage.

Other applications to human modelling include air traffic control [69], controller synthesis of UAVs interacting with human operators [38], human in the loop self-adaptive systems [59], and the design of correct-by-construction Advanced Driver Assistance Systems (ADAS) interacting with a human driver model built using the cognitive architecture ACT-R [35].

2.10   Further Application Areas

In the above, we have identified some of the most well-studied categories of applications for probabilistic model checking. This is of course non-exhaustive, and various other applications have been considered multiple times. These include: *smart grids*, with a focus on performance, resilience or security; *quantum computing*, for example, key distribution algorithms; *blockchain*-based systems, such as bitcoin; and *business process* modelling. See [81] for details.

We conclude this discussion by highlighting a few of the more diverse applications that have appeared in recent years, adding support for our argument that the techniques have very broad applicability.

Sports tactics. One example is the work in [30], which develops a framework for reasoning about effectiveness of team strategies in professional football. The first phase is to learn an MDP model, from event stream data, capturing the probability of moving between areas of the pitch and executing actions, such as passes or shots. Then probabilistic model checking is applied, with LTL used to express outcomes of interest. The analysis yields insights, such as when and where passing and shooting is more effective.

Court Interactions. Another example is an analysis of interactions between participants in cases in the US Supreme Court [32]. Based on a dataset of court transcripts, a DTMC model is constructed that models the dynamics of interactions between people over the course of a trial. Probabilistic model checking is then used to analyse a wide range of properties, from the expected timing of court processes, to the likelihood of various events, such as the decisions taken by judges, and various trends are identified.

Other interesting applications include estimating political affinities through opinion diffusion on Twitter [73], studying the impact of regulations on the performance of public transport [17], planning railway infrastructure through capacity evaluation [36] and assessing graphical user interfaces [16].

3   Conclusions and Outlook

We have taken a retrospective look at some of the successful applications of probabilistic model checking over the last 25 years, identifying commonly studied application domains and discussing why and how they are amenable to this approach. Commonalities exist between the different problem domains, but there is considerable breadth and diversity to the set of applications.

Broadly speaking, we observe that this is largely because probabilistic model checking offers ease of rigorous modelling for many different types of stochastic behaviour, combined with extremely flexible logical formalisms to specify their quantitative characteristics. These go well beyond the classical notions of correctness typically used in formal verification. This is backed up by a wide range of flexible analysis techniques and stable tool support. These tools are continually evolving, incorporating new advances in modelling and verification techniques, which we anticipate leading to further application domains in the future.

Acknowledgements. This project was funded by the ERC under the European Union's Horizon 2020 research and innovation programme (FUN2MODEL, grant agreement No. 834115).

References

1. Alasmari, N., Calinescu, R., Paterson, C., Mirandola, R.: Quantitative verification with adaptive uncertainty reduction. J. Syst. Softw. **188**, 111275 (2022)

2. Alexiou, N., Basagiannis, S., Katsaros, P., Deshpande, T., Smolka, S.A.: Formal analysis of the Kaminsky DNS cache-poisoning attack using probabilistic model checking. In: Proceedings 9th IEEE International Symposim on High Assurance Systems Engineering (HASE 2010), pp. 94–103. IEEE Computer Society (2010)

3. Aljazzar, H., Fischer, M., Grunske, L., Kuntz, M., Leitner, F., Leue, S.: Safety analysis of an airbag system using probabilistic FMEA and probabilistic counter examples. In: Proceedings 6th International Conference on Quantitative Evaluation of Systems (QEST 2009), pp. 299–308. IEEE Computer Society (2009)

4. Andrei, O., Calder, M., Higgs, M., Girolami, M.: Probabilistic model checking of DTMC models of user activity patterns. In: Proceedings of the 11th International Conference on Quantitative Evaluation of Systems (QEST 2014). LNCS, vol. 8657, pp. 138–153. Springer (2014)

5. Andriushchenko, R., et al.: Tools at the frontiers of quantitative verification. In: TOOLympics Challenge 2023, pp. 90–146. Springer (2024)

6. Aziz, A., Sanwal, K., Singhal, V., Brayton, R.: Verifying continuous time Markov chains. In: Proceedings of the 8th International Conference on Computer Aided Verification (CAV 1996). LNCS, vol. 1102, pp. 269–276. Springer (1996)

7. Baier, C., Clarke, E., Hartonas-Garmhausen, V., Kwiatkowska, M., Ryan, M.: Symbolic model checking for probabilistic processes. In: Proc. 24th International Colloquium on Automata, Languages and Programming (ICALP 1997). LNCS, vol. 1256, pp. 430–440. Springer (1997)

8. Baier, C., Haverkort, B., Hermanns, H., Katoen, J.P.: Model-checking algorithms for continuous-time Markov chains. IEEE Trans. Softw. Eng. **29**(6), 524–541 (2003)

9. Baier, C., Haverkort, B., Hermanns, H., Katoen, J.P.: Performance evaluation and model checking join forces. Commun. ACM **53**(9), 76–85 (2010)

10. Baier, C., de Alfaro, L., Forejt, V., Kwiatkowska, M.: Model checking probabilistic systems. In: Handbook of Model Checking, pp. 963–999. Springer (2018)

11. Baier, C., Daum, M., Dubslaff, C., Klein, J., Klüppelholz, S.: Energy-utility quantiles. In: NASA Formal Methods. LNCS, vol. 8430, pp. 285–299. Springer (2014)

12. Baier, C., Dubslaff, C., Klüppelholz, S.: Trade-off analysis meets probabilistic model checking. In: Proc. 23rd International Conference on Computer Science Logic and the 29th Symposium on Logic In Computer Science (CSL-LICS 2014), pp. 1:1–1:10. ACM (2014)

13. Ballarini, P., Guerriero, M.: Query-based verification of qualitative trends and oscillations in biochemical systems. Theoret. Comput. Sci. **411**(20), 2019–2036 (2010)

14. Ballarini, P., Ben-Othman, J., Mokdad, L.: Quantitative verification of WiMAX traffic shaping solutions. In: Proceedings of the 7th International Symposium on Intelligent System Techniques for Ad hoc and Wireless Sensor Networks (IST-AWSN) (2012)

15. Basagiannis, S., Petridou, S.G., Alexiou, N., Papadimitriou, G.I., Katsaros, P.: Quantitative analysis of a certified e-mail protocol in mobile environments: a probabilistic model checking approach. Comput. Secur. **30**(4), 257–272 (2011)

16. Bertolini, C., Mota, A.: Using probabilistic model checking to evaluate GUI testing techniques. In: SEFM, pp. 115–124. IEEE Computer Society (2009)

17. Bertrand, N., Bordais, B., Hélouët, L., Mari, T., Parreaux, J., Sankur, O.: Performance evaluation of metro regulations using probabilistic model-checking. In: RSSRail. LNCS, vol. 11495, pp. 59–76. Springer (2019)

18. Bianco, A., de Alfaro, L.: Model checking of probabilistic and nondeterministic systems. In: Proceedings of the 15th Conference on Foundations of Software Tech-

nology and Theoretical Computer Science (FSTTCS 1995). LNCS, vol. 1026, pp. 499–513. Springer (1995)

19. Bruna, M., Grigore, R., Kiefer, S., Ouaknine, J., Worrell, J.: Proving the Herman-protocol conjecture. In: Proceedings of the 43rd International Colloquium on Automata, Languages, and Programming (ICALP 2016). LIPIcs, vol. 55, pp. 104:1–104:12 (2016)

20. Budd, M., et al.: Probabilistic planning for AUV data harvesting from smart underwater sensor networks. In: Proceedings of the IEEE/RSJ International Conference on Intelligent Robots and Systems (IROS 2022), pp. 12051–12057 (2022)

21. Budde, C.E., et al.: On correctness, precision, and performance in quantitative verification: QComp 2020 competition report. In: Proceedings of the 9th International Symposium on Leveraging Applications of Formal Methods, Verification and Validation (ISoLA 2020). LNCS, vol. 12479, pp. 216–241. Springer (2020)

22. Calinescu, R., et al.: Controller synthesis for autonomous systems with deep-learning perception components. IEEE Trans. Softw. Eng. **50**(6) (2024)

23. Cámara, J., Garlan, D., Schmerl, B.R., Pandey, A.: Optimal planning for architecture-based self-adaptation via model checking of stochastic games. In: Proceedings of the 30th ACM Symposium on Applied Computing (SAC 2015), pp. 428–435. ACM (2015)

24. Chen, M., Mokdad, L., Ben-Othman, J., Fourneau, J.M.: Probabilistic performance evaluation of the class-A device in LoRaWAN protocol on the MAC layer. Perform. Eval. **166**(C) (2024)

25. Chen, S., Feng, L., Rickels, M.R., Peleckis, A., Sokolsky, O., Lee, I.: A data-driven behavior modeling and analysis framework for diabetic patients on insulin pumps. In: International Conference on Healthcare Informatics, pp. 213–222 (2015)

26. Chen, T., Diciolla, M., Kwiatkowska, M.Z., Mereacre, A.: Quantitative verification of implantable cardiac pacemakers. In: Proceedings of the 33rd IEEE Real-Time Systems Symposium (RTSS 2012), pp. 263–272 (2012)

27. Chrszon, P., Dubslaff, C., Klüppelholz, S., Baier, C.: ProFeat: feature-oriented engineering for family-based probabilistic model checking. Formal Aspects Comput. **30**(1), 45–75 (2018)

28. Chrszon, P., et al.: Model checking of spacecraft operational designs: a scalability analysis. Softw. Syst. Model. (2025)

29. Ciocchetta, F., Hillston, J.: Bio-PEPA for epidemiological models. In: Proceedings of the 4th International Workshop on Practical Applications of Stochastic Modelling (PASM 2009). ENTCS, vol. 261, pp. 43–69. Elsevier (2008)

30. Clijmans, J., Roy, M., Davis, J.: Looking beyond the past: Analyzing the intrinsic playing style of soccer teams. In: Machine Learning and Knowledge Discovery in Databases. LNCS, vol. 13718, pp. 370–385 (2023)

31. Courcoubetis, C., Yannakakis, M.: Verifying temporal properties of finite state probabilistic programs. In: Proceedings of the 29th Symposium on Foundations of Computer Science (FOCS 1988), pp. 338–345. IEEE CS Press (1988)

32. Das, S., Sharma, A.: Probabilistic model checking of temporal interaction dynamics in the supreme court. In: Proceedings of the 11th International Symposium From Data to Models and Back (DataMod 2023) (2023)

33. Dombrowski, C., Junges, S., Katoen, J.P., Gross, J.: Model-checking assisted protocol design for ultra-reliable low-latency wireless networks. In: Proceedings of the IEEE 35th Symposium on Reliable Distributed Systems (SRDS 2016), pp. 307–316 (2016)

34. Duflot, M., Kwiatkowska, M., Norman, G., Parker, D.: A formal analysis of Bluetooth device discovery. Int. J. Softw. Tools Technol. Transf. **8**(6), 621–632 (2006)

35. Eiras, F., Lahijanian, M., Kwiatkowska, M.: Correct-by-construction advanced driver assistance systems based on a cognitive architecture. In: Proceedings of the IEEE 2nd Connected and Automated Vehicles Symposium (CAVS 2019), pp. 1–7 (2019)

36. Emunds, T., Nießen, N.: Evaluating railway junction infrastructure: a queueing-based, timetable-independent analysis. Transp. Res. Part C: Emerg. Technol. **165** (2024)

37. Etessami, K., Kwiatkowska, M., Vardi, M., Yannakakis, M.: Multi-objective model checking of Markov decision processes. Log. Methods Comput. Sci. **4**(4), 1–21 (2008)

38. Feng, L., Wiltsche, C., Humphrey, L., Topcu, U.: Controller synthesis for autonomous systems interacting with human operators. In: Proceedings of the ACM/IEEE 6th International Conference on Cyber-Physical Systems (ICCPS 2015), pp. 70–79. ACM (2015)

39. Ghadhab, M., Junges, S., Katoen, J.-P., Kuntz, M., Volk, M.: Model-based safety analysis for vehicle guidance systems. In: Proceedings of the 36th International Conference on Computer Safety, Reliability and Security (SAFECOMP 2017). LNCS, vol. 10488, pp. 3–19. Springer (2017)

40. Glazier, T.J., Cámara, J., Schmerl, B.R., Garlan, D.: Analyzing resilience properties of different topologies of collective adaptive systems. In: SASO Workshops, pp. 55–60. IEEE Computer Society (2015)

41. Greifeneder, J., Frey, J.: Optimizing quality of control in networked automation systems using probabilistic models. In: Proceedings of the 11th IEEE International Conference on Emerging Technologies and Factory Automation (ETFA 2006), pp. 372–379 (2006)

42. Hansson, H., Jonsson, B.: A logic for reasoning about time and reliability. Formal Aspects Comput. **6**(5), 512–535 (1994)

43. Hartmanns, A., Hermanns, H.: The Modest Toolset: An Integrated Environment for Quantitative Modelling and Verification. In: Proceedings of the 20th International Conference on Tools and Algorithms for the Construction and Analysis of Systems (TACAS 2014). LNCS, vol. 8413, pp. 593–598. Springer (2014)

44. Haverkort, B., Cloth, L., Hermanns, H., Katoen, J.P., Baier, C.: Model checking performability properties. In: Proceedings of the International Conference on Dependable Systems and Networks (DSN 2002). IEEE CS Press (2002)

45. Hawes, N., et al.: The STRANDS project: long-term autonomy in everyday environments. IEEE Robot. Autom. Mag. **24**(3), 146–156 (2017)

46. He, F., Baresi, L., Ghezzi, C., Spoletini, P.: Formal analysis of publish-subscribe systems by probabilistic timed automata. In: Proceedings of the Formal Techniques for Networked and Distributed Systems (FORTE 2007). LNCS, vol. 4574, pp. 247–262. Springer (2007)

47. Hensel, C., Junges, S., Katoen, J.P., Quatmann, T., Volk, M.: The probabilistic model checker Storm. Int. J. Softw. Tools Technol. Transf. **24**(4), 589–610 (2022)

48. Herman, T.: Probabilistic self-stabilization. Inf. Process. Lett. **35**(2), 63–67 (1990)

49. Hoque, K.A., Mohamed, O.A., Savaria, Y.: Towards an accurate reliability, availability and maintainability analysis approach for satellite systems based on probabilistic model checking. In: Proceedings of the Design, Automation & Test in Europe Conference & Exhibition (DATE 2015), pp. 1635–1640. ACM (2015)

50. Hucka, M., et al.: The systems biology markup language (SBML): a medium for representation and exchange of biochemical network models. Bioinformatics **1**(19), 524–531 (2003)

51. Junges, S., et al.: Parameter synthesis for Markov models: covering the parameter space. Formal Methods Syst. Design **62**(1), 181–259 (2024)
52. Koziolek, H., Schlich, B., Bilich, C.: A large-scale industrial case study on architecture-based software reliability analysis. In: Proceedings of the IEEE 21st International Symposium on Software Reliability Engineering (ISSRE 2010), pp. 279–288 (2010)
53. Kromer, J.A., Märcker, S., Lange, S., Baier, C., Friedrich, B.M.: Decision making improves sperm chemotaxis in the presence of noise. PLoS Comput. Biol. **14**(4) (2018)
54. Krotsiani, M., Kloukinas, C., Spanoudakis, G.: Validation of service level agreements using probabilistic model checking. In: Proceedings of the 14th IEEE International Conference on Services Computing (SCC 2017), pp. 148–155 (2017)
55. Kwiatkowska, M., Heath, J.: Biological pathways as communicating computer systems. J. Cell Sci. **122**(16), 2793–2800 (2009)
56. Kwiatkowska, M., Norman, G., Parker, D.: PRISM 4.0: Verification of Probabilistic Real-Time Systems. In: Proceedings of the 23rd International Conference on Computer Aided Verification (CAV 2011). LNCS, vol. 6806, pp. 585–591. Springer (2011)
57. Kwiatkowska, M., Norman, G., Parker, D.: Probabilistic model checking and autonomy. Annu. Rev. Control Robot. Auton. Syst. **5**, 385–410 (2022)
58. Lakin, M., Parker, D., Cardelli, L., Kwiatkowska, M., Phillips, A.: Design and analysis of DNA strand displacement devices using probabilistic model checking. J. R. Soc. Interface **9**(72), 1470–1485 (2012)
59. Li, N., Adepu, S., Kang, E., Garlan, D.: Explanations for human-on-the-loop: a probabilistic model checking approach. In: Proceedings of the IEEE/ACM 15th International Symposium on Software Engineering for Adaptive and Self-Managing Systems (SEAMS 2020), pp. 181–187. ACM (2020)
60. McIver, A., Morgan, C.: An elementary proof that Herman's ring is $\Theta(N^2)$. Inf. Process. Lett. **94**(2), 79–84 (2005)
61. Moreno, G.A., Cámara, J., Garlan, D., Schmerl, B.R.: Proactive self-adaptation under uncertainty: a probabilistic model checking approach. In: Proceedings of the 10th Joint Meeting on Foundations of Software Engineering (FSE 2015), pp. 1–12. ACM (2015)
62. Norman, G.: Analysing randomized distributed algorithms. In: Validation of Stochastic Systems: A Guide to Current Research. LNCS (Tutorial Volume), vol. 2925, pp. 384–418. Springer (2004)
63. Norman, G., Parker, D., Kwiatkowska, M., Shukla, S.: Evaluating the reliability of NAND multiplexing with PRISM. IEEE Trans. Comput. Aided Des. Integr. Circuits Syst. **24**(10), 1629–1637 (2005)
64. Norman, G., Parker, D., Sproston, J.: Model checking for probabilistic timed automata. Formal Methods Syst. Design **43**(2), 164–190 (2013)
65. Nuraishah, S., Jawaddi, A., Ismail, A., Sulaiman, M.S., Cardellini, V.: Analyzing energy-efficient and Kubernetes-based autoscaling of microservices using probabilistic model checking. J. Grid Comput. **23**(3) (2024)
66. Parker, D.: Implementation of symbolic model checking for probabilistic systems. Ph.D. thesis, University of Birmingham (2002)
67. Päßler, J., ter Beek, M.H., Damiani, F., Johnsen, E.B., Tapia Tarifa, S.L.: Analysing self-adaptive systems as software product lines. J. Syst. Softw. **222**, 112324 (2025)
68. Raitza, M., et al.: Quantitative characterization of reconfigurable transistor logic gates. IEEE Access **8**, 112598–112614 (2020)

69. Rungta, N., et al.: Aviation safety: modeling and analyzing complex interactions between humans and automated systems. In: Proceedings of the 3rd International Conference on Application and Theory of Automation in Command and Control Systems (ATACCS 2013), pp. 27–37. ACM (2013)
70. Rybinski, M., Szymanska, Z., Lasota, S., Gambin, A.: Modelling the efficacy of hyperthermia treatment. J. Roy. Soc. Interface **10**(88) (2013)
71. Sardar, M.U., Dubslaff, C., Klüppelholz, S., Baier, C., Kumar, A.: Performance evaluation of thermal-constrained scheduling strategies in multi-core systems. In: Proceedings of the 16th European Workshop on Computer Performance Engineering (EPEW 2019). LNCS, vol. 12039, pp. 133–147. Springer (2019)
72. Siddique, U., Hoque, K.A., Johnson, T.T.: Formal specification and dependability analysis of optical communication networks. In: Proceedings of the Design, Automation & Test in Europe Conference & Exhibition (DATE 2017) (2017)
73. Stamatelatos, G., Gyftopoulos, S., Drosatos, G., Efraimidis, P.S.: Revealing the political affinity of online entities through their twitter followers. Inf. Process. Manage. **57**(2), 102172 (2020)
74. Steel, G.: Formal analysis of PIN block attacks. Theoret. Comput. Sci. **367**(1–2), 257–270 (2006)
75. Tavolato, P., Luh, R., Eresheim, S.: Formalizing real-world threat scenarios. In: Proceedings of the 8th International Conference on Information Systems Security and Privacy (ICISSP 2022), pp. 281–289. SciTePress (2022)
76. Vardi, M.: Automatic verification of probabilistic concurrent finite state programs. In: Proceedings of the 26th Symposium on Foundations of Computer Science (FOCS 1985), pp. 327–338. IEEE CS Press (1985)
77. Volk, M., Sher, F., Katoen, J.P., Stoelinga, M.: SAFEST: fault tree analysis via probabilistic model checking. In: Proceedings of the 70th Reliability and Maintainability Symposium (RAMS 2024), pp. 1–7 (2024)
78. Weik, N., Volk, M., Katoen, J., Nießen, N.: DFT modeling approach for operational risk assessment of railway infrastructure. Int. J. Softw. Tools Technol. Transf. **24**(3), 331–350 (2022)
79. Younes, H., Simmons, R.: Probabilistic verification of discrete event systems using acceptance sampling. In: Proceedings of the 14th International Conference on Computer Aided Verification (CAV 2002). LNCS, vol. 2404. Springer (2002)
80. Zheng, X., Bolton, M.L., Daly, C., Biltekoff, E.: The development of a next-generation human reliability analysis: systems analysis for formal pharmaceutical human reliability (SAFPH). Reliab. Eng. Syst. Saf. **202** (2020)
81. PRISM bibliography. prismmodelchecker.org/bib.php

Probabilistic Counterexamples Through the Ages

Erika Ábrahám[1]([✉]), Nils Jansen[2,3], and Simon Jantsch[4]

[1] RWTH Aachen University, Aachen, Germany
abraham@cs.rwth-aachen.de
[2] Ruhr-University Bochum, Bochum, Germany
[3] Radboud University, Nijmegen, The Netherlands
[4] Siemens EDA, Munich, Germany

Abstract. Markov models are ubiquitous in decision-making and reasoning under uncertainty. A major branch of the formal verification for such models is probabilistic model checking. Numerous techniques and tools exist, and Christel Baier has been at the forefront of this area for decades. A particular branch of probabilistic model checking is the generation of counterexamples or explanations. In this paper, we reflect on some milestones in this area and highlight Christel's contribution to shed light on the often opaque model checking results that are limited to numerical values or mere "yes" or "no" answers.

Keywords: Markov chains · Markov decision processes · Model checking · Counterexamples · Witnesses

1 Introduction

We dedicate this paper to Christel Baier, on the occasion of her 60th birthday. All who have the pleasure to know her would confirm that Christel is the queen of formal methods for probabilistic systems. It is impossible to recall her life-work in a single paper, thus we will focus on some aspects that connect our scientific interests: *the definition and generation of counterexamples, witnesses and certificates for probabilistic properties.*

Counterexamples are *the* killer-feature of classical model checking applications in the real world. The ability to generate explanations for why a property fails is invaluable. Explanations should be as concise as possible and focussed on the actual root of the violation. Since no single trace can (usually) explain why a property holds in probabilistic model checking, finding useful explanations is a challenge.

The most direct extension of classical counterexamples to the probabilistic setting is to consider *sets of traces* as counterexamples [13]. Such sets may, however, become extremely large and even infinite, and therefore can be hard to interpret or represent concisely. This triggered research into compact representations, for example, using regular expressions [9]. Another natural way of

© The Author(s), under exclusive license to Springer Nature Switzerland AG 2026
N. Bertrand et al. (Eds.): Christel Baier Festschrift, LNCS 15760, pp. 174–196, 2026.
https://doi.org/10.1007/978-3-031-97439-7_8

compactly representing sets of traces is to use *subsystems* of the system in question [29]. When a subsystem is used to demonstrate that a property holds in a system, we call it a *witnessing subsystem* (or simply a witness). Vice versa, a subsystem that violates a property proves that the property is violated in the original system and, therefore, serves as a counterexample. Such subsystems are called *critical* for the property and were introduced in the DFG project *Counter Example Generation for Stochastic Systems using Bounded Model Checking (CEBug)*. First, a number of approaches exploited the previously mentioned path-based counterexamples to form such critical subsystems [15]. These approaches were extended to symbolic state space representations using decision diagrams [18]. Finally, solving techniques such as sat-modulo-theories (SMT) or mixed-integer linear programming (MILP) provide means to compute *miminal* critical subsystems [24, 30]. The corresponding decision problem is NP-complete, even for acyclic Markov chains and Markov chains with low path-width [12, 21]. Many of the approaches were implemented in the COMICS tool [16]. An overview of these results can be found in [1].

In contrast to classical counterexamples, the aim of *certificates* is not to provide an explanation that is easy to interpret by humans [23]. Rather, a certificate should allow for an independent algorithm, the *certificate checker*, to validate that the result of the computation is correct. Of course, validating the certificate must be considerably easier than computing the answer in the first place. Therefore, a witnessing or critical subsystem is not a certificate: determining whether the property holds in the subsystem is as hard as solving the original verification problem in general.

One approach to implement certifying algorithms for probabilistic model checking uses so called *Farkas certificates* [12, 19]. They can certify all types of lower and upper bounds on optimal reachability probabilities in MDPs. Farkas certificates are solutions of linear real-arithmetic constraint systems derived from variants of the standard linear programs for optimal reachability probabilities in MDPs, based on Farkas' Lemma [11]. Interestingly, Farkas certificates induce witnessing subsystems for the same property, and this observation was used to devise new algorithms for computing small and minimal witnesses. These approaches are implemented in the tool SWITSS [20]. A related definition of certificates for probabilistic model checking are *fixed point certificates* [8]. This work also discusses a formally verified certificate checker and efficient computation methods based on value- and policy-iteration.

In this paper, we give a high-level overview of these different approaches to compute critical (resp. witnessing) subsystems for MDPs, and related certification methods. For Farkas certificates, we highlight some results of the PhD-thesis [19] which were previously unpublished elsewhere. In particular, we present the extension of Farkas certificates [12] to MDPs with non-trivial end components, certifying (non-)existence of end components, and a new method for computing minimal witnesses for invariants and Rabin conditions. The construction for Rabin conditions can be used to compute minimal and small witnessing subsystems for LTL properties. Furthermore, we compare Farkas certificates to the newly proposed *fixed point certificates* [8].

Outline. We start in Sect. 2 with some preliminaries, which we illustrate on a running example called the Christel MDP. We introduce critical subsystems in Sect. 3 and give an informal presentation of our work on minimal critical subsystems in Sect. 4. We present a deeper and more formal presentation of Christel's work on certifying algorithms and the novel extensions in the Sects. 5 and 6. Finally, we conclude the paper in Sect. 7.

2   Preliminaries and the Christel MDP

Discrete-Time Markov Models. Given a non-empty countable set S, a *distribution over* S is a function $\iota : S \to [0,1]$ satisfying $\sum_{s \in S} \iota(s) = 1$. We call $\mathrm{supp}(\iota) = \{s \in S \mid \iota(s) > 0\}$ the *support* of ι, and name ι a *Dirac distribution* if $|\mathrm{supp}(\iota)| = 1$. We denote by $\mathrm{Dist}(S)$ the set of all distributions over S.

A *Markov decision process (MDP)* is a tuple $\mathcal{M} = (S, \mathrm{Act}, s_{in}, P)$, where S is a non-empty countable set of *states* and Act is a function mapping each state to a finite, non-empty set of *actions* $\mathrm{Act}(s)$. With a slight abuse of notation, we also denote by Act the set of all actions $\bigcup_{s \in S} \mathrm{Act}(s)$; it should always be clear from the context which one is meant. The state $s_{in} \in S$ is the *initial state* and $P \colon S \times \mathrm{Act} \times S \to [0,1]$ is a *transition probability function*. We require P to satisfy $\sum_{s' \in S} P(s, \alpha, s') = 1$ for each s and $\alpha \in \mathrm{Act}(s)$, and $P(s, \alpha, s') = 0$ whenever $\alpha \notin \mathrm{Act}(s)$. While we assume Markov models with a single initial state, all results can be extended to initial distributions. A *sub-stochastic* MDP is defined in the same way, with the exception that P must satisfy $\sum_{s' \in S} P(s, \alpha, s') \leq 1$ for each $\alpha \in \mathrm{Act}(s)$.

An *infinite path* of $\mathcal{M}$ is an infinite sequence $s_1 \alpha_1 s_2 \alpha_2 \ldots \in (S \times \mathrm{Act})^\omega$ such that $P(s_i, \alpha_i, s_{i+1}) > 0$ for all $i \geq 1$. A *finite path* is a non-empty finite prefix $s_1 \alpha_1 \ldots s_n \in (S \times \mathrm{Act})^* S$ of an infinite path; we refer to the last state of a finite path π as $\mathrm{last}(\pi)$. We denote the sets of all infinite (respectively finite) paths of $\mathcal{M}$ that start in s_{in} by $\mathrm{Paths}(\mathcal{M})$ (respectively $\mathrm{Paths}_{\mathrm{fin}}(\mathcal{M})$).

Borrowing notation from temporal logic, for any set of states $R \subseteq S$, we define $\lozenge R = \{s_1 \alpha_1 s_2 \alpha_2 \ldots \in \mathrm{Paths}(\mathcal{M}) \mid s_i \in R \text{ for some } i \geq 1\}$ and $\square R = \{s_1 \alpha_1 s_2 \alpha_2 \ldots \in \mathrm{Paths}(\mathcal{M}) \mid s_i \in R \text{ for all } i \geq 1\}$. We will also use $\neg R = S \setminus R$, and if $R = \{s\}$ is a singleton then we may write $\lozenge s$, $\square s$, or $\neg s$.

The Christel MDP: A Typical Working Day. We illustrate the above concepts on a motivating example (Fig. 1), which will serve as a running example throughout this article. Instances of this example will be marked by a gray box.

Christel's usual working day starts at home (🏠). The overall and highest goal of each day is to finish a research paper (✔) with a very high probability. When leaving her home to reach the office (🖥), Christel faces the first difficult decision of the day. Will she take the tram, will she walk, or will she take a taxi? Choosing the tram (🚊) means that she may have to wait for its arrival (🧍). In case she decides to walk, there is a small probability that the walkway may be freezing. Unless she is wearing her

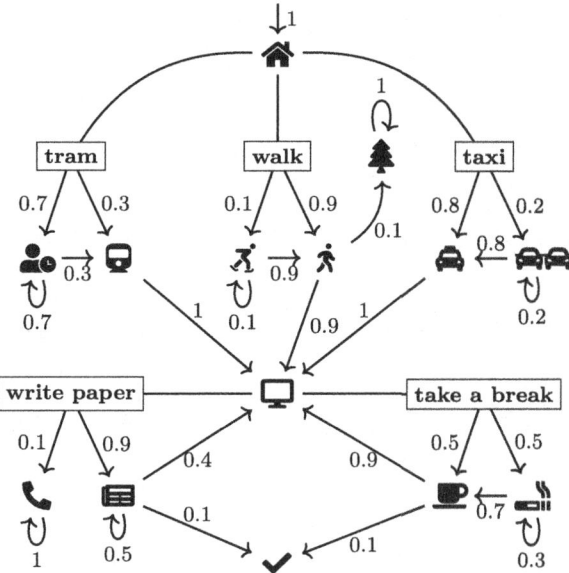

Fig. 1. The Christel MDP.

spiked shoes, she will, with a small probability, slip and skate for some time (🛼) until she can walk (🚶) to the office. Walking bears the risk of creating a desire to breathe more fresh air. In this case, Christel may decide to take a long walk in the nearby forest (🌲), in which case she will not reach the office on this day. Finally, she may make the unpopular but highly effective decision to call for a taxi. In this case, she may either swiftly (🚗) arrive at the office, or she may face a traffic jam (🚗🚗) and be delayed for some time with a small probability.

Let us now assume that Christel has safely arrived at the office (🖥). As in each scientist's life, she has to make the next difficult decision: She may either start to write a paper, or she may take a break.

In case she decides to write the paper, there is a high probability that she can indeed start writing (🖩), but also a small probability that right at that moment, the phone will ring (📞). This will almost surely lead to a long conference call preventing any further work on the paper. When writing, there is a small probability that the paper will get finished (✔), otherwise, after a while, Christel will re-consider the choice between writing and break. If Christel decides to take a break, there is a uniform probability distribution over two possible events. She may have coffee (☕), or she may decide to have another type of break (🚬). In the latter case, there is a likely delay until she decides that she needs a coffee anyway. After coffee, there is a very high probability that she may be back at the computer, and can start fresh. However, there is also the very unlikely event that one of Christel's PhD students took advantage of the break and finished the paper without her.

A *scheduler* for $\mathcal{M} = (S, \mathrm{Act}, s_{in}, P)$ is a function $\sigma : \mathrm{Paths}_{\mathrm{fin}}(\mathcal{M}) \to \mathrm{Dist}(\mathrm{Act})$, where we require that $\mathrm{supp}(\sigma(\pi)) \subseteq \mathrm{Act}(\mathrm{last}(\pi))$ holds for each π. If the chosen probability distribution is always a Dirac distribution, we call the scheduler *deterministic*, and otherwise *randomized*. The scheduler σ is called *memoryless* if for all finite paths π, π' with $\mathrm{last}(\pi) = \mathrm{last}(\pi')$ we have $\sigma(\pi) = \sigma(\pi')$. A scheduler is called *finite-memory* if it can be realized by a finite state machine. Memoryless and deterministic schedulers will be abbreviated as MD-schedulers, and memoryless randomized schedulers will be called MR-schedulers.

Schedulers for the Christel MDP.

Scheduler σ_1. On some days, Christel chooses a randomized finite-memory scheduler for her day: First, she throws a (fair) coin to decide between taking a tram or walking. After taking the tram, she needs a break before starting to write a paper. However, when walking she enjoys structuring her thoughts and hence is eager to start immediately to put her new ideas on paper. Thus, for any paths π, the scheduler makes the following decisions:

$$\sigma_1(\text{🏠})(\textbf{tram}) = 0.5 \qquad\qquad \sigma_1(\text{🏠})(\textbf{walk}) = 0.5$$
$$\sigma_1(\pi\, \text{🖥️}\,\text{🖥️})(\textbf{take a break}) = 1 \quad \sigma_1(\pi\, \text{🏃}\,\text{🖥️})(\textbf{write paper}) = 1$$

Scheduler σ_2. On other days, Christel waives probabilistic decisions but still uses a memory: She definitely takes a walk to her office and tries to start writing immediately. However, when paper writing gets delayed by a call, she needs a break to restructure her thoughts before she starts again. Thus, for any path π, she makes the following decisions:

$$\sigma_2(\text{🏠})(\textbf{walk}) = 1$$
$$\sigma_2(\pi\, \text{🖥️}\,\text{🖥️})(\textbf{write paper}) = 1 \quad \sigma_2(\pi\, \text{🏃}\,\text{🖥️})(\textbf{write paper}) = 1$$
$$\sigma_2(\pi\, \text{🚌}\,\text{🖥️})(\textbf{write paper}) = 1$$
$$\sigma_2(\pi\, \text{🖥️}\,\text{📇}^*\text{🖥️})(\textbf{write paper}) = 1$$
$$\sigma_2(\pi\, \text{🖥️}\,\text{📞}^*\, \text{📇}^*\, \text{🖥️})(\textbf{take a break}) = 1$$

Scheduler σ_3. On other occasions, Christel opts for the simple deterministic and memoryless strategy to walk and to repeatedly take breaks. This allows her PhD students to learn how to write a paper:

$$\sigma_3(\text{🏠})(\textbf{walk}) = 1 \quad \sigma_3(\pi\, \text{🖥️})(\textbf{take a break}) = 1$$

A *discrete-time Markov chain (DTMC)* is an MDP in which exactly one action is enabled in every state. For DTMCs we omit the actions yielding $\mathcal{D} = (S, s_{in}, P)$, with a probability transition function of type $P : S \times S \to [0,1]$ satisfying $\sum_{s' \in S} P(s, s') = 1$ for all $s \in S$ (respectively $\sum_{s' \in S} P(s, s') \leq 1$

for sub-stochastic DTMCs). Each scheduler σ for MDP $\mathcal{M}$ *induces* a DTMC, denoted as $\mathcal{M}^\sigma$. A DTMC induces a probability space over its infinite paths, and we denote by $\mathrm{Pr}_\mathcal{D}(\Pi)$ the probability of the set $\Pi \subseteq S^\omega$, assuming that Π is measurable. All ω-regular sets are measurable, which is enough for our purposes (see [2, Chapter 10] for more details).

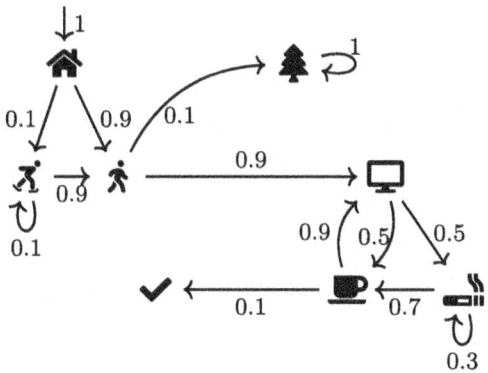

Fig. 2. DTMC induced by the scheduler σ_3 for the Christel MDP.

Induced Markov chains. Christel's deterministic memoryless scheduler σ_3 induces the DTMC depicted in Fig. 2.

Assuming an MDP $\mathcal{M} = (S, \mathrm{Act}, s_{in}, P)$ and an ω-regular set $\Pi \subseteq$ Paths($\mathcal{M}$), we define $\mathbf{Pr}_\mathcal{M}^{max}(\Pi) = \sup_\sigma \mathrm{Pr}_{\mathcal{M}^\sigma}(\Pi)$ and $\mathbf{Pr}_\mathcal{M}^{min}(\Pi) = \inf_\sigma \mathrm{Pr}_{\mathcal{M}^\sigma}(\Pi)$, where σ ranges over all schedulers of $\mathcal{M}$. Since Π is ω-regular, there exist finite-memory schedulers which achieve the supremum, respectively infimum [26]. If we are interested in the optimal probabilities from some state $s \in S$ (i.e., in the MDP (S, Act, s, P)), we write $\mathbf{Pr}_{\mathcal{M},s}^{max}(\Pi)$ and $\mathbf{Pr}_{\mathcal{M},s}^{min}(\Pi)$. If $\mathcal{M}$ is clear from the context we will write $\mathbf{Pr}^{max}(\Pi)$ and $\mathbf{Pr}^{min}(\Pi)$.

Optimal schedulers for reachability properties. In order to maximize the probability of finishing the paper, Christel should avoid walking. Furthermore, even if it sounds weird, the highest chances of success are achieved by avoiding to write the paper. All schedulers that avoid starting to write yield the maximal probability of 1 for the outcome of completing the paper. In other words: Christel should trust her PhD students to complete the paper!

Verifying Probabilistic Reachability Properties. A common specification language for discrete-time Markov models is probabilistic computation tree logic (PCTL). PCTL is rooted in computation tree logic (CTL), but replaces path quantification by probability operators. Whereas the techniques described in this and the following section can handle PCTL properties, we restrict ourselves to *reachability* properties for simplicity. More precisely, we consider *probabilistic reachability constraints* of the form $\mathbf{Pr}^{\mathfrak{m}}(\lozenge T) \sim \lambda$ with $\mathfrak{m} \in \{\min, \max\}$, target set $T \subseteq S$, comparison operator $\sim \in \{<, \leq, =, \geq, >\}$, and probability bound $\lambda \in [0, 1]$.

Probabilistic reachability constraints. The followings are probabilistic reachability constraints for the Christel MDP:

P1: The maximal probability of having a walk through the forest is at most 5% ($\mathbf{Pr}^{\max}(\lozenge \text{🌲}) \leq 0.05$).
P2: The maximal probability that Christel arrives to her office delayed due to being caught in a traffic jam is at most 10% ($\mathbf{Pr}^{\max}(\lozenge \text{🚗🚗}) \leq 0.1$).
P3: The paper gets completed with a probability of at least 50% regardless which decisions Christel makes ($\mathbf{Pr}^{\min}(\lozenge \text{✔}) \geq 0.5$).

Probabilistic reachability constraints can be decided by different model checking methods. For example, reachability probabilities in a DTMC $\mathcal{D} = (S, s_{in}, P)$, can be encoded as a system of linear real arithmetic equations. Hence, a reachability constraint can be decided by solving the equations and comparing the value in the initial state against the bound. More formally, the reachability probabilities p_s to eventually reach a target state from each state s is the unique solution of a linear equation system [2] containing an equation for each state $s \in S$: $p_s = 1$ if $s \in T$, $p_s = 0$ if there is no path from s to any state in T, and $p_s = \sum_{s' \in S} P(s, s') \cdot p_{s'}$ in all other cases.

Computing reachability probabilities. For the Christel DTMC induced by scheduler σ_3, the probabilities to reach the office can be computed as follows:

$$p_{\text{🏠}} = 0.1 p_{\text{🚶}} + 0.9 p_{\text{🚴}} \qquad p_{\text{🚶}} = 0.1 p_{\text{🚶}} + 0.9 p_{\text{🚴}} \qquad p_{\text{🚴}} = 0.1 p_{\text{🚗}} + 0.9 p_{\text{🖥}}$$
$$p_{\text{🚙}} = 0.1 p_{\text{✔}} + 0.9 p_{\text{🖥}} \qquad p_{\text{🚧}} = 0.3 p_{\text{🚧}} + 0.7 p_{\text{🚙}} \qquad p_{\text{🖥}} = 1$$
$$p_{\text{🚗}} = 0 \qquad\qquad\qquad p_{\text{✔}} = 0$$

Solving this system gives us the probability of 0.9 to reach the office from home.

Another model checking approach - which we do not further detail here because it is not directly relevant for the following - is based on iterative computations to determine reachability probabilities within a bounded number of steps, starting with bound 0 and iteratively increasing the path length, so that the values converge to the exact probability values.

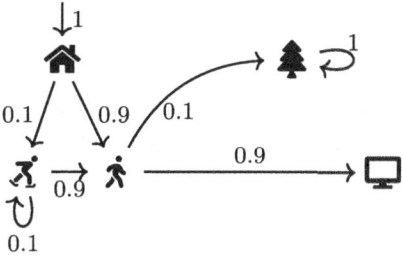

Fig. 3. A small DTMC from the Christel MDP

3 Probabilistic Counterexamples for DTMCs

A major asset in formal verification are *counterexamples*, which provide evidence on *why* a specification is not satisfied. In the classical model checking problem of transition systems against linear temporal logic (LTL) specifications, a counterexample is a single execution of the system violating the specification.

> **Christel does not reach the office.** Consider again the Christel MDP from Fig. 1, and in particular, the small excerpt sub-stochastic DTMC in Fig. 3. We are interested in the property $\lozenge\, \square$, that is, for Christel to eventually reach her office so she can start working on a paper. Model checking would reveal that this property is not almost surely satisfied. Consider the path π_1: 🏠🚶🌲🌲···, as induced by a scheduler that chooses to take the action **walk** from home. This path shows that there is a possibility that Christel may end up in the forest forever after she has been walking. Note that the likelihood of this happening is relatively low (0.09%).

The above example shows that single paths may serve as counterexamples to almost sure reachability. However, a general notion of counterexamples in the probabilistic setting requires to account for (or: collect) a potentially large or even infinite set of paths and take their joint probability mass into account [13].

> **Path-based counterexample.** Consider the probabilistic reachability constraint $\mathbf{Pr}^{\max}(\lozenge\, 🌲) \leq 0.095$ and again the excerpt DTMC in Fig. 3. Basically, we want to ensure that the probability of Christel getting lost in the forest is at most 0.095. The path π_1: 🏠🚶🌲🌲··· from the previous example has a probability of 0.09, so it cannot serve as a counterexample here. However, let us also consider the path π_2: 🏠⛸🚶🌲,··· which has a probability of 0.009. Together, these paths have a probability mass of 0.099, which exceeds the threshold of 0.095. Thus the set $\{\pi_1, \pi_2\}$ of paths is a counterexample to φ.

A significant problem with path-based counterexamples is that one may need many paths. A finite set of paths that can serve as a counterexample may not even exist. As shown in [13], one can easily construct an example that requires an infinite set of paths whose probability mass is computed in the limit.

Infinite sets of paths. Consider again the excerpt DTMC in Fig. 3 and the probabilistic reachability constraint $\Pr(\Diamond \clubsuit) < 0.1$. Note that the only change to the previous example is that the probability threshold is now strict and slightly higher at 0.1. The previous counterexample $\{\pi_1, \pi_2\}$ with π_1: 🏠🚶♣♣··· and π_2: 🏠⛸🚶♣,··· has a joint probability of 0.099. To form a counterexample for $\Pr(\Diamond \clubsuit) \leq 0.1$, we need to add more paths. Let us assume that Christel spends more time at skating ⛸, and add the path π_3: 🏠⛸⛸🚶♣,··· with a probability of 0.0009. Yet, together, we only achieve a mass of 0.0999.

Let us examine how all additional paths may be formed. Basically, we need $\pi(i) =:$ 🏠$(⛸)^i$🚶♣,··· , where $(⛸)^i$ indicates that ⛸ is taken i times for $i \in \mathbb{N}_{>0}$. The probability of all these paths is calculated using the geometric series as

$$\sum_{i=0}^{\infty} 0.1 \cdot (0.1)^i \cdot 0.9 \cdot 0.1 = 0.1 \cdot \frac{1}{1 - 0.1} \cdot 0.9 \cdot 0.1 = 0.01 \ .$$

Together with π_1, we achieve the desired probability mass of 0.1, but only because we consider the infinite set of paths $\pi(i)$.

Subsystems as Counterexamples. The above example shows how a set of paths can be used as a counterexample in probabilistic model checking. To achieve a more compact representation for such counterexamples, we discuss the concept of *subsystems*. Let $\mathcal{M} = (S, \mathrm{Act}, s_{in}, P)$ be an MDP. A subsystem $\mathcal{M}' = (S', \mathrm{Act}, s_{in}, P')$ of $\mathcal{M}$ is a sub-stochastic MDP which satisfies $S' \subseteq S$, $s_{in} \in S'$, and $P'(s, \alpha, s') \leq P(s, \alpha, s')$ for all $s, s' \in S'$ and $\alpha \in \mathrm{Act}(s)$. We say that $\mathcal{M}'$ is a *strict subsystem*, if additionally $P'(s, \alpha, s') > 0 \implies P'(s, \alpha, s') = P(s, \alpha, s')$ holds.

To use subsystems as counterexamples, we observe that for every linear temporal logic property $\Pi \subseteq S^\omega$ we have $\mathbf{Pr}^{\mathfrak{m}}_{\mathcal{M}'}(\Pi) \leq \mathbf{Pr}^{\mathfrak{m}}_{\mathcal{M}}(\Pi)$, for $\mathfrak{m} \in \{\min, \max\}$. Note that the set of enabled actions $\mathrm{Act}(s)$ is preserved in a subsystem for every state, and therefore each scheduler of $\mathcal{M}$ has a corresponding scheduler in $\mathcal{M}'$, and vice versa. Hence, a (preferably small) subsystem $\mathcal{M}'$ satisfying $\mathbf{Pr}^{\mathfrak{m}}_{\mathcal{M}'}(\Pi) \geq \lambda$ is a *witness* for the same constraint holding in the original MDP $\mathcal{M}$. We call such a witness a *critical subsystem* to $\mathbf{Pr}^{\mathfrak{m}}_{\mathcal{M}}(\Pi) < \lambda$ and consider it a counterexample for this upper bound. Given a set of states $R \subseteq S$ with $s_{in} \in R$, we define the subsystem $\mathcal{M}_R$ *induced by* R to be the maximal strict subsystem of $\mathcal{M}$ which contains exactly the states R.

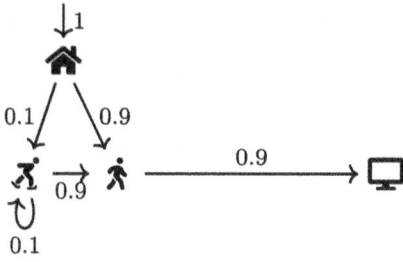

Fig. 4. Critical subsystem for property $\Pr(\Diamond \square) < 0.85$ in the DTMC in Fig. 2.

The central algorithmic question is to find *small* critical subsystems, since they contain the most information on why the property in question is violated. A natural measure of size is the number of states of the subsystem. Another useful measure is to consider a transition labeling function $\Lambda : S \times S \to 2^L$ and aim at subsystems $\mathcal{D}' = (S', s_{in}, P')$ with small $| \cup_{P'(s,s')>0} \Lambda(s,s')|$. One application of such a labeling is to compute *high-level* counterexamples [32], e.g. in terms of the common representation of DTMCs using a guarded command language, where transitions are labelled with all guarded commands that they instantiate.

Deciding whether a critical subsystem of size $\leq k$ exists is NP-complete even for restricted classes of Markov chains, including acyclic Markov chains [12] and Markov chains with low pathwidth [21].

Critical subsystems. Consider again the the DTMC in Fig. 2, as induced by Christel's scheduler σ_3. The sub-stochastic DTMC depicted in Fig. 4 is a critical subsystem for the property $\Pr(\Diamond \square) < 0.85$.
For the property $\Pr(\Diamond \square) < 0.8$, we could further remove the self-loop on the state 🏂.

4 Computing Critical Subsystems for DTMCs

4.1 Path-Based Synthesis of Critical Subsystems

To identify critical subsystems, our first research direction was to use graph algorithms and bounded model checking to enumerate paths from the initial state to the target states, until the joint probability mass of their cylinder sets exceeds the given threshold [5,6].

4.2 Logical Synthesis of Minimal Critical Subsystems

Path-based critical subsystems provide explanations, but they have certain weaknesses. Firstly, the computations are time-consuming and thus the methods come with restricted scalability. Secondly, we cannot provide any quality assurances regarding the size of the critical subsystems.

To achieve (state-)*minimal* critical subsystems, in [29,30] we proposed a technique based on a logical encoding of minimal critical subsystems as optimization problems, and the usage of optimizing solvers to solve the logical encodings and provide this way solutions with formal guarantees for minimality. To formalize the encoding, we need to introduce some further notions.

A state of the DTMC $\mathcal{D} = (S, \mathrm{Act}, s_{in}, P)$ is *relevant* if it lies on a path from the initial state over non-target states to a target state from some set $T \subseteq S$. With other words, a state is relevant if making it absorbing would reduce the probability of reaching the target from the initial state. Let us denote the relevant states as S_{rel}.

A *transition* is a pair $(s, s') \in S \times S$ with source s and target s', such that $P(s, s') > 0$. The transition (s, s') is *relevant* if both s and s' are relevant, and s is not a target state. Let us denote the set of all relevant transitions as E_{rel}.

Using these notions, we encode minimal critical subsystems as counterexamples to $\Pr(\Diamond T) \leq \lambda$ as solutions to optimization problems. The following encoding uses for each state $s \in S$ two variables: (1) a real-valued variable p_s with domain $[0, 1] \subset \mathbb{R}$ to encode the probability to reach the target from s in the critical subsystem, and (2) a variable x_s with domain $\{0, 1\}$ to encode whether the state s is part of the critical subsystem ($x_s = 1$) or not ($x_s = 0$). Using $\oplus$ for the binary XOR operator, the encoding is as follows:

$$minimize \sum_{s \in S_{rel}} x_s \ such \ that \tag{1}$$

$$p_{s_{in}} > \lambda \tag{2}$$

$$\forall s \in S_{rel} \cap T : ((x_s = 0 \land p_s = 0) \oplus (x_s = 1 \land p_s = 1)) \tag{3}$$

$$\forall s \in S_{rel} \setminus T : ((x_s = 0 \land p_s = 0) \oplus (x_s = 1 \land p_s = \sum_{(s,s') \in E_{rel}} P(s, s') \cdot p_{s'})) \tag{4}$$

Consider again the Christel DTMC in Fig. 2. For target state $T = \{🖥\}$, the relevant states are $S_{rel} = \{🏠, ✂, 🏃, 🖥\}$, and the relevant transitions are $E_{rel} = \{(🏠, ✂), (🏠, 🏃), (✂, ✂), (✂, 🏃), (🏃, 🖥)\}$.
For a violated property $\Pr(\Diamond 🖥) \leq \lambda$ with $\lambda \in [0, 1] \subset \mathbb{R}$, we encode minimal critical subsystems as follows:

$$minimize \ x_{🏠} + x_{✂} + x_{🏃} + x_{🖥} \ such \ that \tag{5}$$

$$p_{🏠} > \lambda \tag{6}$$

$$((x_{🖥} = 0 \land p_{🖥} = 0) \oplus (x_{🖥} = 1 \land p_{🖥} = 1)) \tag{7}$$

$$((x_{🏠} = 0 \land p_{🏠} = 0) \oplus (x_{🏠} = 1 \land p_{🏠} = 0.1p_{✂} + 0.9p_{🏃})) \tag{8}$$

$$((x_{✂} = 0 \land p_{✂} = 0) \oplus (x_{✂} = 1 \land p_{✂} = 0.1p_{✂} + 0.9p_{🏃})) \tag{9}$$

$$((x_{🏃} = 0 \land p_{🏃} = 0) \oplus (x_{🏃} = 1 \land p_{🏃} = 0.9p_{🖥})) \tag{10}$$

For $\lambda < 0.81$, solutions will exclude 🏃 from the critical subsystem (i.e., solutions will set $x_{🏃} = 0$). For $0.81 \leq \lambda < 0.9$, in this example we need to include all relevant states, and achieve the minimal critical subsystem from Fig. 4. Note that for the presented encoding, self-loops on relevant non-target states (e.g. on 🏃) cannot be extended from critical subsystems; this needs an additional encoding of transition selection, which we do not detail here.

4.3 Extensions and Implementation

We worked out several extensions of the above ideas. In addition to the above SMT encoding, we also provided an MILP encoding [29,30]. We further developed algorithms for checking general ω-regular properties for DTMCs [28] as well as for MDPs [27]. Markov models with rewards have been covered in [24], hierarchical counterexamples were proposed in [15], and symbolic approaches in [17,18]. The computation of counterexamples formalized in terms of high-level probabilistic guarded commands were the objective of [31,32]. The presented methods have been implemented in the COMICS tool [16].

5 Certifying Algorithms

Since verification tools are used to ensure the correctness of safety-relevant systems, it is essential that their results can be trusted. The two main approaches to ensure the validity of verification results are *formally verified* implementations, usually utilizing deductive theorem provers for higher-order logic, and *certifying algorithms.*

A certifying algorithm accompanies every result with a *certificate*, a token that enables a simple, independent validation of the result. It is essential, of course, that validating the certificate is significantly simpler than producing the result in the first place. Certifying algorithms have the advantage that the implementation of the algorithm itself can still be complex, optimized and even incorrect – as long as it produces certificates. On the other hand, not all problems are naturally amenable to certification. A common approach to combine the benefits of formally verified implementations and certifying algorithms is to use *formally verified certificate checkers.*

This section gives an overview of certifying algorithms in the domain of probabilistic model checking. It covers existing techniques for certifying *probabilistic reachability constraints* using Farkas certificates and certifying the computation of end components [12,19]. A probabilistic reachability constraint is a bound (upper or lower) on the optimal probability (maximal or minimal) of a reachability property in an MDP. Further, we compare these with the recently proposed *fixed point certificates* [8].

5.1   Farkas Certificates

It is well known that optimal reachability probabilities in MDPs can be computed using linear programs [2, 22]. Farkas' Lemma [11] is a standard result in linear algebra and linear programming [25], relating the *unsatisfiability* of a set of linear inequalities to the *satisfiability* of a dual system. This result allows certifying unsatisfiability of linear inequalities and has found numerous application beyond that [10]. In the context of MDPs, the dual formulations of linear programs have also been studied [7, 22]. Our focus is their applicability to defining certifying algorithms, and to computing witnesses and counterexamples. We now give a brief overview of Farkas certificates for probabilistic reachability constraints in MDPs, as proposed in [12, 19]. Assume in the following an MDP $\mathcal{M} = (S, \text{Act}, s_{in}, P)$ and a target set $T \subseteq S$.

Universal Statements. A Farkas certificate for $\mathbf{Pr}^{\max}(\lozenge T) \lesssim \lambda$ (all schedulers achieve probability $\lesssim \lambda$) is a vector $\mathbf{z} \in \mathbb{R}_{\geq 0}^{S \setminus T}$ satisfying

$$\mathbf{z}(s) \geq \sum_{s' \in S} P(s, \alpha, s') \cdot \mathbf{z}(s'), \qquad \text{for all } s \in S \setminus T \text{ and } \alpha \in \text{Act}(s)$$

$$\mathbf{z}(s) = 1 \qquad \text{for all } s \in T$$

and $\mathbf{z}(s_{in}) \lesssim \lambda$. In words, the value of state s is *at least as high* as supported by all possible actions. Such solutions are called *inductive* [14] and indeed imply that $\mathbf{z}(s) \geq \mathbf{Pr}_s^{\max}(\lozenge T)$ for all states.

The certificate condition for $\mathbf{Pr}^{\min}(\lozenge T) \gtrsim \lambda$ (all schedulers achieve probability $\gtrsim \lambda$) is analogous, but requires an additional condition for states in $S_{\min=0} = S \setminus S_{rel}$:

$$\mathbf{z}(s) \leq \sum_{s' \in S} P(s, \alpha, s') \cdot \mathbf{z}(s'), \qquad \text{for all } s \in S \setminus T \text{ and } \alpha \in \text{Act}(s)$$

$$\mathbf{z}(s) = 1 \qquad \text{for all } s \in T$$

$$\mathbf{z}(s) = 0 \qquad \text{for all } s \in S_{\min=0}$$

and $\mathbf{z}(s_{in}) \gtrsim \lambda$.

Existential Statements. Farkas certificates for *existential statements* (there exists a scheduler such that ...) are derived from the dual systems of the certificate conditions for universal statements, using Farkas' Lemma.

A Farkas certificate $\mathbf{y} \in \mathbb{R}_{\geq 0}^{(S \setminus T) \times \text{Act}}$ for $\mathbf{Pr}^{\max}(\lozenge T) \gtrsim \lambda$ satisfies

$$\sum_{\alpha \in \text{Act}} \mathbf{y}(s, \alpha) \leq \delta_{s_{in}}(s) + \sum_{s' \in S \setminus T} \sum_{\beta \in \text{Act}} \mathbf{y}(s', \beta) \cdot P(s', \beta, s)$$

for all $s \in S \setminus T$, where $\delta_{s_{in}}(s)$ is the Dirac distribution concentrated on s_{in}. Values $\mathbf{y}(s, \alpha)$ can be interpreted as the *expected number of times* that a satisfying scheduler visits s and chooses action α before reaching T. The left-hand

Table 1. Overview of Farkas certificates for the different types of probabilistic reachability constraints (with $\lesssim \in \{\leq, <\}$ and $\gtrsim \in \{\geq, >\}$). A constraint holds if and only if the certificate condition is satisfiable.

constraint	certificate dimension	certificate condition
$\mathbf{Pr}^{\max}(\lozenge T) \lesssim \lambda$	$\mathbf{z} \in \mathbb{R}_{\geq 0}^{S \setminus T}$	$\mathbf{Az} \geq \mathbf{t} \wedge \mathbf{z}(s_{in}) \lesssim \lambda$
$\mathbf{Pr}^{\min}(\lozenge T) \gtrsim \lambda$	$\mathbf{z} \in \mathbb{R}_{\geq 0}^{S^*}$	$\mathbf{A}^*\mathbf{z} \leq \mathbf{t}^* \wedge \mathbf{z}(s_{in}) \gtrsim \lambda$
$\mathbf{Pr}^{\max}(\lozenge T) \gtrsim \lambda$	$\mathbf{y} \in \mathbb{R}_{\geq 0}^{(S \setminus T) \times \text{Act}}$	$\mathbf{yA} \leq \delta_{s_{in}} \wedge \mathbf{yt} \gtrsim \lambda$
$\mathbf{Pr}^{\min}(\lozenge T) \lesssim \lambda$	$\mathbf{y} \in \mathbb{R}_{\geq 0}^{S^* \times \text{Act}}$	$\mathbf{yA}^* \geq \delta_{s_{in}} \wedge \mathbf{yt}^* \lesssim \lambda$

side $\sum_{\alpha \in \text{Act}} \mathbf{y}(s, \alpha)$ represents the expected total number of visits of s under that scheduler. The inequality requires this value to be at most the probability of starting in s plus the expected number of visits of incoming transitions to s. With this interpretation, the constraint on the probability of reaching T can be expressed as:

$$\sum_{s \in S \setminus T} \sum_{\alpha \in \text{Act}} \mathbf{y}(s, \alpha) \cdot \sum_{t \in T} P(s, \alpha, t) \gtrsim \lambda$$

The Farkas certificate for $\mathbf{Pr}^{\min}(\lozenge T) \lesssim \lambda$ is based on analogous inequalities. However, it also requires separately considering states in $S_{\min=0}$, which are excluded from the set of inequalities.

To express these conditions in matrix notation, let $S' = S \setminus T$ and define the *system matrix* $\mathbf{A} \in \mathbb{R}^{(S' \times \text{Act}) \times S'}$ of $\mathcal{M}$ as follows: $\mathbf{A}((s, \alpha), t) = -P(s, \alpha, t)$ if $s \neq t$, and else $\mathbf{A}((s, \alpha), s) = 1 - P(s, \alpha, s)$. Further, let $\mathbf{t} \in \mathbb{R}_{\geq 0}^{S \times \text{Act}}$ be defined by $\mathbf{t}(s, \alpha) = \sum_{t \in T} P(s, \alpha, t)$. Let $\mathbf{A}^*$ and $\mathbf{t}^*$ be the restrictions of $\mathbf{A}$ and $\mathbf{t}$ to states in $S^* = S' \setminus S_{\min=0}$. These restrictions ensure that the additional requirements for states in $S_{\min=0}$ are met for certificates involving $\mathbf{Pr}^{\min}$. The four kinds of Farkas certificates in matrix notation are defined in Table 1.

Farkas certificates. Consider the sub-stochastic DTMC in Fig. 4 and the property $\Pr(\lozenge \,\square) \geq 0.8$. Since minimal and maximal probabilities coincide in DTMCs, we may certify it using either certificates for lower bounds using either **z**- or **y**-formulations. The following two vectors are both Farkas certificates for this property:

$$\mathbf{z}(\text{🏠}) = 0.8, \quad \mathbf{z}(\text{🏹}) = 0, \quad \mathbf{z}(\text{🏃}) = 0.9$$
$$\mathbf{y}(\text{🏠}) = 1, \quad \mathbf{y}(\text{🏹}) = 0, \quad \mathbf{y}(\text{🏃}) = 0.9$$

Since this is a DTMC we can use states, rather than state-action-pairs, to index vector **y**.

5.2   Certifying End Components

Variants of the systems of linear inequalities used to define Farkas certificates can also be used to reason about end components. An *end component* of an MDP $\mathcal{M} = (S, \text{Act}, s_{in}, P)$ is a strongly connected strict subsystem of $\mathcal{M}$, i.e., an MDP $\mathcal{M}' = (S', \text{Act}', s'_{in}, P')$ with $S' \subseteq S$, $\text{Act}' \subseteq \text{Act}$ and $s'_{in} \in \mathfrak{S}'$, such that $P'(s, \alpha, s') \in \{0, P(s, \alpha, s')\}$ for all $s, s' \in S'$ and $\alpha \in \text{Act}'$, and such that each pair of states $s, s' \in S'$ is connected by a path $s \ldots s'$ in $\mathcal{M}'$. A *maximal end component* is an end component that cannot be extended.

We now consider the question of certifying (non-)existence of end components in a given sub-stochastic MDP $\mathcal{M}$. Let $\mathbf{A}$ be the system matrix as defined in the previous section for $\mathcal{M}$, with empty target set $T = \varnothing$.

Proposition 1 [19, Lemma 3.28]. *Let $\mathcal{M}$ be a sub-stochastic MDP with states S and system matrix $\mathbf{A}$. Then, $\mathcal{M}$ has no end component iff the system of linear inequalities $\mathbf{A}\mathbf{z} \geq \mathbf{1}$ is satisfiable with $\mathbf{z} \in \mathbb{R}^{S}_{\geq 0}$.*

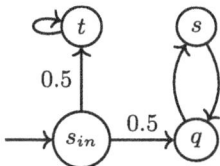

Fig. 5. A DTMC with end components $\{t\}$ and $\{s, q\}$.

Certifying non-existence of end-components. Consider the DTMC $\mathcal{M}$ pictured in Fig. 5. The system $\mathbf{A}\mathbf{z} \geq \mathbf{1}$ for $\mathcal{M}$ includes the inequalities

$$\mathbf{z}(s) \geq 1 + \mathbf{z}(q) \qquad \text{and} \qquad \mathbf{z}(q) \geq 1 + \mathbf{z}(s)$$

which are clearly inconsistent.

Based on Proposition 1, one can define a certificate for the computation of maximal end components. It includes certificates of strong-connectedness of the induced graphs of every sub-MDP in the candidate partition, and a certificate that the induced quotient is indeed free of end components [19, Proposition 3.26]. Applying Farkas' Lemma to Proposition 1 directly yields a condition for the existence of end components:

Proposition 2 [19, Remark 3.29]. *Let $\mathcal{M}$ be a sub-stochastic MDP with states S and system matrix $\mathbf{A}$. Then, $\mathcal{M}$ has an end component iff the system of linear inequalities $\mathbf{y}\mathbf{A} \leq \mathbf{0}$ and $\mathbf{y} \cdot \mathbf{1} > 0$ is satisfiable with $\mathbf{y} \in \mathbb{R}^{S \times \text{Act}}_{\geq 0}$. Moreover, the support of a solution $\mathbf{y}$ induces a set of end components.*

If (s, α) is part of the support of a solution to the system $\mathbf{y}\mathbf{A} \leq \mathbf{0}$, then $\mathbf{y}(s, \alpha)$ can be used to infer the *long-run frequency* of visiting (s, α) under a related scheduler. This observation is used in [7] to optimize for multiple long-run average objectives in MDPs using linear programs.

Certifying existence of end-components. Consider again the DTMC $\mathcal{M}$ in Fig. 5. The inequality for s_{in} in $\mathbf{y}\mathbf{A} \leq \mathbf{0}$ is $\mathbf{y}(s_{in}) \leq 0$, since s_{in} has no incoming transitions. After substituting $\mathbf{y}(s_{in}) = 0$ in $\mathbf{y}\mathbf{A} \leq \mathbf{0}$, the following inequalities remain:

$$\mathbf{y}(t) \leq \mathbf{y}(t) \qquad \mathbf{y}(s) \leq \mathbf{y}(q) \qquad \text{and} \qquad \mathbf{y}(q) \leq \mathbf{y}(s)$$

They are satisfiable together with $\mathbf{y} \cdot \mathbf{1} > 0$ by choosing $\mathbf{y}(s) = \mathbf{y}(q) > 0$ or $\mathbf{y}(t) > 0$. The support of every solution induces a set of end components. Since $\mathbf{y}(q) = \mathbf{y}(s)$ in every solution, the long-run frequencies of visiting s and q always coincide.

5.3   A Comparison with Fixed Point Certificates

Farkas certificates are defined as solutions of systems of linear inequalities, and can therefore be computed by solving a linear program. In contrast, the recently proposed *fixed point certificates* [8] allow using "min" and "max" operators, and are derived using fixed point induction on variants of the standard Bellman operators for MDPs. Further, fixed point certificates always provide bounds on the (optimal) probabilities for every state, the Farkas certificates based on $\mathbf{y}$-formulations are specific to a fixed initial distribution.

Universal Statements. For constraints $\mathbf{Pr}^{\min}(\lozenge T) \gtrsim \lambda$ and $\mathbf{Pr}^{\max}(\lozenge T) \lesssim \lambda$, fixed point certificates and Farkas certificates essentially coincide. The fixed point certificate condition $\mathbf{z}(s_{in}) \lesssim \lambda$ and

$$\mathbf{z}(s) \;\geq\; \max_{\alpha \in \text{Act}} \Big\{ \sum_{s' \in S} P(s, \alpha, s') \, \mathbf{z}(s') \Big\}$$

for $\mathbf{Pr}^{\max}(\lozenge T) \lesssim \lambda$ can be written as a conjunction of $|\text{Act}|$ inequalities, which yields the Farkas certificate condition for the same property. For $\mathbf{Pr}^{\min}(\lozenge T) \gtrsim \lambda$ the case is similar, and both fixed point and Farkas certificates rely on a separate certification of $S_{\min=0}$ (resp. $S_{\min>0}$) in this case.

Existential Statements. The fixed point certificate for $\mathbf{Pr}^{\min}(\lozenge T) \lesssim \lambda$ is defined using inequalities:

$$\mathbf{z}(s) \;\geq\; \min_{\alpha \in \text{Act}} \Big\{ \sum_{s' \in S} P(s, \alpha, s') \, \mathbf{z}(s') \Big\}$$

and $\mathbf{z}(s_{in}) \lesssim \lambda$. This combination of "$\geq$" and "min" cannot be directly rewritten into a conjunction of standard linear inequalities. The Farkas certificate for this property is based on the $\mathbf{y}$-form, and requires an additional certificate for $S_{\min=0}$.

For property $\mathbf{Pr}^{\max}(\lozenge T) \gtrsim \lambda$ the fixed point certificate from [8] specifies a deterministic scheduler, and certifies the bound in the induced DTMC using the certificates for $\mathbf{Pr}^{\min}(\lozenge T) \geq \lambda$. It additionally requires certifying that states in $S_{\min=0}$ have value zero. In total, this amounts to $2|S|$ values in the certificate, where $|S|$ of these are correspond to integer variables for indicating the deterministic scheduler. On the other hand, Farkas certificates for this property require only $|S|$ values, if we restrict ourselves to certificates corresponding to deterministic schedulers. The condition is a set of linear inequalities (without integer variables), and naturally extends to randomized schedulers. However, the Farkas certificate is specific to a fixed initial distribution and does not certify the maximal probability for each state.

Certifying $S_{\min=0}$. The qualitative reachability certificates from [8] can be used to certify that the set $S_{\min=0}$ was computed correctly. This is a useful complement to Farkas certificates for constraints on $\mathbf{Pr}^{\min}(\lozenge T)$, which depend on precomputing $S_{\min=0}$. To this end, consider the set of inequalities:

$$r(s) \leq 1 + \max_{\alpha \in \mathrm{Act}} \left\{ \min_{\substack{s' \in S \\ P(s,\alpha,s')>0}} r(s') \right\} \qquad \text{for } s \in S \setminus T$$

$$r(t) = 0 \qquad\qquad \text{for } t \in T$$

The crucial idea is to allow "infinity" as a value in solutions. If $r \in \mathbb{R}_{\geq 0}^S \cup \{\infty\}$ is a solution to the above inequalities, then, by [8, Remark 1], we have:

$$r(s) = \infty \implies s \in S_{\min=0}$$

This provides a condition for inclusion in $S_{\min=0}$. A related certificate for non-inclusion in $S_{\min=0}$ (i.e. for states satisfying $\mathbf{Pr}_s^{\min}(\lozenge T) > 0$) can also be given [8, Proposition 1]. Together, they can be used to certify $S_{\min=0}$ exactly and thereby complement Farkas certificates for bounds on $\mathbf{Pr}^{\min}$.

> **Example.** Consider the DTMC in Fig. 5, and let $T = \{t\}$. We get inequalities:
>
> $$r(s_{in}) \leq 1+\min\{r(t), r(s)\} \qquad r(t) = 0 \qquad r(s) \leq 1+r(q) \qquad r(q) \leq 1+r(s)$$
>
> The assignment $t \mapsto 0, s_{in} \mapsto 1, s \mapsto \infty, q \mapsto \infty$ is a solution which proves that $\mathbf{Pr}_s^{\min}(\lozenge T) = \mathbf{Pr}_q^{\min}(\lozenge T) = 0$.

6 Farkas Certificates and Witnessing Subsystems

It turns out that Farkas certificates for $\mathbf{Pr}^m(\lozenge T) \gtrsim \lambda$ induce witnessing subsystems for the same property [12,19]. Moreover, *Farkas certificates with few non-zero elements correspond to witnesses with few states.* In the following, let $\mathcal{F}_{\geq}^{m}(\lambda)$ denote the set of Farkas certificates for $\mathbf{Pr}^{m}(\Diamond T) \geq \lambda$. Given a vector $\mathbf{y} \in \mathbb{R}_{\geq 0}^{S \times \mathrm{Act}}$, we will use state-supp$(\mathbf{y}) = \{s \in S \mid \exists \alpha \in \mathrm{Act} . \ \mathbf{y}(s, \alpha) > 0\}$. Recall that $\mathcal{M}_R$ is the subsystem induced by states $R \subseteq S$. The correspondence between witnesses and Farkas certificates is stated formally as:

- There exists $\mathbf{z} \in \mathcal{F}_{\geq}^{\min}(\lambda)$ such that supp$(\mathbf{z}) \subseteq R$ if and only if $\mathcal{M}_R$ is a witness for $\mathbf{Pr}_{\mathcal{M}}^{\min}(\Diamond T) \geq \lambda$.
- There exists $\mathbf{y} \in \mathcal{F}_{\geq}^{\max}(\lambda)$ such that state-supp$(\mathbf{y}) \subseteq R$ if and only if $\mathcal{M}_R$ is a witness for $\mathbf{Pr}_{\mathcal{M}}^{\max}(\Diamond T) \geq \lambda$.

This observation gives rise to algorithms and heuristics for computing small witnessing subsystems.

Farkas certificates and witnesses. We have seen that the following vectors are Farkas certificates for $\mathrm{Pr}(\Diamond \Box) \geq 0.8$ in the DTMC defined in Fig. 4:

$$\mathbf{z}(\text{🏠}) = 0.8, \ \ \mathbf{z}(\text{🏃}) = 0, \ \ \mathbf{z}(\text{🏃}) = 0.9$$
$$\mathbf{y}(\text{🏠}) = 1, \ \ \mathbf{y}(\text{🏃}) = 0, \ \ \mathbf{y}(\text{🏃}) = 0.9$$

Both induce the subsystem which excludes the state 🏃, and thereby prove that this subsystem is a witness for the property.

Mixed-Integer Linear Programming Formulations. Finding solutions of a set of linear inequalities with a minimal number of non-zero elements can be expressed as a mixed-integer linear program (MILP). By the connection to Farkas certificates, minimal witnessing subsystems for $\mathbf{Pr}^{\min}(\Diamond T) \geq \lambda$ correspond to solutions of:

$$\text{minimize} \sum_{s \in S^*} \sigma(s). \quad \mathbf{z} \in \mathcal{F}_{\geq}^{\min}(\lambda) \ \text{and} \ \mathbf{z}(s) \leq \sigma(s) \ \text{for all} \ s \in S^*$$

with $\sigma(s) \in \{0, 1\}$ for all $s \in S^*$. (Recall that $S^* = S \setminus S_{\min=0}$.) The binary indicator variables $\sigma(s)$ are used to "charge" (if $\sigma(s) = 0$) or "discharge" (if $\sigma(s) = 1$) the equation $\mathbf{z}(s) = 0$. This relies on 1 being an upper bound on $\mathbf{z}(s)$.

As discussed, the values of Farkas certificates $\mathbf{y}$ for $\mathbf{Pr}^{\max}(\Diamond T) \geq \lambda$ correspond to *expected number of visits* of a scheduler. Therefore $\mathcal{F}_{\geq}^{\max}(\lambda)$ is unbounded in the presence of end components. One way of working around the missing upper bound is to use *indicator constraints* of the form:

$$\sigma(s, \alpha) = 0 \implies \mathbf{y}(s, \alpha) = 0$$

Algorithms for solving MILPs extended by indicator constraints have been studied [3, 4]. Consider the following MILP with indicator constraints:

$$\text{minimize} \sum_{s \in S \setminus T} \sigma(s). \quad \mathbf{z} \in \mathcal{F}^{\max}_{\geq}(\lambda) \text{ and}$$

$$\sigma(s, \alpha) = 0 \implies \mathbf{y}(s, \alpha) = 0 \quad \text{for all } s, \alpha \in (S \setminus T) \times \text{Act}$$

where $\sigma(s, \alpha) \in \{0, 1\}$ for all s, α. Again, optimal solutions correspond to minimal witnesses of $\mathbf{Pr}^{\max}(\Diamond T) \geq \lambda$. Observe that no special treatment of end components is required.

To avoid indicator constraints, [19] proposes to compute an upper bound $U(s, \alpha)$ on the maximal expected visiting times of s, α under any deterministic and memoryless scheduler that reaches $T \cup \{s \in S \mid \mathbf{Pr}^{\max}_s(\Diamond T) = 0\}$ with probability one. Since $\mathbf{Pr}^{\max}(\Diamond T) \geq \lambda$ can always be witnessed by a scheduler in this class, we can restrict ourselves to certificates corresponding to such schedulers. The indicator constraints can then be replaced by $\mathbf{y}(s, \alpha) \leq \sigma(s, \alpha) \cdot U(s, \alpha)$, for all $s, \alpha \in (S \setminus T) \times \text{Act}$. A generic, exponential bound is $U(s, \alpha) = |S| \cdot \epsilon^{-2|S|}$, where ϵ is the smallest transition probability in $\mathcal{M}$. (This is a standard result, spelled out in [19, Lemma 4.37].) Better bounds can be computed in polynomial time if the size of end components is bounded [19, Lemma 4.38].

The Quotient-Sum Heuristic. The correspondence of Farkas certificates and witnesses implies that any solution of the certificate conditions induces a witness. The *quotient-sum heuristic* [12] iteratively solves a series of linear programs whose optimization functions serve as proxies for the exact minimization criterion.

Consider the property $\mathbf{Pr}^{\min}(\Diamond T) \geq \lambda$. The heuristic starts by computing the linear program: minimize $\sum_{s \in S^*} \mathbf{z}(s). \quad \mathbf{z} \in \mathcal{F}^{\min}_{\geq}(\lambda)$. Let the solution be vector $\mathbf{z}_1$. Now we update the optimization function based on the following hypothesis: the smaller the value of a state in $\mathbf{z}_1$, the more likely it is that we can remove it entirely to achieve a smaller witness. Consequently, such a state should get a high coefficient in the next linear program. We set $\mathbf{o}_1(s) = \frac{1}{\mathbf{z}_1(s)}$ if $\mathbf{z}_1(s) > 0$, else $\mathbf{o}_1(s) = C$ for some large constant C, and compute:

$$\text{minimize} \sum_{s \in S^*} \mathbf{o}_1(s) \cdot \mathbf{z}(s). \quad \mathbf{z} \in \mathcal{F}^{\min}_{\geq}(\lambda)$$

This can be iterated an arbitrary amount of times, but often the solutions stabilize after a few iterations. This approach can yield small witnessing subsystems quickly, which are often close to the optimum [19, 20].

6.1   Probabilistic Constraints on Rabin Conditions

So far we have only considered witnessing subsystems for lower-bounded reachability constraints. We now consider a richer kind of constraints based on Rabin properties. To this end, we extend the construction for computing minimal witnesses for invariants presented in [19, Section 4.4].

Let $\mathcal{M} = (S, \text{Act}, s_{in}, P)$ be an MDP. A Rabin pair (N, M), with $N, M \subseteq S$, represents the path condition $\Box\Diamond M \wedge \Diamond\Box N$. It requires a path to stay within N from some point on, and to visit M infinitely often. A Rabin property is a disjunction of Rabin pairs. In the following, we focus on a solution for a single Rabin pair, and finally argue that it can be extended to Rabin properties.

We can assume that witnessing subsystems for $\mathbf{Pr}^{\min}(\Box\Diamond M \wedge \Diamond\Box N) \geq \lambda$ are closed under end components. This is because states belonging to a partially included end component in a subsystem have minimal probability zero for any Rabin property. Hence, this case can be reduced to reachability in the quotient of maximal end components.

Now consider $\mathbf{Pr}^{\max}(\Box\Diamond M \wedge \Diamond\Box N) \geq \lambda$. The obligations on a scheduler witnessing this property can be divided into *reaching* an end component satisfying the Rabin property with sufficiently high probability, and *realizing* that end component. Previous sections have considered systems of linear inequalities for reachability constraints and end components, and we now combine these ideas.

We construct auxiliary MDPs $\mathcal{T} = (S \cup \{t\}, \text{Act}_\mathcal{T}, s_{in}, P_\mathcal{T})$ and $\mathcal{R} = (S \cap N, \text{Act}, s_{in}, P_\mathcal{R})$, intuitively representing the transient and recurrent parts of the problem. MDP $\mathcal{T}$ is a copy of $\mathcal{M}$ with additional transitions

$$P_\mathcal{T}(s, \tau, t) = 1 \qquad \text{for all } s \in M \cap N.$$

The MDP $\mathcal{R}$ is a copy of $\mathcal{M}$ restricted to states in N, i.e.: $P_\mathcal{R}(s, \alpha, q) = P(s, \alpha, q)$ for all $s, q \in S \cap N$. Let $\mathbf{A}_\mathcal{T}, \mathbf{t}_\mathcal{T}$ be system matrix and target vector of $\mathcal{T}$ with respect to target set $T = \{t\}$, and $\mathbf{A}_\mathcal{R}$ be the system matrix of $\mathcal{R}$ with respect to the empty target set. Consider the inequalities:

$$\mathbf{y}_\mathcal{T}\mathbf{A}_\mathcal{T} \leq \delta_{s_{in}} \quad \wedge \quad \mathbf{y}_\mathcal{T}\mathbf{t}_\mathcal{T} \geq \lambda \tag{11}$$

$$\mathbf{y}_\mathcal{T}(s, \tau) \leq \sum_{\alpha \in \text{Act}} \mathbf{y}_\mathcal{R}(s, \alpha) \qquad \text{for all } s \in M \cap N \tag{12}$$

$$\mathbf{y}_\mathcal{R}\mathbf{A}_\mathcal{R} \leq 0 \tag{13}$$

Equation (11) is the Farkas certificate for $\mathbf{Pr}_\mathcal{T}^{\max}(\Diamond\{t\}) \geq \lambda$ (see Table 1), and thereby intuitively ensures that action τ is eventually chosen with probability at least λ. Equation (12) enforces that if $\mathbf{y}(s, \tau) > 0$ holds, then s is part of the state support of $\mathbf{y}_\mathcal{R}$. Finally, Eq. (13) implies that the support of any solution $\mathbf{y}_\mathcal{R}$ induces a set of end components, by Proposition 2. Observe that any such end component that additionally contains a state in M satisfies the Rabin condition, since $\mathcal{R}$ is restricted to states in N.

Proposition 3. *There exists a solution* $\mathbf{y}_\mathcal{T}, \mathbf{y}_\mathcal{R}$ *to equations (11–13) with*

$$\text{state-supp}(\mathbf{y}_\mathcal{T}) \cup \text{state-supp}(\mathbf{y}_\mathcal{R}) \subseteq R$$

if and only if $\mathcal{M}_R$ *satisfies* $\mathbf{Pr}_{\mathcal{M}_R}^{\max}(\Box\Diamond M \wedge \Diamond\Box N) \geq \lambda$.

This can be proven in the same way as [19, Proposition 4.52] with small adaptations. The construction can be extended to Rabin properties by introducing a separate recurrent component $\mathcal{R}_p$ for each Rabin pair p, together with dedicated actions τ_p for moving from the transient part to the respective component.

To compute minimal witnesses for an LTL property, consider the product of an MDP and a deterministic Rabin automaton and a labeling function Λ that labels a state in the product by its corresponding MDP state. By Proposition 3, solutions of the presented set of inequalities whose state-support intersects a minimal number of labels, are exactly the minimal witnesses to the LTL property. Such solutions can be computed using extensions to the MILPs in this section. The resulting MILP is doubly-exponential in worst case, due to the size of the Rabin automaton. However, it contains only a linear number of binary variables, which correspond to the number of states of the MDP.

To handle invariants of the form $\mathbf{Pr}^{\max}(\Box I) \geq \lambda$, the construction can be adapted by restricting both $\mathcal{T}$ and $\mathcal{R}$ to states in I. Lower-bounded invariant constraints correspond to *upper bounds* on reachability properties, by the equation $\mathbf{Pr}^{\min}(\Diamond T) < \lambda \iff \mathbf{Pr}^{\max}(\Box \overline{T}) \geq 1-\lambda$, and analogously for $\mathbf{Pr}^{\max}(\Diamond T) < \lambda$.

7   Conclusion

In this paper, which we dedicate to Christel Baier on the occasion of her 60th birthday, we gave an overview over related previous works of Christel and the authors, and proposed some extensions for the generation of counterexamples and witnesses for probabilistic properties of Markov models. These works have in common that they provide reasons for satisfaction and violation, but they have different aims: whereas *critical and witnessing subsystems* derived with the methods from Sect. 4 and 6 aim at explanations to humans, the *certificates* presented in the Sect. 5 target automated checks. With this contextual presentation we aimed to describe the relation between our work and the research results of Christel and her group, and continue this line with the presented novel components.

Starting with the ROCKS project, over the years we look back to lots of exciting collaborations and insightful discussions with Christel. We are grateful to Christel for being such an enrichment to all of us, and wish her all the best for the future.

Acknowledgements. This work was supported by the ERC Starting Grant 101077178 (DEUCE) and the DFG project RealySt.

References

1. Ábrahám, E., Becker, B., Dehnert, C., Jansen, N., Katoen, J.-P., Wimmer, R.: Counterexample generation for discrete-time Markov models: an introductory survey. In: Bernardo, M., Damiani, F., Hähnle, R., Johnsen, E.B., Schaefer, I. (eds.)

SFM 2014. LNCS, vol. 8483, pp. 65–121. Springer, Cham (2014). https://doi.org/10.1007/978-3-319-07317-0_3

2. Baier, C., Katoen, J.: Principles of Model Checking. MIT Press (2008)
3. Belotti, P., et al.: On handling indicator constraints in mixed integer programming. Comput. Optim. Appl. **65**(3), 545–566 (2016). https://doi.org/10.1007/s10589-016-9847-8
4. Bonami, P., Lodi, A., Tramontani, A., Wiese, S.: On mathematical programming with indicator constraints. Math. Program. **151**(1), 191–223 (2015). https://doi.org/10.1007/s10107-015-0891-4
5. Braitling, B., Wimmer, R., Becker, B., Jansen, N., Ábrahám, E.: Counterexample generation for Markov chains using SMT-based bounded model checking. In: Bruni, R., Dingel, J. (eds.) FMOODS/FORTE -2011. LNCS, vol. 6722, pp. 75–89. Springer, Heidelberg (2011). https://doi.org/10.1007/978-3-642-21461-5_5
6. Braitling, B., Wimmer, R., Becker, B., Jansen, N., Ábrahám, E.: SMT-based counterexample generation for Markov chains. In: Proceedings of Methoden und Beschreibungssprachen zur Modellierung und Verifikation von Schaltungen und Systemen (MBMV 2011), pp. 19–28. OFFIS-Institut für Informatik (2011)
7. Brázdil, T., Brozek, V., Chatterjee, K., Forejt, V., Kucera, A.: Two views on multiple mean-payoff objectives in Markov decision processes. Log. Methods Comput. Sci. **10**(1) (2014)
8. Chatterjee, K., Quatmann, T., Schäffeler, M., Weininger, M., Winkler, T., Zilken, D.: Fixed point certificates for reachability and expected rewards in MDPs (2025). https://arxiv.org/abs/2501.11467
9. Damman, B., Han, T., Katoen, J.P.: Regular expressions for PCTL counterexamples. In: Proceedings of the 5th International Conference on Quantitative Evaluation of Systems (QEST 2008), pp. 179–188 (2008)
10. Dinh, N., Jeyakumar, V.: Farkas' lemma: three decades of generalizations for mathematical optimization. Trans. Oper. Res. **22**(1), 1–22 (2014)
11. Farkas, J.: Theorie der einfachen Ungleichungen. J. reine angewandte Math. (Crelles J.) **1902**(124), 1–27 (1902)
12. Funke, F., Jantsch, S., Baier, C.: Farkas certificates and minimal witnesses for probabilistic reachability constraints. In: TACAS 2020. LNCS, vol. 12078, pp. 324–345. Springer, Cham (2020). https://doi.org/10.1007/978-3-030-45190-5_18
13. Han, T., Katoen, J.P., Damman, B.: Counterexample generation in probabilistic model checking. IEEE Trans. Softw. Eng. **35**(2), 241–257 (2009)
14. Hartmanns, A., Kaminski, B.L.: Optimistic value iteration. In: Lahiri, S.K., Wang, C. (eds.) CAV 2020. LNCS, vol. 12225, pp. 488–511. Springer, Cham (2020). https://doi.org/10.1007/978-3-030-53291-8_26
15. Jansen, N., Ábrahám, E., Katelaan, J., Wimmer, R., Katoen, J.-P., Becker, B.: Hierarchical counterexamples for discrete-time Markov chains. In: Bultan, T., Hsiung, P.-A. (eds.) ATVA 2011. LNCS, vol. 6996, pp. 443–452. Springer, Heidelberg (2011). https://doi.org/10.1007/978-3-642-24372-1_33
16. Jansen, N., Ábrahám, E., Volk, M., Wimmer, R., Katoen, J.-P., Becker, B.: The COMICS tool – computing minimal counterexamples for DTMCs. In: Chakraborty, S., Mukund, M. (eds.) ATVA 2012. LNCS, pp. 349–353. Springer, Heidelberg (2012). https://doi.org/10.1007/978-3-642-33386-6_27
17. Jansen, N., et al.: Symbolic counterexample generation for discrete-time Markov chains. In: Pǎsǎreanu, C.S., Salaün, G. (eds.) FACS 2012. LNCS, vol. 7684, pp. 134–151. Springer, Heidelberg (2013). https://doi.org/10.1007/978-3-642-35861-6_9

18. Jansen, N., Wimmer, R., Ábrahám, E., Zajzon, B., Katoen, J.P., Becker, B.: Symbolic counterexample generation for large discrete-time Markov chains. Sci. Comput. Program. **91**(A), 90–114 (2014)
19. Jantsch, S.: Certificates and witnesses for probabilistic model checking. Ph.D. thesis, Dresden University of Technology, Germany (2022)
20. Jantsch, S., Harder, H., Funke, F., Baier, C.: SWITSS: computing small witnessing subsystems. In: Proceedings of the 20th Conference on Formal Methods in Computer-Aided Design (FMCAD 2020), vol. 1, pp. 236–244. TU Wien Academic Press (2020)
21. Jantsch, S., Piribauer, J., Baier, C.: Witnessing subsystems for probabilistic systems with low tree width. In: Proceedings of the 12th International Symposium on Games, Automata, Logics, and Formal Verification (GandALF 2021). Electronic Proceedings in Theoretical Computer Science, vol. 346, pp. 35–51. Open Publishing Association (2021)
22. Kallenberg, L.: Linear programming and finite Markovian control problems. Mathematical Centre, Amsterdam (1983)
23. McConnell, R.M., Mehlhorn, K., Näher, S., Schweitzer, P.: Certifying algorithms. Comput. Sci. Rev. **5**(2), 119–161 (2011)
24. Quatmann, T., et al.: Counterexamples for expected rewards. In: Bjørner, N., de Boer, F. (eds.) FM 2015. LNCS, vol. 9109, pp. 435–452. Springer, Cham (2015). https://doi.org/10.1007/978-3-319-19249-9_27
25. Schrijver, A.: Theory of Linear and Integer Programming. Wiley-Interscience Series in Discrete Mathematics and Optimization. Wiley (1999)
26. Vardi, M.Y.: Automatic verification of probabilistic concurrent finite-state programs. In: Proceedings of the 26th Annual Symposium on Foundations of Computer Science (FOCS 1985), pp. 327–338. IEEE (1985)
27. Wimmer, R., Jansen, N., Ábrahám, E., Katoen, J.P., Becker, B.: Minimal counterexamples for refuting ω-regular properties of Markov decision processes. Reports of SFB/TR 14 AVACS 88 (2012)
28. Wimmer, R., Jansen, N., Ábrahám, E., Katoen, J.P., Becker, B.: Minimal critical subsystems as counterexamples for ω-regular DTMC properties. In: Methoden und Beschreibungssprachen zur Modellierung und Verifikation von Schaltungen und Systemen (MBMV 2012), pp. 169–180. Verlag Dr. Kovač (2012)
29. Wimmer, R., Jansen, N., Ábrahám, E., Becker, B., Katoen, J.-P.: Minimal critical subsystems for discrete-time Markov models. In: Flanagan, C., König, B. (eds.) TACAS 2012. LNCS, vol. 7214, pp. 299–314. Springer, Heidelberg (2012). https://doi.org/10.1007/978-3-642-28756-5_21
30. Wimmer, R., Jansen, N., Ábrahám, E., Katoen, J.P., Becker, B.: Minimal counterexamples for linear-time probabilistic verification. Theoret. Comput. Sci. **549**, 61–100 (2014)
31. Wimmer, R., Jansen, N., Vorpahl, A., Ábrahám, E., Katoen, J.-P., Becker, B.: High-level counterexamples for probabilistic automata. In: Joshi, K., Siegle, M., Stoelinga, M., D'Argenio, P.R. (eds.) QEST 2013. LNCS, vol. 8054, pp. 39–54. Springer, Heidelberg (2013). https://doi.org/10.1007/978-3-642-40196-1_4
32. Wimmer, R., Jansen, N., Vorpahl, A., Ábrahám, E., Katoen, J., Becker, B.: High-level counterexamples for probabilistic automata. Log. Methods Comput. Sci. **11**(1) (2015)

PCTL Satisfiability for Infinite Binary Trees

Antonín Kučera[✉][iD]

Masaryk University, Brno, Czechia
tony@fi.muni.cz

Abstract. We show that the problem whether a given PCTL formula has an infinite binary tree model where all transition probabilities are equal to 1/2 is highly undecidable (i.e., beyond the arithmetical hierarchy). This result holds even for the PCTL fragment where the set of modal connectives is restricted to the **F** and **G** operators, and even under the assumption that the PCTL formula on input is either unsatisfiable or it has a model with the aforementioned structure.

Keywords: Markov chains · Probabilistic temporal logics · Probabilistic CTL

1 Introduction

Probabilistic Computational Tree Logic (PCTL) was introduced by Hansson & Jonsson [16] as a probabilistic extension of the logic CTL (see, e.g., [14]). In the PCTL syntax, the existential/universal path quantifiers of CTL are replaced with the *probabilistic operator* that allows for specifying lower/upper bounds on the probability of all runs satisfying a given path formula. PCTL is interpreted over states in discrete Markov chains, and the decidability/complexity of the corresponding model-checking problem has been studied for finite-state Markov chains [2,3] and also for selected classes of infinite-state models [9,23]. PCTL formulae can also be used as objectives in Markov decision processes and two-player stochastic games, where the task is to design a strategy satisfying a given PCTL formula against an arbitrary strategy of the other player [6,8].

The *satisfiability* problem for PCTL and its fragments attracted considerable attention even before the logic had been explicitly defined. Unlike non-probabilistic temporal logics, PCTL does not have the *small model property*, guaranteeing the existence of a bounded-size model for every satisfiable formula. More precisely, there exist satisfiable PCTL formulae without *any* finite model (see, e.g., [7]). Hence, the PCTL satisfiability problem has also been studied in its finitary variant, where the set of eligible models is restricted to finite-state Markov chains (we refer to this problem as the "finite PCTL satisfiability").

The first positive decidability results have been obtained for the *qualitative fragment* of PCTL, where the range of admissible probability constraints

N. Bertrand et al. (Eds.): Christel Baier Festschrift, LNCS 15760, pp. 197–206, 2026.
https://doi.org/10.1007/978-3-031-97439-7_9

is restricted to $=0$, >0, $=1$, or <1. The satisfiability for qualitative PCTL is **EXPTIME**-complete, and the same holds for finite satisfiability [18,19,21]. Furhermore, a finite description of a model for a satisfiable qualitative PCTL formula is constructible in exponential time [7].

The finite satisfiability is decidable also for certain *quantitative* PCTL fragments where the probability constrains may involve arbitrary rational constants [10,12,20]. A recent result of [11] says that the finite satisfiability problem for *unrestricted* PCTL is *undecidable*. The general satisfiability problem is even *highly undecidable*, i.e., beyond the arithmetical hierarchy [13].

Christel Baier has substantially contributed to developing the algorithmic theory of PCTL. Her habilitation thesis [1] provides an excellent introduction into the area of probabilistic verification (the author of this paper is one of the grateful readers), and the book about model-checking [2] written jointly with Joost-Pieter Katoen is the standard entry point into the area.

This work is inspired by the paper [4] written by Christel's close colleagues and collaborators. In [4], it is argued that for practical reasons, it makes sense to restrict the admissible PCTL models to some simple and implementable subclass. Subsequently, it is shown that the existence of a model where all transition probabilities are equal to $\frac{1}{2}$ or 1 and the number of states is bounded by a given $b \in \mathbb{N}$ is **NP**-complete. Compared to the results mentioned above, this complexity is low, and the result is encouraging. In this paper, we elaborate this line of research by restricting the class of admissible models to Markov chains where all transition probabilities are equal to $\frac{1}{2}$ (without imposing any bound on the number of states). Note that each such Markov chain can be unfolded into an infinite binary tree. We show that the problem whether a given PCTL formula has a binary tree model is highly undecidable, even under the assumption that the PCTL formula on input is either unsatisfiable or it has a binary tree model.

2   Basic Definitions

We use $\mathbb{N}$ to denote the sets of non-negative integers. For the rest of this paper, we fix a countably infinite set AP of *atomic propositions*.

2.1   Markov Chains

A (discrete-time) *Markov chain* is a triple $M = (S, P, v)$, where

- S is a finite or countably infinite set of *states*,
- $P\colon S \times S \to [0,1]$ is a stochastic matrix such that $\sum_{t \in S} P(s,t) = 1$ for every $s \in S$,
- $v\colon S \to 2^{AP}$ is a *valuation*.

We say that M is finite if S is a finite set.

For $s, t \in S$, we say that t is an *immediate successor* of s if $P(s,t) > 0$. A *path* in M is a finite sequence $w = s_0, \ldots, s_n$ of states where $n \geq 0$ and $P(s_i, s_{i+1}) > 0$ for all $i < n$. We say that t is *reachable* from s if there is a path from s to t. A *run* in M is an infinite sequence $\pi = s_0, s_1, \ldots$ of states such that every finite prefix of π is a path in M. We also use $\pi(i)$ to denote the state s_i of π.

For every path $w = s_0, \ldots, s_n$, let $Run(w)$ be the set of all runs starting with w, and let $\mathbb{P}(Run(w)) = \prod_{i=0}^{n-1} P(s_i, s_{i+1})$. To every state s, we associate the probability space $(Run(s), \mathcal{F}_s, \mathbb{P}_s)$, where $\mathcal{F}_s$ is the σ-field generated by all $Run(w)$ where w starts in s, and $\mathbb{P}_s$ is the unique probability measure obtained by extending $\mathbb{P}$ in the standard way (see, e.g., [5]).

2.2   The Logic PCTL

The syntax of Probabilistic CTL (PCTL) [16] is obtained by replacing the existential and universal path quantifiers in the standard CTL (see, e.g., [15]) with the probabilistic operator bounding the probability of all runs satisfying a given path formula by a rational constant. More concretely, the syntax of PCTL state and path formulae is defined as follows:

$$\varphi \quad ::= \quad a \mid \neg\varphi \mid \varphi_1 \wedge \varphi_2 \mid P(\Phi) \bowtie r$$
$$\Phi \quad ::= \quad \mathbf{X}\varphi \mid \varphi_1 \mathbf{U} \varphi_2 \mid \varphi_1 \mathbf{U}^k \varphi_2$$

Here, $a \in AP$, $\bowtie \in \{\geq, >, \leq, <, =, \neq\}$, $r \in [0, 1]$ is a rational constant, and $k \in \mathbb{N}$.

The formulae *true, false* and the other Boolean connectives are defined using $\neg$ and $\wedge$ in the standard way. In the following, we abbreviate a formula of the form $P(\Phi) \bowtie r$ by omitting P and adjoining the probability constraint directly to the topmost path operator of Φ. For example, we write $\mathbf{X}_{=1}\varphi$ instead of $P(\mathbf{X}\varphi) = 1$.

Let $M = (S, P, v)$ be a Markov chain. The *validity* of a PCTL state/path formula for a given state/run of M is defined inductively as follows:

$$
\begin{aligned}
s \models a \qquad &\text{iff} \quad a \in v(s), \\
s \models \neg\varphi \qquad &\text{iff} \quad s \not\models \varphi, \\
s \models \varphi_1 \wedge \varphi_2 \qquad &\text{iff} \quad s \models \varphi_1 \text{ and } s \models \varphi_2, \\
s \models P(\Phi) \bowtie r \qquad &\text{iff} \quad \mathbb{P}_s(\{\pi \in Run(s) \mid \pi \models \Phi\}) \bowtie r, \\[4pt]
\pi \models \mathbf{X}\varphi \qquad &\text{iff} \quad \pi(1) \models \varphi \text{ for some } i \in \mathbb{N} \\
\pi \models \varphi_1 \mathbf{U} \varphi_2 \qquad &\text{iff} \quad \text{there is } j \geq 0 \text{ such that } \pi(j) \models \varphi_2 \text{ and} \\
&\qquad \pi(i) \models \varphi_1 \text{ for all } 0 \leq i < j, \\
\pi \models \varphi_1 \mathbf{U}^k \varphi_2 \qquad &\text{iff} \quad \text{there is } 0 \leq j \leq k \text{ such that } \pi(j) \models \varphi_2 \text{ and} \\
&\qquad \pi(i) \models \varphi_1 \text{ for all } 0 \leq i < j.
\end{aligned}
$$

We also use $\mathbf{F}_{\bowtie r}\varphi$ to abbreviate the formula $true\mathbf{U}_{\bowtie r}\varphi$. Furthermore, $\mathbf{G}_{\bowtie r}\varphi$ abbreviates $\mathbf{F}_{\overline{\bowtie} r}\neg\varphi$. The $\mathbf{F}, \mathbf{G}$-*fragment* of PCTL consists of all PCTL formulae containing only the path connectives $\mathbf{F}$ and $\mathbf{G}$.

A PCTL formula φ is *satisfiable* if there exists a Markov chain M such that $s \models \varphi$ for some state s of M.

2.3   Minsky Machines

A *non-deterministic Minsky machine* $\mathcal{M}$ *with two counters* [22] is a finite program

$$1 : Ins_1; \ \cdots \ m : Ins_m;$$

where $m \geq 1$ and every $i : Ins_i$ is a labeled instruction in one of the following forms:

  I. $i : inc \ c_j; \ goto \ u$
 II. $i : if \ c_j{=}0 \ then \ goto \ u \ else \ dec \ c_j; \ goto \ u'$
III. $i : goto \ u \ or \ u'$

Here, $j \in \{1, 2\}$ is a counter index and $u, u' \in \{1, \ldots, m\}$.

A *configuration* of $\mathcal{M}$ is a tuple (i, n_1, n_2) of non-negative integers where $1 \leq i \leq m$ represents the current control position and n_1, n_2 represent the current counter values. A configuration (i', n_1', n_2') is a *successor* of a configuration (i, n_1, n_2), written $(i, n_1, n_2) \mapsto (i', n_1', n_2')$, if $i \leq m$, the pair (n_1', n_2') is obtained from (n_1, n_2) by performing Ins_i (note that $n_j' \neq n_j$ for at most one $j \in \{1, 2\}$), and i' is the target label obtained by performing Ins_i. A *computation* of $\mathcal{M}$ is an infinite sequence $C_1, C_2, \ldots$ of configurations such that $C_1 = (1, 0, 0)$ and $C_i \mapsto C_{i+1}$ for all $i \geq 1$. A computation is *recurrent* if Ins_1 is executed infinitely often along the computation.

The problem whether a given non-deterministic Minsky machine $\mathcal{M}$ has a recurrent computation is hard for the Σ_1^1 level of the analytical hierarchy [17].

3   Binary Trees as PCTL Models

A *binary tree Markov chain (BTMC)* is a Markov chain $M = (S, P, v)$ where

 - $S = \{0, 1\}^*$, i.e., the states are finite words over the alphabet $\{0, 1\}$, including the empty word ε.
 - For every $w \in \{0, 1\}^*$ we have that $P(w, w0) = P(w, w1) = \frac{1}{2}$.

We say that a BTMC M is a model of a PCTL formula φ if $\varepsilon \models \varphi$. Furthermore, we say that a formula φ is *BTMC-faithful* if φ is either unsatisfiable or it has a BTMC model.

Theorem 1. *Let φ be a PCTL formula of the $\mathbf{F}, \mathbf{G}$-fragment. The problem whether there exists a BTMC model of φ is Σ_1^1-hard, even under the assumption that φ is BTMC-faithful.*

The rest of this section is devoted to the proof of Theorem 1. Our starting point is the recent result of [13] saying that the PCTL satisfiability problem is Σ_1^1-hard even for the $\mathbf{F}, \mathbf{G}$-fragment. More precisely, for a given non-deterministic Minsky machine $\mathcal{M}$, an $\mathbf{F}, \mathbf{G}$-formula $\varphi_\mathcal{M}$ is constructed such that $\mathcal{M}$ has a recurrent computation iff $\varphi_\mathcal{M}$ is satisfiable. Our proof of Theorem 1 is obtained by modifying the formula $\varphi_\mathcal{M}$ into a formula $\varphi_\mathcal{M}^*$ such that

 - if $\mathcal{M}$ has a recurrent computation, then $\varphi_\mathcal{M}^*$ has a BTMC model;
 - if $\mathcal{M}$ has no recurrent computation, then $\varphi_\mathcal{M}^*$ is not satisfiable.

Note that $\varphi_\mathcal{M}^*$ is BTMC-faithful.

3.1   The Structure of $\varphi_\mathcal{M}$ Models

In this section, we identify the properties of the formula $\varphi_\mathcal{M}$ presented in [13] that are crucial for constructing the formula $\varphi_\mathcal{M}^*$.

The formula $\varphi_\mathcal{M}$ of [13] takes the form

$$\varphi_\mathcal{M} \equiv Struct \wedge Simulate$$

where $Struct$ defines the basic structure of a model and $Simulate$ enforces the existence of a run encoding a recurrent computation of $\mathcal{M}$.

Let $\mathcal{A}$ be the finite set of all atomic propositions occurring in $\varphi_\mathcal{M}$. Slightly abusing our notation, for a given $\alpha \subseteq \mathcal{A}$ we use the same symbol α to denote the PCTL formula

$$\bigwedge_{a \in \alpha} a \wedge \bigwedge_{a \in \mathcal{A} \setminus \alpha} \neg a \tag{1}$$

saying that precisely the propositions of α are satisfied in a given state.

Suppose that $s \models \varphi_\mathcal{M}$ where s a state of a Markov chain M. Without restrictions, we assume that every state of M is reachable from s.

For every state t of M, we use $\alpha(t)$ denote the unique subset of $\mathcal{A}$ such that $t \models \alpha(t)$. The formula $\varphi_\mathcal{M}$ determines a set $\mathcal{E} \subseteq 2^\mathcal{A}$ such that $\alpha(t) \in \mathcal{E}$ for every state t of M.

The formula $Struct$ is essentially a conjunction of the form

$$Struct \equiv \bigwedge_{\alpha \in \mathcal{E}} Succ_\alpha$$

where $Succ_\alpha$ is either the formula

$$\mathbf{G}_{=1}\big(\alpha \Rightarrow (\mathbf{G}_{=1}\alpha)\big) \tag{2}$$

or a formula of the form

$$\mathbf{G}_{=1}\left(\alpha \Rightarrow \bigvee_{j=1}^{n} \alpha\mathbf{U}_{=1}(\beta_{j,1} \vee \cdots \vee \beta_{j,k_j})\right) \tag{3}$$

where $n \geq 1$ and $\alpha, \beta_{j,1}, \ldots, \beta_{j,k_j}$ are pairwise different elements of $\mathcal{E}$ for all $j \leq n$. Furthermore, the formula (3) is constructed so that for every state t of M such that $t \models \alpha$ there is precisely one $j \leq n$ such that $t \models \alpha\mathbf{U}_{=1}(\beta_{j,1} \vee \cdots \vee \beta_{j,k_j})$.

Note that (3) contains the $\mathbf{U}$ operator. In [13], it is shown how to rewrite (3) into a formula of the $\mathbf{F}, \mathbf{G}$-fragment with essentially the same meaning. For simplicity, we keep the current form of (3).

A state t of M if absorbing or transient depending on whether $Succ_{\alpha(t)}$ is the formula (2) or the formula (3), respectively. Observe that all successors of an absorbing state t are absorbing and satisfy the same subset of atomic propositions as t. Hence, the precise structure of absorbing states is irrelevant, and it does not influence the validity of $\varphi_\mathcal{M}$ in s.

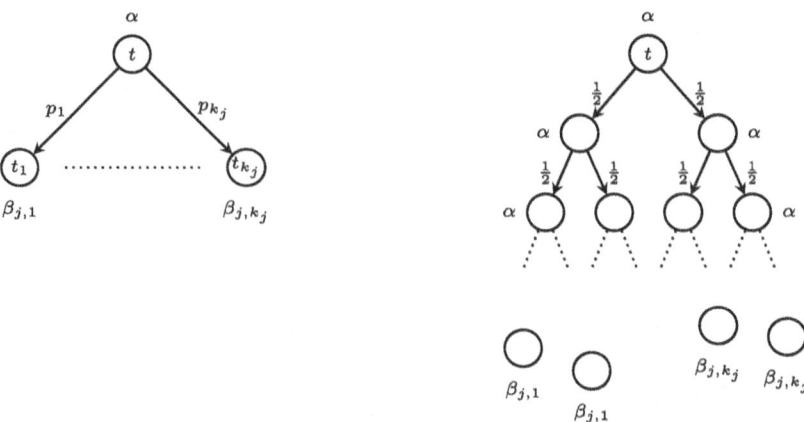

Fig. 1. Transforming a model of $\varphi_{\mathcal{M}}$ into a BTMC model.

Let t be a transient state, and consider the formula (3) associated to $\alpha = \alpha(t)$. Let j be the unique index such that

$$t \models \alpha \mathbf{U}_{=1}(\beta_{j,1} \vee \cdots \vee \beta_{j,k_j})$$

We use $Prob_t = (p_1, \ldots, p_{k_j})$ to denote the tuple of probabilities such that $t \models \alpha \mathbf{U}_{=p_i} \beta_{j,i}$ for every $i \in \{1, \ldots, k_j\}$. Furthermore, we say that a state u is a *relevant descendant* of t if u is transient and there is a finite path from t to u such that $u \models \beta_{j,1} \vee \cdots \vee \beta_{j,k_j}$, and all states in the path except for u satisfy α. We define *Rel* as the least set of states that contains s and is closed under relevant descendants (the state s satisfying $\varphi_{\mathcal{M}}$ is always transient). Note that there may exist transient states that do *not* belong to *Rel*; these are the states visited along paths leading from a relevant state t to its relevant descendants.

The formula *Simulate* enforces additional properties of transient states. Technically,

$$Simulate \equiv Init \wedge \bigwedge_{j=1}^{m} \mathbf{G}_{=1}(\psi_j \Rightarrow \xi_j).$$

The subformula *Init* says that the state s satisfying $\varphi_{\mathcal{M}}$ encodes the initial configuration $(1, 0, 0)$ of $\mathcal{M}$. For every $j \leq m$, the formula ψ_j is a Boolean combination of atomic predicates such that for every absorbing state t we have that $t \not\models \psi_j$. Hence, *Simulate* requires that $t \models \bigwedge_{j=1}^{m}(\psi_j \Rightarrow \xi_j)$ for every transient state t.

3.2   Modifying the Formula $\varphi_{\mathcal{M}}$ into $\varphi_{\mathcal{M}}^*$

Before presenting the modified formula $\varphi_{\mathcal{M}}^*$, let us first explain why we cannot construct a BTMC model directly for $\varphi_{\mathcal{M}}$. In [13], it is shown that if $\mathcal{M}$ has a recurrent computation, then $\varphi_{\mathcal{M}}$ has a model such that

- every absorbing state t has a self-loop with probability 1,
- every transient state t such that $t \models \alpha \mathbf{U}_{=1}(\beta_{j,1} \vee \cdots \vee \beta_{j,k_j})$ has k_j *immediate* successors $t_1, \ldots, t_{k_j}$ where $t_i \models \beta_{j,i}$ for all $i \leq k_j$ (see Fig. 1 (left)).

Observe that every transient state in this model is relevant. A natural idea is to transform this model into a BTMC model by introducing auxiliary non-relevant transient states and arranging them into a binary tree so that the tuple of probabilities $Prob_t$ is preserved (see Fig. 1 (right)). The main problem of this transformation is that, according to *Struct*, each of the newly introduced transient states t' must also satisfy the formula $\bigwedge_{j=1}^{m}(\psi_j \Rightarrow \xi_j)$. However, due to the new branching structure, it may happen that $Prob_{t'} \neq Prob_t$, and this may spoil the validity of $\bigwedge_{j=1}^{m}(\psi_j \Rightarrow \xi_j)$ in t'.

Although the transformation indicated in Fig. 1 does not necessarily preserve the validity of $\varphi_{\mathcal{M}}$, we can make it work by slightly modifying the formula $\varphi_{\mathcal{M}}$. Thus, we obtain the formula $\varphi_{\mathcal{M}}^*$. We construct $\varphi_{\mathcal{M}}^*$ so that the following properties are satisfied:

A. If $\varphi_{\mathcal{M}}^*$ is satisfiable, then there is an infinite run initiated in s encoding a recurrent computation of $\mathcal{M}$.
B. If $\mathcal{M}$ has a recurrent computation, then $\varphi_{\mathcal{M}}^*$ has a BTMC model obtained by applying the transformation of Fig. 1 to the model of $\varphi_{\mathcal{M}}$.

To achieve that, we need two more observations about the formula $\varphi_{\mathcal{M}}$. In [13], it is shown that if $\varphi_{\mathcal{M}}$ is satisfiable, i.e., $s \models Struct \wedge Simulate$, then there is an infinite run initiated in s encoding a recurrent computation of $\mathcal{M}$. However, a proof of this property actually requires the satisfaction of $\bigwedge_{j=1}^{m}(\psi_j \Rightarrow \xi_j)$ *only for all* $t \in Rel$. Furthermore, the satisfaction of $\bigwedge_{j=1}^{m}(\psi_j \Rightarrow \xi_j)$ in the states of Rel depends *only* on the tuples of probabilities $Prob_t$ where $t \in Rel$.

We start by extending $\mathcal{A}$ with a fresh atomic proposition $*$ (this also influences the formula (1) which is now considered for the extended $\mathcal{A}$). For every $\alpha \in \mathcal{E}$, let $\alpha_* = \alpha \cup \{*\}$. The formula $\varphi_{\mathcal{M}}^*$ is obtained by changing every $Succ_\alpha$ of the form (3) into

$$\mathbf{G}_{=1}\left(\alpha \Rightarrow \bigvee_{j=1}^{n}(\alpha \vee \alpha_*)\mathbf{U}_{=1}(\beta_{j,1} \vee \cdots \vee \beta_{j,k_j})\right) \tag{4}$$

and the formula *Simulate* into

$$Simulate \equiv Init \wedge \bigwedge_{j=1}^{m}\mathbf{G}_{=1}((\psi_j \wedge \neg *) \Rightarrow \xi_j). \tag{5}$$

The formula (4) allows to mark the *non-relevant* transient states with $*$. The formula (5) requires the satisfaction of $\bigwedge_{j=1}^{m}(\psi_j \Rightarrow \xi_j)$ only in the states not satisfying $*$.

Observe that $\varphi_{\mathcal{M}}^*$ still guarantees the satisfaction of $\bigwedge_{j=1}^{m}(\psi_j \Rightarrow \xi_j)$ in all states of Rel, because the states of Rel do not satisfy $*$. Thus, we obtain the property A. Furthermore, observe that a model of $\varphi_{\mathcal{M}}$ is also a model of $\varphi_{\mathcal{M}}^*$.

Hence, we can apply the transformation indicated in Fig. 1 to the model of $\varphi_{\mathcal{M}}$ constructed in [13] so that the newly added non-important transient states satisfy α_* instead of α (this is admitted by (4)). Note that the satisfaction of $\bigwedge_{j=1}^{m}(\psi_j \Rightarrow \xi_j)$ is *not* required in these states due to (5). Thus, we achieve B. The details of the transformation are given in the next subsection.

3.3   Constructing a BTMC Model of $\varphi_{\mathcal{M}}^*$

In this section, we show that if $\mathcal{M}$ has a recurrent computation, then the formula $\varphi_{\mathcal{M}}^*$ has a BTMC model.

A *general binary tree Markov chain (gBTMC)* is a directed tree where every inner vertex t has exactly two immediate successors u, v where $P(t, u) = P(t, v) = \frac{1}{2}$, and every leaf t satisfies $p(t, t) = 1$.

We start by unfolding the model of $\varphi_{\mathcal{M}}$ constructed in [13] into an infinite directed tree T. Then, we proceed by gradually transforming T into a BTMC which is a model of $\varphi_{\mathcal{M}}^*$. This is achieved by processing all vertices of T from the root towards its successors in the way indicated in Fig. 1. That is, at each step we transform a vertex t of T by replacing t with a gBTMC τ_t rooted by t. If t is absorbing, then τ_t has no leaves and all inner vertices satisfy α_t. If t is transient with immediate successors $t_1, \ldots, t_{k_j}$ satisfying $\beta_{j,1}, \ldots, \beta_{j,k_k}$, then

- every inner vertex of τ_t except for the root t satisfies $\alpha_*(t)$;
- every leaf of τ_t is labeled by some t_i;
- the probability of reaching a leaf labeled by t_i from the root of τ_t is equal to $Prob_t(i)$ (i.e., the i-th component of $Prob_t$)

Then, every leaf of τ_t labeled by t_i is replaced with the subtree of T rooted by t_i, and the transformation proceeds by processing all of these t_i's.

Note that the gBTMC τ_t for a transient t with the above properties exists, although it may have infinitely many vertices. More concretely, it is constructed so that for all $i \leq k_j$ and $d \geq 1$, the gBTMC τ_t contains at most one leaf labeled by t_i with distance d from the root, and such a leaf exists iff the d-th digit in the binary representation of $Prob_t(i)$ (after the leading 0.) is equal to 1.

4   Conclusions

The presented result shows that the PCTL satisfiability remains undecidable even if the class of eligible models is restricted to BTMC and the PCTL formula of input is BTMC-faithful. Hence, although restricting the class of eligible models may have positive impact on the decidability of the PCTL satisfiability problem, it is not yet clear what structural properties (except for boundedness) can bring more decidability/tractability, and for which PCTL fragments. This is an interesting challenge for future research.

References

1. Baier, C.: On Algorithmic Verification Methods for Probabilistic Systems. Habilitation thesis, University of Mannheim (1998)
2. Baier, C., Katoen, J.P.: Principles of Model Checking. The MIT Press, Cambridge (2008)
3. Baier, C., Kwiatkowska, M.: Model checking for a probabilistic branching time logic with fairness. Distrib. Comput. **11**(3), 125–155 (1998)
4. Bertrand, N., Fearnley, J., Schewe, S.: Bounded satisfiability for PCTL. In: Proceedings of CSL 2012. Leibniz International Proceedings in Informatics, vol. 16, pp. 92–106. Schloss Dagstuhl–Leibniz-Zentrum für Informatik (2012)
5. Billingsley, P.: Probability and Measure. Wiley, Hoboken (1995)
6. Brázdil, T., Brožek, V., Forejt, V., Kučera, A.: Stochastic games with branching-time winning objectives. In: Proceedings of LICS 2006, pp. 349–358. IEEE Computer Society Press (2006)
7. Brázdil, T., Forejt, V., Křetínský, J., Kučera, A.: The satisfiability problem for probabilistic CTL. In: Proceedings of LICS 2008, pp. 391–402. IEEE Computer Society Press (2008)
8. Brázdil, T., Forejt, V., Kučera, A.: Controller synthesis and verification for Markov decision processes with qualitative branching time objectives. In: Aceto, L., Damgård, I., Goldberg, L.A., Halldórsson, M.M., Ingólfsdóttir, A., Walukiewicz, I. (eds.) ICALP 2008. LNCS, vol. 5126, pp. 148–159. Springer, Heidelberg (2008). https://doi.org/10.1007/978-3-540-70583-3_13
9. Brázdil, T., Kučera, A., Stražovský, O.: On the decidability of temporal properties of probabilistic pushdown automata. In: Diekert, V., Durand, B. (eds.) STACS 2005. LNCS, vol. 3404, pp. 145–157. Springer, Heidelberg (2005). https://doi.org/10.1007/978-3-540-31856-9_12
10. Chakraborty, S., Katoen, J.: On the satisfiability of some simple probabilistic logics. In: Proceedings of LICS 2016, pp. 56–65 (2016)
11. Chodil, M., Kučera, A.: The finite satisfiability problem for PCTL is undecidable. In: Proceedings of LICS 2024, Article No. 22, pages 1–14. ACM Press (2024)
12. Chodil, M., Kučera, A.: The satisfiability problem for a quantitative fragment of PCTL. J. Comput. Syst. Sci. **139**, 103478 (2024)
13. Chodil, M., Kučera, A.: The satisfiability and validity problems for probabilistic computational tree logic are highly undecidable. In: Proceedings of ICALP 2025. Leibniz International Proceedings in Informatics, vol. 334. Schloss Dagstuhl–Leibniz-Zentrum für Informatik (2025)
14. Clark, E., Grumberg, O., Peled, D.: Model Checking. The MIT Press, Cambridge (1999)
15. Emerson, E.: Temporal and modal logic. Handb. Theor. Comput. Sci. **B**, 995–1072 (1991)
16. Hansson, H., Jonsson, B.: A logic for reasoning about time and reliability. Formal Aspects Comput. **6**, 512–535 (1994)
17. Harel, D.: Effective transformations on infinite trees with applications to high undecidability dominoes, and fairness. J. Assoc. Comput. Mach. **33**(1) (1986)
18. Hart, S., Sharir, M.: Probabilistic temporal logic for finite and bounded models. In: Proceedings of POPL'84, pp. 1–13. ACM Press (1984)
19. Kraus, S., Lehmann, D.: Decision procedures for time and chance (extended abstract). In: Proceedings of FOCS'83, pp. 202–209. IEEE Computer Society Press (1983)

20. Křetínský, J., Rotar, A.: The satisfiability problem for unbounded fragments of probabilistic CTL. In: Proceedings of CONCUR 2018. Leibniz International Proceedings in Informatics, vol. 118, pp. 32:1–32:16. Schloss Dagstuhl–Leibniz-Zentrum für Informatik (2018)
21. Lehmann, D., Shelah, S.: Reasoning with time and chance. Inf. Control **53**, 165–198 (1982)
22. Minsky, M.: Computation: Finite and Infinite Machines. Prentice-Hall, Upper Saddle River (1967)
23. Winkler, T., Gehnen, C., Katoen, J.-P.: Model checking temporal properties of recursive probabilistic programs. In: FoSSaCS 2022. LNCS, vol. 13242, pp. 449–469. Springer, Cham (2022). https://doi.org/10.1007/978-3-030-99253-8_23

CTL^* Verification and Synthesis Using Existential Horn Clauses

Mishel Carelli$^{(\boxtimes)}$ and Orna Grumberg

Technion - Israel Institute of Technology, Haifa, Israel
mishel.carelli@cispa.de, orna@cs.technion.ac.il

Abstract. This work proposes a novel approach for automatic verification and synthesis of infinite-state reactive programs with respect to CTL^* specifications, based on translation to Existential Horn Clauses (EHCs).

CTL^* is a powerful temporal logic, which subsumes the temporal logics LTL and CTL, both widely used in specification, verification, and synthesis of complex systems.

EHCs with its solver E-HSF, is an extension of Constrained Horn Clauses, which includes existential quantification as well as the power of handling well-foundedness.

We develop the translation system Trans, which given a verification problem consisting of a program P and a specification ϕ, builds a set of EHCs which is satisfiable iff P satisfies ϕ. We also develop a synthesis algorithm that given a program with holes in conditions and assignments, fills the holes so that the synthesized program satisfies the given CTL^* specification.

We prove that our verification and synthesis algorithms are both sound and relative complete. Finally, we present case studies to demonstrate the applicability of our algorithms for CTL^* verification and synthesis.

1 Introduction

This work proposes a novel approach for automatic verification and synthesis of reactive infinite-state programs with respect to CTL^* specifications, based on translation to Existential Horn Clauses (EHC).

CTL* is a powerful temporal logic, which subsumes the temporal logics LTL and CTL, both widely used in specification, verification, and synthesis of complex systems in industry and in academic research. In addition to specifying the behavior of a system over time, CTL* is capable of describing the system's branching structure.

Constrained Horn Clauses (CHCs) is a fragment of first-order logic which has been recently highly successful in automated verification of infinite-state programs with respect to safety properties [5,6]. CHCs include uninterpreted predicates that represent, for instance, loop invariants within the checked program. CHCs are accompanied with powerful solvers that can determine their

© The Author(s), under exclusive license to Springer Nature Switzerland AG 2026
N. Bertrand et al. (Eds.): Christel Baier Festschrift, LNCS 15760, pp. 207–233, 2026.
https://doi.org/10.1007/978-3-031-97439-7_10

satisfiability. That is, they can find an appropriate interpretation for the uninterpreted predicates, so that the CHCs are satisfied. Given a verification problem consisting of a program and a safety property, the problem can be translated to a set of CHCs, which is satisfiable iff the program satisfies the property.

While being successful in proving safety properties, CHCs lack the power of proving properties such as termination and general liveness or branching temporal properties. For the former, a notion of well-foundedness is needed. For the latter, an existential quantifier is needed for choosing one of several branches.

Existential Horn Clauses (EHCs) with their solver E-HSF [3], is an extension of CHCs, which include existential quantification as well as the power of handling well-foundedness.

Our work is built on two pillars. One, is the seminal paper by Kesten and Pnueli [15], which establishes a compositional approach to deductive verification of First-Order CTL^*. The other is the EHC logic with its solver E-HSF [3] applicable to linear programs, which together provides a strong formal setting for verification problems that can be determined by EHC satisfiability.

The compositional approach to CTL^* in [15] provides proof rules for basic (assertional) state formulas of the form Qc, where Q is a path quantifier and c is an assertion. Formulas including temporal operators are reduced to the basic form by applying rules for successive eliminations of temporal operators. The reductions involve building program-like (testers) structures that are later composed with each other and with the verified program.

Our work adopts the idea of transforming a specification to basic state formulas while eliminating temporal operators. However, it takes a significantly different approach. We develop a translation system Trans, which given a verification problem consisting of a program P and a specification ϕ, builds a set of EHCs which is satisfiable iff P satisfies ϕ. Trans consists of 8 translation rules. The first replaces each basic subformula in ϕ with a boolean variable. The second replaces non-fair path quantifiers Q with fair path quantifiers Q_f. The next three rules eliminate temporal path subformulas of the form $Next\ (Xc)$, $Globally\ (Gx)$, and $Until\ (c_1 U c_2)$, using fresh boolean variables and fresh uninterpreted predicates. Additional constraints require that the program satisfies the appropriate temporal property.

It should be noted, that our programs include fairness conditions that restrict the paths under consideration. This is because, the elimination of both *Globally* and *Until* require such conditions. Moreover, handling fairness as part of the program and not as part of a specification is much more efficient. Typically, the program is large and the specification is relatively small. However, specifications describing program fairness might be of size of the program.

The sixth and seventh translation rules in Trans deviate the most from [15]. Rule 6 handles the verification of $A_f c$, i.e., along every fair path, assertion c holds. For that purpose, the translation imposes the requirement that if a path does not satisfy c, then it is not fair. Well-foundedness is needed for proving this fact. Rule 7 handles $E_f c$, i.e., for every initial state there is a fair path starting at it. Here we need both well-foundedness and an existential quantification, both supplied by EHCs. In our translation rules we also use negations and disjunctions that are not allowed by CHCs, but are expressible by EHCs.

We prove that our translation Trans is sound and relative complete.

Next, we turn to exploiting our approach for synthesis. We are given a *partial program*, in which some of the conditions (e.g. in while loops) and assignments are missing (*holes*). We are also given a CTL^* specification. Our goal is to synthesize a "filling" to the holes, so that the synthesized program satisfies the CTL^* specification.

We obtain this by first filling each hole with an uninterpreted predicate. Additional constraints are added to guarantee that predicates associated with a condition or an assignment indeed behave accordingly. Finally, we build EHCs to describe the new program and send it to the EHC solver, which will return interpreted predicates to serve as hole fillings. Here again we prove the soundness and relative completeness of our approach.

Finally, we present case studies to demonstrate the applicability of our algorithms for CTL^* verification and synthesis.

Remark: A preliminary version of this work appeared in ATVA 2024 [8]. The current version includes the full set of proofs and also a few comments referring to new insights we had while preparing the final version.

Related Work. Several works [1,4] exploit the approach presented in [15] to perform CTL+FO verification via EHC-solving. CTL+FO has the same syntax as CTL, except that the atomic propositins are first-order (FO) formulas. We significantly extend their result to perform full CTL^* verification for CTL^* with first-order atomic propositions. We also show that EHCs can be further exploited for CTL^* synthesis.

In [2], EHCs are used to solve reachability, safety and LTL games on infinite graphs. This tool can perform CTL^* verification, since CTL^* can be expressed through μ-calculus [11], μ-calculus verification can be performed through parity-games [19], which are a special case of LTL games. However, the translation of CTL^* to μ-calculus is doubly exponential [11]. As a result, a direct model checking of CTL^* according to our approach is more efficient than first translating and then solving the resulting LTL game.

In [21], condition synthesis is performed, given a program with holes in conditions and specifications as assertions in the code. The program behaviors are modeled "backwards" using CHCs. The CHC solver infers conditions so that the synthesized program satisfies the assertions. In our work we synthesize both conditions and assignments with respect to full CTL^*. Program behaviors are modeled "forward". However, we use EHCs rather than the more standard modeling via CHCs to model the problem.

Many works use synthesis to fill holes in partial program. Sketch [14,22] synthesizes programs to satisfy input-output specifications. In template-based synthesis [23] the specification is in the form of assertions in the code. Overview on synthesis and on synthesis of reactive programs can be found in [13] and [12]. Condition synthesis can also be used to *repair* programs [18,21,26,27].

Termination has been proved in [10]. Several works are able to synthesize *finite-state* full systems for CTL* specifications with *propositional* atomic formulas. [17] translates propositional CTL* formulas to hesitant alternating automata. [16] proposes bounded synthesis for CTL*. The paper bounds the number of states in the synthesized system to some k. The bound is increased if no such system exists. Clearly, such an algorithm is complete only if the goal of the synthesis is a finite-state system. Moreover, it cannot determine unrealizability. [7] shows how to apply CTL* synthesis via LTL synthesis. Both logics are propositional. It exploits translation to a hesitant automaton, which is then sent to an SMT solver.

In contrast, our work handles CTL* specifications with First-Order atomic formulas. It can synthesize infinite-state systems and can determine unrealizability. However, we are given as input a partial (sketch-like) program, from which we synthesize a full program.

Another type of Horn-like clauses, very similar to EHCs, called pfwCSP, has been introduced in [25]. pfwCSP does not allow existential quantification in the heads of clauses but allows disjunctions and can require that an uninterpreted predicate behaves like a function. Instead of disjunctively well-founded clauses, pfwCSP can require well-foundedness.

The authors of [25] present a fully automated CEGIS-based method for solving pfwCSP and implement it for the theory of quantifier-free linear arithmetic. Furthermore, the authors describe how relational verification concerning different relational verification problems, including k-safety, co-termination, termination-sensitive non-interference, and generalized non-interference for infinite-state programs, can be performed using pfwCSP.

The authors conjecture that EHCs and pfwCSP are inter-reducible.

In future work, our rules can be redesigned to reduce CTL^* verification to pfwCSP satisfiability by replacing the only occurrence of the existential quantifier in rule 7 with a Skolem function using a functional predicate and substituting disjunctive well-foundedness with well-foundedness.

Another use of pfwCSP is presented in [24]. The authors define a first-order fixpoint logic with background theories called μCLP and reduce the validity checking of μCLP to pfwCSP solving. Moreover, they present a modular primal-dual method that improves the solving of the pfwCSP, generated for μCLP formulas.

The authors mention that several problems, such as LTL, CTL, and even full μ-calculus model checking, can be encoded using μCLP for infinite-state systems in a sound and complete manner. Similar techniques were used in [25] to perform μ-calculus verification in a sound but incomplete way.

Another sound but incomplete procedure for CTL^* verification of infinite-state systems has been presented in [9].

2 Preliminaries

2.1 Forall-Exists Horn-Like Clauses (EHC)

Let $\mathcal{T}$ be a first-order theory over signature Σ. Then, *Forall-exists Horn-like clauses*, also called *Existential Horn Clauses* (EHCs), are first-order formulas of two types: implications and dwf-clauses.

1. Implications are formulas with uninterpreted predicates of the form:

$$c(v_0) \wedge q_1(v_1) \wedge \cdots \wedge q_n(v_n) \rightarrow \exists w : b(w_0) \wedge p_1(w_1) \wedge \ldots p_m(w_m)$$

   where $q_1, \ldots, q_n, p_1, \ldots, p_m$ are uninterpreted predicate symbols, not included in Σ; $v_0 \ldots v_n, w_0, \ldots, w_m$ are tuples of variables, not necessarily disjoint; c and b are quantifier-free formulas from $\mathcal{T}$ over variables v_0 and w_0, respectively, and $w \subseteq \bigcup_{i=0}^{m} w_i$.
   The left-hand side of the clause is called the *body* of the clause and the right-hand side is called the *head*.
2. Disjunctively well-founded clauses or dwf-clauses are of the form:

$$dwf(q)$$

   where $q(v, v')$ is an uninterpreted predicate of arity $2n$ for some $n \in \mathbb{N}$.

The Semantics of EHCs. Let $\mathcal{I}$ be an interpretation, defined over a domain $\mathcal{E}$. $\mathcal{I}$ associates with each uninterpreted predicate of arity k a k-ary relation over $\mathcal{E}$. Given a formula ϕ, $\mathcal{I}(\phi)$ is the formula obtained by replacing each occurrence of an uninterpreted predicate p with its interpretation, $\mathcal{I}(p)$. We say that clause $f(u) = body(u) \rightarrow head(u)$ is satisfied under $\mathcal{I}$, if $\mathcal{I}(body(u)) \models_{\mathcal{T}} \mathcal{I}(head(u))$.

   To define the semantics of dwf-clauses we need the notion of a disjunctively well-founded relation.

Definition 1. *A relation $r(v, v')$ over a set X is* well-founded *if there is no infinite sequence of elements in X, $x_1, x_2, \ldots$ such that for every i: $r(x_i, x_{i+1})$. A relation $r(v, v')$ is* disjunctively well-founded *if it is included in a finite union of well-founded relations. Formally, $r(v, v') \rightarrow r_1(v, v') \vee \cdots \vee r_l(v, v')$ for some well-founded relations $r_1, \ldots, r_l$.*

   An interpretation $\mathcal{I}$ satisfies $dwf(q)$ for a predicate q of arity $2n$, if $\mathcal{I}(q)$ is a disjunctively well-founded relation over the set $\mathcal{E}^n$.

   Note that, our notion of satisfiability is semantic, referring only to the domain over which the interpretation $\mathcal{I}$ is defined, ignoring the syntax over which relations associated with uninterpreted predicates are expressed.

   Theorem 1 from [20] can be reformulated as follows: A relation is *well-founded* if its transitive closure is disjunctively well-founded. We refer to this formulation when we use this theorem in our work.

Expressing Disjunction and Negation with EHC. For our purposes, we need to express a disjunction in the heads of EHC clauses. A clause of the form $body \rightarrow head_1 \vee head_2$ is equivalent to the following three forall-exists Horn-like clauses, using a new boolean variable a:

$$body \rightarrow \exists a : head(a); \quad head(0) \rightarrow head_1; \quad head(1) \rightarrow head_2$$

Given a predicate $p(v)$, we define a predicate $q(v)$ such that $q(v) \equiv \neg p(v)$ using the following two clauses:

$$p(v) \wedge q(v) \rightarrow \bot; \quad \top \rightarrow p(v) \vee q(v)$$

If both of them are satisfied, then $q(v) \equiv \neg p(v)$.

From now on, we use disjunctions and negations as part of the forall-exists Horn-like clauses, implying that we express them as described above.

2.2 CTL^* Verification

In this section we present the verification problems that we solve.

Program. Let us fix a first-order theory $\mathcal{T}$. We view a program as a transition system with fairness conditions. A program P consists of a tuple of program variables v, and two formulas $next(v, v')$ and $init(v)$, describing the transition relation and the set of initial states, respectively. In addition, it includes a finite set J of assertions in $\mathcal{T}$, describing the set of fairness conditions.

The states of P are the valuations of v. A *path* of P is a sequence of states $s_1, s_2, \ldots$, such that $s_i, s_{i+1} \models_{\mathcal{T}} next(v, v')$ for every i. An infinite path $s_1, s_2, \ldots$ is *fair* if for every $\phi \in J$ we have $s_i \models_{\mathcal{T}} \phi$ for infinitely many i's. If $J = \emptyset$ then every infinite path is fair. Note that the transition relation is not required to be total. Nevertheless, only infinite sequences of states are considered as program paths in the semantics of CTL^* defined later.

Syntax and semantics of CTL^*. We define the syntax of CTL^* formulas in negation normal form (NNF), where negations are applied only to atomic assertions. Let c range over assertions in $\mathcal{T}$. State formulas ϕ and path formulas ψ can be defined by the following grammar.

$$\phi := c|\phi \wedge \phi|\phi \vee \phi|E\psi|A\psi|E_f\psi|A_f\psi$$

$$\psi := \phi|G\psi|X\psi|\psi U\psi|\psi \wedge \psi|\psi \vee \psi$$

CTL^* is the set of state formulas defined by this grammar.

We use Q to denote a non-fair quantifier (A or E), ranging over *all* infinite program paths. Q_f denotes a fair quantifier (A_f or E_f), ranging only over fair paths.

A CTL^* state formula of the form $Q\psi$ or $Q_f\psi$ is called *basic*, if ψ does not contain path quantifiers.

Let P be a program, s range over the states of P, and π range over the paths of P. We use π^i to denote the suffix of π, starting at the i'th state.

The semantics of CTL^* is defined with respect to a program P, where boolean and temporal operators have their usual semantics: $\pi \models X\psi$ iff ψ holds on π^1; $\pi \models G\psi$ iff ψ holds on every suffix of π; $\pi \models \psi_1 U\psi_2$ iff ψ_2 eventually holds on a suffix π^j of π and ψ_1 holds in all preceding suffixes.

$P, s \models c$ iff $s \models_\mathcal{T} c$

$P, s \models E\psi$ iff there exists an infinite path π starting at s such that $P, \pi \models \psi$

$P, s \models A\psi$ iff for every infinite path π starting at s : $P, \pi \models \psi$

$P, s \models E_f\psi$ iff there exists a fair path π starting at s such that $P, \pi \models \psi$

$P, s \models A_f\psi$ iff for every fair path π starting at s : $P, \pi \models \psi$

$P, \pi \models \phi$ iff s is the first state of π and $P, s \models \phi$

The CTL^* Verification Problem. We say that P satisfies a CTL^* formula ϕ, denoted $P \models \phi$, if ϕ holds in every initial state of P. That is, for every state s of P: $init(s) \rightarrow P, s \models \phi$.

We view the verification problem as the tuple $(v, init(v), next(v, v'), J, \phi)$, where the first four elements are the parameters of the program and the last one is the CTL^* state formula that the program must satisfy.

3   Reduction of CTL^* Verification Problem to EHC Satisfiability

Our goal is to generate a set of EHCs for a given verification problem, such that the set of EHCs is satisfiable iff the program meets the specification.

We denote the desired set of EHCs as $Clauses(v, init(v), next(v, v'), J, \phi)$ and define it recursively on the structure of ϕ. We prove that this recursive definition is always sound (Theorem 2) and relatively complete (Theorem 3).

Below, we present our *translation system* Trans, which consists of eight rules, and explain the intuition behind them.

We use the notation $\phi_1(\phi_2)$ to describe a formula ϕ_1 that contains one or more occurrences of a state subformula ϕ_2. $\phi_1(\gamma)$ then stands for the formula ϕ_1 in which every occurrence of ϕ_2 is replaced by the state formula γ.

Denote $D = (v, init(v), next(v, v'), J)$. Recall that if $D \models \phi$ then for all s, $init(s) \rightarrow \phi$.

1. Consider the CTL^* state formula $\phi_2(\phi_1)$ with a basic state subformula ϕ_1 and let aux be a fresh uninterpreted predicate of arity $|v|$, then

$$Clauses(D, \phi_2(\phi_1))$$

$$= Clauses(v, aux(v), next(v, v'), J, \phi_1) \cup Clauses(D, \phi_2(aux)(v)).$$

Explanation of Rule 1: The fresh predicate aux represents an underapproximation of the set of states in which ϕ_1 is satisfied. Since our formulas are in NNF, where negations are applied only to atomic assertions, we can conclude that $\phi_2(aux(v))$ is an underapproximation of $\phi_2(\phi_1)$. Thus, by $init(v) \rightarrow \phi_2(aux)$ we also have $init(v) \rightarrow \phi_2(\phi_1)$.

2. Let $Q\phi$ be a CTL^* basic state formula, where Q is a non-fair path quantifier (E or A). Let Q_f be the fair version of Q (E_f or A_f, respectively), then

$$Clauses(D, Q\phi) = Clauses(v, init(v), next(v, v'), \emptyset, Q_f\phi).$$

Explanation of Rule 2: Since the only path quantifier in $Q\phi$ is non-fair, it does not depend on the fairness conditions. Hence, we can eliminate those conditions from the program and replace Q with its fair version.

3. Let c be an assertion in T. Suppose we have the CTL^* basic state formula $Q_f\phi(Xc)$. Let x_X be a new boolean variable, then

$$Clauses(D, A_f\phi(Xc))$$

$$= Clauses(v \cup \{x_X\}, init(v), next(v, v') \wedge (x_X = c(v')), J, A_f\phi(x_X)).$$

In the case of an existential path quantifier we also need a fresh uninterpreted predicate aux of arity $|v| + 1$:

$$Clauses(D, E_f\phi(Xc)) = \{init(v) \rightarrow \exists x_X : aux(v, x_X)\}\cup$$

$$\cup\, Clauses(v \cup \{x_X\}, aux(v, x_X), next(v, v') \wedge (x_X = c(v')), J, E_f\phi(x_X)).$$

Explanation of Rule 3: The new variable x_X represents the value of $c(v)$ in the next state of the path. This is achieved through the new constraint added to $next(v, v')$. Hence, we can replace Xc by x_X in the verified formula. For the existential quantifier E_f, we add a new auxiliary predicate aux over (v, x_X) that, if initially holds for a certain value of x_X, it is guaranteed that $E_f\phi(x_X)$ holds as well. The additional clause $init(v) \rightarrow \exists x_X : aux(v, x_X)$, when interpreted together with the rest of the clauses, guarantees that for every initial state there is a value of x_X, such that $aux(v, x_X)$ initially holds. The reason why we need this auxiliary predicate is that we do not know the value of x_X in the beginning of the path. For example, if $\phi(Xc) = FXc$, then a path that satisfies this formula may or may not have Xc in the first state. Therefore, we want to express that we can consider paths with any initial value of x_X. The second set of clauses guarantees that if initially $aux(v, x_X)$ holds for some value x_X, then $F_f\phi(x_X)$ holds as well.

The same approach is applied in rules 4 and 5 with the existential quantifier.

4. Let c be an assertion in T. Suppose we have the CTL^* basic state formula $Q_f\phi(Gc)$. Let x_G be a new boolean variable, then

$$Clauses(D, A_f\phi(Gc))$$

$$= Clauses(v \cup \{x_G\}, init(v), next(v, v') \wedge (x_G = (c(v) \wedge x'_G)), J \cup \{x_G \vee \neg c(v)\}, A_f\phi(x_G)).$$

In the case of an existential path quantifier, similarly to Rule 3, we also need a fresh uninterpreted predicate aux of arity $|v| + 1$:

$$Clauses(D, E_f\phi(Gc)) = \{init(v) \rightarrow \exists x_G : aux(v, x_G)\}\cup$$

$$Clauses(v \cup \{x_G\}, aux(v, x_G), next(v, v') \wedge (x_G = c(v) \wedge x'_G), J \cup \{x_G \vee \neg c(v)\}, E_f\phi(x_G))$$

Explanation of Rule 4: If the new variable x_G is true in some state, then the new condition added to the transition relation ensures that $c(v)$ and x_G are true in every consecutive state. Thus, x_G represents the value of Gc along the path. The new fairness condition ensures that if for every state of the path $c(v)$ is true, then x_G must be true infinitely often along this path.

5. Let c_1 and c_2 be assertions in T. Suppose we have the CTL^* basic state formula $Q_f\phi(c_1Uc_2)$. Let x_U be a new boolean variable, then

$$Clauses(D, A_f\phi(c_1Uc_2)) = Clauses(v \cup \{x_U\}, init(v),$$

$$next(v, v') \wedge (x_U = c_2(v) \vee (c_1(v) \wedge x'_U)), J \cup \{\neg x_U \vee c_2(v)\}, A_f\phi(x_U)).$$

In the case of an existential path quantifier, similarly to Rule 3, we need a fresh uninterpreted predicate aux of arity $|v| + 1$:

$$Clauses(D, E_f\phi(c_1Uc_2))$$

$$= Clauses(v \cup \{x_U\}, aux(v, x_U), next(v, v') \wedge (x_U = (c_2(v) \vee (c_1(v) \wedge x'_U))),$$

$$J \cup \{\neg x_U \vee c_2(v)\}, E_f\phi(x_U)) \cup \{init(v) \rightarrow \exists x_U : aux(v, x_U)\}.$$

Explanation of Rule 5: The new variable x_U represents the value of c_1Uc_2 along the path. The new condition added to the transition relation defines U recursively. The new fairness condition ensures that if for every state of the path x_U holds, then $c_2(v)$ must hold infinitely often along this path.

6. Let c be an assertion in T. Suppose we have a basic CTL^* state formula A_fc. Let p be a fresh uninterpreted predicate of arity $|v|$ and let r, t be fresh uninterpreted predicates of arity $2|v|$. Let $v_0, \ldots, v_{|J|}$ be $|J| + 1$ copies of the set of program variables v. Suppose $J \neq \emptyset$, then,

$$Clauses(D, A_fc) = \{init(v) \wedge \neg c(v) \rightarrow p(v), \ next(v, v') \wedge p(v) \rightarrow p(v'),$$

$$p(v_0) \wedge \bigwedge_{i=1}^{|J|}(t(v_{i-1}, v_i) \wedge J_i(v_i)) \rightarrow r(v_0, v_{|J|}), \ dwf(r),$$

$$next(v, v') \rightarrow t(v, v'), \ t(v, v') \wedge next(v', v'') \rightarrow t(v, v'')\},$$

where J_i is the i-th element of J.

If $J = \emptyset$, then,

$$Clauses(D, A_fc) = \{init(v) \wedge \neg c(v) \rightarrow p(v), \ next(v, v') \wedge p(v) \rightarrow p(v'),$$

$$p(v) \wedge t(v, v') \rightarrow r(v, v'), \ dwf(r),$$

$$next(v, v') \rightarrow t(v, v'), \ t(v, v') \wedge next(v', v'') \rightarrow t(v, v'')\}$$

Explanation of Rule 6: The new predicate $p(v)$ represents states that are reachable from initial states that do not satisfy $c(v)$. The predicate t represents the transitive closure of the transition relation $next$. The predicate $r(v, v')$ represents a pair of states from p that can be connected by a path that satisfies every fairness constraint at least once.

In order to verify that the program satisfies $A_f c$ we need to prove that every initial state either satisfies $c(v)$ or there is no fair path starting in that state. The latter is guaranteed by showing that $r(v, v')$ is well-founded, which implies that not all fairness conditions hold infinitely often.

If $J = \emptyset$, then every infinite path is fair, since the program does not have any fairness constraints. Hence, we want to avoid infinite paths that start in initial states that do not satisfy c. In that case, we set $r(v, v')$ to represent pairs of states from p that can be connected by a path. Since r is disjunctively well-founded, there is no infinite path from a state that satisfies p.

7. Let c be an assertion in $\mathcal{T}$ and let $E_f c$ be a basic CTL^* state formula. Let $q_1, \ldots, q_{|J|}$ be fresh uninterpreted predicates of arity $|v|$ and $r_1, \ldots, r_{|J|}$ be fresh uninterpreted predicates of arity $2|v|$. Suppose $J \neq \emptyset$. Then,

$$Clauses(D, E_f c) \;=\; \{init(v) \to c(v) \wedge q_1(v)\} \; \cup$$

$$\cup \; \{q_i(v) \to \exists v' : next(v, v') \wedge ((J_i(v) \wedge q_{(i\%|J|)+1}(v')) \vee (r_i(v, v') \wedge q_i(v'))),$$

$$dwf(r_i), \; r_i(v, v') \wedge r_i(v', v'') \to r_i(v, v'') \mid i \leq |J|\},$$

where J_i is the i-th element of J. $i\%|J|$ is the remainder of i modulo $|J|$. If $J = \emptyset$, let q be a fresh predicate of arity $|v|$, then

$$Clauses(D, E_f c) \;=\; \{init(v) \to c(v) \wedge q(v), \; q(v) \to \exists v' : next(v, v') \wedge q(v')\}$$

Explanation of Rule 7: In order to verify that the program satisfies $E_f c$ we need to prove that every initial state s_0 satisfies $c(v)$ and there is a fair path starting at s_0.

The clauses $dwf(r_i)$, $r_i(v, v') \wedge r_i(v', v'') \to r_i(v, v'')$ ensure that r_i is a disjunctively well-founded and transitive relation. Hence, it is well-founded. $r_i(v, v')$ means that v' is closer to a state that satisfies J_i than v. Since r_i is well-founded it takes only finitely many steps to get to a state that satisfies J_i.

The new predicate q_i represents the states from which it is possible to get to a state s that satisfies J_i. In addition, there is a transition from s to a state satisfying q_j for $j = (i\%|J|) + 1$, meaning that eventually, J_j holds too.

We use $(i\%|J|) + 1$ instead of $i + 1$ to emphasize that when a state that satisfies $J_{|J|}$ is found, we continue to look for a state that satisfies J_1, so every J_i will be satisfied infinitely many times along the path.

If $J = \emptyset$, then every infinite path is a fair path. Hence, we just prove that every initial state satisfies $c(v)$ and is the start of an infinite path. Predicate q represents the set of states, that lie on an infinite path.

8. Let c be an assertion in $\mathcal{T}$. Suppose we have the CTL^* state formula c, then

$$Clauses(v, init(v), next(v, v'), J, c) = \{init(v) \to c(v)\}.$$

Remark 1. Applying Trans to a verification problem may result in a non-unique set of clauses. This may occur when the specification is a basic state formula that contains more than one path subformula. For example, $E_f(Xc_1 \vee Gc_2)$. In such cases, these subformulas can be eliminated (using rules 3 and 4) in any order. As the soundness and relative completeness theorems prove, all sets produced by Trans correctly represent the given verification problem.

Unless otherwise stated, $Clauses(v, init(v), next(v, v'), J, \phi)$ will refer to *any* set of clauses that may be produced by Trans.

For every verification problem, Trans terminates, since every rule calls procedure *Clauses* for a problem with a smaller verification formula, except of rule 2, which can be called only once for every non-fair path quantifier in the specification formula.

Theorem 1 (complexity). *Let (D, ϕ) be a verification problem, with a program of size m, with k fairness conditions, and a specification formula of size n. Then Trans produces a set $Clauses(D, \phi)$ consisting of $O(n(n + k))$ clauses with a total size of $O(n(n + k)(n + m))$.*

Proof. At first, we prove that in the case when ϕ is a basic state formula, the number of clauses and the overall size are $O(n + k)$ and $O((n + k)(n + m))$ respectively. At the end of the generation process Trans applies rule 6 or rule 7. Rule 6 produces constant number of clauses, rule 7 produces $O(k)$ clauses and in both cases overall size is $O(n + mk)$.

Before the last step Trans applies rules 3, 4 and 5, replacing assertions inside the formula with boolean variables and extending the set of fairness conditions (by one element on each step) and transition relation (linearly from the size of the replaced assertion). Hence, by the last step of the algorithm, the sizes of the set of fairness constraints and the program are $O(n + k)$ and $O(n + m)$ respectively, hence the number of the clauses in the resulting set and the overall size are $O(n + k)$ and $O((n + k)(n + m))$. Also, during the implication of rules 3, 4 and 5 Trans may produce one clause per step of the size m, hence not more than n clauses with overall size $n(n + m)$.

Rule 2 makes a recursive call of *Clauses* for the problem of the same size, hence the bounds are preserved.

To deal with the case of non-basic specification formula ϕ we need to examine Rule 1. It takes a state basic subformula of the specification formula of the size l and makes two recursive calls. One for the problem with sizes of the parameters $m, n - l, k$ and one for the problem with sizes of the parameters m, l, k and with the basic state specification formula. Eventually, after numerous implications of rule 1 Trans reduces our problem to not more than n subproblems with the same sizes of program and the set of fairness constraints and smaller specification state formulas. Hence, Trans produces $O(n(n + k))$ clauses with the overall size $O(n(n + k)(n + m))$.   □

Note that the verification problems we deal with are undecidable in general, and so are the satisfiability problems of sets of EHC clauses. Nevertheless,

Theorem 1 is meaningful since it shows that our translation increases the representation of the problem at most polynomially. Thus, its solution, if exists, remains feasible.

3.1   Soundness

Theorem 2 (soundness). *For every set $Clauses(v, init(v), next(v, v'), J, \phi)$ of clauses produced by* Trans, *and for every interpretation $\mathcal{I}$, if the set of clauses is satisfiable by $\mathcal{I}$, then the program $D = \mathcal{I}((v, init(v), next(v, v'), J))$ satisfies $\mathcal{I}(\phi)$.*

Note that initially, D and ϕ do not include uninterpreted predicates and therefore $\mathcal{I}$ does not influence the result. However, inductive calls do include uninterpreted predicates, and therefore $\mathcal{I}$ should be taken into account.

Proof. We prove the theorem by induction on the size of the formula ϕ. Recall that by definition, CTL* formulas are state formulas. As such, they are true in a program D if all initial states of D satisfy the formula. That is, $init(v) \rightarrow \phi$. We will use this fact in the proof.

Base: Rule 8 forms the basis of the induction. For an assertion c in $\mathcal{T}$, $Clauses(D, c) = \{init(v) \rightarrow c(v)\}$. If this set of clauses is satisfiable then, by definition of CTL* satisfiability, $D \models c$.

Induction step:
Rule 1:
Assume a set of clauses $Clauses(D, \phi_2(\phi_1))$ is satisfiable by an interpretation $\mathcal{I}$. Then, $\mathcal{I}$ satisfies $Clauses(v, aux(v), next(v, v'), J, \phi_1)$ and by the induction hypothesis, $(v, \mathcal{I}(aux)(v), next(v, v'), J) \models \phi_1$. Thus, $I(aux)(v) \rightarrow \phi_1$. Also, $\mathcal{I}$ satisfies $Clauses(D, \phi_2(aux)(v))$ and by the induction hypothesis $D \models \phi_2(I(aux))$. Since ϕ_2 is in Negation Normal Form (NNF), replacing all occurrences of $I(aux)$ in ϕ_2 with ϕ_1 results in a formula that is also true in D. We thus conclude that $D \models \phi_2(\phi_1)$, as required.

Rule 2:
Since the formula $Q\phi$ does not depend on the fairness conditions, $(v, init(v), next(v, v'), J) \models Q\phi$ iff $(v, init(v), next(v, v'), \emptyset) \models Q_f\phi$.

Assume by way of contradiction that a set of clauses produced by Trans for $Clauses(v, init(v), next(v, v'), J, Q\phi)$ is satisfiable but $(v, init(v), next(v, v'), J) \not\models Q\phi$. Then $(v, init(v), next(v, v'), \emptyset) \not\models Q_f\phi$. By the induction hypothesis, since we consider $Q_f\phi$ to be smaller than $Q\phi$, every set of clauses produced by Trans for $Clauses(v, init(v), next(v, v'), \emptyset, Q_f\phi)$ is unsatisfiable. Recall that Rule 2 defines $Clauses(v, init(v), next(v, v'), J, Q\phi) = Clauses(v, init(v), next(v, v'), \emptyset, Q_f\phi)$. Consequently, every set of clauses produced by Trans for $Clauses(v, init(v), next(v, v'), J, Q\phi)$ is unsatisfiable as well. A contradiction.

We partition the proof for each of the rules 3, 4 and 5 into two cases – universal and existential quantifiers.

Rule 3, Universal:

We prove here that if $D = (v, init(v), next(v, v'), J)$ does not satisfy $A_f\phi(Xc)$ then for every set of clauses, produced by Trans for $Clauses(v \cup \{x_X\}, init(v), next(v, v') \wedge (x_X = c(v')), J, A_f\phi(x_X))$, the set is unsatisfiable.

Suppose D does not satisfy $A_f\phi(Xc)$. Then there is a fair path $\pi = s_1, s_2, \ldots$ such that $D, s_1 \models init(v)$ and $D, \pi \not\models \phi(Xc)$.

Consider the path $\pi_X = s'_1, s'_2, \ldots$ of the program $D_X = (v \cup \{x_X\}, init(v), next(v, v') \wedge (x_X = c(v')), J)$, where $s'_i(v) = s_i(v)$ and $s'_i(x_X) = s_{i+1}(c(v))$. π_X is a fair path of D_X, since the fairness conditions of D and D_X are identical and defined only over v (and not over x_X). In addition, $D_X, \pi_X \not\models \phi(x_X)$, since x_X and Xc are true on exactly the same locations of the paths π_X and π, respectively.

Hence, D_X does not satisfy $A_f\phi(x_X)$. By the induction hypothesis, for every set of clauses produced by Trans for $Clauses(v \cup \{x_X\}, init(v), next(v, v') \wedge (x_X = c(v')), J, A_f\phi(x_X))$, the set is unsatisfiable, as required.

Rule 3, Existential:

Denote $D = (v, init(v), next(v, v'), J)$. Suppose a set of clauses $Clauses(D, E_f\phi(Xc)) = Clauses(v \cup \{x_X\}, aux(v, x_X), next(v, v') \wedge (x_X = c(v')), J, E_f\phi(x_X)) \cup \{init(v) \to \exists x_X : aux(v, x_X)\}$ is satisfied by interpretation I.

Denote $D_X = (v \cup \{x_X\}, I(aux)(v, x_X), next(v, v') \wedge (x_X = c(v')), J)$. By the induction hypothesis, $D_X \models E_f\phi(x_X)$.

Let s be a state of D such that $D, s \models init(v)$. Since the clause $init(v) \to \exists x_X : I(aux)(v, x_X)$ is true, there is a state s_X of D_X, such that $s_X(v) = s(v)$ and $D_X, s_X \models I(aux)(v, x_X)$. Thus, there is a fair path $\pi_X = s_X, s_2^X, s_3^X, \ldots$ of D_X such that $D_X, \pi_X \models \phi(x_X)$. Consider the path of D defined as $\pi = s, s_2, s_3, \ldots$, where each s_i is the restriction of s_i^X to the variables from v. Then $D, \pi \models \phi(Xc)$. This is because x_X and Xc are true on exactly the same locations of the paths π_X and π, respectively. We conclude that $D \models E_f\phi(Xc)$.

Rule 4, Universal:

Suppose $D = (v, init(v), next(v, v'), J)$ does not satisfy $A_f\phi(Gc)$. Then there is a fair path $\pi = s_1, s_2, \ldots$ such that $D, s_1 \models init(v)$ and $D, \pi \not\models \phi(Gc)$.

Consider the path $\pi_G = s'_1, s'_2, \ldots$ of the program $D_G = (v \cup \{x_G\}, init(v), next(v, v') \wedge (x_G = c(v) \wedge x'_G), J \cup \{x_G \vee \neg c(v)\})$, where $s'_i(v) = s_i(v)$ and $s'_i(x_G) = true$ iff for every $j \geq i : s'_j \models c(v)$. Note that, once x_G is set to true on π_G, $c(v)$ is set to true. Moreover, they both remain true forever from that point on.

π_G is a fair path of D_G since the fairness conditions of D do not depend on x_G. The additional fairness condition of D_G holds as well, since if $c(v)$ holds from a certain point on along π_G then x_G holds from that point on. Thus, $x_G \vee \neg c(v)$ holds infinitely often along π_G.

Also $D_G, \pi_G \not\models \phi(x_G)$, since x_G is true on the same locations of path π_G as the locations of π where Gc is true.

Hence, D_G does not satisfy $A_f\phi(x_G)$. By the induction hypothesis, every set of clauses produced by Trans for $Clauses(D_G, A_f\phi(x_X))$ is unsatisfiable. Since

$Closes(D, A_f\phi(Gc)) = Clauses(D_G, A_f\phi(x_X))$, this implies that every set of clauses produced by Trans for $Closes(D, A_f\phi(Gc))$ is unsatisfiable, as required.

Rule 4, Existential:

Suppose a set $Clauses(v, init(v), next(v, v'), J, E_f\phi(Gc))$ is satisfied by interpretation $\mathcal{I}$. Denote $D_G = (v \cup \{x_G\}, I(aux)(v, x_G), next(v, v') \wedge (x_G = c(v) \wedge x'_G), J \cup \{x_G \vee \neg c(v)\})$. Then, by Rule 4, $Clauses(D_G, E_f\phi(x_G))$ is also satisfied by $\mathcal{I}$. By the induction hypothesis $D_G \models E_f\phi(x_G)$.

Let s be a state of $D = (v, init(v), next(v, v'), J)$, such that $D, s \models init(v)$. Since the clause $\{init(v) \rightarrow \exists x_G : aux(v, x_G)\}$ is satisfied by interpretation $\mathcal{I}$, then there is a state s_G of D_G such that $s_G(v) = s(v)$ and $D_G, s_G \models \mathcal{I}(aux)(v, x_G)$. Moreover, there is a fair path $\pi_G = s_G, s'_2, s'_3, \ldots$ of D_G such that $D_G, \pi_G \models \phi(x_G)$. Consider the path of D defined by $\pi = s, s_2, s_3 \ldots$, where $s_i(v)$ is a restriction of s'_i to the variables from v. Then $D, \pi \models \phi(Gc)$. This is because x_G is true on the same places of the path π_G as the places of π where Gc is true. Moreover, π is a fair path. Thus, D satisfies $E_f\phi(Gc)$.

Rule 5, Universal:

Suppose $D = (v, init(v), next(v, v'), J)$ does not satisfy $A_f\phi(c_1 U c_2)$, then there is a fair path $\pi = s_1, s_2, \ldots$, such that $D, s_1 \models init(v)$ and $D, \pi \not\models \phi(c_1 U c_2)$.

Consider the path $\pi_U = s'_1, s'_2, \ldots$ of the program $D_U = (v \cup \{x_U\}, init(v), next(v, v') \wedge (x_U = c_2(v) \vee (c_1(v) \wedge x'_U)), J \cup \{\neg x_U \vee c_2(v)\})$, where $s'_i(v) = s_i(v)$ and $s'_i(x_U) = true$ iff π^i satisfies $c_1 U c_2$. π_U is a fair path of D_U since the fairness conditions from D do not depend on the evaluation of x_U. Further, if x_U is true from a certain point on along π_U, then c_2 must hold infinitely often. Thus, π_U satisfies the additional fairness condition as well.

Also $D_U, \pi_U \not\models \phi(x_U)$, since x_U is true on the same places of the path π_U as the places of π where $c_1 U c_2$ is true.

Hence D_U does not satisfy $A_f\phi(x_U)$. By the induction hypothesis, every clause produced by Trans for $Clauses(v \cup \{x_U\}, init(v), next(v, v') \wedge (x_U = c_2(v) \vee (c_1(v) \wedge x'_U)), J \cup \{\neg x_U \vee c_2(x)\}, A_f\phi(x_U))$ is unsatisfiable, as required.

Rule 5, Existential:

Suppose a set $Clauses(v, init(v), next(v, v'), J, E_f\phi(c1Uc2))$ is satisfied by interpretation I.

Denote $D_U = (v \cup \{x_U\}, init(v), next(v, v') \wedge (x_U = c_2(v) \vee (c_1(v) \wedge x'_U)), J \cup \{\neg x_U \vee c_2(x)\})$. By the induction hypothesis $D_U \models E_f\phi(x_U)$.

Let s be a state of $D = (v, init(v), next(v, v'), J)$, such that $D, s \models init(v)$. Since the clause $\{init(v) \rightarrow \exists x_X : aux(v, x_G)\}$ is satisfied by interpretation I, then there is a state s_U of D_U such that $s_U(v) = s(v)$ and $D_U, s_U \models I(aux)(v, x_U)$. Moreover, there is a fair path $\pi_U = s_U, s'_2, s'_3, \ldots$ of D_U such that $D_U, \pi_U \models \phi(x_U)$. Consider the path of D defined by $\pi = s, s_2, s_3, \ldots$, where s_i is the restriction of s'_i to the variables from v.

Then $D, \pi \models \phi(c_1 U c_2)$ since x_U is true on the same places of path π_U as the places of π where $c_1 U c_2$ is true. Hence, D satisfies $E_f\phi(c_1 U c_2)$.

Rule 6:

Denote $D = (v, init(v), next(v, v'), J)$. Suppose by way of contradiction that a set $Clauses(D, A_f c)$ is satisfied by an interpretation I but the program D does not satisfy the formula $A_f c$. Then, there is a fair path $\pi = s_0, s_1, s_2, \ldots$, such that $D, s_0 \models init(v)$ and $D, \pi \not\models c$.

By the clauses $init(v) \wedge \neg c(v) \to I(p)(v)$, $next(v, v') \wedge I(p)(v) \to I(p)(v')$ we know that for every $i : s_i \models I(p)(v)$.

Since π is fair, each J_j holds infinitely often along π. Moreover, there is an infinite increasing sequence of indexes $0 < i_1 < i_2 < i_3 <, \ldots$, such that for every $k : s_{i_k} \models J_{k\%|J|+1}$.

By the clause $I(p)(v_0) \wedge \bigwedge_{i=1}^{|J|}(I(t)(v_{i-1}, v_i) \wedge J_i(v_i)) \to I(r)(v_0, v_{|J|})$ we get that $s_{i_{n_1|J|}}, s_{i_{n_2|J|}} \models r(v, v')$ for every pair of natural numbers n_1, n_2, provided $n_1 < n_2$. Thus, $I(r)$, which is disjunctive well-founded, contains an infinite transitive chain, but that is impossible due to Theorem 1 from [20]. A contradiction.

Rule 7:

Suppose a set of clauses $Clauses(v, init(v), next(v, v'), J, E_f c)$ is satisfied by interpretation I.

Denote $D = (v, init(v), next(v, v'), J)$. In order to prove that D satisfies $E_f c$ we need to show that every state that satisfies $init(v)$ also satisfies $c(v)$ and is a start of a fair path.

Suppose for some state s of D, $s \models init(v)$, then by clause $init(v) \to c(v) \wedge I(q_1)(v)$ we have $s \models c(v)$ and $s \models I(q_1)(v)$.

By clauses $\{dwf(I(r_i)), I(r_i)(v, v') \wedge I(r_i)(v', v'') \to I(r_i)(v', v'') \mid i \leq |J|\}$ we know that for every $i \leq |J|$, relation $I(r_i)$ is disjunctively well-founded and transitive, hence it is a well-founded relation by Theorem 1 from [20].

For every $i \leq |J|$, by clause $I(q_i)(v) \to \exists v' : next(v, v') \wedge ((J_i(v') \wedge I(q_{i\%|J|)+1})(v')) \vee (I(r_i)(v, v') \wedge I(q_i)(v')))$ and the fact that $I(r_i)$ is a well-founded relation we can conclude that for every state t of D if $t \models I(q_i)(v)$ then there is a path in D from t to some state t' such that $t' \models J_i(v)$ and there is a transition from t' to some state t'' such that $t'' \models I(q_{(i\%|J|)+1})(v)$.

Thus, there is a fair path, that starts in s.   □

3.2   Relative Completeness

Next we show that Trans is relative complete. That is, if a verification problem holds, i.e., the program meets its specification, then every set of clauses, produced by Trans for that problem, is satisfied by some interpretation of the uninterpreted predicates. However, this interpretation might not be expressible by a first-order formula in our logic.

Theorem 3 (relative completeness). *If $D = (v, init(v), next(v, v'), J)$ satisfies ϕ then every set produced by Trans for $Clauses(D, \phi)$ is satisfiable by some interpretation of the uninterpreted predicates.*

Proof. We prove the theorem by induction on the size of the verified formula.
Base: Rule 8 forms the basis of the induction. For assertion c in $\mathcal{T}$, if $D \models c$

then $Clauses(D, c) = \{init(v) \rightarrow c(v)\}$ is satisfiable by definition of CTL*
satisfiability.

Induction Step:
Rule 1:
Assume $D \models \phi_2(\phi_1)$. For the uninterpreted predicate aux, we choose as interpretation the predicate $\mathcal{I}(aux)$, which is true in exactly those states where ϕ_1 is true. Then, $(v, \mathcal{I}(aux)(v), next(v, v'), J)$ meets ϕ_1 and $(v, init(v), next(v, v'), J)$ meets $\phi_2(\mathcal{I}(aux))$. By the induction hypothesis, the corresponding sets of clauses are satisfiable. Consequently, $Clauses(D, \phi_1(\phi_2))$ is satisfiable.

Rule 2:
Since this verification problem does not depend on the fairness conditions, the following holds: $(v, init(v), next(v, v'), J) \models Q\phi$ iff $(v, init(v), next(v, v'), \emptyset) \models Q\phi$ iff $(v, init(v), next(v, v'), \emptyset) \models Q_f\phi$. By the induction hypothesis (since we consider $Q_f\phi$ to be smaller than $Q\phi$), every set of clauses produced by Trans for $Clauses(v, init(v), next(v, v'), \emptyset, Q_f\phi)$ is satisfiable. The same holds for $Clauses(v, init(v), next(v, v'), J, Q\phi)$, since by Rule 2 is identical.

We partition the proof for each of the rules 3, 4 and 5 into two cases – universal and existential quantifiers.

Rule 3, Universal:
If $D = (v, init(v), next(v, v'), J)$ meets $A_f\phi(Xc)$, then $D_X = (v \cup \{x_X\}, init(v), next(v, v') \wedge (x_X = c(v')), J)$ meets $A_f\phi(x_X)$. This is because for every fair path of D_X, the formula x_X is true in the exact the same places where Xc is true on the corresponding path of D. By the induction hypothesis, every set of clauses produced by Trans for $Clauses(D_X, A_f\phi(x_X))$ is satisfiable. Consequently, every set of clauses produced by Trans for $Clauses(D, A_f\phi(Xc))$ is satisfiable, as required.

Rule 3, Existential:
Assume $D = (v, init(v), next(v, v'), J)$ satisfies $E_f\phi(Xc)$. We need to prove that every set of clauses produce by Trans for $Clauses(D, E_f\phi(Xc)) = Clauses(v \cup \{x_X\}, aux(v, x_X), next(v, v') \wedge (x_X = c(v')), J, E_f\phi(x_X)) \cup \{init(v) \rightarrow \exists x_X : aux(v, x_X)\}$ is satisfiable.

For $D_X = (v \cup \{x_X\}, \mathcal{I}(aux)(v), next(v, v') \wedge (x_X = c(v')), J)$, we choose the predicate $\mathcal{I}(aux)$, which is true in exactly those states of D_X that satisfy $E_f\phi(x_X)$.

Clearly, D_X satisfies $E_f\phi(x_X)$. By the induction hypothesis, every set of clauses produced by Trans for $Clauses(D_X, E_f\phi(x_X))$ is satisfiable.

For every fair path of D_X, the variable x_X is true in exactly the same places where Xc is true on the corresponding fair path of D.

Hence, $init(v) \rightarrow \exists x_X \mathcal{I}(aux)(v, x_X)$, because if $s \models init(v)$ is true for some state s of D, then there is a fair path π, starting at s that satisfies $\phi(Xc)$. Hence, if Xc is true on π, then $s \models \mathcal{I}(aux)(v, 1)$ and if Xc is false on π then $s \models \mathcal{I}(aux)(v, 0)$.

Thus, for every set of clauses produced by Trans for $Clauses(v \cup \{x_X\}, aux(v, x_X), next(v, v') \wedge (x_X = c(v')), J, E_f\phi(x_X)) \cup \{init(v) \rightarrow \exists x_X : aux(v, x_X)\}$, we defined an interpretation that satisfy this set.

The two cases for rules 4 and 5 can be proved similarly to Rule 3 and are omitted.

Rule 6:

Suppose the program $D = (v, init(v), next(v, v'), J)$ satisfies $A_f c$. Recall that, $D \models A_f c$ implies that every initial state of D either satisfies c or has no fair path starting in it. We define an interpretation $\mathcal{I}$ as follows:

- $s \models \mathcal{I}(p)(v)$ iff there is no fair path starting at s.
- $s, s' \models \mathcal{I}(t)(v, v')$ iff there is a path from s to s'.
- $s, s' \models \mathcal{I}(r)(v, v')$ if $s \models \mathcal{I}(p)(v)$, $s' \models J_{|J|}(v')$ and there is a path from s to s' which visits a state satisfying J_1, then a state satisfying J_2 and so on until it ends in s' that satisfies $J_{|J|}$.

For every set of state $s, s', s'', s_0, \ldots, s_{|J|}$ of D, we check that all clauses defined by Rule 6 are satisfied by $\mathcal{I}$.

- $s \models init(v) \wedge \neg c(v) \rightarrow \mathcal{I}(p)(v)$, because if an initial state does not satisfy c, then there is no fair path that starts in it. This is because D satisfies $A_f c$.
- $s, s' \models next(v, v') \wedge \mathcal{I}(p)(v) \rightarrow \mathcal{I}(p)(v')$, because if there is no fair path that starts in s and there is a transition from s to s', then there is no fair path that starts in s'.
- $s_0, \ldots, s_{|J|} \models I(p)(v_0) \wedge \bigwedge_{i=1}^{|J|}(t(v_{i-1}, v_i) \wedge J_i(v_i)) \rightarrow \mathcal{I}(r)(v_0, v_{|J|})$ by definition of $\mathcal{I}(r)$.
- $s, s' \models next(v, v) \rightarrow \mathcal{I}(t)(v, v')$ and $s, s', s'' \models next(v, v') \wedge \mathcal{I}(t)(v', v'') \rightarrow \mathcal{I}(t)(v, v'')$, since $\mathcal{I}(t)$ is the transitive closure of $next$.
- Relation $\mathcal{I}(r)$ is well-founded, since if there was an infinite chain of $\mathcal{I}(r)$, then this chain would represent a fair path on states satisfying $\mathcal{I}(p)$. But such states cannot occur on fair paths. Hence $dwf(\mathcal{I}(r))$.

This concludes the proof of relative completeness for Rule 6.

Rule 7:

Suppose the program $D = (v, init(v), next(v, v'), J)$ satisfies $E_f c$. This means that every initial state s of D satisfies c, and, in addition, s is the start of a fair path. We say that a state s of D is *fair* if there is a fair path starting at s. We define the interpretation $\mathcal{I}$ as follows. For every $i \in \{1, \ldots, |J|\}$:

- $s \models \mathcal{I}(q_i)(v)$ iff s is a fair state.
- $s, s' \models \mathcal{I}(r_i)(v, v')$ iff s' is a fair state and s satisfies J_i or s' is the successor of s on the shortest path π leading from s to a fair state s'' satisfying J_i.

For every two states s, s' in D, we check that the clauses defined by Rule 7 are satisfied by $\mathcal{I}$.

- $s \models init(v) \rightarrow c(v) \wedge \mathcal{I}(q_1)(v)$ since for every initial state s, $c(v)$ holds in s and there is a fair path starting at s. That is, s is fair.

- $s \models \mathcal{I}(q_i)(v) \rightarrow \exists v' : next(v, v') \wedge ((J_i(v) \wedge \mathcal{I}(q_{(i\%|J|)+1})(v')) \vee (\mathcal{I}(r_i)(v, v') \wedge \mathcal{I}(q_i)(v')))$. This is because if $s \models \mathcal{I}(q_i)(v)$, then s is a fair state. On a fair path every J_j holds infinitely often. Thus, either $J_i(v)$ holds in s or there is a shortest path π from s to a fair state where J_i holds.

  In the first case we define s' as the successor of s on a fair path. In the second case we define s' as the successor of s on π. Hence, we either have $s \models J_i(v)$ and $s' \models \mathcal{I}(q_{(i\%|J|)+1})(v')$ (this is because fair states satisfy all q_j) or $s' \models \mathcal{I}(q_i)(v)$ and $s, s' \models \mathcal{I}(r_i)(v, v')$.

- $s \models \mathcal{I}(r_i)(v, v') \wedge \mathcal{I}(r_i)(v', v'') \rightarrow \mathcal{I}(r_i)(v, v'')$, since the relation is transitive by definition.

- $\mathcal{I}(r_i)$ is a well-founded relation because if on each step we move to a state, which is closer along the shortest path to a state satisfying $(J_i \wedge q_{(i\%|J|)+1})$, we will eventually get to such a state. Hence $dwf(\mathcal{I}(r_i))$.

This concludes the proof of relative completeness for Rule 7.   □

4   Case Study 1: CTL* Verification of a Robot System

```
l₁: while(true){
l₂:     robot_id := *;
l₃:     a := *;
l₄:     b := *;
l₅:     if(robot_id == 1){
l₆:         x₁ := x₁ + 2a + b;
        }
l₇:     if(robot_id == 2){
l₈:         x₂ := x₂ + a + b;
l₉:         y₂ := y₂ - 2a - 2b;
        }
l₁₀:    if(robot_id == 3){
l₁₁:        x₃ := x₃ + a + b;
l₁₂:        y₃ := y₃ - a - b;
        }
    }
```

Fig. 1. Robots program R

Suppose we are given a system with three robots placed on rational points of a two-dimensional plane. A user makes an ongoing interaction with the system by sending commands. The system processes the commands and moves the robots accordingly.

We wish to verify two properties. The first property states that the three robots are never located in the same position at the same time. The second property states that for each pair of robots, the user can manipulate the robots in such a way that this pair meets infinitely often.

Program R, presented in Fig. 1, describes the interaction between the user and the system. Program R consists of a while-loop. At the beginning of each iteration, three variables *robot_id*, a and b are assigned non-deterministically.

These are the inputs from the user. Variable *robot_id* represents the identification of the robot that the user wishes to move. Variables a and b represent the moving instruction, given by the user to the selected robot.

Variables x_i and y_i represent the current position of robot i on the plane.

If the user chooses robot i, then the system changes its position according to the linear transformation defined for robot i, based on a and b. For each robot, a different linear transformation is used. For example, for robot 2, the transformation is $(a + b, -2a - 2b)$.

Let us denote the set of variables of R (including the variable pc) as v and the transition relation of R as $next(v, v')$. The initial positions of the robots are given by the initial condition: $init(v) = (x_1 = y_1 = x_2 = y_2 = x_3 = 0) \wedge (y_3 = 2)$. The property we wish to verify is given by the CTL^* formula ϕ.

$$\phi = AG(safe(v)) \wedge EG(Fmeet_{1,2}(v)) \wedge EG(Fmeet_{2,3}(v)) \wedge EG(Fmeet_{1,3}(v)),$$

where the formula $safe(v)$ states that the three robots are not all in the same position on the plane. Formula $meet_{i,j}(v)$ states that robot i and robot j are in the same position[1].

The first conjunct of ϕ ensures that the three robots are never in the same position at the same time. The second conjunct ensures that for every pair of robots, there is a path along which they meet infinitely often.

Next, we demonstrate how to generate the set of EHCs for the verification problem of $R \models \phi$, using Trans.

We need to use rule 1 four times, one for each basic state subformula. As a result, we get the following

$$Clauses(R, \phi) = Clauses(v, aux_1(v), next(v, v'), \emptyset, AG(safe(v)))$$

$$\cup\, Clauses(v, aux_2(v), next(v, v'), \emptyset, EG(Fmeet_{1,2}(v)))$$

$$\cup\, Clauses(v, aux_3(v), next(v, v'), \emptyset, EG(Fmeet_{1,3}(v)))$$

$$\cup\, Clauses(v, aux_4(v), next(v, v'), \emptyset, EG(Fmeet_{2,3}(v)))$$

$$\cup\, Clauses(R, aux_1(v) \wedge aux_2(v) \wedge aux_3(v) \wedge aux_4(v))$$

The first set requires that in every state where $aux_1(v)$ is true, subformula $AG(safe(v))$ is satisfied. Similar requirements are presented in the next three sets. The last set guarantees that in the initial states of R all four aux_i predicates are satisfied.

To deal with the first set, we can use rule 2 to turn A to A_f and then rule 4 to eliminate the temporal operator. Finally, we get the following result.

$$Clauses(v, aux_1(v), next(v, v'), \emptyset, AG(safe(v)))$$

$$= Clauses(v \cup \{x_G\}, aux_1(v), next(v, v') \wedge (x_G = safe(v) \wedge x'_G),$$

$$\{x_G \vee \neg safe(v)\}, A_f x_G).$$

---

[1] $safe(v) = (x_1 \neq x_2) \vee (x_1 \neq x_3) \vee (x_2 \neq x_3) \vee (y_1 \neq y_2) \vee (y_1 \neq y_3) \vee (y_2 \neq y_3)$ and $meet_{i,j}(v) = (x_i = x_j) \wedge (y_i = y_j)$.

Now we need to use rule 6 to get the explicit set of clauses.

$Clauses(v \cup \{x_G\}, aux_1(v), next(v, v') \wedge (x_G = safe(v) \wedge x'_G), \{x_G \vee \neg safe(v)\}, A_f x_G)$

$= \{aux_1(v) \wedge \neg x_G \rightarrow p(v, x_G),\ next(v, v') \wedge (x_G = safe(v) \wedge x'_G) \wedge p(v, x_G) \rightarrow p(v', x'_G),$

$p(v, x_G) \wedge t(v, x_G, v', x'_G) \wedge (x'_G \vee \neg safe(v')) \rightarrow r(v, x_G, v', x'_G),$

$dwf(r),\ next(v, v') \wedge (x_G = safe(v) \wedge x'_G) \rightarrow t(v, x_G, v', x'_G),$

$t(v, x_G, v', x'_G) \wedge next(v', v'') \wedge (x'_G = safe(v') \wedge x''_G) \rightarrow t(v, x_G, v'', x''_G)\}.$

The resulting set of clauses guarantees that $aux_1(v)$ is evaluated in such a way that from every state in which $aux_1(v)$ is true, the user cannot direct three robots to the same point at the same time (since $safe(V)$ holds).

Next, we explain informally why the set of clauses above is satisfiable. Suppose from a state s the user can apply a sequence of moves so that eventually all three robots arrive at position $(x; y)$. Hence $y = s(y_1)$, since the second coordination of robot 1 cannot be changed. For the second and third robots the sums $2x_2 + y_2$ and $x_3 + y_3$ cannot be changed. Hence

$$2x + s(y_1) = 2x + y = 2\,s(x_2) + s(y_2),$$

$$x = (2\,s(x_2) + s(y_2) - s(y_1))/2$$

and from the condition for the last robot, we need to have $(2s(x_2) + s(y_2) - s(y_1))/2 + s(y_1) = s(x_3) + s(y_3)$. By these equations, we can conclude that $aux_1(v)$ can be evaluated as

$$((2x_2 + y_2 + y_1)/2) \neq x_3 + y_3$$

and $init(v)$ implies this condition.

For the second, third and fourth sets the generation of clauses in similar. We will look only at the fourth set.

Let us rewrite the formula to replace F with U.

$$EG(F\,meet_{2,3}(v)) = EG(True\,U\,meet_{2,3}(v))$$

First, we need to apply rule 2 to turn E to E_f. Then we need to apply rule 5 to eliminate the temporal operator U.

$Clauses(v, aux_4(v), next(v, v'), \emptyset, EG(True\,U\,meet_{2,3}(v))) = \{aux_4(v) \rightarrow \exists x_U : aux_5(v, x_U)\}$

$\cup\ Clauses(v \cup \{x_G\}, aux_5(v, x_U), next(v, v') \wedge (x_U = meet_{2,3}(v) \vee x'_U, \{\neg x_U \vee meet_{2,3}(v)\}, EG(X_U))$

As the last two steps, we need to apply rule 4 to eliminate the temporal operator G and rule 7 to get the explicit set of clauses.

5 CTL^* Synthesis from Partial Programs

In this section, we investigate the problem of CTL^* synthesis from partial programs, which include holes of two types – condition holes and assignment holes.

As in the previous part, a program is presented as a transition system with fairness conditions. Additionally, we have the variable pc, ranging over program locations. The set of program variables is now presented as $v = \{pc\} \cup v_r$, where v_r is the set of all program variables except pc.

A *partial program* is a pair $\langle P, H \rangle$, where P is a program and H is a finite set of holes in P. There are two types of holes.

A *condition hole* is a triple of locations $\langle l, l_t, l_f \rangle$ – the first element represents the location where we would like to insert a condition. If this condition is satisfied by a certain state at location l, then the synthesized program moves from location l to location l_t, otherwise, it moves to location l_f.

An *assignment hole* is a pair of locations $\langle l, l' \rangle$. In this case, we want to synthesize an assignment in the form of a relation $\alpha(v_r, v_r')$ over pairs of states such that the synthesized program moves from location l to location l', assigning variables according to α.

A *resolving function* for a partial program $\langle P, H \rangle$ is a function Ψ that maps every condition hole to a subset of $\mathcal{E}^{|v_r|}$, and every assignment hole to a subset of $\mathcal{E}^{|v_r|} \times \mathcal{E}^{|v_r|}$ [2], such that for every assignment hole $\langle l, l' \rangle$ the following holds: $\forall v_r \exists v_r' : \Psi(\langle l, l' \rangle)(v_r, v_r')$.

The constraint above characterizes a non-deterministic multiple assignment. Using constraints to characterize the assignment holes gives us flexibility in the synthesis. For example, we could alternatively require that the assignment changes exactly one variable, while all others preserve their previous value.

For a partial program $\langle P, H \rangle$ and a resolving function Ψ the *synthesized program* is a program $P_\Psi = \langle v, init(v), next_\Psi(v, v'), \mathcal{J} \rangle$, where

$$next_\Psi(v, v') = \bigvee_{\langle l, l_t, l_f \rangle \in H} (pc = l \wedge pc' = l_t \wedge \Psi(\langle l, l_t, l_f \rangle)(v_r) \wedge v_r' = v_r) \vee$$

$$\bigvee_{\langle l, l_t, l_f \rangle \in H} (pc = l \wedge pc' = l_f \wedge \neg \Psi(\langle l, l_t, l_f \rangle)(v_r) \wedge v_r' = v_r) \vee$$

$$\bigvee_{\langle l, l' \rangle \in H} (pc = l \wedge pc' = l' \wedge \Psi(\langle l, l' \rangle)(v_r, v_r')) \vee next(v, v').$$

So basically P_Ψ is P with the holes filled according to Ψ.

The CTL^* **synthesis problem** is given a partial program $\langle P, H \rangle$ and a CTL^* formula ϕ. A solution to the CTL^* synthesis problem is a resolving function Ψ, such that the synthesized program P_Ψ satisfies ϕ. If such a resolving function exists we say that the problem is *realizable*. Otherwise, the problem is *unrealizable*.

5.1 Reduction of CTL^* Synthesis to EHC Satisfiability

Our goal is to generate a set of EHCs for a given synthesis problem, such that the set of EHCs is satisfiable if and only if there is a solution to the synthesis problem, exactly as we did for the verification problem.

---

[2] Recall that $\mathcal{E}$ is the domain over which interpretations are defined.

Actually, the verification problem is a special case of the synthesis problem, so it is not surprising that we use an extension of our approach to CTL^* verification to perform CTL^* synthesis. The main idea of our algorithm is to fill holes with uninterpreted predicates, apply our reduction for CTL^* verification and add some additional clauses to the resulted set of EHCs, in order to set some constraints on uninterpreted predicates in holes.

Formally, for a given synthesis problem consisting of a partial program $\langle P, H \rangle$ and a CTL^* formula ϕ, we define the set of uninterpreted predicates U as follows: $U := \{u_l^c(v_r)|\langle l, l_t, l_f \rangle \in H\} \cup \{u_l^a(v_r, v_r')|\langle l, l' \rangle \in H\}$.

Basically, U contains a predicate over v_r for every condition hole and a predicate over v_r, v_r' for every assignment hole.

Intuitively, u_l^c and $\neg u_l^c$ represent the sets of states from which the program moves from l to l_t and l_f, respectively.

The predicate u_l^a intuitively represents the set of pairs of states s, s' (with locations l and l'), such that the program moves from s to s'. We need to require that for every state s in l there exists a state s' in l', such that the program moves from s to s'. For that purpose, we introduce the following set of clauses $\Delta_a(P, H) := \{\top \rightarrow \exists v_r' : u_l^a(v_r, v_r') \mid \langle l, l' \rangle \in H\}$.

Let us define the program P_U, which is the program P with holes filled with uninterpreted predicates. Formally $P_U = \langle v, init(v), next_U(v, v'), \; \mathrm{J} \rangle$, where[3]

$$next_U(v, v') = next(v, v') \vee \bigvee_{\langle l, l_t, l_f \rangle \in H} (pc = l \wedge pc' = l_t \wedge u_l^c(v_r) \wedge v_r' = v_r)$$
$$\vee \bigvee_{\langle l, l_t, l_f \rangle \in H} (pc = l \wedge pc' = l_f \wedge \neg u_l^c(v_r) \wedge v_r' = v_r) \vee \bigvee_{\langle l, l' \rangle \in H} (pc = l \wedge pc' = l' \wedge u_l^a(v_r, v_r')).$$

Finally, the desired set of clauses is $\Delta(P, H, \phi) := \Delta_a(P, H) \cup Clauses(P_U, \phi)$.

Theorem 4 (soundness). *Suppose for a given CTL^* synthesis problem $\langle P, H, \phi \rangle$, a set of clauses $\Delta(P, H, \phi)$ produced by Trans is satisfiable by interpretation $\mathcal{I}$. Let Ψ be a resolving function for $\langle P, H \rangle$ defined by:*
*   For every condition hole $\langle l, l_t, l_f \rangle \in H$, $\Psi(\langle l, l_t, l_f \rangle)(v_r) = \mathcal{I}(u_l^c)(v_r)$.
*   For every assertion hole $\langle l, l' \rangle \in H$, $\Psi(\langle l, l' \rangle)(v_r, v_r') = \mathcal{I}(u_l^a)(v_r, v_r')$.
Then P_Ψ satisfies ϕ.

Proof. First, we check that Ψ is a resolving function. That is, for every assignment hole $\langle l, l' \rangle \in H$, the formula $\forall v_r \exists v_r' : \Psi(\langle l, l' \rangle)(v_r, v_r')$ is true. But this follows from the fact that interpretation I satisfies clauses $\Delta_a(P, H)$.

Next, we partition I to I_1 and I_2, where I_1 is a restriction of I to the set of predicates U and I_2 is restricted to the predicates not in U.

We denote by $I_1(next_U)$ the formula $next_U$, in which every predicate u_l^c or u_l^a is interpreted according to I_1. Note that, by definition of $next_U, next_\Psi$ and Ψ, we get $I_1(next_U)(v, v') = next_\Psi(v, v')$. Similarly, we denote by $I_1(P_U)$ the program P_U, in which the transition relation $next_U$ is replaced by its interpreted version $I_1(next_U)$. Note that, $I_1(P_U) = P_\Psi$.

---

[3] Recall that we can express the negation of a predicate using EHCs. See Sect. 2.1.

Applying I_1 to the predicates inside $Clauses(P_U, \phi)$ we get

$$I_1(Clauses(P_U, \phi)) = Clauses(I_1(P_U), \phi) = Clauses(P_\Psi, \phi).$$

Since the set $Clauses(P_U, \phi)$ is satisfied by interpretation I and $I = I_1 \cup I_2$, then $Clauses(P_\Psi, \phi) = I_1(Clauses(P_U, \phi))$ is satisfied by I_2. Thus, by Theorem 2 the program P_Ψ satisfies ϕ, as required. $\square$

Theorem 5 (Relative completeness). *Given a realizable CTL^* synthesis problem $\langle P, H, \phi \rangle$, then every set of clauses $\Delta(P, H, \phi)$ produced by Trans is satisfiable.*

Proof. Suppose $\langle P, H, \phi \rangle$ is realizable by resolving function Ψ.
As before, we partition the satisfying interpretation I that we would like to define into two interpretations, I_1 and I_2, where the first interprets the predicates from U and the second interprets the rest. We define interpretation I_1 on U as follows:
For every condition hole $\langle l, l_t, l_f \rangle \in H$, $I_1(u_l^c)(v_r) := \Psi(\langle l, l_t, l_f \rangle)(v_r)$.
For every assertion hole $\langle l, l' \rangle \in H$, $I_1(u_l^a)(v_r, v_r') := \Psi(\langle l, l' \rangle)(v_r, v_r')$.
We adopt the notations of $I_1(next_U), I_1(P_U)$ and $I_1(Clauses(P_U, \phi))$ from the proof of Lemma 4. Note that, $I_1(next_U)(v, v') = next_\Psi(v, v')$, hence $P_\Psi = I_1(P_U)$. Thus, by Theorem 3 the set $Clauses(P_\Psi, \phi) = Clauses(I_1(P_U), \phi) = I_1(Clauses(P_U, \phi))$ is satisfiable by some interpretation I_2, since P_Ψ satisfies ϕ. Hence, $I := I_1 \cup I_2$ is a satisfying interpretation for $Clauses(P_U, \phi)$.
Since the formula $\forall v_r \exists v_r' : \Psi(\langle l, l' \rangle)(v_r, v_r')$ is true, then $\Delta_a(P, H)$ is satisfied by I_1, and therefore also by I. Thus, $\Delta(P, H, \phi)$ is satisfied by I. $\square$

6   Case Study 2: Synthesizing a Bank-Client Interaction

Suppose we would like to write a program that will determine the rules for an ongoing interaction between a client and a bank. The goal is to define a process that is satisfactory for both the client and the bank. Specifically, we would like to complete the given sketch on Fig. 2(a) according to a given specification written as a CTL^* formula (see below).

The interaction between the client and the bank is an ongoing iterative process, defined over the variables req (for request), bal (for balance), exp (for expenses) and pro (for profit). At each iteration the client may request to apply either a withdrawal (in which case $req \geq 0$) or a deposit (where $req < 0$) to his account. bal represents the balance (amount of money) in the client's account. exp represents the accumulative withdrawals made by the client up to this iteration, and pro holds the profit accumulated by the bank up to this iteration.

Initially, $bal = exp = pro = 0$.

According to the sketch in Fig. 2(a), after an unspecified check on line l_c, on line l_3 the client's balance is updated according to req. Then the computation is split to the case where request is a deposit (line l_{a_1}) or a withdrawal (line l_5). In both cases the associated assignments are unspecified, but in the latter case, exp is updated according to the current withdrawal. Finally, on line l_6 the bank's profit is updated, decreasing it by \$0.1, regardless of the action that has been applied. This is because the bank has to transfer a fee of \$0.1 to the central bank on each transaction it makes.

The CTL* specification formula ϕ is given as follows:

$$\phi = A(G(exp < 1000) \vee FG(pro > 50)) \wedge EF((exp \geq 100) \wedge pro < 7).$$

The first conjunct of ϕ requires that for every path either exp is always less than \$1000 or from some moment on, pro is greater than \$50 forever. It assures that if the amount of money withdrew by the client is sufficiently large, then pro is also large.

The second conjunct assures that the client can withdraw at least \$100 from his account, while not paying the bank more than \$7.

The entire ϕ ensures that the interaction between the client and the bank, performed according to the synthesized program, is acceptable for the bank (first conjunct) and for the client (second conjunct).

Note that the first conjunct also implies that the fees of \$0.1 transferred from the bank to the central bank (line l_6), are collected from the client. Otherwise, the bank would not be able to maintain $pro > 50$, as it might lose profit on client's actions.

The set of variables of B is $v = \{pc, req, bal, exp, pro\}$, the transition relation is $next(v, v')$ and the initial condition is $init(v) = (bal = exp = pro = 0)$. The set of holes is $H = \{l_c, l_{a_1}, l_{a_2}\}$. We use Trans to generate the set of clauses $\Delta(B, H, \phi)$ for the synthesis problem $\langle B, H, \phi \rangle$.

The set of uninterpreted predicates is $U = \{u_{l_c}^c(v_r), u_{l_{a_1}}^a(v_r, v_r'), u_{l_{a_2}}^a(v_r, v_r')\}$. The program $B_U = \langle v, init(v), next_U(v), \emptyset \rangle$, where

$$next_U(v, v') = (pc = l_c \wedge pc' = l_3 \wedge u_{l_c}^c(v_r) \wedge v_r' = v_r)$$

$$\vee (pc = l_c \wedge pc' = l_2 \wedge \neg u_{l_c}^c(v_r) \wedge v_r' = v_r)$$

$$\vee (pc = l_{a_1} \wedge pc' = l_6 \wedge u_{l_{a_1}}^a(v_r, v_r')) \vee (pc = l_{a_2} \wedge pc' = l_6 \wedge u_{l_{a_2}}^a(v_r, v_r')) \vee next(v, v').$$

By definition, $\Delta(B, H, \phi) = \Delta_a(B, H) \cup Clauses(B_U, \phi)$. Since we have two assignment hole, $\Delta_a(P_T, H) = \{\top \to \exists v_r' : u_{l_{a_1}}^a(v_r, v_r'), \top \to \exists v_r' : u_{l_{a_2}}^a(v_r, v_r')\}$. The holes can be filled as presented in Fig. 2(b), resulting in program B_S.

The synthesized condition on line l_c assures that the client has enough money to pay for his request and for the cost of transaction, \$0.1, so that the bank would not lose its profit.

On lines l_{a_1} and l_{a_2} of B_S the bank collects \$0.1 from the client's account. Additionally, on line l_{a_2} the bank collects 6% of req, thus pro is at least 6% of exp and it never decreases. The first conjunct of ϕ is then satisfied.

To see that the second conjunct holds, consider a scenario in which the client deposits \$110 and then withdraws \$100. After this transaction $exp = \$100$ and $pro = \$6$. Hence, the second conjunct of ϕ is satisfied. Theorem 5 now guarantees that the set $\Delta(B, H, \phi)$ is satisfiable.

```
l₁: while ( true ) {            l₁: while ( true ) {
l₂:    req := *;               l₂:    req := *;
lc:    if ( ?_{c} ) {          lc:    if ( bal−req >= 0.1 ) {
l₃:       bal := bal−req ;     l₃:       bal := bal−req ;
l₄:       if ( req < 0 ) {     l₄:       if ( req < 0 ) {
la₁:          ?_{a_1};         la₁:          bal := bal −0.1;
                                             pro := pro +0.1;
          } else {                       } else {
l₅:          exp := exp+req ;  l₅:          exp := exp+req ;
la₂:          ?_{a_2};         la₂:          bal := bal −0.1−0.06 req ;
                                             pro := pro +0.1+0.06 req ;
          }                             }
l₆:       pro := pro −0.1;     l₆:       pro := pro −0.1;
       }                              }
    }                              }
}                              }

    (a) Partial program B          (b) Synthesized program B_S
```

Fig. 2. Synthesizing bank-client interaction

7 Conclusion

In this work, we exploit the power of EHCs for the verification and synthesis of infinite-state programs with respect to CTL^* specification. The algorithms are both sound and relative complete. The relative completeness is based on semantic interpretation that might not be expressible in a specific syntax. However, this is the most general way to describe the desired interpretation. For synthesis, for instance, when we conclude that the synthesis problem is unrealizable, then indeed no suitable solution exists, regardless of the syntax.

Our translation is independent of the solver or the underlying theory. If, in the future, a new EHC solver for any theory is designed, it can be combined with our translation system to perform CTL^* verification and partial synthesis.

References

1. Beyene, T.A., Brockschmidt, M., Rybalchenko, A.: CTL+FO verification as constraint solving (2014)
2. Beyene, T.A., Chaudhuri, S., Popeea, C., Rybalchenko, A.: A constraint-based approach to solving games on infinite graphs. In: Jagannathan, S., Sewell, P. (eds.) The 41st Annual ACM SIGPLAN-SIGACT Symposium on Principles of Programming Languages, POPL 2014, San Diego, CA, USA, 20–21 January 2014, pp. 221–234. ACM (2014). https://doi.org/10.1145/2535838.2535860
3. Beyene, T.A., Popeea, C., Rybalchenko, A.: Solving existentially quantified horn clauses. In: Sharygina, N., Veith, H. (eds.) Computer Aided Verification, pp. 869–882. Springer, Heidelberg (2013)
4. Beyene, T.A., Popeea, C., Rybalchenko, A.: Efficient CTL verification via horn constraints solving. In: Gallagher, J.P., Rümmer, P. (eds.) Proceedings 3rd Workshop on Horn Clauses for Verification and Synthesis, HCVS@ETAPS 2016, Eindhoven, The Netherlands, 3rd April 2016. EPTCS, vol. 219, pp. 1–14 (2016). https://doi.org/10.4204/EPTCS.219.1

5. Bjørner, N., Gurfinkel, A., McMillan, K., Rybalchenko, A.: Horn clause solvers for program verification, pp. 24–51. Springer, Cham (2015)
6. Bjørner, N., McMillan, K., Rybalchenko, A.: On solving universally quantified horn clauses. In: Logozzo, F., Fähndrich, M. (eds.) SAS 2013. LNCS, vol. 7935, pp. 105–125. Springer, Heidelberg (2013). https://doi.org/10.1007/978-3-642-38856-9_8
7. Bloem, R., Schewe, S., Khalimov, A.: CTL* synthesis via LTL synthesis. Electron. Proc. Theor. Comput. Sci. **260**, 4–22 (2017). https://doi.org/10.4204/eptcs.260.4
8. Carelli, M., Grumberg, O.: CTL^* verification and synthesis using existential horn clauses. In: 22nd International Symposium on Automated Technology for Verification and Analysis, ATVA 2024, Kyoto, Japan. Springer (2024)
9. Cook, B., Khlaaf, H., Piterman, N.: On automation of CTL* verification for infinite-state systems. In: Kroening, D., Păsăreanu, C.S. (eds.) Computer Aided Verification, pp. 13–29. Springer, Cham (2015)
10. Cook, B., Podelski, A., Rybalchenko, A.: Terminator: beyond safety. In: Ball, T., Jones, R.B. (eds.) Computer Aided Verification, pp. 415–418. Springer, Heidelberg (2006)
11. Dam, M.: CTL* and ECTL* as fragments of the modal μ-calculus. Theor. Comput. Sci. **126**(1), 77–96 (1994). https://doi.org/10.1016/0304-3975(94)90269-0, https://www.sciencedirect.com/science/article/pii/0304397594902690
12. Finkbeiner, B.: Synthesis of reactive systems, pp. 72–98 (2016). https://doi.org/10.3233/978-1-61499-627-9-72
13. Gulwani, S., Polozov, O., Singh, R.: Program synthesis. Found. Trends® Program. Lang. **4**(1-2), 1–119 (2017). https://doi.org/10.1561/2500000010
14. Guo, D., Svyatkovskiy, A., Yin, J., Duan, N., Brockschmidt, M., Allamanis, M.: Learning to complete code with sketches (2022)
15. Kesten, Y., Pnueli, A.: A compositional approach to CTL* verification. Theor. Comput. Sci. **331**, 397–428 (2005). https://doi.org/10.1016/j.tcs.2004.09.023
16. Khalimov, A., Bloem, R.: Bounded synthesis for Streett, Rabin, and CTL*. In: Majumdar, R., Kunčak, V. (eds.) CAV 2017, Part II. LNCS, vol. 10427, pp. 333–352. Springer, Cham (2017). https://doi.org/10.1007/978-3-319-63390-9_18
17. Kupferman, O., Vardi, M.Y., Wolper, P.: An automata-theoretic approach to branching-time model checking. J. ACM **47**(2), 312–360 (2000). https://doi.org/10.1145/333979.333987
18. Long, F., Rinard, M.: Staged program repair with condition synthesis, pp. 166–178 (2015). https://doi.org/10.1145/2786805.2786811
19. Niwiński, D., Walukiewicz, I.: Games for the μ-calculus. Theoret. Comput. Sci. **163**(1–2), 99–116 (1996)
20. Podelski, A., Rybalchenko, A.: Transition invariants. In: 2004 Proceedings of the 19th Annual IEEE Symposium on Logic in Computer Science, pp. 32–41 (2004). https://doi.org/10.1109/LICS.2004.1319598
21. Rothenberg, B.C., Grumberg, O., Vizel, Y., Singher, E.: Condition synthesis realizability via constrained horn clauses. In: Rozier, K.Y., Chaudhuri, S. (eds.) NASA Formal Methods, pp. 380–396. Springer, Cham (2023)
22. Solar-Lezama, A., Tancau, L., Bodik, R., Seshia, S., Saraswat, V.: Combinatorial sketching for finite programs. SIGARCH Comput. Archit. News **34**(5), 404–415 (2006). https://doi.org/10.1145/1168919.1168907
23. Srivastava, S., Gulwani, S., Foster, J.S.: Template-based program verification and program synthesis. Int. J. Softw. Tools Technol. Transf. **15**(5), 497–518 (2012). https://www.microsoft.com/en-us/research/publication/template-based-program-verification-program-synthesis/

24. Unno, H., Terauchi, T., Gu, Y., Koskinen, E.: Modular primal-dual fixpoint logic solving for temporal verification. Proc. ACM Program. Lang. **7**(POPL) (2023). https://doi.org/10.1145/3571265

25. Unno, H., Terauchi, T., Koskinen, E.: Constraint-based relational verification. In: Silva, A., Leino, K. (eds.) Computer Aided Verification, pp. 742–766. Springer, Cham (2021)

26. Xiong, Y., Wang, J., Yan, R., Zhang, J., Han, S., Huang, G., Zhang, L.: Precise condition synthesis for program repair. In: 2017 IEEE/ACM 39th International Conference on Software Engineering (ICSE), pp. 416–426 (2017). https://doi.org/10.1109/ICSE.2017.45

27. Xuan, J., et al.: Nopol: automatic repair of conditional statement bugs in java programs. IEEE Trans. Softw. Eng. **43**(1), 34–55 (2017). https://doi.org/10.1109/TSE.2016.2560811

Featured Message Sequence Graphs

Clemens Dubslaff[1,2]($\boxtimes$) (iD)

[1] Eindhoven University of Technology, Eindhoven, The Netherlands
c.dubslaff@tue.nl
[2] Centre for Tactile Internet with Human-in-the-Loop (CeTI), Dresden, Germany

Abstract. Message sequence charts (MSCs) are widely used in computer science to describe communication scenarios. Collections of communication scenarios are specified by graphs over MSCs, so-called message sequence graphs (MSGs). While nowadays almost all software is configurable, including communicating systems, MSGs have not yet been considered in a configurable setting.

In this paper, we strive towards configurable communication scenarios by introducing *featured MSGs*. Following a feature-oriented approach, a configuration is given through a set of features, each encapsulating an incremental or optional unit of functionality. Verification of (featured) MSGs is challenging, since even for standard MSGs most model-checking problems are known to be undecidable. We show that model checking MSGs against action computation tree logic (aCTL) is undecidable. However, if the MSG follows the syntactic criterion of local synchronization, MSG admit a finite transition-system semantics. For MSGs as well as featured MSGs, local synchronization renders aCTL model checking decidable.

1 Introduction

Message sequence charts (MSCs) [36] provide an intuitive graphical formalism to describe communication scenarios, widely used for the documentation and early stage specification of protocols. They comprise time lines of the communicating processes and arrows between them to describe data transfer. Their semantics is naturally provided by a partial order on events, imposed by the total time-line ordering of each process and the requirement that each send event must precede its corresponding receive event. Repetitive behaviors and choice points between different scenarios are usually modeled using *message sequence graphs (MSGs)*, which are automata over a finite alphabet of MSCs. Figure 1 shows an example MSG $\mathcal{G}_{\text{email}}$ that specifies a simple emailing system.[1] The basic communication

---

[1] In reminiscence of perspicuity, all graphics are drawn by hand.

This work was partially supported by the DFG under the projects TRR 248 (see https://perspicuous-computing.science, project ID 389792660) and EXC 2050/1 (CeTI, project ID 390696704, as part of Germany's Excellence Strategy), and by the NWO through Veni grant VI.Veni.222.431.

N. Bertrand et al. (Eds.): Christel Baier Festschrift, LNCS 15760, pp. 234–252, 2026.
https://doi.org/10.1007/978-3-031-97439-7_11

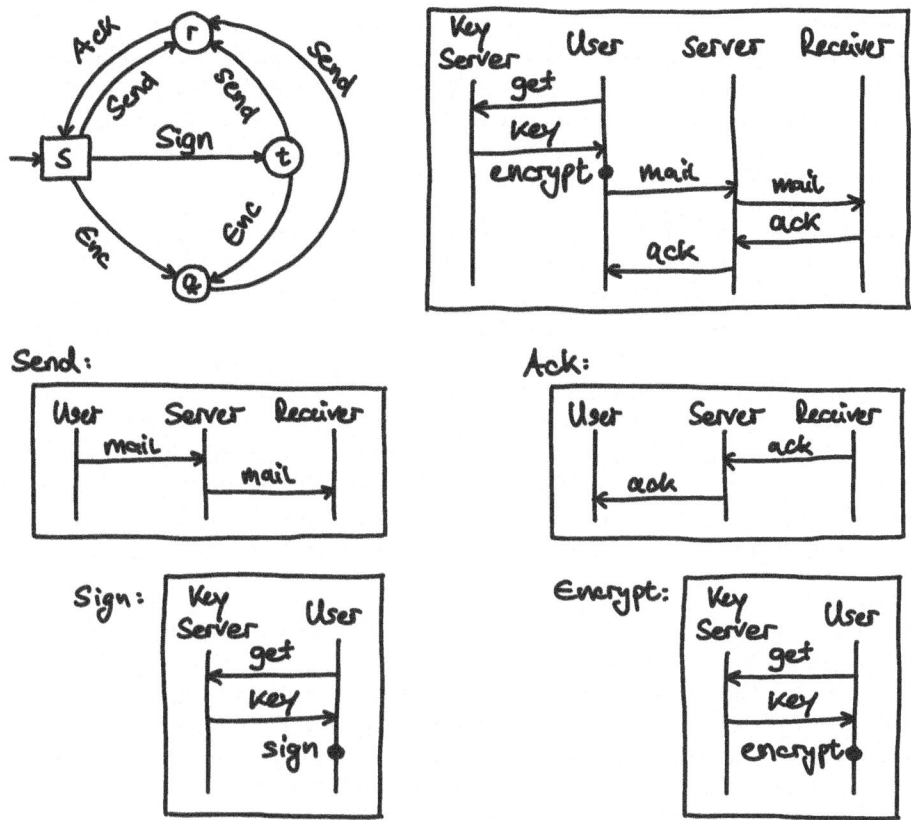

Fig. 1. Example MSG $\mathcal{G}_{\text{email}}$ of an email system (top left), its basic MSCs (bottom), and an MSC associated to an accepting path in the MSG (top right)

scenarios in form of sending, signing, encrypting, and acknowledging reception of an email are modeled by MSCs *Send*, *Sign*, *Encrypt*, and *Ack*, respectively. Every accepting path in $\mathcal{G}_{\text{email}}$ specifies a possible higher-order communication scenario. For instance, the path from s to q, r, and back to s describes the communication scenario depicted on the top right of Fig. 1 as composed MSC, sending an encrypted email and waiting for the acknowledgment of reception.

Branching-Time Semantics of MSGs. The set of MSCs specified by an MSG comprises all those MSCs that arise from composing MSC along accepting paths in the MSG [38]. The standard composition operation for MSCs is the *weak sequential composition* [48] that, intuitively spoken, glues successive time lines of each process together. To this end, the standard semantics of MSGs is a partial-order semantics describing the linear-time behavior of MSGs (cf., e.g., surveys [40,42]). Towards a more implementation-directed usage of MSGs, an operational semantics of MSGs has been standardized in [35], choosing a process-algebraic *delayed choice* operator [4]. Delayed choice ensures that the choice between communication scenarios is made only if it is inevitable. As a consequence, the standard operational behavior of an MSG solely relies on its partial-order semantics and is thus an intrinsic linear-time semantics [24,25]. Any developer using MSGs to specify protocols has to obey this fact and regard MSGs more as an automata-like formalism to specify collections of MSCs, not influenced by the branching structure of the MSG.

Reasoning About MSGs. As MSGs are first and foremost used in early stages of protocol development, verification of their correctness is of utter importance to avoid costly and timely redesign steps. In the linear-time setting, many decision problems on MSGs are undecidable (see, e.g., [1,40,43]), including model checking against formulas in *linear temporal logic (LTL)* [45]. Positive results are mainly obtained by imposing syntactic restrictions on the MSG, e.g., *locally synchronized MSGs* [1,41] or *globally cooperative MSGs* [30]. According to [40], the latter class of MSGs is "perhaps the most general of the classes of MSGs amenable to algorithmic analysis". Due to the intrinsic linear-time nature of delayed choice, many of the (un)decidability results for model-checking problems can be adopted to the branching-time setting of delayed-choice MSGs.

Quantitative MSGs. To improve communication systems in their quantitative properties such as reliability or energy consumption, quantitative MSCs were introduced with a semantics as *continuous-time Markov chains (CTMCs)* [50]. In this model, the rates of message sends and receives or their energy consumption were directly attached to communication events. An analysis of quantitative MSCs was conducted using a simulation-based approach. The notion of quantitative MSCs has been extended to *quantitative MSGs* [24] along with probabilistic model-checking algorithms, e.g., to reason about resilience [7,8].

Feature-Oriented Systems. Almost all software systems nowadays are configurable, let it be through low-level compiler flags or through preference panes in running software. One well-established approach to model configurable systems is by using the notion of *features*, i.e., incremental or optional elements of functionality [3]. In feature-oriented modeling, a system variant is described through its configuration as a set of features. Due to the configurations being exponential in the number of features, analyzing all system variants in a feature-oriented system easily becomes infeasible due the need of analyzing exponentially many systems. However, using symbolic representations such as binary decision diagrams (BDDs) [13], commonalities between the behavior of system variants can be exploited. This allows for scalable symbolic verification of feature-oriented systems [19]. These ideas have proven also to be successful in the quantitative setting [26,27], ultimately leading to the feature probabilistic model-checking tool PROFEAT [16,17].

Featured MSGs. While also communication systems are configurable in practice, configurability of MSGs has not yet been considered in the literature. Take for instance the email system in Fig. 1, where it is assumed that all systems have the functionality of encrypting emails. However, encryption might not always be possible, e.g., if the user does not have encryption keys stored on a public key server. In this case, it is desirable to not having to fully specify the communication scenarios without encryption once again by modeling a fresh MSG. Inspired by the area of software product lines [2], commonalities in communication scenarios should be also modeled, e.g., including variants with and without encryption in one specification. In this paper, we introduce *featured MSGs* following a feature-oriented modeling approach. Common communication scenarios are encapsulated into abstract features, rendering MSGs configurable and allowing for feature-oriented techniques also being applied to MSGs.

Contributions. Christel Baier and her group contributed to all the aforementioned fields of MSGs [24,25], quantitative analysis through CTMCs [5,6,9,32], and feature-oriented systems [16,17,26,27]. The goal of this paper is twofold. First, provide an overview on MSGs, feature-oriented formalisms, and analysis techniques Christel Baier shaped in the past. Second, combine these techniques to cumulate into a featured extension of MSGs. In particular, we

- prove that MSGs have a finite-state delayed-choice semantics if they are locally synchronized;
- show that the aCTL model-checking problem for MSGs is undecidable in general but decidable in the locally synchronized case;
- introduce featured MSGs and weak and strong local synchronization criteria
- show decidability of the aCTL model-checking problem for featured MSGs that are weakly or strongly locally synchronized.

2   Background

We fix our notations used throughout the paper. Let P denote a finite set of processes and Δ a finite set of data labels. For each process $p \in P$, we assume to have given a finite set of actions Act_p, comprising local actions $p(d)$, send actions $p!q(d)$, and receive actions $p?q(d)$ with $q \in P$ being a process different than p and $d \in \Delta$ a data label. The set of all actions is denoted by $Act = \bigcup_{p \in P} Act_p$.

Languages and Automata. Given an alphabet Σ, the sets of all finite, non-empty finite, and infinite words are denoted by $\Sigma^\star$, Σ^+, and Σ^ω, respectively. By $w[i]$ we denote the $(i+1)$-th symbol of a word $w \in \Sigma^\star$ and by $|w|$ the length of w. The empty word is denoted by ε. An *automaton* over Σ is a tuple $\mathcal{A} = (Q, \Sigma, \rightarrow, \iota, F)$, where Q is a countable set of states, Σ is an alphabet, $\rightarrow \subseteq Q \times (\Sigma \cup \{\varepsilon\}) \times Q$ is a transition relation, $\iota \in Q$ is an initial state, and $F \subseteq Q$ is a set of final states. We write $p \xrightarrow{\alpha} q$ for $(p, \alpha, q) \in \rightarrow$. A sequence $\eta = q_0 \alpha_0 q_1 \cdots \alpha_{|w|-1} q_{|w|} \in Q \times (\Sigma \times Q)$ is a *path* of $\mathcal{A}$ for a word $w \in \Sigma^\star$ if for all $0 \le i < |w|$ we have $\alpha_i = w[i]$ and $q_i \xrightarrow{\alpha_i} q_{i+1}$. The path η is called *accepting* for w if $q_{|w|} \in F$. We denote the set of all words with accepting paths starting in $q \in Q$ by $L_q(\mathcal{A}) \subseteq \Sigma^\star$, omitting the subscript if $q = \iota$. $\mathcal{A}$ is *finite* if the set of states reachable by some path in $\mathcal{A}$ is finite. Further, $\mathcal{A}$ is *deterministic* if $q \xrightarrow{\alpha} q'$ and $q \xrightarrow{\alpha} q''$ implies $q' = q''$ for all $q, q', q'' \in Q$, $\alpha \in \Sigma$. A language L is *regular* if there is a finite automaton $\mathcal{A}$ with $L = L(\mathcal{A})$.

Propositional Logic. By $\mathbb{B}(X)$ we denote the set of *Boolean expressions* over X, given by the grammar $\phi ::= \mathtt{tt} \mid x \mid \neg\phi \mid \phi \wedge \phi \mid \phi \vee \phi$ where variables x range over X. We use well-known Boolean connectives such as $\rightarrow$, $\leftrightarrow$, etc. from which a Boolean expression can be easily obtained using standard syntactic transformations such as De Morgan's rule. Further, we define $\mathtt{ff} = \neg\mathtt{tt}$. The satisfaction relation $\models \subseteq 2^X \times \mathbb{B}(X)$ is defined in the usual way, where for $Y \subseteq X$ and $\phi \in \mathbb{B}(X)$ we have $Y \models \phi$ if ϕ evaluates to $\mathtt{tt}$ when all variables in Y are assumed to be $\mathtt{tt}$ and all variables in $X \backslash Y$ are $\mathtt{ff}$, respectively.

2.1 Pomset Families

A *poset* over Act is a tuple $(E, \leqslant_E, \lambda_E)$ where $\leqslant_E$ is a partial order over *events* E and $\lambda_E \colon E \to Act$ is a labeling function. We usually identify a poset with its domain, i.e., write E, inherit set operations, and omit subscript domains if clear from the context, e.g., write $\leqslant$ for $\leqslant_E$. E is *total* if for all $e, e' \in E$ with $e \neq e'$ we have either $e < e'$ or $e > e'$. By $\prec$ we denote the direct predecessor relation, i.e., $e \prec e'$ if $e \neq e'$ and for all $f \in E$ where $e < f$ and $f \leqslant e'$ we have $f = e'$. We define the upward E-closure of a set of events $F \subseteq E$ as $F \uparrow E = \{e \in E : \exists e' \in F. e' \leqslant e\}$. A *suffix* of E is a upward E-closed poset, i.e., an $F \subseteq E$ where $F = F \uparrow E$. The set of suffixes of E is denoted by $\mathrm{Suff}^\star(E)$ and $\mathrm{Suff}_\alpha(E)$ is the set of suffixes F where $E \backslash F = \{e\}$ with $\lambda(e) = \alpha$. We define the process function $\mathrm{proc} \colon E \to P$ by $\mathrm{proc}(e) = p$ iff $\lambda(e) \in Act_p$, which naturally extends to sets of events $F \subseteq E$ by $\mathrm{proc}(F) = \bigcup_{e \in F} \mathrm{proc}(e)$. A poset E can be used as a representative of a *pomset* $[E]$, i.e., the class of posets isomorphic to E. Notations for posets naturally extend to pomsets through its representative. We define the *concatenation* of pomsets $[E]$ and $[F]$ represented by disjoint E and F by weak sequential composition [25,38,48] towards $[E] \odot [F] = [(E \cup F, \leqslant, \lambda_E \cup \lambda_F)]$ where $\leqslant = \left(\leqslant_E \cup \leqslant_F \cup \{(e, f) \in E \times F : \mathrm{proc}(e) = \mathrm{proc}(f)\}\right)^\star$. Concatenation is extended to words π over posets and pomsets inductively by $\odot\varepsilon = \varnothing$ and $\odot(\pi E) = \odot(\pi[E]) = (\odot\pi) \odot [E]$. Note that concatenation always yields a pomset and that disjoint domains can always be ensured by renaming. A *pomset family* is a set of pomsets. We extend our notations from pomsets to pomset families by element-wise application, e.g., $\mathcal{E} \odot \mathcal{F} = \{[E] \odot [F] : [E] \in \mathcal{E}, [F] \in \mathcal{F}\}$ for pomset families $\mathcal{E}$ and $\mathcal{F}$. Suffixes of a pomset family $\mathcal{E}$ are extended by $\mathrm{Suff}_\alpha(\mathcal{E}) = \bigcup_{E \in \mathcal{E}} \mathrm{Suff}_\alpha(E)$ for an action $\alpha \in Act$. This also enables the definition of word-suffixes of a poset E, inductively defined by $\mathrm{Suff}_\varepsilon(E) = \{E\}$ and $\mathrm{Suff}_{\alpha w}(E) = \mathrm{Suff}_\alpha(\mathrm{Suff}_w(E))$, and languages $L \subseteq Act^\star$ by $\mathrm{Suff}_L(E) = \bigcup_{w \in L} \mathrm{Suff}_w(E)$. We then write $\mathrm{Suff}^\star(\mathcal{E}) = \mathrm{Suff}_{Act^\star}(\mathcal{E})$. The *language* $L(\mathcal{E})$ of a pomset family $\mathcal{E}$ is the set of words $w \in Act^\star$ for which $\varnothing \in \mathrm{Suff}_w(\mathcal{E})$. Given a language L over posets or pomsets, we denote the pomset family of L by $\odot L = \{\odot\pi : \pi \in L\}$. An automaton over a finite set of pomsets over Act is called *pomset automaton*, accepting the pomset family $\odot L(\mathcal{A})$.

2.2 Transition Systems and Computation Tree Logic

Transition Systems. A *transition system* over the action alphabet Act is a tuple $\mathcal{T} = (S, Act, \longrightarrow, \iota)$ where S is a countable set of states, $\longrightarrow \subseteq S \times (Act \cup$

$$\mathcal{T}, s \models \texttt{tt}$$

$\mathcal{T}, s \models \neg\Phi$ iff $\mathcal{T}, s \not\models \Phi$

$\mathcal{T}, s \models \Phi \wedge \Psi$ iff $\mathcal{T}, s \models \Phi$ and $\mathcal{T}, s \models \Psi$

$\mathcal{T}, s \models \exists(\phi)$ iff there is $\pi \in \mathrm{Runs}(\mathcal{T}, s)$ such that $\mathcal{T}, \pi \models \phi$

$\mathcal{T}, s \models \forall(\phi)$ iff for all $\pi \in \mathrm{Runs}(\mathcal{T}, s)$ we have $\mathcal{T}, \pi \models \phi$

$\mathcal{T}, \pi \models \mathbf{X}_A\Phi$ iff for $\pi = s_0\alpha_0 s_1 \cdots$ we have $\alpha_0 \in A$ and $\mathcal{T}, s_1 \models \Phi$

$\mathcal{T}, \pi \models \Phi\mathbf{U}_A\Psi$ iff for $\pi = s_0\alpha_0 s_1 \cdots$ there is $j \in \mathbb{N}$ such that $\mathcal{T}, s_j \models \Psi$

       and for all $i < j$ we have $\mathcal{T}, s_i \models \Phi$ and $\alpha_i \in A$

$\mathcal{T}, \pi \models \Box_A\Phi$ iff for $\pi = s_0\alpha_0 s_1 \cdots$ and all $i \in \mathbb{N}$ we have $\mathcal{T}, s_i \models \Phi$ and $\alpha_i \in A$

Fig. 2. Semantics of aCTL

$\{\varepsilon\}) \times S$ a transition relation, and $\iota \in S$ an initial state. Here, we indicate by ε a *silent action* modeling internal moves. An infinite sequence $\pi = s_0\alpha_0 s_1\alpha_1 \ldots \in S \times ((Act \cup \{\varepsilon\}) \times S)^\omega$ where $s_i \xrightarrow{\alpha_i} s_{i+1}$ for all i is called *run* for $\alpha_0\alpha_1 \ldots$ in s_0. For a state $s \in S$, we denote by $\mathrm{Runs}(\mathcal{T}, s)$ the set of runs in s. Note that an infinite run can be a run for a finite sequence of actions. Transition systems induce a language semantics, denoted by $L(\mathcal{T})$, which is the set of words $\alpha_0\alpha_1 \ldots \in \Sigma^\star$ for which there is a run starting in ι. $\mathcal{T}$ is *finite* if the set of states reachable by a run in s_0 is finite. It is *deterministic* if for all $s, t, t' \in S$ and $\alpha \in Act \cup \{\varepsilon\}$ with $s \xrightarrow{\alpha} t$ and $s \xrightarrow{\alpha} t'$ we have $t = t'$.

Computation Tree Logic. To reason about transition systems with action and state labels, we rely on *action computation tree logic (aCTL)*, an action-based variant of computation tree logic (CTL) [18,21]. An aCTL formula Φ over actions Act is defined through the grammar

$$\Phi ::= \texttt{tt} \mid \neg\Phi \mid \Phi \wedge \Phi \mid \exists(\phi) \mid \forall(\phi)$$
$$\phi ::= \mathbf{X}_A\Phi \mid \Phi\mathbf{U}_A\Phi \mid \Box_A\Phi$$

where $A \subseteq Act \cup \{\varepsilon\}$. When $A = Act \cup \{\varepsilon\}$, we omit the subscript action set A. Further standard operators can be derived, e.g., $\Diamond_A\Phi := \texttt{tt}\,\mathbf{U}_A\Phi$. The semantics of an aCTL formula over Act is defined for transition systems $\mathcal{T} = (S, Act, \longrightarrow, \iota)$. Denoting by $s \in S$ a state, $A \subseteq Act \cup \{\varepsilon\}$ an action set, and Φ, Ψ aCTL formulas over Act, the satisfaction relation $\models$ is inductively defined as provided in Fig. 2. We define the *satisfaction set* of an aCTL formula Φ by $[\![\Phi]\!] = \{s \in S : \mathcal{T}, s \models \Phi\}$. We write $\mathcal{T} \models \Phi$ for $\mathcal{T}, \iota \models \Phi$.

The *aCTL model-checking problem* for $\mathcal{T}$ and Φ amounts to decide whether $\mathcal{T} \models \Phi$. It is well known that the CTL model-checking problem is decidable in polynomial time [10] and so is also its action-based variant [21].

3   Principles of Message Sequence Graphs

The ITU-T standard [33] introduced *message sequence charts (MSCs)* as a visual formalism to describe communication scenarios. Sequential, alternative, and repetitive behaviors are specified using *message sequence graphs (MSGs)*, which essentially are graphs over MSCs [34]. In this section, we define MSCs based on pomsets [38,47], introduce a more general variant of MSGs by means of pomset automata over MSCs, provide their delayed-choice semantics [4,25], and establish other basic properties. Recall that we assume fixed sets of communicating processes P and communication actions Act.

Definition 1 (MSC). *An* MSC *over Act is a non-empty pomset $[M]$ where for $M_\dagger = \{e \in M : \lambda(e) = p\dagger q(d)\}$ with $\dagger \in \{!,?\}$ there is a bijection $\xi \colon M_! \to M_?$ with $\lambda(e) = p!q(d)$ implies $\lambda(\xi(e)) = q?p(d)$ for all $e \in M_!$, and for all $p \in P$ there is a total order $\leqslant_p$ on $M_p = \{e \in M : \lambda(e) \in Act_p\}$ such that $\leqslant = \left(\{(e, \xi(e)) : e \in M_!\} \cup \bigcup_{p \in P} \leqslant_p \right)^\star$.*

Intuitively, the bijection ξ in the above definition guarantees that every receive event has a corresponding send event and the requirement of every process $p \in P$ admitting a totally ordered M_p formalizes the time-line ordering of events of p. Note that for any two MSCs M_1 and M_2, $M_1 \odot M_2$ is again an MSC. One important property of MSCs is that there is a one-to-one correspondence of them to their languages [40]. This result also holds for suffixes of MSCs.

Lemma 1. *For $i = 1, 2$ let M_i be an MSC over Act and $S_i \in \mathrm{Suff}^\star(M_i)$. Then $L(S_1) = L(S_2)$ iff $[S_1] = [S_2]$.*

Proof. ($\Rightarrow$): Since for $i = 1, 2$ we have that S_i is upward-closed as it is a suffix, for each process $p \in P$ we have that S_{ip} is totally ordered and comprises the last $|S_{ip}|$ events of M_i only. By $L(S_1) = L(S_2)$, $|S_{1p}| = |S_{2p}|$ for all $p \in P$ and hence, there is a uniquely defined bijection $\xi \colon S_1 \to S_2$ mapping the j-th event of S_{1p} to the j-th event of S_{2p} for each $p \in P$ such that $\lambda_1(e) = \lambda_2(\xi(e))$ for all $e \in S_1$. Further, there is a uniquely defined $\omega_i \colon L(S_i) \times S_i \to \{0, \ldots, |S_i| - 1\}$ that maps each word $w \in L(S_i)$ and event $e \in S_i$ to the position $\omega(w, e) = j$ of the label $\lambda(e) = w[j]$ corresponding to the event. Towards a contradiction, assume that ξ is not an isomorphism, i.e., there are w.l.o.g. $e, e' \in S_1$ such that $e <_1 e'$ but $\xi(e) \not<_2 \xi(e')$. Then let $w = x\alpha y\beta z \in L(S_1)$ such that $\omega(w, e) = |x|$ and $\omega(w, e') = |x\alpha y|$ are the positions of the α- and β-events in w. If $\xi(e) >_2 \xi(e')$, then $w \notin L(S_2)$. Otherwise, due to $e <_1 e'$, we have that $x\beta y\alpha z \notin L(S_1)$ but $x\beta y\alpha z \in L(S_2)$. Both cases contradict $L(S_1) = L(S_2)$.

($\Leftarrow$): Clear by definition of languages of pomsets. $\square$

Definition 2 (MSG). *A* message sequence graph (MSG) *over $\mathbb{M}$, a finite set of MSCs over Act, is a pomset automaton $\mathcal{G} = (Q, \mathbb{M}, \hookrightarrow, \iota, F)$ over $\mathbb{M}$.*

Usually, pomset automata and thus also MSGs are given graphically as illustrated in Fig. 1. In the following, let us fix an MSG $\mathcal{G}$ as above.

Semantics of MSGs. The *pomset semantics* of $\mathcal{G}$ is the pomset family $\odot L(\mathcal{G})$ of MSCs arising by composition of MSCs along accepting paths in $\mathcal{G}$ [25]. The standard branching-time semantics for MSGs is governed by *delayed choice*, a process-algebraic operator [4] where choices are delayed until they become inevitable. The connection between both semantics was made by interpreting delayed choice over pomsets, leading to a transition-system semantics over suffixes [24, 25]. Intuitively, each state comprises all the possible futures as pomsets, starting with the initial state as the MSC family $\odot L(\mathcal{G})$. A transition with action $\alpha \in Act_p$ removes minimal α-events from the pomsets in the current state and disregards those pomsets that do not have such an α-event. The latter implements delayed choice by ruling out possible futures if they do not contain an α-event next on the time line of process p.

Definition 3. *The transition-system semantics* $\mathcal{T}(\mathcal{G}) = (S, Act, \longrightarrow, \odot L(\mathcal{G}))$ *of the MSG* $\mathcal{G}$ *is defined by* $S = [\mathrm{Suff}^\star(\odot L(\mathcal{G}))]$ *and* $\longrightarrow$ *being the smallest transition relation where* $s \xrightarrow{\alpha} [\mathrm{Suff}_\alpha(s)]$ *for* $\alpha \in Act$ *and* $s \xrightarrow{\varepsilon} s$ *for* $\varnothing \in s$.

Note that for a state $s \in S$, $[\mathrm{Suff}_\alpha(s)] = \varnothing$ if there is no pomset $M \in s$ with minimal α. In such a case, s does not have an α-transition since $\varnothing \notin S$. Further, $\varnothing \in s$ stands for all events of a pomset future having been executed. This allows for infinitely staying in this state and not executing any further events. Note that $\mathcal{T}(\mathcal{G})$ is deterministic due to the totally ordered time lines of each process and that $\mathcal{T}(\mathcal{G})$ can be constructed on-the-fly as every state can be described by a regular pomset family [25].

3.1 Local Synchronization

In general, the transition system $\mathcal{T}(\mathcal{G})$ for an MSG $\mathcal{G}$ can be infinite if $\mathcal{G}$ contains a cycle with non-empty MSCs. This is for instance the case for the communication scenario of the email system of Fig. 1. Towards a syntactic criterion that ensures finite descriptions of the behavior of MSGs, Alur and Yannakakis [1] and Muscholl and Peled [41] independently introduced *locally synchronized* MSGs. A cycle in $\mathcal{G}$ is a path $\eta = q_0 b_0 q_1 b_1 \ldots b_n q_n \in Q \times (\mathbb{M} \times Q)^\star$ in $\mathcal{G}$ where $q_0 = q_n$. For an MSC M, the *communication graph* $\mathcal{H}(M)$ is the directed graph $(\mathrm{proc}(M), \rightsquigarrow)$, where $p \rightsquigarrow q$ iff there is an event $e \in M_!$ where $\lambda(e) = p!q(d)$ or $p = q$ and $\lambda(e) = p(d)$. The MSG $\mathcal{G}$ is *locally synchronized* if for all cycles η as above the communication graph $\mathcal{H}(b_0 \odot b_1 \odot \ldots \odot b_n)$ consists of a single strongly connected component. Note that the MSG in Fig. 1 is locally synchronized as every message send needs to be acknowledged.

Lemma 2 ([29, 31, 40]). *Checking whether an MSG* $\mathcal{G}$ *is locally synchronized is coNP-complete.*

For locally synchronized MSGs $\mathcal{G}$, it was shown that there is a finite automaton that accepts the language of $\odot L(\mathcal{G})$ [1, 41]. Let $\mathcal{A}_\mathcal{G} = (Z, Act, \rightarrow, \iota_Z, F_Z)$ denote a deterministic automaton over Act for which $L(\mathcal{A}_\mathcal{G}) = L(\odot L(\mathcal{G}))$. This result can be extended to show that $\mathcal{T}(\mathcal{G})$ is finite if $\mathcal{G}$ is locally synchronized.

Theorem 1. *If* $\mathcal{G}$ *is locally synchronized, then* $\mathcal{T}(\mathcal{G})$ *is finite.*

Proof. Assume $\mathcal{T}(\mathcal{G})$ to be infinite. By König's lemma, there is an infinite path $s_0\alpha_0 s_1\alpha_1 \ldots$ in $\mathcal{T}(\mathcal{G})$ that does not visit any state twice, i.e., for all $i \neq j \in \mathbb{N}$ we have $s_i \neq s_j$. Since $L(\mathcal{T}(\mathcal{G})) = L(\mathcal{A}_\mathcal{G})$ [25] there is a path $z_0\alpha_0 z_1\alpha_1 \ldots$ in $\mathcal{A}_\mathcal{G}$ such that $L(s_i) = L_{z_i}(\mathcal{A}_\mathcal{G})$ for all $i \in \mathbb{N}$. As $\mathcal{A}_\mathcal{G}$ is finite, there are indices $i \neq j \in \mathbb{N}$ where $z_i = z_j$. Thus, $L_{z_i}(\mathcal{A}_\mathcal{G}) = L_{z_j}(\mathcal{A}_\mathcal{G})$, which yields $L(s_i) = L(s_j)$. Since $s_i = \text{Suff}_{\alpha_0\alpha_1\ldots\alpha_{i-1}}(\circledcirc L(\mathcal{G}))$ and $s_j = \text{Suff}_{\alpha_0\alpha_1\ldots\alpha_{j-1}}(\circledcirc L(\mathcal{G}))$, Lemma 1 applied on sets of suffixes then yields $s_i = s_j$, a contradiction. □

The converse does not hold, i.e., an MSG with finite transition-system semantics does not need to be locally synchronized. This can be seen by the following MSG $\mathcal{G}_{ab}$ over two MSCs M and M'. When M and M' contain only one local event $p(d)$ and $p'(d)$ with $p \neq p'$, respectively, and $\mathcal{G}_{ab}$ has a single accepting and initial state ι with $\iota \xrightarrow{M} \iota$ and $\iota \xrightarrow{M'} \iota$, then $L(\mathcal{G}_{ab}) = \{M, M'\}^\star$ and $\mathcal{T}(\mathcal{G}_{ab})$ is a single-state transition system.

3.2 ACTL Model Checking

Formulas in aCTL allow to specify many interesting properties of communication scenarios. For instance, taking the example MSG of Fig. 1 into account, we could specify

(success) $\forall\square\forall\lozenge_{\text{Server!Receiver}(mail)}\forall\lozenge_{\text{Receiver?Server}(mail)}\texttt{tt}$:
  whenever a mail is send, a mail is also eventually received.
(acknowledge) $\forall\lozenge_{\text{Receiver?Server}(mail)}\exists\lozenge_{\text{User?Server}(ack)}\texttt{tt}$:
  whenever a mail is received, there is a possibility of acknowledging reception.

Unfortunately, stemming from many undecidability results for standard MSGs, aCTL model checking of MSGs is undecidable.

Theorem 2. *The aCTL model-checking problem for MSGs is undecidable.*

Proof. Let us provide a reduction from the Post Correspondence Problem (PCP), similar to the reduction presented for the MSG intersection problem [42]. The input for the PCP is a finite sequence of pairs $(u_i, v_i) \in \{0,1\}^+ \times \{0,1\}^+$ for $i = 0, \ldots, n$. A PCP solution is a finite sequence of indices $k_1, k_2, \ldots, k_m$ such that $u_{k_1} u_{k_2} \ldots u_{k_m} = v_{k_1} v_{k_2} \ldots v_{k_m}$. We construct an MSG $\mathcal{G}$ as depicted in Fig. 3 and consider the aCTL formula $\Phi = \exists\lozenge(\exists\mathbf{X}_{q(l)}\texttt{tt} \wedge \exists\mathbf{X}_{q(r)}\texttt{tt})$, i.e., there is a path in $\mathcal{T}(\mathcal{G})$ such that both local events $q(l)$ and $q(r)$ can be executed by process q. Intuitively, the left part of the MSG encodes an index i and the words u_i through local events of independent processes p and q in the MSC $M_{u,i}$, respectively. The right part does the same for v_i through the MSC $M_{v,i}$. Both the left and right can be finalized by taking the N_l or N_r transition, respectively. After executing $q?p(\bot)$ the local events $q(l)$ or $q(r)$ can be executed as final event, depending whether the left or right branch has been taken, respectively. We show that $\mathcal{T}(\mathcal{G}) \models \Phi$ iff the PCP has a solution.
($\Rightarrow$): If $\mathcal{T}(\mathcal{G}) \models \Phi$, then there is a word $w \in L(\mathcal{T}(\mathcal{G}))$ such that $\{e_l\}, \{e_r\} \in \text{Suff}_w(\circledcirc L(\mathcal{G}))$ with $\lambda(e_l) = q(l)$ and $\lambda(e_r) = q(r)$. Since an MSC with e_l (e_r)

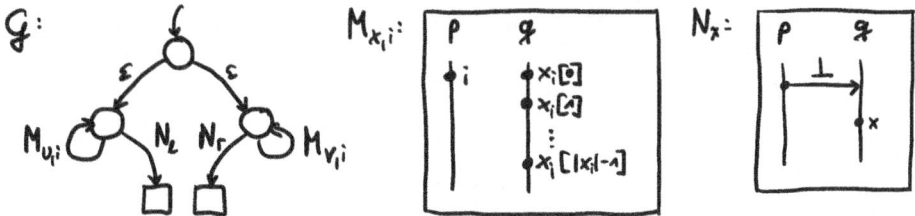

Fig. 3. Reduction from the PCP to the aCTL model checking problem for MSGs

can only be reached based on a left-branch path (right-branch path, respectively), there are $m, m' \in \mathbb{N}$ and sequences $k_1, k_2, \ldots, k_m$ and $k'_1, k'_2, \ldots, k'_{m'}$ such that $wq(l) \in L(M_{u,k_1} \circledcirc M_{u,k_2} \circledcirc \ldots \circledcirc M_{u,k_m} \circledcirc N_l)$ and $wq(r) \in L(M_{v,k'_1} \circledcirc M_{v,k'_2} \circledcirc \ldots \circledcirc M_{v,k_{m'}} \circledcirc N_r)$. Since $w|_{Act_p} = p(k_1)p(k_2)\ldots p(k_m)p(\bot) = p(k'_1)p(k'_2)\ldots p(k'_{m'})p(\bot)$ we obtain $m = m'$ and $k_i = k'_i$ for all $0 < i \leqslant m$. Hence,

$$M = M_{u,k_1} \circledcirc M_{u,k_2} \circledcirc \ldots \circledcirc M_{u,k_m} = M_{v,k_1} \circledcirc M_{v,k_2} \circledcirc \ldots \circledcirc M_{v,k_m} \quad (1)$$

That is, the MSC pomsets arising from concatenating MSCs along the left- and right-branch in the MSG are isomorphic and choices between both MSG branches are delayed due to delayed choice. Thus,

$$w|_{Act_q} = u_{k_1}u_{k_2}\ldots u_{k_m}q?p(\bot) = v_{k_1}v_{k_2}\ldots v_{k_m}q?p(\bot)$$

and $k_1, k_2, \ldots, k_m$ is a solution to the PCP.

($\Leftarrow$): The other way around, if there is a solution $k_1, k_2, \ldots, k_m$ of the PCP, then $u_{k_1}u_{k_2}\ldots u_{k_m} = v_{k_1}v_{k_2}\ldots v_{k_m}$ and thus, (1) holds. By the definition of $\mathcal{T}(\mathcal{G})$, we obtain $\{e_l\}, \{e_r\} \in \mathrm{Suff}_w(\circledcirc L(\mathcal{G}))$ for

$$w = p(k_1)p(k_2)\ldots p(k_m)u_{k_1}u_{k_2}\ldots u_{k_m}p!q(\bot)q?p(\bot)$$

Thus, $\mathcal{T}(\mathcal{G}) \models \Phi$.

Note that while $\mathcal{G}$ is non-deterministic, this is not a restriction as it is well possible to add an initialization of two independent MSCs M_1 and M_2 where one branch is initialized with $M_1 M_2$ and the other with $M_2 M_1$. Since $M_1 \circledcirc M_2 = M_2 \circledcirc M_1$, this renders the left and right branches in the proof result into the same pomset suffixes and hence the same states in the transition system due to delayed choice. □

One natural consequence of an MSG being locally synchronized is the ability to model check branching-time properties such as given by an aCTL formula Φ over Act. For this, the finite transition system $\mathcal{T}(\mathcal{G})$ is constructed, followed by the standard aCTL model-checking procedure [21,44].

4 Principles of Feature-Oriented Systems

Nowadays, almost every software system is configurable. A common approach to model configurable systems is by using the notion of *features*, i.e., optional or incremental functionalities of a system that can be configured to be included or excluded [2,11]. In this section, we recall feature-oriented modeling through feature models and featured transition systems (FTSs) [19].

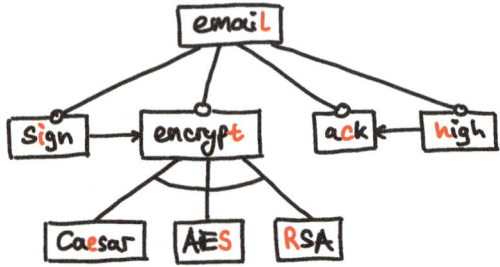

Fig. 4. Feature diagram of a simple email system product line

Let us identify the abstract set of features with a set of Boolean variables X. A *configuration* is then a subset of features $C \subseteq X$. Feature models describe which configurations can lead to a valid system variant, i.e., they specify the set of *valid configurations* Conf $\subseteq 2^X$. *Feature diagrams* [37] constitute the most commonly used feature model, providing a symbolic tree-based representation of valid configurations. Figure 4 depicts an example of a feature diagram specifying an email system product line over features $X = \{\mathbf{c}, \mathbf{h}, \mathbf{r}, \mathbf{i}, \mathbf{s}, \mathbf{t}, \mathbf{e}, \mathbf{l}\}$. The semantics of a feature diagram is given as a Boolean expression over features, e.g., $\phi \in \mathbb{B}(X)$ in Fig. 4 for the email system example. Essentially, this semantics interprets the hierarchical structure of the diagram and its decompositions: (1) the root feature is always included (2) child features imply that parent features are included (3) parent features imply a child group where OR groups are solid connected, ONE-HOT groups are line connected, and AND groups are without branch connections (4) features with a circle at the top are optional (5) arrows between features indicate additional implications as cross-tree constraints. In Fig. 4, the l feature has an AND group comprising optional $\mathbf{i}, \mathbf{t}, \mathbf{c}$, and $\mathbf{h}$, where the latter imposes a ONE-HOT group between $\mathbf{e}, \mathbf{s}$, and $\mathbf{r}$. The high-feature is identified with $\mathbf{h}$ in Fig. 4 and stands for the option to give an email high priority while the option to require an acknowledgement is identified with the feature $\mathbf{c}$. Then, cross-tree constraints can formalize implications indicated as connecting arrows between features. For instance, the high-feature $\mathbf{h}$ implies requesting an acknowledgement $\mathbf{c}$, leading to all valid configurations including $\mathbf{h}$ to also include $\mathbf{c}$.

Featured Transition Systems. To specify operational behaviors of configurable systems, *featured transition systems (FTSs)* were introduced as transition systems with transitions guarded by Boolean expressions over features. We here focus on FTSs over communication actions, i.e., tuples $\mathsf{T} = (S, Act, \Longrightarrow, \iota)$ where S is a countable set of states, $\Longrightarrow\, \subseteq S \times \mathbb{B}(X) \times (Act \cup \{\varepsilon\}) \times S$ a featured transition relation, and $\iota \in S$ an initial state. We write $s \overset{\phi:\alpha}{\Longrightarrow} s'$ for a transition $(s, \phi, \alpha, s') \in \Longrightarrow$ that is guarded by an expression $\phi \in \mathbb{B}(X)$. The *product semantics* of an FTS T w.r.t. a valid configuration $C \in$ Conf is the transition system $\mathsf{T}|_C = (S, Act, \longrightarrow_C, \iota)$ whose transition relation is defined by $s \overset{\alpha}{\longrightarrow}_C s'$ iff there is $s \overset{\phi:\alpha}{\Longrightarrow} s'$ where $C \models \phi$. Note that even if T does not have terminal states, products might have. An FTS is deterministic if $\mathsf{T}|_C$ is deterministic for all $C \in$ Conf. This semantics can be extended to a *family semantics*, comprising all product transition systems by $[\![\mathsf{T}]\!] = \{(C, \mathsf{T}|_C) : C \in \text{Conf}\}$.

The *lifted semantics* $\uparrow\mathsf{T}$ of T is a transition system that encodes configurations in the state space of the FTS with an initial *seeding phase* to select configurations. This approach is usually taken for practical applications for model checking FTSs. Formally, $\uparrow\mathsf{T} = (\text{Conf} \times S, Act \cup \text{Conf}, \longrightarrow, (\varnothing, \iota))$ where $\longrightarrow$ is the smallest transition relation where for all $C \in$ Conf we have $(\varnothing, \iota) \overset{C}{\longrightarrow} (C, \iota)$ and $s \overset{\alpha}{\longrightarrow}_C s'$ implies $(C, s) \overset{\alpha}{\longrightarrow} (C, s')$.

Featured Automata. The definition of FTSs naturally extends to automata, i.e., a *featured automaton* over Σ and X is a tuple $\mathsf{A} = (Q, \Sigma, \Rightarrow, \iota, F)$, where Q, Σ, ι, and F are as for automata and $\Rightarrow \subseteq Q \times \mathbb{B}(X) \times (\Sigma \cup \{\varepsilon\}) \times Q$ is a featured transition relation. Analogously, the product semantics $\mathsf{A}|_C$ for a valid configuration $C \in$ Conf is defined as the automaton $\mathsf{A}|_C = (Q, \Sigma, \rightarrow, \iota, F)$ where $q \overset{\alpha}{\rightarrow} q'$ iff there is $q \overset{\phi:\alpha}{\Rightarrow} q'$ with $C \models \phi$. The family semantics of featured automata comprises all product automata, i.e., $[\![\mathsf{A}]\!] = \{(C, \mathsf{A}|_C) : C \in \text{Conf}\}$.

Featured Pomsets. Let us define *featured posets* as posets whose events are guarded by feature expressions, i.e., $\mathsf{E} = (E, \leqslant_E, \lambda_E, g_E)$ where $(E, \leqslant_E, \lambda_E)$ is a poset and $g_E \colon E \to \mathbb{B}(X)$. The definition naturally extends to featured pomsets, for which we also define a concatenation operation $\circledcirc$ on disjoint featured pomsets E and F as for standard pomsets by $[\mathsf{E}] \circledcirc [\mathsf{F}] = ([E] \circledcirc [F], g_E \cup g_F)$. The product semantics is defined as $\mathsf{E}|_C = (F, \leqslant_F, \lambda_F)$ for valid configurations $C \in$ Conf where $F = \{e \in E : \forall f \in E, f \leqslant e.C \models g_E(f)\}$ is the maximal set of downward-closed events whose feature guards are satisfied in configuration C. The family semantics for featured posets maps valid configurations to posets, i.e., $[\![\mathsf{E}]\!] = \{(C, \mathsf{E}|_C) : C \in \text{Conf}\}$. Suffixes are defined as for standard pomsets. Given an action $\alpha \in Act$, we define the feature guards of α of the featured pomset E by $\Phi_\alpha(\mathsf{E}) = \{g_E(e) : e \in E \backslash \text{Suff}_\alpha(E)\}$.

5   Featured Message Sequence Graphs

We are now ready to combine the presented principles for MSGs and FTSs towards *featured MSGs*, allowing to specify families of communication scenarios using a feature-oriented approach.

Definition 4. *A* featured MSG *(FMSG) over a finite set of MSCs* $\mathbb{M}$ *and features* X *is a featured automaton* $\mathsf{G} = (Q, \mathbb{M}, \Rightarrow, \iota, F)$ *over* $\mathbb{M}$ *and* X.

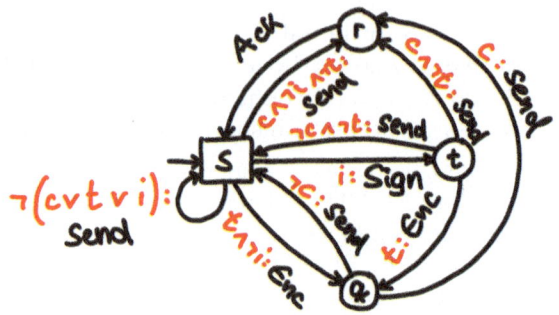

Fig. 5. Featured MSG of the simple email system product line

Figure 5 shows an FMSG of our simple email system product line with a set of valid configurations specified by the feature diagram Fig. 4. In the following, let us fix an FMSG as above and a set of valid configurations $\text{Conf} \subseteq 2^X$.

Semantics of FMSGs. The pomset semantics for MSGs naturally extends to FMSGs, mapping each valid configuration to the pomset semantics of the corresponding product MSG: $[\![G]\!] = \{(C, \odot L(G|_C)) : C \in \text{Conf}\}$. Defining a proper featured transition-system semantics is more intricate as feature guards in FMSGs are effective at the level of pomsets and not on events. We therefore push the global feature guards at FMSG level inside MSCs towards MSGs over featured MSCs: For an MSC M and a feature guard $\phi \in \mathbb{B}(X)$, we define the featured MSC $\phi(M) = (M, \leqslant_M, \lambda_M, g_M)$ by $g_M(e) = \phi$ for all $e \in M$. From G we now define an MSG $\underline{G} = (Q, \mathbb{F}, \hookrightarrow, \iota, F)$ over featured MSCs

$$\mathbb{F} = \{\phi(M) : \exists q, q' \in Q, M \in \mathbb{M}. q \overset{\phi:M}{\Rightarrow} q'\}$$

where $q \overset{\phi(M)}{\hookrightarrow} q'$ iff $q \overset{\phi:M}{\Rightarrow} q'$. While this enables us to see the featured pomset semantics of an FMSG G as the pomset semantics $\odot L(\underline{G})$ of $\underline{G}$, shifting to event-level feature guards may neglect accepting paths at MSG level:

Lemma 3. *For all valid configurations $C \in \text{Conf}$:*

$$[\![G]\!](C) = \odot L(G|_C) \subseteq (\odot L(\underline{G}))|_C$$

Proof. The first equality holds by definition. Let $\eta = q_0(\phi_0{:}M_0)q_1 \ldots q_n$ be an accepting path in G such that $C \models \phi_i$ for all $i < n$. Hence, $M = \odot(M_0 M_1 \ldots M_{n-1}) \in \odot L(G|_C)$. Then, $\eta' = q_0(\phi_0(M_0))q_1 \ldots q_n$ is an accepting path in $\underline{G}$ by definition of $\underline{G}$ and thus, $M' = \odot(\phi_0(M_0)\phi_1(M_1)\ldots\phi_{n-1}(M_{n-1})) \in \odot L(\underline{G})$. By definition of product semantics of featured pomsets, $[\phi_i(M_i)|_C] = [M_i]$ and thus, $M'|_C = M \in (\odot L(\underline{G}))|_C$. Note that the converse does not hold, e.g., there could be $C' \in \text{Conf}$ such that $C \models \phi_0$ but $C \not\models \phi_1$. Then, $M_0 \in (\odot L(\underline{G}))|_C$ but $M_0 \notin \odot L(G|_C)$. $\qquad\square$

Towards an FTS semantics of FMSGs, delayed choice has hence to be interpreted with taking feature guards into account. There are several possibilities to achieve a semantics compatible with delayed choice. A naive approach would be to define the FTS that combines transition systems for each valid configuration following a product construction, guarding the actions with the characteristic Boolean expression for all configurations the action could be executed. However, this would require semantic evaluation, change the feature guards compared to the original FMSG specification, and renders an on-the-fly construction infeasible due to the potential exponentially many valid configurations. We propose to define the FTS semantics similar to the delayed-choice semantics for MSGs but now on the MSG $\underline{G}$ over featured MSCs.

Definition 5. *The FTS semantics* $\mathsf{T}(\mathsf{G}) = (S, Act, \Longrightarrow, \odot L(\underline{G}))$ *of the FMSG* G *is defined by* $S = [\mathrm{Suff}^\star(\odot L(\underline{G}))]$ *and* $\Longrightarrow$ *being the smallest transition relation where* $s \xrightarrow{\phi:\alpha} [\mathrm{Suff}_\alpha(s)]$ *for* $\alpha \in Act$ *with* $\phi = \bigvee \Phi_\alpha(s)$ *and* $s \xrightarrow{\mathtt{tt}:\varepsilon} s$ *for* $\varnothing \in s$.

Recall that $\Phi_\alpha(\mathsf{E})$ for a featured pomset E is the set of all feature guards that guard a minimal α-event in E. Note that $\mathsf{T}(\underline{G})$ is deterministic. In Lemma 3 we showed that there might be pomsets in $(\odot L(\underline{G}))|_C$ that do not have an accepting counterpart in $\mathsf{G}|_C$. This phenomenon can also arise in the FTS semantics, leading to parts in $\mathsf{T}(\underline{G})$ that might contain terminal states. However, transition systems, and hence also FTSs, are usually interpreted over infinite runs and could potentially only lead to finite action sequences through ε transitions. Due to $\mathcal{T}(\mathsf{G})$ and $\mathsf{T}(\mathsf{G})$ being deterministic where internal actions only appear in self-loops, this can be interpreted as compatibility between featured and product transition system semantics of FMSGs:

Theorem 3. *For all configurations* $C \in \mathrm{Conf}$:

$$L\big(\mathcal{T}(\mathsf{G}|_C)\big) \;=\; L\big(\mathsf{T}(\mathsf{G})|_C\big)$$

Proof. ($\subseteq$): Let $s_0\alpha_0 s_1\alpha_1 \ldots \in \mathrm{Runs}(\mathcal{T}(\mathsf{G}|_C), \iota)$ be a run for a finite word $w = \alpha_0\alpha_1 \ldots \in Act^\star$. Then there is an accepting path $q_0 M_0 q_1 M_1 \ldots M_{n-1} q_n$ in $\mathsf{G}|_C$ such that $\varnothing \in \mathrm{Suff}_w(M_0 \odot M_1 \odot \ldots \odot M_{n-1})$. By definition of product semantics for FMSGs, there are hence accepting paths $q_0(\phi_0{:}M_0)q_1(\phi_1{:}M_1)\ldots (\phi_{n-1}{:}M_{n-1})q_n$ in G and $q_0(\phi_0(M_0))q_1(\phi_1(M_1))\ldots (\phi_{n-1}(M_{n-1}))q_n$ in $\underline{G}$ where $C \models \phi_i$ for all $i < n$. Thus, there is a run $s_0'(\phi_0'{:}\alpha_0)s_1' \ldots \in \mathrm{Runs}(\mathsf{T}(\mathsf{G}), \iota)$ in $\mathsf{T}(\mathsf{G})$ where $\phi_i' = \bigvee \Phi_{\alpha_i}(s_i')$ and $\phi_i \in \Phi_{\alpha_i}(s_i')$ for all $i < n$. Since $C \models \phi_i$ implies $C \models \phi_i'$ for all $i < n$, it follows $s_0'\alpha_0 s_1'\alpha_1 \ldots \in \mathrm{Runs}(\mathsf{T}(\mathsf{G})|_C, \iota)$ and thus $w \in L(\mathsf{T}(\mathsf{G})|_C)$. ($\supseteq$): Let $s_0'\alpha_0 s_1'\alpha_1 \ldots \in \mathrm{Runs}(\mathsf{T}(\mathsf{G})|_C, \iota)$ be a run for $w = \alpha_0\alpha_1 \ldots \in Act^\star$. Since the run is infinite, there is a $k \in \mathbb{N}$ such that $\alpha_i = \varepsilon$ for all $i \geq k$. W.l.o.g., let k be the first index where $\alpha_k = \varepsilon$ and hence $k = |w|$. Then $\varnothing \in \mathrm{Suff}_w(\odot L(\underline{G}))$ and thus, there are accepting paths $q_0(\phi_0(M_0))q_1(\phi_1(M_1))\ldots (\phi_{n-1}(M_{n-1}))q_n$ in $\underline{G}$ and $q_0(\phi_0{:}M_0)q_1(\phi_1{:}M_1)\ldots (\phi_{n-1}{:}M_{n-1})q_n$ in G where $C \models \phi_i$ for all $i < n$ and $w \in L(M_0 \odot M_1 \odot \ldots \odot M_{n-1})$. Thus, also $q_0 M_0 q_1 M_1 \ldots M_{n-1} q_n$ is an accepting path in $\mathsf{G}|_C$ and $\varnothing \in \mathrm{Suff}_w(M_0 \odot M_1 \odot \ldots \odot M_{n-1})$, which leads to $w \in L(\mathcal{T}(\mathsf{G}|_C))$. $\square$

5.1 Locally Synchronized FMSGs

We consider two variants of local synchronization in FMSGs: weakly and strongly locally synchronized. An FMSG G is *weakly locally synchronized* w.r.t. a feature model given as Boolean expression $\Psi \in \mathbb{B}(X)$ if for all valid configurations $C \models \Psi$ we have that $\mathsf{G}|_C$ is locally synchronized. It is *strongly locally synchronized* if the MSG $\underline{\mathsf{G}}$ is locally synchronized. Put differently, the weak variant takes feature guards into account, while the strong variant disregards feature guards and requires local synchronization on the underlying MSG. Interestingly, checking for weak local synchronization is of the same complexity as checking for the feature agnostic criterion of strong local synchronization:

Lemma 4. *Let G be an FMSG and $\Psi \in \mathbb{B}(X)$ a feature model. Then both, checking whether G is weakly locally synchronized w.r.t. Ψ, and checking whether G is strongly locally synchronized is coNP-complete.*

Proof. Containment in coNP is immediate: guessing a configuration $C \subseteq X$ and a cycle in $\mathsf{G}|_C$ where checking whether $C \not\models \Psi$ or the communication graph of the cycle is a single strongly connected component can be done in polynomial time (see also Lemma 2). For showing completeness, let us provide a reduction from UNSAT. Assume we have given a Boolean expression $\phi \in \mathbb{B}(X)$ and consider an MSC M that contains two events labeled by local actions $p(d)$ and $q(d)$ with $p \neq q$, respectively. The encoding FMSG $\mathsf{G}_{pq} = (\{\iota\}, \{M\}, \Rightarrow, \iota, \{\iota\})$ where $\iota \overset{\phi:M}{\Rightarrow} \iota$ is weakly locally synchronized w.r.t. tt iff ϕ is not satisfiable. The statement about strong local synchronization follows directly due to Lemma 2.   □

Weak local synchronization requires to evaluate the semantics of feature guards while strong local synchronization is a mere syntactic criterion. This is also reflected in the FTS semantics, which could be infinite even though weakly locally synchronized. The latter is witnessed by an MSG that has two independent loops guarded by complementary feature guards.

Theorem 4. *Let G be an FMSG and $\Psi \in \mathbb{B}(X)$ a feature model.*

(a) $\mathsf{T}(\mathsf{G})|_C$ is finite for all $C \models \Psi$ if G is weakly locally synchronized w.r.t. Ψ.
(b) $\mathsf{T}(\mathsf{G})$ is finite if G is strongly locally synchronized.

Proof. (a) We first lift Lemma 1 towards featured pomset suffixes by a syntactic interpretation of feature guards and actions in pomsets: let $\Omega = \{\phi \in \mathbb{B}(X) : \exists q, q' \in Q, M \in \mathbb{M}.q \overset{\phi:M}{\Rightarrow} q'\}$ the set of feature guards in G. Then for a suffix $s \in \mathsf{Suff}^\star(\odot L(\underline{\mathsf{G}}))$ we define $L(s)$ as the language of s interpreted as pomset over $\Omega \times Act$. With this transformation, the proof of Lemma 1 is verbatim and also holds for featured suffixes. Since for all $C \models \Psi$ we have that $\mathsf{G}|_C$ is locally synchronized, $\mathcal{T}(\mathsf{G}|_C)$ is finite by Theorem 1. By Theorem 3 we have $L(\mathsf{T}(\mathsf{G})|_C) = L(\mathcal{T}(\mathsf{G}|_C))$ and thus, there is a finite automata $\mathcal{A}_C$ accepting $L(\mathsf{T}(\mathsf{G})|_C)$. Using $\mathcal{A}_C$ in the proof for Theorem 1 in combination with the featured variant of Lemma 1, we obtain that $\mathsf{T}(\mathsf{G})|_C$ is finite.
(b) Immediate consequence of Theorem 1.   □

5.2   ACTL Model Checking FMSGs

One key consequence of Theorem 4 is that for a given configuration $C \in \text{Conf}$, the aCTL model-checking problem is decidable using the standard approach on finite transition systems. However, in practice one is interested in solving the aCTL model-checking problem for all configurations $C \in \text{Conf}$. The naive one-by-one approach [49], where for each configuration $C \in \text{Conf}$ the transition system $T(G)|_C$ is constructed and checked, suffers from an exponential blowup in the number of features. This issue is commonly addressed by following an all-in-one approach, performing an analysis directly on the FTS in a single analysis run. One way to achieve an all-in-one model checking on FTSs is by adapting model-checking procedures by incorporating feature guards to the satisfiability sets over states [20]. However, this requires a reimplementation of the highly optimized existing model checkers and thus is usually avoided. In practice, the lifting approach [46] is employed, encoding feature configurations into the state space of a transition system and directly resolving the feature guards [22,23, 26,27]. The exponential blowup in the number of features is then mitigated by using BDD-based symbolic model checkers [20,26]. In the setting of FMSGs, this amounts to model check the finite lifted transition system $\uparrow T(G)$ (see Sect. 4).

6   Discussion and Future Work

In this paper we made contributions to the field of branching-time verification of MSGs and their configurability by introducing *featured MSGs* and related aCTL model-checking problems. Given the existing literature on FTSs [19,26] as well as branching-time semantics [24,25], extending their concepts to MSGs seemed to be straight-forward. However, delayed choice as a non-trivial branching operator required several technicalities to be obeyed. Our main result of this paper is that the FTS semantics of featured MSGs is finite if a strong variant of local synchronization holds, while a weaker variant ensures finiteness of the product FTSs. We further showed that checking for weak or strong local synchronization is both coNP-complete and thus has the same complexity for both variants. Since practical model-checking approaches for feature-oriented systems usually rely on lifting configurations into the state-space of a transition system [46] and exploit symbolic data structures [22,23], we conclude that weak local synchronization is the criterion to target for the verification of featured MSGs.

There are several extensions to be imagined. First, the lifted approach for encoding configurations in the state space could be improved by a purely symbolic approach similar to the synthesis of configurable monitors [28] and configuration monitors [39]. These approaches encode feature guards into the state space and ensure most general feature guards in states, allowing for state-merging and more concise FTSs and featured automata. In practice, monitoring communication systems and establishing featured MSGs through combining techniques from configuration monitors [39] and MSG learning [12] would be relevant for specification and documentation of communication scenarios in configurable systems. Including information on communication roles such as "key server" and "email server" both providing communication scenarios following a server role, would allow for a further separation of concerns and improve the featured MSG formalism even further [14]. A natural and straight-forward extension of featured MSGs is by annotating quantitative information, following the approach of quantitative MSGs [24]. This would enable for a quantitative analysis of configurable communication scenarios using feature-oriented probabilistic model checking [17,26]. Role-based variants of quantitative MSGs would be a natural next step for future work using techniques from quantitative role-based systems [15].

References

1. Alur, R., Yannakakis, M.: Model checking of message sequence charts. In: Baeten, J.C.M., Mauw, S. (eds.) CONCUR 1999. LNCS, vol. 1664, pp. 114–129. Springer, Heidelberg (1999). https://doi.org/10.1007/3-540-48320-9_10
2. Apel, S., Batory, D.S., Kästner, C., Saake, G.: Feature-Oriented Software Product Lines - Concepts and Implementation. Springer, Cham (2013)
3. Apel, S., Kästner, C.: An overview of feature-oriented software development. J. Object Technol. **8**, 49–84 (2009)
4. Baeten, J.C.M., Mauw, S.: Delayed choice: an operator for joining message sequence charts. In: Formal Description Techniques VII. IAICT, pp. 340–354. Springer, Boston, MA (1995). https://doi.org/10.1007/978-0-387-34878-0_27
5. Baier, C., Haverkort, B., Hermanns, H., Katoen, J.-P.: Model-checking algorithms for continuous-time Markov chains. IEEE TSE **29**(6), 524–541 (2003)
6. Baier, C., Cloth, L., Haverkort, B.R., Kuntz, M., Siegle, M.: Model checking Markov chains with actions and state labels. IEEE TSE **33**(4), 209–224 (2007)
7. Baier, C., Dubslaff, C., Klüppelholz, S., Leuschner, L.: Energy-utility analysis for resilient systems using probabilistic model checking. In: Ciardo, G., Kindler, E. (eds.) PETRI NETS 2014. LNCS, vol. 8489, pp. 20–39. Springer, Cham (2014). https://doi.org/10.1007/978-3-319-07734-5_2
8. Baier, C., Dubslaff, C., Korenčiak, L., Kučera, A., Řehák, V.: Synthesis of optimal resilient control strategies. In: D'Souza, D., Narayan Kumar, K. (eds.) ATVA 2017. LNCS, vol. 10482, pp. 417–434. Springer, Cham (2017). https://doi.org/10.1007/978-3-319-68167-2_27
9. Baier, C., Dubslaff, C., Korenčiak, L. U., Kučera, A., Řehák, V.: Mean-payoff optimization in continuous-time Markov chains with parametric alarms. ACM Trans. Model. Comput. Simul. **29**(4) (2019)
10. Baier, C., Katoen, J.-P.: Principles of Model Checking. The MIT Press, Cambridge (2008)
11. Batory, D.: Feature models, grammars, and propositional formulas. In: Obbink, H., Pohl, K. (eds.) SPLC 2005. LNCS, vol. 3714, pp. 7–20. Springer, Heidelberg (2005). https://doi.org/10.1007/11554844_3
12. Bollig, B., Katoen, J.-P., Kern, C., Leucker, M.: Learning communicating automata from MSCs. IEEE Trans. Softw. Eng. **36**(3), 390–408 (2010). https://doi.org/10.1109/TSE.2009.89
13. Bryant, R.E.: Graph-based algorithms for Boolean function manipulation. IEEE Trans. Comput. **35**, 677–691 (1986)
14. Chrszon, P., Baier, C., Dubslaff, C., Klüppelholz, S.: From features to roles. In: Proceedings of the 24th Systems and Software Product Line Conference (SPLC), pp. 1–11. ACM (2020)

15. Chrszon, P., Baier, C., Dubslaff, C., Klüppelholz, S.: Interaction detection in configurable systems - a formal approach featuring roles. J. Syst. Softw. **196**, 111556 (2023)
16. Chrszon, P., Dubslaff, C., Klüppelholz, S., Baier, C.: Family-based modeling and analysis for probabilistic systems - featuring PROFEAT. In: Stevens, P., Wasowski, A. (eds.) FASE 2016. LNCS, vol. 9633, pp. 287–304. Springer, Heidelberg (2016). https://doi.org/10.1007/978-3-662-49665-7_17
17. Chrszon, P., Dubslaff, C., Klüppelholz, S., Baier, C.: PROFEAT: feature-oriented engineering for family-based probabilistic model checking. Formal Aspects Comput. **30**, 45–75 (2018)
18. Clarke, E.M., Allen Emerson, E., Sistla, A.P.: Automatic verification of finite-state concurrent systems using temporal logic specifications. Trans. Program. Lang. Syst. **8**, 244–263 (1986)
19. Classen, A., Cordy, M., Schobbens, P.-Y., Heymans, P., Legay, A., Raskin, J.-F.: Featured transition systems: foundations for verifying variability-intensive systems and their application to LTL model checking. Trans. Softw. Eng. **39**, 1069–1089 (2013)
20. Classen, A., Heymans, P., Schobbens, P-Y., Legay, A.: Symbolic model checking of software product lines. In: Proceedings of the 33rd Conference on Software Engineering (ICSE), pp. 321–330. ACM (2011)
21. De Nicola, R., Vaandrager, F.: Action versus state based logics for transition systems. In: Guessarian, I. (ed.) LITP 1990. LNCS, vol. 469, pp. 407–419. Springer, Heidelberg (1990). https://doi.org/10.1007/3-540-53479-2_17
22. Dubslaff, C.: Compositional feature-oriented systems. In: Ölveczky, P.C., Salaün, G. (eds.) SEFM 2019. LNCS, vol. 11724, pp. 162–180. Springer, Cham (2019). https://doi.org/10.1007/978-3-030-30446-1_9
23. Dubslaff, C.; Quantitative Analysis of Configurable and Reconfigurable Systems. PhD thesis, TU Dresden, Institute for Theoretical Computer Science (2021)
24. Dubslaff, C., Baier, C.: Quantitative analysis of communication scenarios. In: Sankaranarayanan, S., Vicario, E. (eds.) FORMATS 2015. LNCS, vol. 9268, pp. 76–92. Springer, Cham (2015). https://doi.org/10.1007/978-3-319-22975-1_6
25. Dubslaff, C., Baier, C.: Delayed-choice semantics for pomset families and message sequence graphs. In: Katoen, J.-P., Langerak, R., Rensink, A. (eds.) ModelEd, TestEd, TrustEd. LNCS, vol. 10500, pp. 64–84. Springer, Cham (2017). https://doi.org/10.1007/978-3-319-68270-9_4
26. Dubslaff, C., Baier, C., Klüppelholz, S.: Probabilistic model checking for feature-oriented systems. In: Chiba, S., Tanter, É., Ernst, E., Hirschfeld, R. (eds.) Transactions on Aspect-Oriented Software Development XII. LNCS, vol. 8989, pp. 180–220. Springer, Heidelberg (2015). https://doi.org/10.1007/978-3-662-46734-3_5
27. Dubslaff, C., Klüppelholz, S., Baier, C.: Probabilistic model checking for energy analysis in software product lines. In: Proceedings of the 13th Conference on Modularity (MODULARITY), pp. 169–180. ACM (2014)
28. Dubslaff, C., Köhl, M.A.: Configurable-by-construction runtime monitoring. In: Margaria, T., Steffen, B. (eds.) ISoLA 2022. LNCS, vol. 13701, pp. 220–241. Springer, Cham (2022). https://doi.org/10.1007/978-3-031-19849-6_14
29. Genest, B., Kuske, D., Muscholl, A.: A Kleene theorem and model-checking algorithms for existentially bounded communicating automata. Inf. Comput. **204**(6)
30. Genest, B., Muscholl, A., Seidl, H., Zeitoun, M.: Infinite-state high-level MSCs: model-checking and realizability. In: Widmayer, P., Eidenbenz, S., Triguero, F., Morales, R., Conejo, R., Hennessy, M. (eds.) ICALP 2002. LNCS, vol. 2380, pp. 657–668. Springer, Heidelberg (2002). https://doi.org/10.1007/3-540-45465-9_56

31. Gunter, E.L., Muscholl, A., Peled, D.A.: Compositional message sequence charts. In: Margaria, T., Yi, W. (eds.) TACAS 2001. LNCS, vol. 2031, pp. 496–511. Springer, Heidelberg (2001). https://doi.org/10.1007/3-540-45319-9_34
32. Haverkort, B., Cloth, L., Hermanns, H., Katoen, J.-P., Baier, C.: Model checking performability properties. In: Proceedings International Conference on Dependable Systems and Networks, pp. 103–112 (2002)
33. ITU-T. Message sequence chart (msc). Recommendation Z.120, Edition 1.0 (1993)
34. ITU-T. Message sequence chart (msc). Recommendation Z.120, Edition 2.0 (1996)
35. ITU-T. Annex b: Formal semantics of message sequence charts. Z.120, v2.2 (1998)
36. ITU-T. Message Sequence Chart (MSC). Z.120, Edition 5.0 (2011)
37. Kang, K.C., Cohen, S.G., Hess, J.A., Novak, W.E., Spencer Peterson, A.: Feature-oriented domain analysis (foda) feasibility study. Technical report, Carnegie-Mellon University Software Engineering Institute (1990)
38. Katoen, J.P., Lambert, L.: Pomsets for message sequence charts. In: König, H., Langendörfer, P., (eds.) 8. GI/ITG-Fachgespraech, pp. 197–207. Shaker Verlag (1998)
39. Köhl, M.A., Dubslaff, C., Hermanns, H.: Configuration monitor synthesis. In: Akshay, Aina Niemetz, S., Sankaranarayanan, S., (eds.) Automated Technology for Verification and Analysis, pp. 3–27. Springer, Cham (2025)
40. Narayan Kumar, K.: The Theory of Message Sequence Charts, pp. 289–323. Co-Published with Indian Institute of Science (IISc), Bangalore (2012)
41. Muscholl, A., Peled, D.: Message sequence graphs and decision problems on Mazurkiewicz traces. In: Kutyłowski, M., Pacholski, L., Wierzbicki, T. (eds.) MFCS 1999. LNCS, vol. 1672, pp. 81–91. Springer, Heidelberg (1999). https://doi.org/10.1007/3-540-48340-3_8
42. Muscholl, A., Peled, D., Su, Z.: Deciding properties for message sequence charts. In: Nivat, M. (ed.) FoSSaCS 1998. LNCS, vol. 1378, pp. 226–242. Springer, Heidelberg (1998). https://doi.org/10.1007/BFb0053553
43. Muscholl, A., Peled, D., Su, Z.: Deciding properties for message sequence charts. In: Nivat, M. (ed.) FoSSaCS 1998. LNCS, vol. 1378, pp. 226–242. Springer, Heidelberg (1998). https://doi.org/10.1007/BFb0053553
44. Pecheur, C., Raimondi, F.: Symbolic model checking of logics with actions. In: Edelkamp, S., Lomuscio, A. (eds.) MoChArt 2006. LNCS (LNAI), vol. 4428, pp. 113–128. Springer, Heidelberg (2007). https://doi.org/10.1007/978-3-540-74128-2_8
45. Pnueli, A.: The temporal logic of programs. In: 18th Annual Symposium on the Foundations of Computer Science (FOCS-77), pp. 46–57. IEEE. Providence, Rhode Island (1977)
46. Post, H., Sinz, C.: Configuration lifting: verification meets software configuration. In: Proceedings of the 23rd Conference on Automated Software Engineering (ASE), pp. 347–350. IEEE (2008)
47. Pratt, V.: Modeling concurrency with partial orders. Int. J. Parallel Prog. **15**, 33–71 (1986)
48. Rensink, A., Wehrheim, H.: Weak sequential composition in process algebras. In: Jonsson, B., Parrow, J. (eds.) CONCUR 1994. LNCS, vol. 836, pp. 226–241. Springer, Heidelberg (1994). https://doi.org/10.1007/978-3-540-48654-1_20
49. Thüm, T., Apel, S., Kästner, C., Schaefer, I., Saake, G.: A classification and survey of analysis strategies for software product lines. Comput. Surv. **47**, 6:1–6:45 (2014)
50. Zhou, Z., Sheldon, F.T., Potok, T.E.: Modeling with stochastic message sequence charts. In: IIIS Proceedings of the International Conference on Computer, Communication, and Control Technology (CCCT 2003) (2003)

A Note on Runtime Verification
of Concurrent Systems

Martin Leucker[✉]

Institute for Software Engineering and Programming Languages, University of
Lübeck, Lübeck, Germany
leucker@isp.uni-luebeck.de

Abstract. To maximize the information gained from a single execution
when verifying a concurrent system, one can derive all concurrency-aware
equivalent executions and check them against linear specifications. This
paper offers an alternative perspective on verification of concurrent systems by leveraging trace-based logics rather than sequence-based formalisms. Linear Temporal Logic over Mazurkiewicz Traces (LTrL) operates on partial-order representations of executions, meaning that once
a single execution is specified, all equivalent interleavings are implicitly
considered. This paper introduces a three valued version of LTrL, indicating whether the so-far observed execution of the concurrent system is
one of correct, incorrect or inconclusive, together with a suitable monitor synthesis procedure. To this end, the paper recalls a construction of
trace-consistent Büchi automata for LTrL formulas and explains how to
employ it in well-understood monitor synthesis procedures. In this way,
a monitor results that yields for any linearization of an observed trace
the same verification verdict.

Keywords: Runtime Verification · Concurrency · Mazurkiewicz Traces

1 Introduction

Following [14], *runtime verification* is a lightweight verification technique that
checks whether a system execution complies with a formally specified correctness property. Given such a property, a *monitor* is synthesized, typically in the
form of an automaton, which observes the system's execution either offline or
online (in real-time). In online runtime verification, the monitor incrementally
processes observed events as the system runs and determines whether the execution satisfies or violates the correctness property. Unlike exhaustive verification
techniques such as model checking, runtime verification provides *on-the-fly analysis*, making it particularly useful for detecting issues in complex, concurrent, or
distributed systems that are difficult to analyse statically.

Concurrent systems exhibit behaviours that are influenced by their *environment*, including factors such as a *scheduler*, *system timing*, and *resource availability*. In other words, unlike sequential systems, concurrent systems consist of

© The Author(s), under exclusive license to Springer Nature Switzerland AG 2026
N. Bertrand et al. (Eds.): Christel Baier Festschrift, LNCS 15760, pp. 253–265, 2026.
https://doi.org/10.1007/978-3-031-97439-7_12

independent threads or processes that may share a processor, leading to executions where their actions occur in *different orders* depending on scheduling decisions. This *non-determinism* means that the same program can exhibit multiple behaviours across different executions. A fundamental property of concurrency is *independence*: if two actions a and b are independent, observing execution a followed by b implies that an alternative execution order, b followed by a, was also possible. This notion of *equivalence between interleavings* is essential in reasoning about *correctness* in concurrent systems.

In a series of papers [3,4,15], runtime verification of concurrent systems was explored along the following lines. A correctness property[1] φ is specified, which identifies when a single sequence is incorrect. Additionally, independence relations in concurrent systems were studied, allowing for the generalization from a single observation to potential alternative interleavings of the observed actions, as permitted by the given equivalence relation. The monitor reports success (or failure) if any of the considered interleavings are identified by the monitor.[2]

The general idea is sound; however, we propose to specify the correctness property while explicitly considering the concurrent nature of the system. Rather than defining correctness based on a particular execution order, the specification should be *interleaving-independent*, ensuring that it captures the intended concurrent behaviour regardless of how independent actions are scheduled. The monitor, in turn, must account for all possible interleavings permitted by the system's concurrency model. By doing so, the verification process remains robust, correctly identifying violations or confirmations of the correctness property across all valid execution orders. This paper provides an example of how to address the monitoring of concurrent systems using a trace logic that treats interleaving as a first-class citizen.

We base our explanation on LTrL. Thiagarajan and Walukiewicz introduced LTrL, which is interpreted over partial-order representations of traces [23,24]. More specifically, it is defined over *Mazurkiewicz traces* [16], a special class of partial orders that respect independence between actions, as made precise in subsequent sections. It has been shown that LTrL is expressively equivalent to the first-order theory of traces when interpreted over (finite and) infinite traces.

Each trace T can be linearized in multiple ways, and we define $\lin(T)$ as the set of all such linearizations. Clearly, $\lin(T)$ is *trace-closed*, meaning that it consists of all sequences obtained by taking one linearization of T and applying all possible permutations of independent actions. Furthermore, LTrL provides a characterization of so-called *trace-consistent* (or robust) LTL specifications. In other words, the models of any LTrL formula φ define trace-closed languages.

As such, our approach proceeds as follows:

---

[1] Correctness and incorrectness are dual notions here, so there is no significant difference in which one is explicitly defined.

[2] [3,4,15] use the term *predictive runtime verification* as further observations are *predicted* from one observation. However, the term *predictive runtime verification* was used with a different meaning in [28], where it referred to using the underlying program to predict how the currently observed behaviour might evolve.

- Use LTrL to specify correctness properties to be monitored.
- Synthesize a monitor that accepts all linearizations of traces satisfying the given correctness property.

While the general scheme from [12] is applicable to monitoring various linear-time temporal logics, particularly those that employ automata-based techniques to capture the models of a given formula, we explicitly provide the individual steps for clarity. Specifically, we build on existing methods to construct Büchi automata that accept all linearizations satisfying a given LTrL formula [7,8].

Furthermore, we analyse the complexity of this approach. We show the complexity of monitors in non-elementary in the nesting of until-formulas, but, at the same time, optimal. Moreover, we discuss potential practical optimizations.

The paper is organized as follows: We recall the concepts of words and automata in the next section. Mazurkiewicz traces are recalled in Sect. 3 while LTrL for Mazurkiewicz traces is given in Sect. 4. The main contribution of the paper is provided in Sect. 5 which introduces a three-valued version of the LTrL together with a suitable monitor synthesis procedure. Section 6 provides a short discussion on Mazurkiewicz trace logics in verification. In Sect. 7 we draw the conclusion of our approach and give directions for future work.

2 Words and Automata

For the remainder of this paper, let us fix an alphabet, i.e. a non-empty finite set, Σ. We write a, a_i for any single element of Σ and sometimes call it an action. Finite words over Σ are elements of Σ^*, and are usually denoted by $u, u', v, v', u_1, u_2, \ldots$, whereas infinite words are elements of Σ^ω, usually denoted by $w, w', w_1, w_2, \ldots$. We let Σ^∞ denote the union of finite and infinite words. The empty word is denoted by ϵ. Finally, we take $\mathrm{prf}(w)$ to be the set of finite prefixes of w and let $alph(w)$ denote the set of actions occurring in w.

A (nondeterministic) Büchi automaton (NBA) is a tuple $\mathcal{A} = (\Sigma, Q, Q_0, \delta, F)$, where Σ is a finite alphabet, Q is a finite non-empty set of states, $Q_0 \subseteq Q$ is a set of initial states, $\delta : Q \times \Sigma \to 2^Q$ is the transition function, and $F \subseteq Q$ is a set of accepting states. We extend the transition function $\delta : Q \times \Sigma \to 2^Q$ to sets of states and (input) words as usual. A *run* of an automaton $\mathcal{A}$ on a word $w = a_1 \ldots \in \Sigma^\omega$ is a sequence of states and actions $\rho = q_0 a_1 q_1 \ldots$, where q_0 is an initial state of $\mathcal{A}$ and for all $i \in \mathbb{N}$ we have $q_{i+1} \in \delta(q_i, a_i)$. For a run ρ, let $\mathrm{Inf}(\rho)$ denote the states visited infinitely often. ρ is called *accepting* iff $\mathrm{Inf}(\rho) \cap F \neq \emptyset$. A nondeterministic *finite automaton* (NFA) $\mathcal{A} = (\Sigma, Q, Q_0, \delta, F)$, where Σ, Q, Q_0, δ, and F are defined as for a Büchi automaton, operates on finite words. A *run* of $\mathcal{A}$ on a word $u = a_1 \ldots a_n \in \Sigma^*$ is a sequence of states and actions $\rho = q_0 a_1 q_1 \ldots q_n$, where q_0 is an initial state of $\mathcal{A}$ and for all $i \in \mathbb{N}$ we have $q_{i+1} \in \delta(q_i, a_i)$. The run is called accepting if $q_n \in F$. A NFA is called *deterministic* and denoted DFA, iff for all $q \in Q$, $a \in \Sigma$, $|\delta(q, a)| = 1$, and $|Q_0| = 1$.

As usual, the language accepted by an automaton (NBA/NFA/DFA), denoted by $\mathcal{L}(\mathcal{A})$, is given by its set of accepted words.

A *Moore machine* (also *finite-state machine*, FSM) is a finite state automaton enriched with output, formally denoted by a tuple $(\Sigma, Q, Q_0, \delta, \Delta, \lambda)$, where Σ, Q, $Q_0 \subseteq Q$, δ is as before and Δ is the output alphabet, $\lambda : Q \to \Delta$ the output function. As before, δ extends to the domain of words as expected. Moreover, we denote by λ also the function that applied to a word u yields the output in the state reached by u rather than the sequence of outputs.

3   Mazurkiewic Traces

Following [8], a *(Mazurkiewicz) trace alphabet* is a pair (Σ, I), where Σ is an alphabet and $I \subseteq \Sigma \times \Sigma$ is an irreflexive and symmetric *independence relation*. Usually, Σ consists of the *actions* performed by a distributed system while I captures a static notion of causal independence between actions. We define $D = (\Sigma \times \Sigma) - I$ to be the *dependency relation*, which is then reflexive and symmetric.

For the rest of the section, we fix a trace alphabet (Σ, I). We will use aIb to denote that the actions a and b are independent, i.e. that $(a, b) \in I$, and use similar notation for $(a, b) \in D$. We extend the notion to sets of actions $X, Y \subseteq \Sigma$, and let XIY denote that each pair of actions $a \in X$ and $b \in Y$ is independent. Moreover, XDY will denote that X is dependent on Y, i.e. that there exists a pair of actions $a \in X$ and $b \in Y$ with a and b dependent. For convenience, we will write $\{a\}IY$ as aIY etc.

For the purpose of interpreting a linear temporal logice over traces, we will adopt the viewpoint that traces are restricted labelled partial orders of events and hence have an explicit representation of causality and concurrency.

Let $T = (E, \leq, \lambda)$ be a Σ-labelled poset, i.e. $(E, \leq)$ is a poset and $\lambda : E \to \Sigma$ is a labelling function. λ can be extended to subsets of E in the straightforward manner. For $e \in E$, we define $\downarrow e = \{x \in E \mid x \leq e\}$ and $\uparrow e = \{x \in E \mid e \leq x\}$. We let $\lessdot$ be the *covering relation* given by $x \lessdot y$ iff $x < y$ and for all $z \in E$, $x \leq z \leq y$ implies $x = z$ or $z = y$.

A *(Mazurkiewicz) trace* over (Σ, I) is a Σ-labelled poset $T = (E, \leq, \lambda)$ satisfying:

- $\downarrow e$ is a finite set for each $e \in E$.
- For every $e, e' \in E$, $e \lessdot e'$ implies $\lambda(e) D \lambda(e')$.
- For every $e, e' \in E$, $\lambda(e) D \lambda(e')$ implies $e \leq e'$ or $e' \leq e$.

We let (Σ, I) denote the class of traces over (Σ, I). A trace language L is a subset of traces, i.e. $L \subseteq (\Sigma, I)$. Throughout the paper, we will not distinguish between isomorphic elements in (Σ, I). We will refer to members of E as *events*.

Let $T = (E, \leq, \lambda)$ be a trace over (Σ, I). A *configuration* of T is a finite subset of events $c \subseteq E$ with $\downarrow c = c$ where $\downarrow c = \bigcup_{e \in c} \downarrow e$. The set of configurations of T will be denoted $\mathcal{C}_T$. Trivially, $\emptyset \in \mathcal{C}_T$ for any trace T. $\mathcal{C}_T$ can be equipped with a transition relation $\longrightarrow_T \subseteq \mathcal{C}_T \times \Sigma \times \mathcal{C}_T$ given by $c \xrightarrow{a}_T c'$ iff there exists an $e \in E$ such that $\lambda(e) = a$, $e \notin c$, and $c' = c \cup \{e\}$. Configurations of $\mathcal{C}_T$ are the trace-theoretic analogues of finite prefixes of words. As will become apparent in Sect. 4, the formulas of our logic are to be interpreted at configurations of traces.

In its original formulation [17], Mazurkiewicz introduced traces as certain equivalence classes of words, and this correspondence turns out to be essential for our developments here. Let $T = (E, \leq, \lambda) \in (\Sigma, I)$. Then $w \in \Sigma^\infty$ is a *linearisation* of T iff there exists a map $\rho : \mathrm{prf}(w) \to \mathcal{C}_T$ such that the following conditions are met:

- $\rho(\varepsilon) = \emptyset$.
- $\rho(v) \overset{a}{\longrightarrow}_T \rho(va)$ for each $va \in \mathrm{prf}(w)$.
- For every $e \in E$, there exists some $u \in \mathrm{prf}(w)$ such that $e \in \rho(u)$.

The function ρ will be called a *run map* of the linearisation w. Note that the run map of a linearisation is unique. In what follows, we shall take $\mathrm{lin}(T)$ to be the *set of linearisations* of the trace T.

A set $p \subseteq \Sigma$ is called a *D-clique* iff $p \times p \subseteq D$. The equivalence relation $\approx \subseteq \Sigma^\infty \times \Sigma^\infty$ induced by I is given by: $w \approx w'$ iff $w{\restriction}p = w'{\restriction}p$ for every D-clique p. Here and elsewhere, if $X \subseteq \Sigma$, $w{\restriction}X$ is the sequence obtained by erasing from w all occurrences of letters in $\Sigma - X$. We take $[w]_\approx$ to denote the $\approx$-equivalence class of $w \in \Sigma^\infty$.

It is not hard to show that elements of (Σ, I) and $\approx$-equivalence classes are two representations of the same object: A labelled partial-order $T \in (\Sigma, I)$ is represented by $\mathrm{lin}(T)$ and vice versa (see also[11]). We exploit this duality of representation and let T_w denote the (unique) trace corresponding to $[w]_\approx$. Moreover, for each $v \in \mathrm{prf}(w)$ we will use c_v to denote the configuration of $\mathcal{C}_{T_w}$ given by $\rho(v)$.

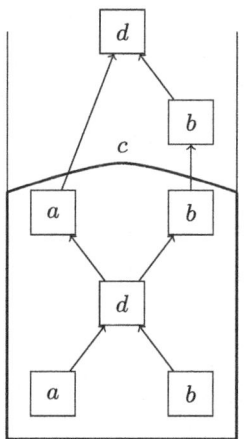

Fig. 1. A trace over (Σ, I).

To illustrate these concepts, consider the trace alphabet (Σ, I) with $\Sigma = \{a, b, d\}$ and $I = \{(a, b), (b, a)\}$. An example trace T over (Σ, I) is depicted in Fig. 1 with smaller elements (with respect to $\leq$) appearing below larger elements.

Furthermore, it can easily be verified that $abdbabd \in \text{lin}(T)$ so $T = T_{abdbabd}$, but $adabbbd \notin \text{lin}(T)$. The configuration $c \in \mathcal{C}_T$ consists of the first two a's, first d, first two b's, and is also denoted by c_{abdab}, which is identical to c_{badab} as $abdab \approx badab$.

We transfer considering traces as equivalence classes to the level of languages and call a word language $L \subseteq \Sigma^\omega$ *trace closed* iff for all words $w, w' \in \Sigma^\omega$ with $w \approx w'$, it holds $w \in L$ iff $w' \in L$.

Finally, we call a Büchi automaton, NFA, DFA etc. $\mathcal{A}$ *trace closed* if its language $\mathcal{L}(\mathcal{A})$ is trace closed.

4   LTL for Mazurkiewicz Traces

In this section, we bring out the syntax and semantics of the linear temporal logic LTL, which will be our basic object of study. It was originally introduced for words by Pnueli [21]. It was later equipped with a trace semantics [23] and some additional operators, and termed LTrL. Diekert and Gastin [10] have shown that LTL with the same trace semantics but without these additional operators is already expressively equivalent to first-order logic for traces. As such, we will use this version in the following. Note that we use LTrL to highlight that we consider Mazurkiewicz traces but LTL when considering words. However, formally one can combine the two approaches using a parameterized version of LTL:

The formulas of LTL are parameterised by a trace alphabet (Σ, I) and are defined inductively as follows:

$$\text{LTL}(\Sigma, I) := \text{tt} \mid \neg\varphi \mid \varphi \vee \psi \mid \langle a \rangle \varphi \mid \varphi \mathcal{U} \psi, a \in \Sigma.$$

Formulas of $\text{LTL}(\Sigma, I)$ are interpreted over configurations of traces over (Σ, I). More precisely, given a trace $T \in (\Sigma, I)$, a configuration $c \in \mathcal{C}_T$, and a formula $\varphi \in \text{LTL}(\Sigma, I)$, the notion of $T, c \models \varphi$ is defined inductively via:

-   $T, c \models \text{tt}$.
-   $T, c \models \neg\varphi$ iff $T, c \not\models \varphi$.
-   $T, c \models \varphi \vee \psi$ iff $T, c \models \varphi$ or $T, c \models \psi$.
-   $T, c \models \langle a \rangle \varphi$ iff there exists a $c' \in \mathcal{C}_T$ such that $c \xrightarrow{a}_T c'$ and $T, c' \models \varphi$.
-   $T, c \models \varphi \mathcal{U} \psi$ iff there exists a $c' \in \mathcal{C}_T$ with $c \subseteq c'$ such that $T, c' \models \psi$ and all $c'' \in \mathcal{C}_T$ with $c \subseteq c'' \subset c'$ satisfy φ.

We will freely use the standard abbreviations such as e.g. $\text{ff} = \neg\text{tt}$, $\varphi \wedge \psi = \neg(\neg\varphi \vee \neg\psi)$. Furthermore, we sometimes abbreviate $T, \emptyset \models \varphi$ by $T \models \varphi$. All models of a formula $\varphi \in \text{LTL}(\Sigma, I)$ constitute a subset of (Σ, I), thus a language. It is denoted by $\mathcal{L}(\varphi)$ and is called the language *defined* by φ. Furthermore, every formula defines an ω-language *viz* the set $\bigcup\{\text{lin}(T) \mid T \models \varphi\}$, which is also indicated by $\mathcal{L}(\varphi)$.

A simple example of a formula of LTL is $\varphi = \langle a \rangle \langle b \rangle \psi$. Note that for the trace from Fig. 1, it holds that $T \models \varphi$ if and only if $T, c_{ab} \models \psi$. Moreover, φ is equivalent to $\varphi' = \langle b \rangle \langle a \rangle \psi$ over this particular trace alphabet because aIb, i.e. the models of φ and φ' coincide.

We note that, in case of the empty independence relation, $\mathrm{LTL}(\Sigma, I)$ and LTL interpreted over words coincide in the expected manner. Thus, we identify LTL over words with $\mathrm{LTL}(\Sigma, \emptyset)$ and save the work of formally introducing LTL over words.[3]

Note that an LTL formula φ (considered over the alphabet $(\Sigma, \emptyset)$) is called *trace consistent wrt. the trace alphabet* (Σ, I), if $\mathcal{L}(\varphi)$ is trace closed wrt. (Σ, I). In other words, φ is called trace consistent iff it cannot distinguish between two equivalent linearizations of a trace. All $\mathrm{LTL}(\Sigma, I)$ formulas are trace consistent by definition. In simple words, every $\mathrm{LTL}(\Sigma, I)$ formula respects the given concurrency relation.

We conclude this section recalling the result that each $\mathrm{LTL}(\Sigma, I)$ formula φ can be translated into a corresponding (alternating) Büchi automaton accepting all linearizations of all traces satisfying φ. As such, the language is especially trace closed.[4]

Theorem 1 *([8]).* *Let φ be a formula of $\mathrm{LTL}(\Sigma, I)$. There is an (alternating) Büchi automaton $\mathcal{A}_\varphi$ such that $\mathcal{L}(\mathcal{A}_\varphi) = \mathcal{L}(\varphi)$. The size of $\mathcal{A}_\varphi$ is non-elementary in the size of φ.*

5   Runtime Verification for LTrL

Let us now introduce runtime verification for LTrL following the anticipatory approach originally introduced in [5,6]. To this end, let us recall the 3-valued approach for LTL over words first.

Monitoring LTL_3 over words. Let us recall our 3-valued semantics, denoted by LTL_3, over the set of truth values $\mathbb{B}_3 = \{\bot, ?, \top\}$ from [5,6]: Let $u \in \Sigma^*$ denote a finite word. The *truth value* of a LTL_3 formula φ wrt. u, denoted by $[u \models \varphi]$, is an element of $\mathbb{B}_3$ defined by

---

[3] Typically, LTL is introduced using atomic propositions rather than labelled actions. This difference should not bother us any further.

[4] While a non-deterministic automaton maps a given state and an input action to a set of possible next states, an alternating automaton maps to a Boolean combination of states [9]. It is easy to see that every alternating Büchi automaton can be translated into a non-deterministic Büchi automaton accepting the same language, involving an exponential blow-up [18]. While in the seminal work by Vardi and Wolper [25], an LTL formula is translated into a non-deterministic Büchi automaton directly, Vardi presented a translation via alternating automata [26], which, arguably, conceptually simplifies the translation into two steps: Translating LTL into an automaton that supports conjunctions of states, and eliminating conjunction when translating alternation into non-determinism. [8] follows the latter route, but adapted to LTrL. Hence, strictly speaking [8] only states the translation result into alternating machines, which can subsequently be translated to non-determinis Büchi automata.

$$[u \models \varphi] = \begin{cases} \top & \text{if } \forall \sigma \in \Sigma^\omega : u\sigma \models \varphi \\ \bot & \text{if } \forall \sigma \in \Sigma^\omega : u\sigma \not\models \varphi \\ ? & \text{otherwise.} \end{cases}$$

Monitor synthesis was introduced for LTL$_3$ in [5,6], and the ideas were later generalized in a systematic way to classes of linear temporal logics satisfying several additional properties in [13]. Since LTrL meets all these criteria, we can obtain a monitor for LTrL using this approach. However, to simplify the presentation, we directly illustrate how the construction for LTrL aligns with that of LTL$_3$, rather than recalling the more general framework from [13].

As such, let us first recall the monitor synthesis approach along the lines of [5,6]: For a given formula $\varphi \in$ LTL, we construct a finite Moore machine (FSM), $\mathcal{B}_{\hat{\mathcal{P}}}^\varphi$ that reads finite words $u \in \Sigma^*$ and outputs $[u \models \varphi] \in \mathbb{B}_3$.

For an NBA $\mathcal{A}$, we denote by $\mathcal{A}(q)$ the NBA that coincides with $\mathcal{A}$ except for Q_0, which is defined as $Q_0 = \{q\}$. Fix $\varphi \in$ LTL for the rest of this paragraph and let $\mathcal{A}^\varphi = (\Sigma, Q^\varphi, Q_0^\varphi, \delta^\varphi, F^\varphi)$ denote the NBA, which accepts all models of φ, and let $\mathcal{A}^{\neg\varphi} = (\Sigma, Q^{\neg\varphi}, Q_0^{\neg\varphi}, \delta^{\neg\varphi}, F^{\neg\varphi})$ denote the NBA, which accepts all counter examples of φ. The corresponding construction is standard [25].

For $u \in \Sigma^*$ and $\delta(Q_0^\varphi, u) = \{q_1, \ldots, q_l\}$, we have $[u \models \varphi] \neq \bot$ iff $\exists q \in \{q_1, \ldots, q_l\}$ such that $\mathcal{L}(\mathcal{A}^\varphi(q)) \neq \emptyset$. Likewise, for the NBA $\mathcal{A}^{\neg\varphi}$, we have for $u \in \Sigma^*$, and $\delta(Q_0^{\neg\varphi}, u) = \{q_1, \ldots, q_l\}$ that $[u \models \varphi] \neq \top$ iff $\exists q \in \{q_1, \ldots, q_l\}$ such that $\mathcal{L}(\mathcal{A}^{\neg\varphi}(q)) \neq \emptyset$.

Following [6], for $\mathcal{A}^\varphi$ and $\mathcal{A}^{\neg\varphi}$, we now define a function $\mathcal{F}^\varphi : Q^\varphi \to \mathbb{B}$ respectively $\mathcal{F}^{\neg\varphi} : Q^{\neg\varphi} \to \mathbb{B}$ (where $\mathbb{B} = \{\top, \bot\}$), assigning to each state q whether the language of the respective automaton starting in state q is not empty. Thus, if $\mathcal{F}^\varphi(q) = \top$ holds, then the automaton $\mathcal{A}^\varphi$ starting at state q accepts a non-empty language and each finite prefix u leading to state q can be extended to an (infinite) run satisfying φ.

Using $\mathcal{F}^\varphi$ and $\mathcal{F}^{\neg\varphi}$, we turn from automata over infinite words to automata over finite words: We define two NFAs $\hat{\mathcal{A}}^\varphi = (\Sigma, Q^\varphi, Q_0^\varphi, \delta^\varphi, \hat{F}^\varphi)$ and $\hat{\mathcal{A}}^{\neg\varphi} = (\Sigma, Q^{\neg\varphi}, Q_0^{\neg\varphi}, \delta^{\neg\varphi}, \hat{F}^{\neg\varphi})$ where $\hat{F}^\varphi = \{q \in Q^\varphi \mid \mathcal{F}^\varphi(q) = \top\}$ and $\hat{F}^{\neg\varphi} = \{q \in Q^{\neg\varphi} \mid \mathcal{F}^{\neg\varphi}(q) = \top\}$. Then, we have for all $u \in \Sigma^*$:

$$u \in \mathcal{L}(\hat{\mathcal{A}}^\varphi) \text{ iff } [u \models \varphi] \neq \bot \quad \text{and} \quad u \in \mathcal{L}(\hat{\mathcal{A}}^{\neg\varphi}) \text{ iff } [u \models \varphi] \neq \top$$

Hence, we can evaluate $[u \models_{\hat{\mathcal{P}}} \varphi]$ as follows: We have $[u \models \varphi] = \top$ if $u \notin \mathcal{L}(\hat{\mathcal{A}}^{\neg\varphi})$, $[u \models \varphi] = \bot$ if $u \notin \mathcal{L}(\hat{\mathcal{A}}^\varphi)$, and $[u \models \varphi] =?$ if $u \in \mathcal{L}(\hat{\mathcal{A}}^\varphi)$ and $u \in \mathcal{L}(\hat{\mathcal{A}}^{\neg\varphi})$.

As a final step, we now define a (deterministic) FSM $\mathcal{B}^\varphi$ that outputs for each finite string u and formula φ its three valued semantics. Let $\tilde{\mathcal{A}}^\varphi$ and $\tilde{\mathcal{A}}^{\neg\varphi}$ be the deterministic versions of $\hat{\mathcal{A}}^\varphi$ and $\hat{\mathcal{A}}^{\neg\varphi}$, which can be computed in the standard manner by power-set construction. Now, we define the FSM in question as a product of $\tilde{\mathcal{A}}^\varphi$ and $\tilde{\mathcal{A}}^{\neg\varphi}$:

Definition 1 (Monitor $\mathcal{B}^\varphi$ for LTL-formula φ). Let $\tilde{\mathcal{A}}^\varphi = (\Sigma, Q^\varphi, \{q_0^\varphi\}, \delta^\varphi, \tilde{F}^\varphi)$ and $\tilde{\mathcal{A}}^{\neg\varphi} = (\Sigma, Q^{\neg\varphi}, \{q_0^{\neg\varphi}\}, \delta^{\neg\varphi}, \tilde{F}^{\neg\varphi})$ be the DFAs which correspond to the two NFAs $\hat{\mathcal{A}}^\varphi$ and $\hat{\mathcal{A}}^{\neg\varphi}$ as defined before. Then we define the *monitor*

$\mathcal{B}^\varphi = \tilde{\mathcal{A}}^\varphi \times \tilde{\mathcal{A}}^{\neg\varphi}$ for φ as the minimized version of the FSM $(\Sigma, \bar{Q}, \bar{q}_0, \bar{\delta}, \bar{\lambda})$, where $\bar{Q} = Q^\varphi \times Q^{\neg\varphi}$, $\bar{q}_0 = (q_0^\varphi, q_0^{\neg\varphi})$, $\bar{\delta}((q, q'), a) = (\delta^\varphi(q, a), \delta^{\neg\varphi}(q', a))$, and $\bar{\lambda} : \bar{Q} \to \mathbb{B}_3$ is defined by

$$\bar{\lambda}((q, q')) = \begin{cases} \top & \text{if } q' \notin \tilde{F}^{\neg\varphi} \\ \bot & \text{if } q \notin \tilde{F}^\varphi \\ ? & \text{if } q \in \tilde{F}^\varphi \text{ and } q' \in \tilde{F}^{\neg\varphi}. \end{cases}$$

We sum up our entire construction in Fig. 2 and conclude with the following correctness theorem.

Theorem 2 ([5, 6]). *Let $\varphi \in \mathrm{LTL}$, and let $\mathcal{B}^\varphi = (\Sigma, \bar{Q}, \bar{q}_0, \bar{\delta}, \bar{\lambda})$ be the corresponding monitor. Then, for all $u \in \Sigma^*$: $[u \models \varphi] = \bar{\lambda}(\bar{\delta}(\bar{q}_0, u))$.*

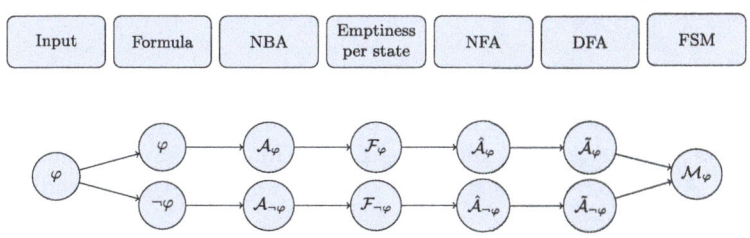

Fig. 2. The procedure for getting $[u \models \varphi]$ for a given φ.

Monitoring LTrL₃ Traces over Mazurkiewicz Traces.

Let us now define our 3-valued semantics, denoted by LTrL₃, with the set of truth values $\mathbb{B}_3 = \{\bot, ?, \top\}$ similar as above:

Definition 2 (3-valued semantics of LTrL). Let $u \in \Sigma^*$ denote a finite word. The *truth value* of a LTrL₃ formula φ wrt. u, denoted by $[u \models^t \varphi]$, is an element of $\mathbb{B}_3$ defined by

$$[u \models^t \varphi] = \begin{cases} \top & \text{if } \forall \sigma \in \Sigma^\omega : T_{u\sigma} \models \varphi \\ \bot & \text{if } \forall \sigma \in \Sigma^\omega : T_{u\sigma} \not\models \varphi \\ ? & \text{otherwise.} \end{cases}$$

Lemma 1. *3-valued LTrL coincides with 3-valued LTL when all actions are independent.*

Assume we have observed a sequence of actions u. This yields the configuration c_u of any trace T with $c_u \subset T$. As $c_u = c_v$ for any $u \approx v$, we get that, for any $w \in \Sigma^\omega$, that $T_{uw} = T_{vw}$ and thus $T_{uw} \models \varphi$ iff $T_{vw} \models \varphi$. In other words, we get that LTrL₃ gives the same verdict of all linearizations of c_u and thus for all words equivalent to u.

Theorem 3. *Let* $\varphi \in LTrL_3$ *and* $u \in \Sigma^*$,

$$[u \models^t \varphi] = [v \models^t \varphi]$$

for all $v \in [u]_\approx$.

In other words, any monitor following the semantics of $LTrL_3$ yields the same value for v if seen u, when $v \approx u$.

Now, we turn our focus to the monitoring procedure. Our main goal is to understand that we can follow the approach for LTL_3 but starting with the Büchi automaton for an $LTrL_3$ formula according to Theorem 1. To this extent, let us consider the automaton for any $LTrL$ formula in more detail.

Reading a finite linearization u yields sets of states in the automaton. The languages accepted from that states characterize together with the prefix u the traces satisfying φ with having c_u as configuration.

We call this language L_u.

Lemma 2. $L_u = L_v$, *if* u *and* v *are equivalent.*

Proof. Assume the contrary: Then there is w such that, wlog, w is in L_u but not in L_v. As such $uw \in \mathcal{L}(\mathcal{A})$ but $vw \notin \mathcal{L}(\mathcal{A})$. But, as $u \approx v$, we have $T_{uw} \models \varphi$ iff $T_{vw} \models \varphi$. As $\mathcal{A}$ is trace closed, $uw \in \mathcal{L}(\mathcal{A})$ iff $vw \in \mathcal{L}(\mathcal{A})$. Contradiction.

Note that the idea of solely relying on the states reached when reading a prefix u, was called *forgettable past* in [13].

Corollary 1. $L_u = \emptyset$ *iff for all* w *we have* $T_{uw} \not\models \varphi$.

In other words, we can follow the schema for LTL_3 to build a monitor for $LTrL$. We follow the construction shown in Fig. 2 and conclude, using the notation before, with the following correctness theorem.

Theorem 4. *Let* $\varphi \in LTrL_3$, *and let* $\mathcal{B}^\varphi = (\Sigma, \bar{Q}, \bar{q}_0, \bar{\delta}, \bar{\lambda})$ *be the corresponding monitor. Then, for all* $u \in \Sigma^*$: $[u \models^t \varphi] = \bar{\lambda}(\bar{\delta}(\bar{q}_0, u))$.

Since the Büchi automaton for an LTL formula over traces is necessarily non-elementary [27], the resulting monitor is also non-elementary. When the independence alphabet is empty, the construction from [8] aligns with traditional LTL methods, producing a Büchi automaton of single exponential size and, consequently, a monitor of double exponential size. Regardless of the case, since the final FSM is minimized, all monitors obtained through this approach are optimal in terms of size.

6   Discussion

Trace-consistent properties have been successfully applied in model checking to enable *partial order reduction*. The core idea of partial order reduction in model checking is to consider only a single execution (or a small representative subset) for all executions that are equivalent up to permutation of independent actions. Simply put, when actions a and b are independent, one considers only executions where a happens before b, rather than also including the reverse order. Fixing such an order not only reduces the number of executions that need to be explored in the underlying system but also allows for pruning the automaton that accepts models of the underlying correctness specification.

In runtime verification, where the underlying system is not under control, the monitor must be prepared to handle any possible ordering of independent actions. In other words, model checking is a *white-box* technique that allows for more optimizations, whereas runtime verification operates as a *black-box* technique, limiting the available reductions. Still, it might be worthwhile to consider potential optimizations following ideas of reordering actions.

An alternative approach to the one presented here[5] is to use an algorithm that checks whether a specification is Mazurkiewcz-trace closed. Then, one can simply use LTL and traditional automata constructions. Note that, as the final monitor is a minimized, unique Moore machine, the originating LTL formula may be non-elementary longer than an equivalent LTrL formula. In practice, readability may be of great importance and it is unclear whether compact formulas may be easier or more difficult to understand. A notable reference for checking whether an LTL specification is trace closed and also considers more general equivalence relations is [19,20].

The method proposed here may work similarly for different notions of concurrency and corresponding notions of equivalences, suitable logics, and automata-based decision procedures, e.g [1,2,22].

7 Conclusion

In this paper, we have explored the use of (Mazurkiewicz) trace logics for monitoring concurrent systems. By incorporating concurrency as a first-class citizen within temporal logic, we have demonstrated that it is possible to obtain the same verdict based on any linearization that could have occurred instead of the observed one, thanks to the independence of actions.

Several directions remain for future research. A key next step is the practical implementation and evaluation of our approach in real-world systems to assess its efficiency and applicability. Additionally, a more detailed comparison with existing methods that rely on generalizations of executions rather than built-in concurrency will help to clarify the advantages of our approach. Beyond this, we aim to explore past-time logics for Mazurkiewicz traces, as well as logics that extend beyond Mazurkiewicz traces to model concurrency in a broader range of systems.

---

[5] This remark was communicated by one of the anonymous reviewers of this paper.

References

1. Alur, R., McMillan, K., Peled, D.: Deciding global partial-order properties. In: Larsen, K., Skyum, S., Winskel, G. (eds.) Proceedings of 25th International Colloquium on Automata, Languages and Programming (ICALP'98). LNCS, vol. 1443, pp. 41–52 (1998). http://link.springer.de/link/service/series/0558/tocs/t1443.htm
2. Alur, R., Peled, D., Penczek, W.: Model checking of causality properties. In: Proceedings of the 10th Annual IEEE Symposium on Logic in Computer Science (LICS'95), pp. 90–100. IEEE Computer Society Press, San Diego, California (1995)
3. Ang, Z., Mathur, U.: Predictive monitoring against pattern regular languages. Proc. ACM Program. Lang. **8**(POPL), pp. 2191–2225 (2024). https://doi.org/10.1145/3632915
4. Ang, Z., Mathur, U.: Predictive monitoring with strong trace prefixes. In: Gurfinkel, A., Ganesh, V. (eds.) Computer Aided Verification - 36th International Conference, CAV 2024, Montreal, QC, Canada, July 24-27, 2024, Proceedings, Part II. LNCS, vol. 14682, pp. 182–204. Springer (2024). https://doi.org/10.1007/978-3-031-65630-9_9
5. Bauer, A., Leucker, M., Schallhart, C.: Monitoring of Real-Time Properties. In: Arun-Kumar, S., Garg, N. (eds.) FSTTCS 2006. LNCS, vol. 4337, pp. 260–272. Springer, Heidelberg (2006). https://doi.org/10.1007/11944836_25
6. Bauer, A., Leucker, M., Schallhart, C.: Runtime verification for LTL and TLTL. ACM Trans. Softw. Eng. Methodol. **20**(4), pp. 14:1–14:64 (2011). https://doi.org/10.1145/2000799.2000800
7. Bollig, B., Leucker, M.: Deciding LTL over Mazurkiewicz traces. In: Eigth International Symposium on Temporal Representation and Reasoning, TIME-01, Civdale del Friuli, Italy, June 14-16, 2001, pp. 189–197. IEEE Computer Society (2001). https://doi.org/10.1109/TIME.2001.930717
8. Bollig, B., Leucker, M.: Deciding LTL over Mazurkiewicz traces. Data Knowl. Eng. **44**(2), 219–238 (2003)
9. Chandra, A.K., Kozen, D.C., Stockmeyer, L.J.: Alternation. J. ACM **28**(1), 114–133 (1981)
10. Diekert, V., Gastin, P.: LTL Is Expressively Complete for Mazurkiewicz Traces. In: Montanari, U., Rolim, J.D.P., Welzl, E. (eds.) ICALP 2000. LNCS, vol. 1853, pp. 211–223. Springer, Heidelberg (2000). https://doi.org/10.1007/3-540-45022-X_18
11. Diekert, V., Rozenberg, G. (eds.): The Book of Traces. World Scientific, Singapore (1995)
12. Dong, W., Leucker, M., Schallhart, C.: Impartial anticipation in runtime-verification. In: Cha, S.D., Choi, J., Kim, M., Lee, I., Viswanathan, M. (eds.) Automated Technology for Verification and Analysis, 6th International Symposium, ATVA 2008, Seoul, Korea, October 20-23, 2008. Proceedings. LNCS, vol. 5311, pp. 386–396. Springer (2008). https://doi.org/10.1007/978-3-540-88387-6_33
13. Dong, W., Leucker, M., Schallhart, C.: Impartial anticipation in runtime-verification. In: Cha, S.D., Choi, J., Kim, M., Lee, I., Viswanathan, M. (eds.) Automated Technology for Verification and Analysis, 6th International Symposium, ATVA 2008, Seoul, Korea, October 20-23, 2008. Proceedings. LNCS, vol. 5311, pp. 386–396. Springer (2008). https://doi.org/10.1007/978-3-540-88387-6_33
14. Leucker, M., Schallhart, C.: A brief account of runtime verification. J. Log. Algebraic Methods Program. **78**(5), 293–303 (2009). https://doi.org/10.1016/j.jlap.2008.08.004

15. Mathur, U., Pavlogiannis, A., Viswanathan, M.: Optimal prediction of synchronization-preserving races. Proc. ACM Program. Lang. **5**(POPL), pp. 1–29 (2021). https://doi.org/10.1145/3434317

16. Mazurkiewicz, A.: Basic notions of trace theory. In: de Bakker, J.W., de Roever, W.P., Rozenberg, G. (eds.) Proceedings of the School/Workshop on Linear Time, Branching Time and Partial Order in Logics and Models for Concurrency. LNCS, vol. 354, pp. 364–397. Springer (1988)

17. Mazurkiewicz, A.: Concurrent program schemes and their interpretations. DAIMI Rep. PB 78, Aarhus University, Aarhus (1977)

18. Miyano, S., Hayashi, T.: Alternating finite automata on ω-words. Theoret. Comput. Sci. **32**, 321–330 (1984)

19. Peled, D., Wilke, T., Wolper, P.: An algorithmic approach for checking closure properties of ω-regular languages. In: Proceedings of the 7th International Conference on Concurrency Theory (CONCUR'96). LNCS, vol. 1119, pp. 596–610. Springer, Pisa, Italy (1996)

20. Peled, D., Wilke, T., Wolper, P.: An algorithmic approach for checking closure properties of ω-regular languages. Theor. Comput. Sci. **195**(2), 183–203 (1998). a preliminary version appeared in [19]

21. Pnueli, A.: The temporal logic of programs. In: Proceedings of the 18th IEEE Symposium on the Foundations of Computer Science (FOCS-77), pp. 46–57. IEEE Computer Society Press, Providence, Rhode Island (Oct 31–Nov 2 1977)

22. Thiagarajan, P.S.: A trace based extension of linear time temporal logic. In: Proceedings, Ninth Annual IEEE Symposium on Logic in Computer Science, pp. 438–447. IEEE Computer Society Press, Paris, France (4–7 Jul 1994)

23. Thiagarajan, P.S., Walukiewicz, I.: An expressively complete linear time temporal logic for Mazurkiewicz traces. In: Proceedings, Twelth Annual IEEE Symposium on Logic in Computer Science, pp. 183–194. IEEE Computer Society Press, Warsaw, Poland (29 Jun–2 Jul 1997)

24. Thiagarajan, P.S., Walukiewicz, I.: An expressively complete linear time temporal logic for Mazurkiewicz traces. Inf. Comput. **179**(2), 230–249 (2002). https://doi.org/10.1006/INCO.2001.2956

25. Vardi, M.Y., Wolper, P.: An automata-theoretic approach to automatic program verification. In: Symposium on Logic in Computer Science (LICS'86), pp. 332–345. IEEE Computer Society Press, Washington, D.C., USA (1986)

26. Vardi, M.Y.: An Automata-Theoretic Approach to Linear Temporal Logic, LNCS, vol. 1043, pp. 238–266. Springer, New York, NY, USA (1996)

27. Walukiewicz, I.: Difficult configurations - on the complexity of LTrL. In: Larsen, K., Skyum, S., Winskel, G. (eds.) Proceedings of 25th International Colloquium on Automata, Languages and Programming (ICALP'98). LNCS, vol. 1443, pp. 140–151 (1998). http://www.brics.dk/~igw/icalp98.ps

28. Zhang, X., Leucker, M., Dong, W.: Runtime verification with predictive semantics. In: Goodloe, A., Person, S. (eds.) NASA Formal Methods - 4th International Symposium, NFM 2012, Norfolk, VA, USA, April 3-5, 2012. Proceedings. LNCS, vol. 7226, pp. 418–432. Springer (2012). https://doi.org/10.1007/978-3-642-28891-3_37

Embedding Monitoring of First-Order Temporal Logic in a Programming Language

Klaus Havelund[1(✉)], Moran Omer[2], and Doron Peled[2]

[1] Jet Propulsion Laboratory, California Institute of Technology, Pasadena, USA
Klaus.Havelund@jpl.nasa.gov
[2] Bar Ilan University, Ramat Gan, Israel

Abstract. Runtime verification (RV) consists of verifying that an execution trace satisfies a specification. The specification can be written in a formal logic, or it can be written as code in a general purpose programming language. In this paper we present an RV method, and a corresponding tool, that explores the boundary between formal logic and general purpose programming. The tool, called PyDejaVu, supports a two-phase approach to writing properties, where a property can be expressed in a combination of an operational phase and a declarative phase. The operational phase is expressed in an internal Python DSL (Domain-Specific Language), whereas the declarative phase is expressed in the external DSL Qtl (Quantified Temporal Logic) for first-order past time temporal logic. This approach benefits from the expressiveness of Python and the succinctness and efficiency of monitoring Qtl. Our tool builds on the previous runtime verification tool DejaVu monitoring Qtl properties.

1 Introduction

Runtime Verification (RV) allows monitoring the execution of a system, usually in the form of a trace of events, and checking it against a formal specification. The provided verdict can be used to avert a problematic behaviour, or just report the violation. Three key parameters in applying RV are expressiveness of the formalism that is used for writing specifications, the succinctness of the formalism, and the efficiency of the monitors. Classical formalisms that are used for RV and model checking techniques are temporal logics and various kind of automata.

RV techniques (as opposed to techniques used in model checking, which is a more comprehensive formal method, hence posses some further complications) have been expanded to allow temporal specification of traces of events that

The research performed by the first author was carried out at Jet Propulsion Laboratory, California Institute of Technology, under a contract with the National Aeronautics and Space Administration. The research performed by the second and third authors was partially funded by Israeli Science Foundation grant 2454/23: "Validating and controlling software and hardware systems assisted by machine learning".

N. Bertrand et al. (Eds.): Christel Baier Festschrift, LNCS 15760, pp. 266–283, 2026.
https://doi.org/10.1007/978-3-031-97439-7_13

carry data, using first-order predicate logic constructs, including universal and existential quantification over the data values. Online RV monitors a sequence of events that are emitted by the checked system as it executes. In order to allow the online RV to cope with the speed of the intercepted events, a carefully constructed compact representation is required to keep a summary of the trace observed so far, which is updated with each new intercepted event. This entails a tradeoff between expressiveness, succinctness, and efficiency, which is typical for formal methods.

One tradeoff in selecting the specification formalism for RV is the use of a "programming language" style formalism versus a "logic-based" formalism. The use of a programming language as formalism allows a high degree of flexibility in describing the desired property. It permits using programming tricks to implement the runtime verification checks. On the other hand, a formal logic specification is usually very succinct, and can benefit from a standard efficient implementation, rather than using an ad-hoc implementation.

In previous work [15], we showed a method and tool, called TP-DEJAVU, for splitting the RV specification into two parts for achieving optimized expressiveness, succinctness, and efficiency. A *declarative part* is used for expressing a purely (first-order) temporal specification, which is devoid of operations on data, including e.g. arithmetic operations. An *operational part* is used to process intercepted events from the system under observation, and modify them, based on some local summary, before passing them to the declarative part. The operational part can embed arithmetic operations, e.g., addition, comparison and calculating extrema values. In the tool TP-DEJAVU, the operational part was implemented based on a syntax that allows intercepting events and performing computations on the data, sending modified events to the RV tool DEJAVU [18]. The tool DEJAVU provides the declarative part based on a first-order past time temporal logic.

In this work, we describe a new tool, PYDEJAVU, that further enhances the capabilities of RV on traces of events with data. The contribution is to implement the operational part as an *internal DSL* (Domain-Specific Language) that is embedded in Python. Then, the declarative part is embedded in Python as an *external DSL* based on the DEJAVU system (which is implemented in Scala). The tool can also be seen as a Python interface to DEJAVU. This allows a much higher flexibility than the small DSL for the operational part in TP-DEJAVU, and immediately opens new capabilities that were not originally allowed in TP-DEJAVU, e.g., adding further arithmetic operations, and generally using builtin and programmed functions, as well as using more complex data structures, such as vectors, when processing the observed events. The contribution of the tool is as follows:

– The combination of a programmable internal DSL, working in tandem with a powerful and efficient external DSL provides a large step up in the expressiveness of the RV specification. This includes functions (builtin and user defined), the use of various data structures, and even the integration of other Python libraries such as Neural Networks (PyTorch [10]) and Databases (SQL [9]).

- We benefit from the succinctness and power of the external DSL part as an efficient monitor, optimized for runtime verification of events with data, based on first-order past time temporal logic.
- We demonstrate the new tool with examples and experiments. The tool is available for download from GitHub[1].

The paper is organized as follows. Section 2 provides an overview of various flavours of DSLs, and shows where the different DSLs mentioned in this paper fit in. Section 3 defines the QTL past-time temporal logic, which is the foundation for the tools presented in the paper. Section 4 gives an overview of the previous tools developed, namely DEJAVU and TP-DEJAVU. Section 5 introduces the tool PYDEJAVU, which is the key contribution of this paper. Section 6 presents experiments. Finally Sect. 7 concludes the paper.

Related Work

Some early tools supported data comparison and computations as part of the logic [2]. The version of the tool MONPOLY in [3] supports comparisons and aggregate operations such as sum and maximum/minimum within a first-order LTL formalism. It uses a database-oriented implementation. Other tools supporting automata based limited first-order capabilities include [4,25]. In [7], a framework that lifts the monitor synthesis for a propositional temporal logic to a temporal logic over a first-order theory, using an SMT solver, is described. A number of pure internal DSLs (libraries in a programming language) for RV have been developed [1,5,11,13,14], which offer the full power of the host programming language for writing monitors and therefore allow for arbitrary comparisons and computations on data to be performed. Of interest in this context is the internal Python DSL PYCONTRACT [5], which is inspired by the internal Scala DSL DAUT [13]. The concept of phasing monitoring such that one phase produces output to another phase has been explored in other frameworks, including early frameworks such as [23] with a fixed number of phases (two), but with propositional monitoring, and later frameworks with arbitrary user defined phases [2,12]. In particular, stream processing systems support this idea [6,21]. Another related work on increasing the expressive power of temporal logic is the extension of DEJAVU with rules described in [17].

2   Classification of Domain-Specific Languages

In this section, we provide an overview of the various types of Domain-Specific Languages (DSLs), highlighting their advantages and disadvantages. DSLs are specialized languages designed to simplify the expression of domain-specific tasks by providing constructs that closely align with particular domain concepts. This focus allows DSLs to enhance productivity and reduce errors compared to general-purpose programming languages.

---

[1] https://github.com/moraneus/pydejavu.

DSLs play a significant role in runtime verification by offering tailored constructs that make it easier to specify and monitor system behaviors. They can be categorized into three main types based on their interaction with a host language. A host language refers to a general-purpose programming language, such as Python, Java, or Scala, that provides the underlying environment in which a DSL operates. The three types of DSLs are: external DSLs, which are completely separate from the host language they are implemented in (i.e., no host-language code is written by a user); internal DSLs, which are embedded directly within the host language (the specification language is the host language); and hybrid DSLs, which combine elements of both, offering a balance between flexibility and integration. Figure 1 provides a visual overview of this classification.

Before we provide a more detailed explanation of these concepts, we can reveal that PYDEJAVU falls under the category of *hybrid closed DSLs*. It leverages the strengths of both external and internal approaches to provide a structured yet adaptable way to express domain-specific rules.

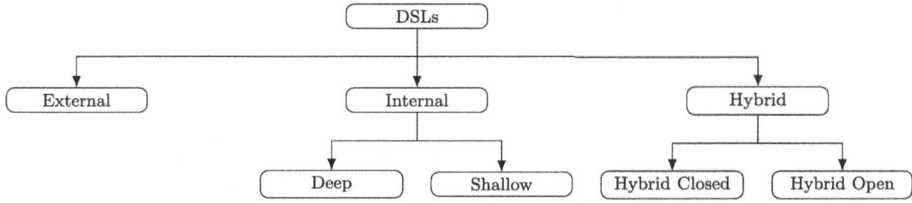

Fig. 1. DSL categorization.

2.1   External DSLs

External DSLs are stand alone languages with dedicated grammars, parsers, and interpreters or compilers, designed to address specific domain requirements independently of any host language. Systems like DEJAVU and TP-DEJAVU exemplify this approach by providing specialized syntax for runtime verification. For example, TP-DEJAVU implements its operational phase using a distinct grammar, as shown in Fig. 4, offering a domain-specific abstraction separate from general-purpose programming languages.

These languages provide users with a concise, domain-focused syntax that simplifies analysis, optimization, and transformations. They enhance reliability by making specifications more formal and interpretable. However, external DSLs come with increased implementation complexity due to the need for dedicated parsing and interpretation machinery. Additionally, their rigid structure makes it challenging to adapt to evolving requirements, a phenomenon known as "requirement creep", where new or unforeseen needs necessitate extending the language's capabilities. As discussed in [16], this can e.g. involve adding support for complex operations like arithmetic comparisons or timing constraints, features without which the language may become too limited for practical use.

2.2   Internal DSLs

Internal DSLs represent a powerful approach to domain-specific language implementation that leverages the existing infrastructure of a host programming language. Unlike external DSLs that require dedicated parsers and interpreters, internal DSLs are implemented as libraries within the host language, making them more accessible and maintainable. These DSLs come in two distinct varieties.

Deep internal DSLs represent specifications or programs as data structures within the host language. This design choice means that a formula by the user is specified as an object of a data type. An external DSL is usually parsed from text into an abstract syntax tree, which is then processed further. In a deep internal DSL, we skip the parsing of text, and the specification writer manually creates the abstract syntax tree, potentially using various auxiliary functions. The implementation in [14] demonstrates this approach, presenting a runtime verification DSL that is mostly deep in nature but extends the paradigm by allowing code as part of the specification. Such deep DSLs share most of the qualities of external DSLs in that they are easier to analyze, optimize, and transform, though they provide less succinct notation (than external DSLs) for users and are limited in expressiveness (as external DSLs).

Shallow internal DSLs, in contrast, embrace the full expressiveness of the host programming language. The DAUT Scala library [19] is an example of a shallow internal DSL. Written in Scala, it takes advantage of Scala's powerful pattern-matching capabilities and object-oriented features. Consider the example in Fig. 2. The monitor class CommandMonitor inherits from the class Monitor[Event], demonstrating how the internal DSL seamlessly integrates with the host language's type system and inheritance mechanisms. The implementation leverages Scala's pattern matching for event processing and condition checking. This implementation has been successfully used in practice, including in mission-critical applications[2].

Deep DSLs offer better analytical properties and optimization potential but sacrifice expressiveness. Shallow DSLs provide full computational power but make static analysis and optimization more challenging, often requiring sophisticated meta-programming techniques to reason about the code. Their integration into existing development environments and tool chains makes them particularly attractive for scenarios where seamless interoperability with existing code is prioritized over domain-specific syntactic optimization.

2.3   Hybrid DSL

Hybrid DSLs represent a language design approach that combines features of both external and internal DSLs to leverage their respective strengths while

---

[2] The monitor verifies that whenever (always) a command dispatch is observed with a task id, a command number, and a time, then within 20 time units a completion must be observed, with no other dispatch of the same task id and command number observed in between.

```
class CommandMonitor extends Monitor[Event] {
  always {
    case Dispatch(taskId , cmdNum, time1) =>
      hot {
        case Dispatch('taskId ', 'cmdNum', _) => error
        case Complete('taskId ', 'cmdNum', time2)
          if time2 - time1 <= 20 => ok
      }
  }
}
```

Fig. 2. Example - shallow DSL.

mitigating their limitations. We categorize them into two distinct types: Hybrid Closed DSLs and Hybrid Open DSLs.

Hybrid Closed DSLs are external DSLs that are used/called from within a programming language. An illustrative example is Python's MySQL library [9], which allows a Python program to execute MySQL statements provided as text strings. This is a hybrid solution since the DSL is invoked from a programming language, and closed since from within the DSL there is no reference back to the programming language. One may refer to such DSLs as *programming language first* DSLs. A PyDejaVu example is shown in Fig. 6.

Hybrid Open DSLs go in the opposite direction by allowing an external DSL to contain statements in a programming language, typically within special brackets. One may refer to such DSLs as *programming language second* DSLs. The UNIX yacc parser generator [20] exemplifies this approach, where the grammar is specified in an external DSL while semantic actions are implemented in C witin curley brackets. This is a hybrid solution since the programming language is invoked from the DSL, and open since from within the DSL there is reference back to the programming language. The aspect-oriented AspectJ language [22] is another example of a hybrid open DSL, since aspects (external DSL) can contain Java code.

3   QTL Syntax and Semantics

To motivate the development of our PyDejaVu tool, we first introduce QTL (Quantified Temporal Logic), the formalism underlying DejaVu. This logic allows the specification of trace properties involving data and, by limiting its use to past-time operators, is suitable for interpretation over finite traces.

3.1   QTL Syntax

The formulas of the QTL logic are defined using the following grammar, where p stands for a *predicate* symbol, a is a *constant* and x is a *variable*. For simplicity of the presentation, we define here the QTL logic with unary predicates, but this is not due to a principal restriction, and in fact DEJAVU supports predicates over multiple arguments, including zero arguments, corresponding to propositions.

$$\varphi ::= true \mid p(a) \mid p(x) \mid (\varphi \wedge \varphi) \mid \neg\varphi \mid (\varphi \, \mathcal{S} \, \varphi) \mid \ominus\varphi \mid \exists x \, \varphi$$

A formula can be interpreted over multiple types (domains), e.g., natural numbers or strings. Accordingly, each variable, constant and parameter of a predicate is defined over a specific type. Type matching is enforced, e.g., between $p(a)$ and $p(x)$, where the types of the parameter of p and of a must be the same. We denote the type of a variable x by $type(x)$. *Propositional* past time linear temporal logic is obtained by restricting the predicates to be parameterless, essentially Boolean propositions. In this case no variables, constants and quantification are needed either.

QTL subformulas have the following informal meaning: $p(a)$ is true if the last event in the trace is $p(a)$. The formula $p(x)$, for some variable x, holds if x is bound to a constant a such that $p(a)$ is the last event in the trace. The formula $(\varphi \, \mathcal{S} \, \psi)$, which reads as φ *since* ψ, means that ψ occurred in the past (including now) and since then (beyond that state) φ has been true. (The *since* operator is the past dual of the future time *until* modality in the commonly used future time temporal logic.) The property $\ominus \, \varphi$ means that φ is true in the trace that is obtained from the current one by omitting the last event. The formula $\exists x \, \varphi$ is *true* if there exists a value a such that φ is true with x bound to a. We can also define the following additional derived operators: $false = \neg true$, $(\varphi \vee \psi) = \neg(\neg\varphi \wedge \neg\psi)$, $(\varphi \rightarrow \psi) = (\neg\varphi \vee \psi)$, $\diamondsuit \, \varphi = (true \, \mathcal{S} \, \varphi)$ ("previously"), $\boxminus \varphi = \neg \diamondsuit \neg\varphi$ ("always in the past" or "historically"), and $\forall x \, \varphi = \neg\exists x \, \neg\varphi$.

3.2 QTL Formal semantics

A QTL formula is interpreted over a *trace*, which is a finite sequence of *events*. Each event consists of a predicate symbol and parameters (no variables), e.g., $p(a), q(7)$. Formally, let $sub(\varphi)$, be the set of subformulas of φ. Let $free(\varphi)$ be the set of free (i.e., unquantified) variables of φ. The bookkeeping of which variables are mapped to what values is recorded in *assignments*, which map variables to values. Let ϵ be the empty assignment. Let γ be an assignment to the variables $free(\varphi)$. We write $[v \mapsto a]$ to denote the assignment that consists of a single variable v mapped to value a. We denote by $\gamma[v \mapsto a]$ the assignment that differs from γ only by associating the value a to v. Let σ be a trace of events of length $|\sigma|$ and i a natural number, where $i \leq |\sigma|$. Then $(\gamma, \sigma, i) \models \varphi$ denotes that φ holds for the prefix of length i of σ with the assignment γ. We denote by $\gamma|_{free(\varphi)}$ the restriction (projection) of an assignment γ to the free variables appearing in φ. The formal semantics of QTL is defined as follows, where $(\gamma, \sigma, i) \models \varphi$ is defined when γ is an assignment over $free(\varphi)$, and $i \geq 1$.

- $(\epsilon, \sigma, i) \models true$.
- $(\epsilon, \sigma, i) \models p(a)$ if $\sigma[i] = p(a)$.

- $([x \mapsto a], \sigma, i) \models p(x)$ if $\sigma[i] = p(a)$.
- $(\gamma, \sigma, i) \models (\varphi \wedge \psi)$ if $(\gamma|_{free(\varphi)}, \sigma, i) \models \varphi$ and $(\gamma|_{free(\psi)}, \sigma, i) \models \psi$.
- $(\gamma, \sigma, i) \models \neg \varphi$ if not $(\gamma, \sigma, i) \models \varphi$.
- $(\gamma, \sigma, i) \models (\varphi \, \mathcal{S} \, \psi)$ if for some $1 \leq j \leq i$, $(\gamma|_{free(\psi)}, \sigma, j) \models \psi$ and for all $j < k \leq i$, $(\gamma|_{free(\varphi)}, \sigma, k) \models \varphi$.
- $(\gamma, \sigma, i) \models \ominus \varphi$ if $i > 1$ and $(\gamma, \sigma, i - 1) \models \varphi$.
- $(\gamma, \sigma, i) \models \exists x \, \varphi$ if there exists $a \in type(x)$ such that $(\gamma[x \mapsto a], \sigma, i) \models \varphi$.

4 Previous QTL Tools

In the following section, we discuss existing tools including DEJAVU, implemented in Scala, and its extension TP-DEJAVU, which enhances functionality with a two-phase verification process. Building upon these, we present in Sect. 5 the tool PYDEJAVU, which addresses limitations of DEJAVU and TP-DEJAVU.

4.1 DEJAVU

DEJAVU [8, 18] is a runtime verification tool designed to monitor systems against QTL properties. By incrementally analyzing events, DEJAVU detects violations of temporal properties in real time. It utilizes Binary Decision Diagrams (BDDs) to efficiently represent and manipulate sets of assignments, allowing for compact storage and fast evaluation of logical formulas. DEJAVU supports both offline trace analysis and live monitoring, handling data-driven specifications by mapping variable bindings to Boolean values through BDDs.

The DEJAVU tool uses keyword characters to express QTL formulas; it employs the following notation: forall and exists stand for $\forall$ and $\exists$, respectively. P, H, S, and @ correspond to $\diamondsuit$, $\boxminus$, $\mathcal{S}$, and $\ominus$, respectively. Additionally, |, &, and ! represent $\vee$, $\wedge$, and $\neg$, respectively.

Example 1. We will illustrate DEJAVU with an example. This example will be used later to demonstrate the tools TP-DEJAVU and PYDEJAVU. Consider a simple filesystem mechanism that handles several key events. The filesystem allows opening a file with the event open(F, f, m, s), carrying as arguments the folder name F, filename f, access mode m (read or write), and maximum writable size s in bytes. The close(f) event indicates that a file f has been closed. The write event write(f, d) contains the filename and the data d (a string) being written. Additionally, the system supports create(F) and delete(F) events, which represent the creation or deletion of a folder.

The requirement we are verifying states that if data is written to a file, the file must have been opened in write mode, not closed since, and must reside in a folder that has been created and not deleted since. The formalization of this property is shown below in a mathematical QTL-like format. In Fig. 3 the property is shown using DEJAVU syntax.

$$\forall f \,\forall d \;\mathsf{write}(f,d) \;\longrightarrow\; \left(\begin{array}{l} \exists F\, \exists s \\[4pt] \big((\neg\,\mathsf{close}(f)\; \mathcal{S}\; \mathsf{open}(F,f,"w",s))\big) \\[4pt] \wedge\; (\neg\,\mathsf{delete}(F)\; \mathcal{S}\; \mathsf{create}(F))\big) \end{array} \right)$$

```
prop example1:
  forall f . forall d .
    write(f, d) ->
      (exists F . Exists s .
        ((! close(f) S open(F, f, "w", s)) & (! delete(F) S create(F)))))
```

DEJAVU Syntax

Fig. 3. Example 1 - DEJAVU.

Suppose we now wanted to express a property about the number of data bytes written over time. Expressing such dynamic file size conditions cannot be accomplished with DEJAVU, as it requires arithmetic operations, such as calculating totals (sum), which DEJAVU does not support. TP-DEJAVU was created for this purpose.

4.2   TP-DEJAVU

The DEJAVU tool provides a powerful and efficient RV algorithm for monitoring events with data, based on the QTL logic. However, as mentioned, it has some deficiencies in supporting specifications that include e.g., arithmetic operations. It does allow using basic arithmetic comparisons, but their use tames down the strong optimization of the DEJAVU monitoring, which is based on enumerating data values, representing them as bit-vectors and using BDD operations on sets of bit-vectors. The reason is that comparisons are performed between the original values, not the bit-vector representation of enumerations. This deficiency is related to the particular DEJAVU representation, which focuses on logical operators rather than arithmetic operators.

This gave the motivation for a "two phase" RV monitoring, as implemented in TP-DEJAVU [26]. The first phase is *operational*. It performs *preprocessing* of the monitored events. The specification of the operational part is given as a collection of small procedures in a programming language like syntax. The procedures maintain a *summary* of the trace of events seen so far, based on the argument values of the intercepted (observed) events. An event can be modified, e.g., augmented with newly calculated values, before forwarding it to the second phase. The operational phase is in particular effective for comparing values, and for performing calculations, such as aggregation, e.g., summing up observed data values or calculating the maximal value.

The *operational* phase of a TP-DEJAVU specification includes an *initiate* section defining the initial state, including variables, their types and their initial values. This is followed by *on* clauses that specify actions to be executed when various events are observed. Within each *on* clause, variables can be updated, and the *output* keyword is used to generate modified events for further processing by DEJAVU in the second phase. The syntax allows for arithmetic operations, Boolean logic, and data manipulations, enabling TP-DEJAVU to preprocess events and potentially modify them for temporal runtime verification checks performed by DEJAVU.

The *declarative* phase expresses the properties in terms of temporal specifications over the sequence of modified events forwarded by the operational part. This phase is supported by the DEJAVU tool. In TP-DEJAVU, the two phases, operational and declarative, are implemented together as one monolithic tool, programmed in the programming language Scala. The operational part is written as an extension of the DEJAVU tool. An additional feature allows the operational part to use the intermediate verdicts calculated by the declarative part on the trace intercepted so far in preprocessing the next event. This makes the interaction between the two phases bi-directional.

The supported formalism is a purely external DSL, hence it is naturally susceptible to a fixed set of predefined features that can be used, including operators, functions, data structures and programming constructs. If a need to extend the capabilities of the tool occurs, e.g., adding the use of collections (vectors, maps, sets) of data that appear in observed events, some non-trivial implementation effort would be required.

Example 2. As in Example 1, our goal is to ensure that all requirements for writing data into a file are met. Additionally, we can leverage the operational phase to verify that the total number of bytes written to all files together does not exceed a specified max value. To achieve this, we need to implement a counting mechanism. The TP-DEJAVU specification is shown below in Fig. 4.

```
initiate
    total_size : int := 0
    max: int := 1000000

on open(F: str, f: str, m: str, s: int)
    output open(F, f, m)

on write(f: str, d: str)
    total_size : int := total_size + b.length
    ok: bool := total_size <= max
    output write(f, ok)
```

Operational Phase

```
prop example2:
    forall f .
        !write(f, "false") &
        (write(f, "true") ->
            (exists F . (
                (!close(f) S open(F, f, "w")) &
                (!delete(F) S create(F)))))
```

Declarative Phase

Fig. 4. Example 2 - TP-DEJAVU.

In this specification the variable total keeps track of the total number of bytes written. Upon observing an open event, the open event is resubmitted to DEJAVU without the size argument since it is not needed. Upon observing a write event, the total is augmented and a write event is returned to DEJAVU with a Boolean second argument being true only if the total number of bytes is still below the limit. The DEJAVU formula requires that this value must never be FALSE.

5   PYDEJAVU

PYDEJAVU [24] is a Python3 library providing a bridge between Python and the Scala-based DEJAVU [8,18] runtime verification tool. It is, as TP-DEJAVU, designed for two-phase processing, but combining the flexibility and expressiveness of Python with the rigorous and efficient, declarative monitoring capabilities of the DEJAVU tool. Simply stated, PYDEJAVU replaces the operational phase DSL in TP-DEJAVU with Python. Specially decorated functions provide the interface between the operational part and the declarative part. Although DEJAVU is implemented in Scala, we chose Python as the front end language due to its widespread use[3], thereby potentially reaching a larger audience for runtime verification.

The communication between the Python component and DEJAVU is facilitated by the third-party library *pyjnius*. Pyjnius is a Python library that uses the Java Native Interface (JNI) to enable Java code running in a Java Virtual Machine (JVM) to invoke and be invoked by Python applications. Adopting this approach necessitated some changes to DEJAVU, including configuration of DEJAVU from within Python, result feedback to Python, and creation of a result file.

PYDEJAVU embodies the integration of an external DSL (DEJAVU's QTL) with an internal DSL, with Python hosting the internal DSL. By using Python, users can perform any desirable operations on events, including arithmetic, string manipulations, and Boolean logic. Additionally, Python's extensive data structures and objects can be leveraged for data storage and complex calculations. This flexibility allows PYDEJAVU to manage a wide range of operational tasks during runtime, giving users the ability to customize event processing and analysis to suit their specific needs. Figure 5 shows the evolution of the three tools DEJAVU, TP-DEJAVU, and PYDEJAVU presented in this paper, all based on the QTL logic.

Example 3. This example demonstrates how PYDEJAVU provides additional expressiveness, compared to DEJAVU (Example 1) and TP-DEJAVU (Example 2). We augment the original property in Example 1 with the property that there is a limit on how many bytes can be written *to each file*. For this we

---

[3] Python is by several sources evaluated to be the most popular programming language at the time of writing, see e.g. https://spectrum.ieee.org/top-programming-languages-2024.

need to store the available space for each file. This is done in the specification below by maintaining a global variable available_space, which is a Python dictionary, mapping file names to their available space. This cannot be expressed in TP-DEJAVU without extending that system, or taking care of it outside the system.

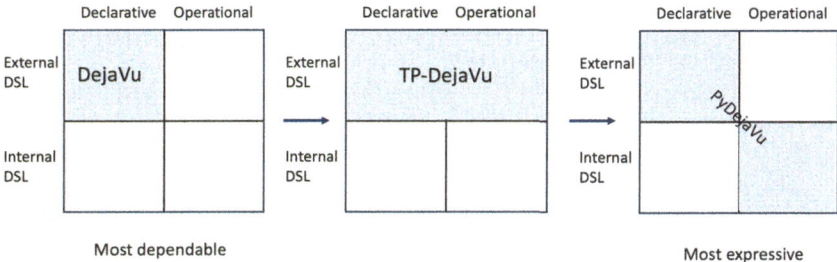

Fig. 5. The evolution of QTL tools.

The monitor in Fig. 6 is initialized as an instance of the Monitor class, part of the PYDEJAVU library, which must be imported. This monitor takes as argument a string representing a specification in QTL. Serving as the core engine, the monitor tracks events issued from the operational part in Python during program execution and ensures they align with the QTL specification. The event handler functions, such as open, close, and write, are decorated with the @event annotation[4], which binds them to specific events. That is, if a function is annotated with @event("e") then that function will be called on all events with name e. Each such annotated event handler function returns a list, where the first item is the event name and subsequent items represent values associated with that event. The returned list is then submitted to DEJAVU as an event. For example, in the open event, the function returns ["open", F, f, m], where F, f and m are the arguments passed to the handler, representing the folder where the file located, the file being opened and the mode in which it is opened. Similarly, the write function returns ["write", f, ok], where f is the file being written to, and ok is a Boolean flag being true iff. the resulting file size is within the size limit provided when the file was opened.

Additionally, if an event handler returns None, it means that nothing is forwarded to the declarative part, and the system waits for new events. Event handlers are not required for every event, e.g., for create and delete. In these cases, events submitted to the monitor are directly passed to DEJAVU, ensuring that the default processing behavior is maintained without additional modifications.

---

[4] In Python, one can define a function D that takes a function as an argument and returns a new function, we call D a decorator. If a function g is decorated with D using the @-sign (i.e., @D), then g is effectively replaced by $D(g)$. In the case of @event("open"), the call event("open") returns a decorator that modifies the function defined below it.

6   Experiments

We present experimental results on the relative efficiency of runtime monitoring QTL specifications using the TP-DejaVu tool compared to the combined internal and external RV tool PyDejaVu. The experiments were performed on an Apple MacBook Pro laptop with an M1 Core processor, 16 GB RAM,

```python
from pydejavu.core.monitor import Monitor, event

specification = """
  prop example3:
    forall f .
      ! write(f, " false") &
      (write(f, "true") −>
        (exists F . ((! close(f) S open(F, f, "w")) & (!delete(F) S create(F)))))
"""

monitor = Monitor(specification)
available_space : dict[str, int] = {}

@event("open")
def open(F: str, f: str, m: str, s: int):
    global available_space
    if mode == "w":
        available_space[f] = s
    return ["open", F, f, m]

@event("close")
def close(f: str):
    global available_space
    del available_space[f]
    return ["close", f]

@event("write")
def write(f: str, d: str):
    global available_space
    if f not in available_space:
        available_space[f] = 0
    data_len = len(d)
    ok = available_space[f] >= data_len
    if ok:
        available_space[f] −= data_len
    return ["write", f, ok]

# −−−−−−−−−−−−−−−−−−−−−−
# Applying monitor to an example trace:
# −−−−−−−−−−−−−−−−−−−−−−

events = [
    {"name": "create", "args": ["tmp"]},
    {"name": "open", "args": ["tmp", "f1", "w", "10"]},
    {"name": "write", "args": ["f1", "some text"]}
]

for e in events:
    monitor.verify(e)
monitor.end()
```

Fig. 6. Example 3 - PyDejaVu.

and 512 GB SSD storage, running the macOS Sonoma operating system. In all cases, both specifications produced identical outputs. Our benchmarks focused solely on time and memory consumption during the evaluation phase, excluding compilation time.

Properties. The QTL properties used in our experiments are shown in Fig. 7. Their adaptations for TP-DEJAVU and PYDEJAVU are shown in Figs. 8 and 9, respectively.

---

1. **forall** x . (p(x) -> **exists** y . (P q(x, y) & y > 10))
2. **forall** x . **forall** y . ((p(x) & @q(y) & x < y) -> P r(x, y))
3. **forall** x . (p(x) -> (**forall** y . (@P q(y) -> x > y) & **exists** z . @ P q(z)))

Fig. 7. Experimental Properties.

```
on q(x: int, y: int)
  in_bound: bool := y > 10
  output ite(in_bound,
    q(x), skip)

─────────────────────

forall x .
  ( p(x) -> P q(x) )
```

```
initiate
  last_q: bool := false
  y: int := 0

on p(x: int)
  x_lt_y: bool :=
    last_seen_q && x < y
  last_q: bool := false
  output p_q(x, y, x_lt_y)

on q(y: int)
  last_q: bool := true
  output skip

on r(x: int, y: int)
  last_q: bool := false
  output r(x,y)

─────────────────────

forall x . forall y .
  (p_q(x, y, "true") ->
    P r(x, y))
```

```
initiate
  max_y: int := −1

on p(x: int)
  x_gt_y: bool := x > max_y
  output p(x, x_gt_y)

on q(y: int)
  is_max: bool := max_y < y
  max_y: int :=
    ite(is_max, y, max_y)
  output q(y)

─────────────────────

forall x . forall y .
  (p(x, y) ->
    (p(x, "true") &
    exists z . @ P q(z)))
```

Property 1   Property 2   Property 3

Fig. 8. TP-DEJAVU properties: operational (above) and declarative (below).

```
@event("q")
    def handle_q(arg_x: int, arg_y: int):
        if arg_y > 10:
            return "q", arg_x
        return None
```

```
forall x .
  ( p(x) -> P q(x) )
```

Property 1

```
last_seen_q  = False
y = 0

@event("p")
def handle_p(arg_x: int):
    nonlocal y, last_seen_q
    x_lt_y  = last_seen_q and arg_x < y
    last_seen_q  = False
    return "p_q", arg_x, y, x_lt_y

@event("q")
def handle_q(arg_y: int):
    nonlocal y, last_seen_q
    y = arg_y
    last_seen_q  = True

@event("r")
def handle_r(arg_x: int, arg_y: int):
    nonlocal last_seen_q
    last_seen_q  = False
    return "r", arg_x, arg_y
```

```
forall x . forall y .
  (p_q(x, y, "true") ->
      P r(x, y))
```

Property 2

```
max_y = -1

@event("p")
def handle_p(arg_x: int):
    nonlocal max_y
    x_gt_y  = arg_x > max_y
    return "p", arg_x, x_gt_y

@event("q")
def handle_q(arg_y: int):
    nonlocal max_y
    max_y = max(max_y, arg_y)
    return "q", arg_y
```

```
forall x . forall y .
  (p(x, y) ->
    (p(x, "true") &
    exists z . @ P q(z)))
```

Property 3

Fig. 9. PyDejaVu properties matching the TP-DejaVu properties in Fig. 8.

Traces. In our experiment, we utilized traces containing varying events to test our approach's scalability and performance. The event sequences and their corresponding values were generated using Python scripts that employed a mix of random methods and minimal modeling, with constraints to ensure variety, control over trace characteristics, and a diverse range of scenarios for all properties. Specifically, we used traces with 10K, 100K, 500K, 1M, and 5M events. This range allowed us to observe the behavior of the system under different levels of complexity and load.

Results. Table 1 presents the results from the experiments, indicating that TP-DEJAVU is slightly faster than PYDEJAVU. Notably, DEJAVU was unable to evaluate any of the traces except one, as its execution times consistently exceeded 1000 s. These results are somewhat predictable, given that Python is likely less efficient than Scala. We actually anticipated that TP-DEJAVU would be even faster relative to PYDEJAVU: while TP-DEJAVU executes its operational phase and declarative phase in the same language (Scala), PYDEJAVU involves two languages, Python and Scala. This dependency introduces additional overhead due to the communication between the operational phase in Python and the declarative phase in Scala.

Table 1. DEJAVU vs. TP-DEJAVU vs. PYDEJAVU: Time and memory usage.

Property	Method	Trace 10K	Trace 100K	Trace 500K	Trace 1M	Trace 5M
$P1$	DEJAVU	11.35 s	∞	∞	∞	∞
		1.21 GB	-	-	-	-
	TP-DEJAVU	0.39 s	0.91 s	2.91 s	5.45 s	44.86 s
		105 MB	311 MB	646 MB	1.34 GB	4.01 GB
	PYDEJAVU	0.50 s	1.21 s	4.55 s	8.85 s	63.90 s
		174 MB	435 MB	1.24 GB	2.26 GB	3.88 GB
$P2$	DEJAVU	∞	∞	∞	∞	∞
		-	-	-	-	-
	TP-DEJAVU	0.39 s	0.98 s	3.02 s	5.48 s	59.14 s
		110 MB	297 MB	707 MB	1.06 GB	4.23 GB
	PYDEJAVU	0.62 s	1.38 s	4.75 s	9.26 s	84.14 s
		155 MB	431 MB	957 MB	1.37 GB	3.43 GB
$P3$	DEJAVU	∞	∞	∞	∞	∞
		-	-	-	-	-
	TP-DEJAVU	0.41 s	0.95 s	2.52 s	4.74 s	31.68 s
		143 MB	311 MB	672 MB	960 MB	3.91 GB
	PYDEJAVU	0.53 s	1.47 s	4.89 s	9.33 s	59.16 s
		172 MB	434 MB	1.00 GB	1.69 GB	3.67 GB

7   Conclusion

The true strength of PyDEJAVU, which combines both internal and external DSLs, lies in its flexibility for executing complex calculations, especially in scenarios where the specifications are challenging or impossible to express in TP-DEJAVU. This flexibility stems largely from Python's clear, familiar, and expressive syntax, enabling developers to easily construct intricate logic. Additionally, the embedded declarative component leverages DEJAVU's capabilities to perform complex relational calculations, particularly for handling the first-order past-time aspects of the desired specifications. A further advantage of embedding RV in Python is the possibility of integrating other libraries such as e.g. SQL and PyTorch. Future work can include implementing a Scala interface to DEJAVU. Another line of work includes investigating a hybrid open DSL, where a QTL formula can directly refer to functions in the host programming language.

References

1. Barringer, H., Havelund, K.: TraceContract: a scala DSL for trace analysis. In: Butler, M., Schulte, W. (eds.) FM 2011. LNCS, vol. 6664, pp. 57–72. Springer, Heidelberg (2011). https://doi.org/10.1007/978-3-642-21437-0_7
2. Barringer, H., Rydeheard, D., Havelund, K.: Rule systems for run-time monitoring: from Eagle to RuleR. In: Sokolsky, O., Taşıran, S. (eds.) RV 2007. LNCS, vol. 4839, pp. 111–125. Springer, Heidelberg (2007). https://doi.org/10.1007/978-3-540-77395-5_10
3. Basin, D., Klaedtke, F., Marinovic, S., Zălinescu, E.: Monitoring of temporal first-order properties with aggregations. Formal Methods Syst. Des. **46**(3), 262–285 (2015). https://doi.org/10.1007/s10703-015-0222-7
4. Colombo, C., Gauci, A., Pace, G.J.: LarvaStat: monitoring of statistical properties. In: Barringer, H., et al. (eds.) RV 2010. LNCS, vol. 6418, pp. 480–484. Springer, Heidelberg (2010). https://doi.org/10.1007/978-3-642-16612-9_38
5. Dams, D., Havelund, K., Kauffman, S.: A Python library for trace analysis. In: Dang, T., Stolz, V. (eds.) RV 2022. LNCS, vol. 13498, pp. 264–273. Springer, Cham (2022). https://doi.org/10.1007/978-3-031-17196-3_15
6. D'Angelo, B., et al.: Lola: runtime monitoring of synchronous systems. In: 12th International Symposium on Temporal Representation and Reasoning (TIME 2005), pp. 166–174 (2005)
7. Decker, N., Leucker, M., Thoma, D.: Monitoring modulo theories. Int. J. Softw. Tools Technol. Transf. **18**(2), 205–225 (2016)
8. DejaVu tool source code. https://github.com/havelund/dejavu
9. Andy Dustman. Python MySQL (2024). https://pypi.org/project/MySQL-python/
10. Github. Pytorch (2024). https://github.com/pytorch/pytorch
11. Gorostiaga, F., Sánchez, C.: HStriver: a very functional extensible tool for the runtime verification of real-time event streams. In: Huisman, M., Păsăreanu, C., Zhan, N. (eds.) FM 2021. LNCS, vol. 13047, pp. 563–580. Springer, Cham (2021). https://doi.org/10.1007/978-3-030-90870-6_30
12. Halle, S., Villemaire, R.: Runtime enforcement of web service message contracts with data. **5**, 192–206 (2012)

13. Havelund, K.: Data automata in Scala. In: 2014 Theoretical Aspects of Software Engineering Conference, TASE 2014, Changsha, China, 1–3 September 2014, pp. 1–9. IEEE Computer Society (2014)
14. Havelund, K.: Rule-based runtime verification revisited. Softw. Tools Technol. Transfer (STTT) **17**(2), 143–170 (2015)
15. Havelund, K., Katsaros, P., Omer, M., Peled, D., Temperekidis, A.: TP-DejaVu: combining operational and declarative runtime verification. In: Dimitrova, R., Lahav, O., Wolff, S. (eds.) VMCAI 2024. LNCS, vol. 14500, pp. 249–263. Springer, Cham (2024). https://doi.org/10.1007/978-3-031-50521-8_12
16. Havelund, K., Omer, M., Peled, D.: Operational and declarative runtime verification (keynote). In: Proceedings of the 7th ACM International Workshop on Verification and Monitoring at Runtime Execution, VORTEX 2024, pp. 3–12. Association for Computing Machinery, New York (2024)
17. Havelund, K., Peled, D.: An extension of first-order LTL with rules with application to runtime verification. Int. J. Softw. Tools Technol. Transfer **23**(4), 547–563 (2021). https://doi.org/10.1007/s10009-021-00626-y
18. Havelund, K., Peled, D., Ulus, D.: First-order temporal logic monitoring with BDDs. Formal Methods Syst. Des. **56**(1–3), 1–21 (2020)
19. Haveund, K.: Daut - Monitoring Data Streams with Data Automata (2024). https://github.com/havelund/daut
20. Johnson, S.C.: YACC: Yet Another Compiler-Compiler. https://en.wikipedia.org/wiki/Yacc
21. Kallwies, H., Leucker, M., Schmitz, M., Schulz, A., Thoma, D., Weiss, A.: TeSSLa – an ecosystem for runtime verification. In: Dang, T., Stolz, V. (eds.) RV 2022. LNCS, vol. 13498, pp. 314–324. Springer, Cham (2022). https://doi.org/10.1007/978-3-031-17196-3_20
22. Kiczales, G., Hilsdale, E., Hugunin, J., Kersten, M., Palm, J., Griswold, W.G.: An overview of AspectJ. In: Knudsen, J.L. (ed.) ECOOP 2001. LNCS, vol. 2072, pp. 327–354. Springer, Heidelberg (2001). https://doi.org/10.1007/3-540-45337-7_18
23. Kim, M., Kannan, S., Lee, I., Sokolsky, O., Viswanathan, M.: Java-MaC: a runtime assurance tool for Java programs. In: Havelund, K., Rosu, G. (eds.) Workshop on Runtime Verification, RV 2001, in connection with CAV 2001, Paris, France, 23 July 2001. Electronic Notes in Theoretical Computer Science, vol. 55, pp. 218–235. Elsevier (2001)
24. PyDejaVu tool source code. https://github.com/moraneus/pydejavu
25. Reger, G., Cruz, H.C., Rydeheard, D.: MarQ: monitoring at runtime with QEA. In: Baier, C., Tinelli, C. (eds.) TACAS 2015. LNCS, vol. 9035, pp. 596–610. Springer, Heidelberg (2015). https://doi.org/10.1007/978-3-662-46681-0_55
26. TPDejaVu tool source code. https://doi.org/10.5281/zenodo.8322559

Causality Monitoring for MIMOS

Chengzi Huang[1] , Behnam Khodabandeloo[1], Duc Anh Nguyen[1],
Wang Yi[1]([✉]) , Jie An[2] , and Zhenya Zhang[3]

[1] Uppsala University, Uppsala, Sweden
{chengzi.huang,behnam.khodabandeloo,ducanh.nguyen,yi}@it.uu.se
[2] Institute of Software, Chinese Academy of Sciences, Beijing, China
anjie@iscas.ac.cn
[3] Kyushu University, Fukuoka, Japan
zhang@ait.kyushu-u.ac.jp

Abstract. MIMOS is a tool environment developed based on an asynchronous and deterministic data flow model for the design and evolution of real-time systems, aiming at safety-critical applications. Currently MIMOS supports modeling, simulation, verification, scheduling, timing analysis and (multi-core) code generation as well as dynamic updates of systems after deployment. The asynchronous yet deterministic design paradigm supported by MIMOS enables developers to build robust applications that can evolve without compromising their critical safety properties. This paper presents the run-time monitoring feature of MIMOS, with a focus on the causality of specifications in Signal Temporal Logic (STL). A causality monitor of MIMOS, determines and visualizes not only whether an executing trace violates a given STL specification but also how relevant the incremental changes of the trace is to the violation, with quantitative information providing insights into the system's behavior. We present a case study to illustrate the power and usefulness of causality monitoring in MIMOS.

Keywords: MIMOS · real-time systems · monitoring · causality

1 Introduction

MIMOS is a tool environment designed for the development and evolution of embedded real-time systems, specifically tailored for safety-critical applications. It provides a platform for modeling, simulation, verification, scheduling, timing analysis and multi-core code generation as well as dynamic system updates. Unlike traditional synchronous design tools, MIMOS utilizes an asynchronous yet deterministic data-flow design model to facilitate incremental system modifications and dynamic updates after deployment without compromising safety guarantees.

With this work, we congratulate Christel on her outstanding scientific achievements and thank her for inspiring so many in our community. Wang would also like to thank Christel for her many years of friendship.

In this paper, we present the *monitoring* feature of MIMOS. Online monitoring (a.k.a. runtime verification) [1] is an effective approach for checking whether a trace of system execution satisfies a given specification at runtime. By extending Temporal Logic [10] with real-time constraints and real-valued predicates, Signal Temporal Logic (STL) [9] has been widely utilized to specify properties of real-time systems. In STL monitoring, an online monitor analyzes partial execution traces at runtime, evaluating the satisfaction of an STL formula φ based on the partial trace observed up to each time instant. Typically, the monitor reports results according to the STL robust semantics. However, since the trace is incomplete, a standard online monitor—such as the *online robust monitor* described in [3]—outputs a robustness interval. This interval represents the range of possible robustness values that could be achieved under any future suffix trace. From this interval, the satisfaction of φ can be inferred, for example, φ is violated if the upper bound of the robustness is negative, as Robustness values represent how much the given trace satisfies or violates the STL specification.

A main limitation arises from the definition of robust semantics is that the upper and lower bounds are monotonically decreasing and increasing; this has the consequence that the robustness interval at a given step is "masked" by the history of previous robustness intervals, and, e.g., it is not possible to detect mitigation of the violation severity. This issue has been acknowledged in the literature [11,15] as a significant drawback of these monitoring approaches. To address the problem of information masking, Zhang et al. [13,14] recently introduced the *causality monitoring* of STL. Instead of relying directly on robustness, it considers the causality of violation or satisfaction. This approach not only determines whether an executing trace violates the specification but also evaluates how relevant each incremental update of the trace is to the violation. By doing so, causality monitoring avoids monotonicity and provides richer insights into the system's behavior. It can identify the specific time intervals relevant to the violation or satisfaction of the specification and thus even count the number of violations, a feature particularly valuable for system engineers in practical applications. Furthermore, the causality monitor serves as a refinement of the classic robust monitor, meaning the latter can be straightforwardly derived from the former.

We shall introduce the causality monitoring of STL to the MIMOS platform as one of the tool features, and illustrate the power and usefulness of causality monitor through a case study using MIMOS.

The paper is organized as follows: Sect. 2 recalls the MIMOS model and tool features. Section 3 introduces Signal Temporal Logic, including syntax, robust semantics, online robust monitoring, and causality monitoring of STL. Section 4 presents a case study of an autopilot control system for marine vessels, demonstrating how causality monitoring can detect and localize specification violations. Finally, Sect. 5 concludes the paper and discusses potential future directions for MIMOS and causality monitoring in safety-critical applications.

2   An Overview of MIMOS

This section presents an overview of the semantic of MIMOS [12], and the tool chain developed based on the model.

2.1   MIMOS Model

MIMOS extends the Kahn Process Network (KPN) model [7] by integrating timing semantics, achieved through modeling its processes as periodic real-time tasks. This enables MIMOS to represent embedded systems with precise timing requirements while ensuring deterministic behavior, guaranteeing that for any given set of timed input streams, the model produces a unique set of timed output streams.

A MIMOS system may comprise a collection of process networks. Each network with the system can be represented as a simple directed graph $G = (N, L)$, where

- Nodes (N): A set of real-time tasks. Each task N_i is characterized by:
  - A local state and a set of input and output ports.
  - A function mapping inputs and its current state to outputs.
  - A release pattern dictating its activation pattern.
  - A relative deadline D_i, specifying its latest allowable completion time after release.
- Channels (L): A set of directed edges, where each channel $L_{i,j}$ connects node N_i to node N_j, enabling inter-task communication. Two types of data exchange mechanisms are supported:
  - FIFO channels: Provide ordered buffering, ensuring tokens are retrieved in a first-in-first-out manner.
  - Register channels: Used for sampling time-dependent data sources (like sensors). They store only the most recently written value, overwriting previous ones.
- Two sets of nonnegative integers, W and R, which define the token flow between nodes:
  - $W_{i,j}$ specifies the number of tokens produced by node N_i and added to channel $L_{i,j}$ at the end of each (absolute) deadline.
  - $R_{i,j}$ specifies the number of tokens consumed from channel $L_{i,j}$ when node N_j starts execution.

A node N_i is eligible to start execution at its activation time if and only if all its incoming channels $L_{j,i}$ satisfy their specific enablement rules:

- FIFO Channels (Blocking Mode): If an incoming FIFO channel $L_{j,i}$ is configured in blocking mode, then this channel is enabled if it stores at least $R_{j,i}$. Upon activation, exactly $R_{j,i}$ tokens are removed from this channel.
- FIFO Channels (Non-Blocking Mode – Upto-N): If an incoming FIFO channel $L_{j,i}$ operates in "Upto-N" mode, then this channel is always enabled. At runtime, the task may consume anywhere between 0 and $R_{j,i}$ tokens from this channel based on its availability.

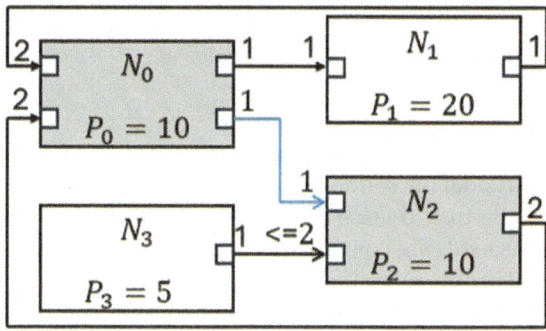

Fig. 1. An example for MIMOS

– Register Channels: If an incoming channel $L_{j,i}$ is a register channel, this channel is always enabled. N_i can always read from it at the start of execution, retrieving the latest value written by its predecessor N_j. Unlike FIFO channels, reading from a register channel does not consume tokens, and the stored value remains available for subsequent activations.

Once a task N_i is released, if it meets the activation conditions at its release time, it can execute, consuming tokens from its incoming channels according to their respective rules. However, if the activation conditions are not met, the task remains inactive and does not execute in this cycle, waiting until its next release time for another activation opportunity.

After execution begins, the task runs until completion, and at the end of its absolute deadline, it produces tokens for each outgoing channel, making them available for downstream tasks. Additionally, the MIMOS model employs read-after-write semantics for scenarios where read and write operations on the same channel logically occur simultaneously. Under these semantics, the read operation always takes place after the write operation, ensuring that the most recently written token is consistently obtained.

In this paper, we assume that each node N_i follows the release pattern of *implicit-deadline periodic tasks* [8], where P_i denotes the minimum separation between consecutive activations of N_i, and $D_i = P_i$. This assumption ensures that every task is activated periodically and must complete its execution within the same period before the next activation occurs.

Example 1. To illustrate the MIMOS model, Fig. 1 presents an example where black arrows represent FIFO channels and the blue arrow represents a register channel. In this example, node N_1 has a period of 20 time units and adds one token to channel $L_{1,0}$ at the end of each period. Meanwhile, node N_0, which has a period of 10 time units, requires at least two tokens from channel $L_{2,0}$ to begin execution and, once it completes execution, produces one token to channel $L_{0,2}$. Additionally, the notation "≤ 2" indicates that the corresponding incoming channel $L_{3,2}$ of node N_2 operates in "Upto-N" mode, meaning that this channel is always enabled. N_2 may consume between 0 and 2 tokens, depending on the number of tokens present at the time of its release.

2.2   Signals and Timed Data Streams

While the MIMOS model introduced earlier emphasizes discrete, token-based interactions at periodic task release events, it can also be beneficial to conceptualize channel states as evolving continuously over time. Adopting this continuous-time perspective naturally introduces the concepts of *signals* and *timed data streams*, which provide complementary ways of representing real-time behavior.

Definition 1 (Signal). Given a positive real number $T \in \mathbb{R} > 0$, referred to as the *time horizon*, and a positive integer $d \in \mathbb{N} > 0$ denoting the dimension, a *signal* is defined as a function:

$$\mathbf{v} : [0, T] \rightarrow \mathbb{R}^d$$

For each time $t \in [0, T]$, $\mathbf{v}(t)$ provides a d-dimensional vector describing the instantaneous state of a channel. If the condition $\mathbf{v}(t) \in \Omega \subseteq \mathbb{R}^d$ holds for every $t \in [0, T]$, the signal $\mathbf{v}$ is said to be *spatially bounded* by the region Ω.

Definition 2 (Timed Data Stream). A *timed data stream* is a finite or infinite sequence of timestamped data points:

$$\langle (x_1, t_1), (x_2, t_2), \ldots \rangle,$$

where each data item $x_i \in \mathbb{R}^d$ is paired with a timestamp $t_i \in [0, T]$. The sequence of timestamps is non-decreasing, satisfying $t_1 \leq t_2 \leq \ldots$, and typically remains bounded within the global time horizon T.

In many real-time modeling frameworks, signals and timed data streams serve as interchangeable abstractions to represent system behavior. Specifically, a continuous signal $\mathbf{v}(t)$ can be sampled at discrete time points $t_1 \leq t_2 \leq \ldots$ to yield a corresponding timed data stream:

$$\langle (\mathbf{v}(t_1), t_1), (\mathbf{v}(t_2), t_2), \ldots \rangle.$$

Conversely, a timed data stream can be converted into a continuous-time signal through suitable interpolation or by holding the last sampled value constant between timestamps.

From this signal-oriented viewpoint, each MIMOS channel $L_{i,j}$ is naturally described by a function

$$\gamma_{i,j} : [0, T] \rightarrow \mathbb{R}^d$$

that characterizes the evolution of the channel state or token count continuously over the interval $[0, T]$. Traditional discrete operations, such as token reads and writes, become operations of sampling from or updating the function $\gamma_{i,j}$. This continuous-time interpretation, bolstered by our assumption of *holding the last sampled value constant*, enhances the analytical capabilities and simulation accuracy of the MIMOS model, providing a cohesive framework that unifies discrete token interactions and continuous-time semantics.

2.3   MIMOS Tool Features

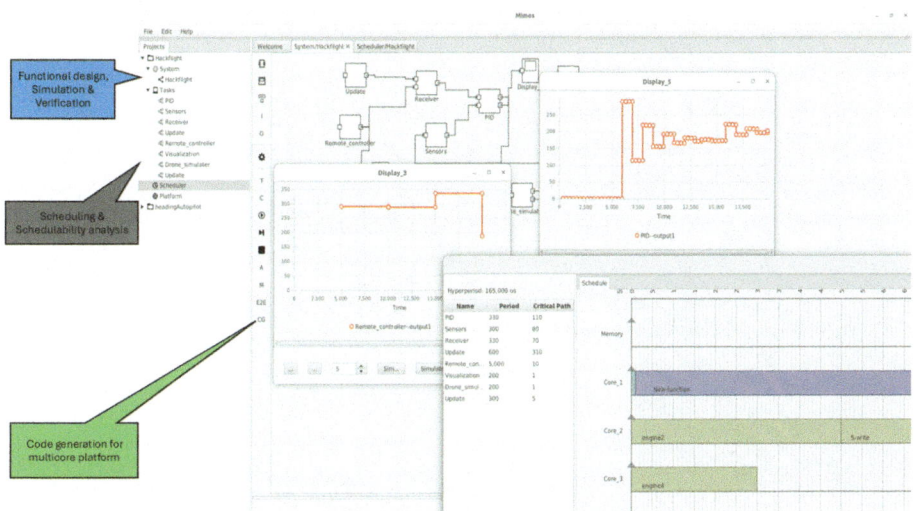

Fig. 2. Main interface of the MIMOS tool.

This subsection outlines the key features, architecture, and core components of the MIMOS tool. Figure 2 illustrates the primary user interface, highlighting the functionality and user interactions supported by the tool.

The design and analysis capabilities of MIMOS are structured into three distinct layers:

- *Functional Layer*: Provides a graphical interface for designing MIMOS models as networks of computational nodes, each performing periodic computations defined by user-specified periods and deadlines. This platform-independent layer supports for modeling of functional behaviors, functional simulation and verification, and also monitoring of time-variant properties.
- *Software Layer*: Abstracts the functional designs into real-time tasks represented as periodic Directed Acyclic Graphs (DAGs), supporting rigorous schedulability and timing analysis, including estimation of end-to-end latency bounds.
- *Hardware Layer*: Allows precise specification of multi-core processor platforms, including details such as cores, caches, and memory. Integration with the Gem5 simulator enables Worst-Case Execution Time (WCET) analysis through cycle-accurate simulation.

Currently the tool supports the following features:

- **Function Editor (see Fig. 3)** Provides a graphical interface to model a system as a network of nodes connected via unidirectional channels, following the MIMOS model (described in Sect. 2). Each node represents a periodic real-time task characterized by input/output ports and mapping functions. Users can configure node parameters, define mapping functions in supported languages (C/C++, Java, and Python), and set assertions on ports for verification.

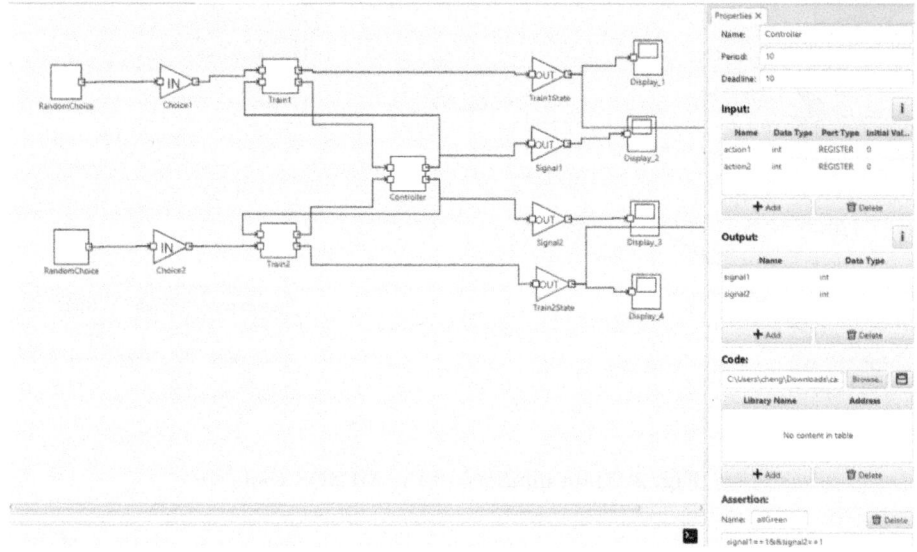

Fig. 3. System modeling in MIMOS.

- **Discrete Event Simulator (Fig. 4a)**: Dynamically visualizes execution behaviors and timing properties. Simulation traces can be displayed step-by-step, continuously up to a predefined step, or indefinitely. Visualization speed is adjustable on a scale from 1 to 10. This simulator also includes monitoring functionalities to track and validate system conditions at each signal (see Fig. 4b).
- **Analyzer (Fig. 5)** Checks whether the tasks associated with a system model satisfy their timing requirements. The analysis suite includes schedulability analysis under fixed priority, earliest deadline first (EDF) scheduling, and partitioned or semi-partitioned scheduling strategies. Execution can also be visualized as Gantt charts.
- **Code Generator**: The Multi-core Code Generator in MIMOS enables the automatic generation of executable C code from system models, supporting multi-core deployment. It translates the system model into a set of communicating tasks connected via one-to-one channels, ensuring efficient execution on

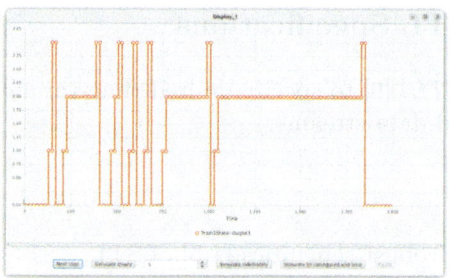
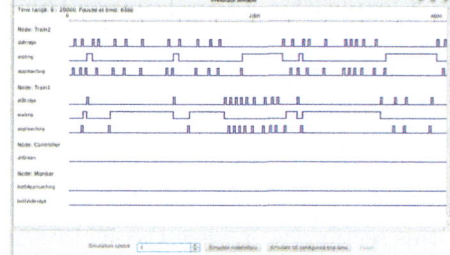

(a) Signal Value Trace in MIMOS.   (b) Boolean State Monitoring in MIMOS.

Fig. 4. Discrete event simulation in MIMOS

multi-core platforms. The compiler applies deterministic semantic refinement, guaranteeing that the generated code preserves both functional correctness and timing constraints when deployed on the target architecture.

In the following section, we present a framework for monitoring the time-variant properties of the input and output signals and timed streams of an MIMOS model.

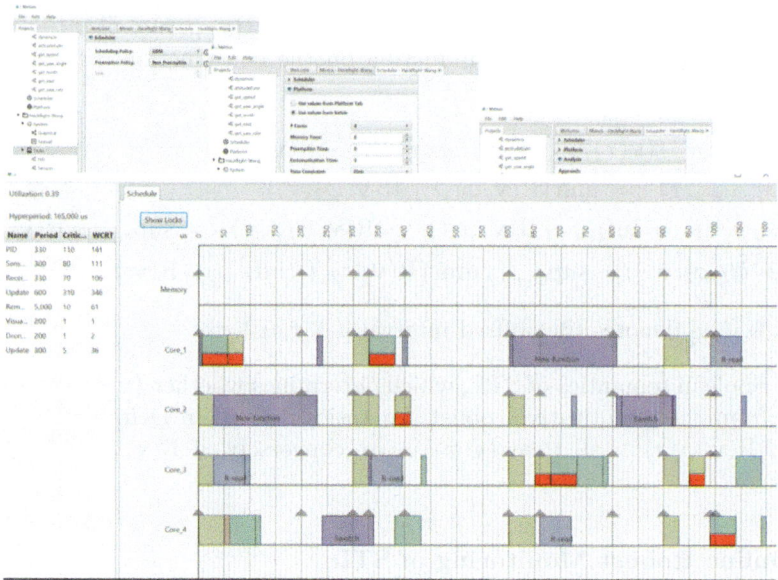

Fig. 5. Schedulability analysis in MIMOS.

3   Causality Monitoring for STL Specifications

We shall use the *Signal temporal logic (STL)* [9] to specify desired time-variant properties over MIMOS signals and timed data streams.

3.1   Signal Temporal Logic

In this section, we review the syntax and robust semantics [4,5] of STL.

Definition 3 (STL syntax). In STL, the *atomic propositions* α and the *formulas* φ are respectively defined as follows:

$$\alpha ::\equiv f(w_1, \ldots, w_K) > 0 \qquad \varphi ::\equiv \alpha \mid \bot \mid \neg\varphi \mid \varphi \wedge \varphi \mid \Box_I \varphi \mid \Diamond_I \varphi \mid \varphi \, \mathcal{U}_I \, \varphi$$

Here f is a K-ary function $f : \mathbb{R}^K \to \mathbb{R}$, $w_1, \ldots, w_K \in \mathbf{Var}$, and I is a closed non-singular interval in $\mathbb{R}_{\geq 0}$, i.e., $I = [l, u]$, where $l, u \in \mathbb{R}$ and $l < u$. $\Box, \Diamond$ and $\mathcal{U}$ are temporal operators, which are known as *always, eventually* and *until*, respectively. The always operator $\Box$ and eventually operator $\Diamond$ are two special cases of the until operator $\mathcal{U}$, where $\Diamond_I \varphi \equiv \top \, \mathcal{U}_I \, \varphi$ and $\Box_I \varphi \equiv \neg\Diamond_I \neg\varphi$. Other common connectives such as $\vee, \to$ are introduced as syntactic sugar: $\varphi_1 \vee \varphi_2 \equiv \neg(\neg\varphi_1 \wedge \neg\varphi_2)$, $\varphi_1 \to \varphi_2 \equiv \neg\varphi_1 \vee \varphi_2$.

Definition 4 (STL robust semantics). Let $\mathbf{v}$ be a signal, φ be an STL formula and $\tau \in \mathbb{R}_+$ be an instant. The *robustness* $R(\mathbf{v}, \varphi, \tau) \in \mathbb{R} \cup \{+\infty, -\infty\}$ of $\mathbf{v}$ w.r.t. φ at τ is defined by induction on the construction of formulas, as follows,

$$R(\mathbf{v}, \alpha, \tau) := f(\mathbf{v}(\tau)) \qquad R(\mathbf{v}, \neg\varphi, \tau) := -R(\mathbf{v}, \varphi, \tau)$$
$$R(\mathbf{v}, \varphi_1 \wedge \varphi_2, \tau) := \min\left(R(\mathbf{v}, \varphi_1, \tau), R(\mathbf{v}, \varphi_2, \tau)\right)$$
$$R(\mathbf{v}, \Box_I \varphi, \tau) := \inf_{t \in \tau + I} R(\mathbf{v}, \varphi, t) \qquad R(\mathbf{v}, \Diamond_I \varphi, \tau) := \sup_{t \in \tau + I} R(\mathbf{v}, \varphi, t)$$
$$R(\mathbf{v}, \varphi_1 \, \mathcal{U}_I \, \varphi_2, \tau) := \sup_{t \in \tau + I} \min\left(R(\mathbf{v}, \varphi_2, t), \inf_{t' \in [\tau, t)} R(\mathbf{v}, \varphi_1, t')\right)$$

where $\tau + [l, u]$ denotes the shifted interval $[l + \tau, u + \tau]$.

The Boolean semantics of STL, which determines whether $(\mathbf{v}, \tau) \models \varphi$, can be derived from the quantitative robust semantics defined in Definition 4. Specifically, if $R(\mathbf{v}, \varphi, \tau) > 0$, then $(\mathbf{v}, \tau) \models \varphi$; conversely, if $R(\mathbf{v}, \varphi, \tau) < 0$, then $(\mathbf{v}, \tau) \not\models \varphi$.

3.2   Online Robust Monitoring of STL

Online monitoring concerns the satisfaction of a *partial signal* $\mathbf{v}_{0:b} : [0, b] \to \mathbb{R}^d$ w.r.t. an STL formula φ. We define a *completion* of $\mathbf{v}_{0:b}$ as a signal $\mathbf{v} : [0, T] \to \mathbb{R}^d$ $(b \leq T)$ such that $\forall t \in [0, b], \mathbf{v}(t) = \mathbf{v}_{0:b}(t)$. A completion $\mathbf{v}$ can be written as the concatenation of $\mathbf{v}_{0:b}$ with a *suffix signal* $\mathbf{v}_{b:T}$, i.e., $\mathbf{v} = \mathbf{v}_{0:b} \cdot \mathbf{v}_{b:T}$.

Definition 5 (Online robust monitor [3]). Let $\mathbf{v}_{0:b}$ be a partial signal, and let φ be an STL formula. We denote by R_{max}^{α} and R_{min}^{α} the possible *maximum* and *minimum bounds* of the robustness $R(\mathbf{v}, \alpha, \tau)^1$. Then, an *online robust monitor* returns a sub-interval $[R](\mathbf{v}_{0:b}, \varphi, \tau) \subseteq [R_{min}^{\alpha}, R_{max}^{\alpha}]$ at instant b, which is defined as follows, by induction on the construction of formulas.

$$[R](\mathbf{v}_{0:b}, \alpha, \tau) := \begin{cases} [f(\mathbf{v}_{0:b}(\tau)), f(\mathbf{v}_{0:b}(\tau))] & \text{if } \tau \in [0, b] \\ [R_{min}^{\alpha}, R_{max}^{\alpha}] & \text{otherwise} \end{cases}$$

$$[R](\mathbf{v}_{0:b}, \neg\varphi, \tau) := -[R](\mathbf{v}_{0:b}, \varphi, \tau)$$

$$[R](\mathbf{v}_{0:b}, \varphi_1 \wedge \varphi_2, \tau) := \min\left([R](\mathbf{v}_{0:b}, \varphi_1, \tau), [R](\mathbf{v}_{0:b}, \varphi_2, \tau)\right)$$

$$[R](\mathbf{v}_{0:b}, \Box_I \varphi, \tau) := \inf_{t \in \tau + I}\left([R](\mathbf{v}_{0:b}, \varphi, t)\right)$$

$$[R](\mathbf{v}_{0:b}, \varphi_1 \mathcal{U}_I \varphi_2, \tau) := \sup_{t \in \tau + I} \min\left([R](\mathbf{v}_{0:b}, \varphi_2, t), \inf_{t' \in [\tau, t)} [R](\mathbf{v}_{0:b}, \varphi_1, t')\right)$$

Here, f is defined as in Definition 3, and the arithmetic rules over intervals $I = [l, u]$ are defined as follows: $-I := [-u, -l]$ and $\min(I_1, I_2) := [\min(l_1, l_2), \min(u_1, u_2)]$.

We denote by $[R]^U(\mathbf{v}_{0:b}, \varphi, \tau)$ and $[R]^L(\mathbf{v}_{0:b}, \varphi, \tau)$ the upper and lower bounds of $[R](\mathbf{v}_{0:b}, \varphi, \tau)$ respectively. Intuitively, this interval $[R](\mathbf{v}_{0:b}, \varphi, \tau)$ represents the range of possible robustness values that can be reached by the completion of $\mathbf{v}_{0:b}$ with any suffix signal $\mathbf{v}_{b:T}$. This interval enables a 3-valued verdict for a given $\mathbf{v}_{0:b}$ in relation to the specification φ: if $[R]^L(\mathbf{v}_{0:b}, \varphi, \tau) > 0$, it implies that $\mathbf{v}_{0:b}$ satisfies φ (thus returning the verdict of **true**); if $[R]^U(\mathbf{v}_{0:b}, \varphi, \tau) < 0$, it implies $\mathbf{v}_{0:b}$ violates φ (thus returning the verdict of **false**); otherwise, it does not imply either of the cases, and so it returns **unknown**.

3.3   Overview of Causality Monitoring

The information masking issue of robust monitors (as described in Definition 5) has been identified as a serious problem in [11,13,15]. This problem stems from the inherent *monotonicity* of robust monitors, i.e., as the signal $\mathbf{v}_{0:b}$ evolves, $[R]^U(\mathbf{v}_{0:b}, \varphi, \tau)$ monotonically decreases and $[R]^L(\mathbf{v}_{0:b}, \varphi, \tau)$ monotonically increases. See [13] for a formal description of the problem.

Online Causality Monitoring. *Causality monitoring* is proposed in [13] to address the information masking issue. Specifically, instead of monitoring robustness that indicates whether a partial trace violates the specification, this approach assesses whether each incremental update to the trace can serve as a *cause* of the violation of the specification. To determine whether an update of a signal at a moment is a cause, we follow the *trace diagnostics* technique [2,15] that returns a (violation or satisfaction) *epoch*, i.e., a set of signal segments that *sufficiently* contribute to either the violation or satisfaction of the specification. Intuitively,

---

[1] $R(\mathbf{v}, \alpha, \tau)$ is bounded because of the bound Ω of $\mathbf{v}$. In practice, if Ω is unknown, we just need to set R_{max}^{α} and R_{min}^{α} to be ∞ and $-\infty$ respectively.

Table 1. The definitions of violation and satisfaction causation distances

$$[\mathscr{R}]^{\ominus}(\mathbf{v}_{0:b}, \alpha, \tau) := \begin{cases} f(\mathbf{v}_{0:b}(\tau)) & \text{if } b = \tau \\ \mathrm{R}^{\alpha}_{\max} & \text{otherwise} \end{cases} \qquad [\mathscr{R}]^{\ominus}(\mathbf{v}_{0:b}, \neg\varphi, \tau) := -[\mathscr{R}]^{\oplus}(\mathbf{v}_{0:b}, \varphi, \tau)$$

$$[\mathscr{R}]^{\ominus}(\mathbf{v}_{0:b}, \varphi_1 \wedge \varphi_2, \tau) := \min\left([\mathscr{R}]^{\ominus}(\mathbf{v}_{0:b}, \varphi_1, \tau), [\mathscr{R}]^{\ominus}(\mathbf{v}_{0:b}, \varphi_2, \tau)\right)$$

$$[\mathscr{R}]^{\ominus}(\mathbf{v}_{0:b}, \Box_I \varphi, \tau) := \inf_{t \in \tau + I}\left([\mathscr{R}]^{\ominus}(\mathbf{v}_{0:b}, \varphi, t)\right)$$

$$[\mathscr{R}]^{\ominus}(\mathbf{v}_{0:b}, \varphi_1 \, \mathcal{U}_I \, \varphi_2, \tau) := \inf_{t \in \tau + I}\left(\max\left(\min\left(\begin{matrix} \inf_{t' \in [\tau, t)} [\mathscr{R}]^{\ominus}(\mathbf{v}_{0:b}, \varphi_1, t') \\ [\mathscr{R}]^{\ominus}(\mathbf{v}_{0:b}, \varphi_2, t) \end{matrix}\right)\right)\right)$$
$$[\mathrm{R}]^{\mathsf{U}}(\mathbf{v}_{0:b}, \varphi_1 \, \mathcal{U}_I \, \varphi_2, \tau)$$

$$[\mathscr{R}]^{\oplus}(\mathbf{v}_{0:b}, \alpha, \tau) := \begin{cases} f(\mathbf{v}_{0:b}(\tau)) & \text{if } b = \tau \\ \mathrm{R}^{\alpha}_{\min} & \text{otherwise} \end{cases} \qquad [\mathscr{R}]^{\oplus}(\mathbf{v}_{0:b}, \neg\varphi, \tau) := -[\mathscr{R}]^{\ominus}(\mathbf{v}_{0:b}, \varphi, \tau)$$

$$[\mathscr{R}]^{\oplus}(\mathbf{v}_{0:b}, \varphi_1 \wedge \varphi_2, \tau) := \max\left(\begin{matrix} \min\left([\mathscr{R}]^{\oplus}(\mathbf{v}_{0:b}, \varphi_1, \tau), [\mathrm{R}]^{\mathsf{L}}(\mathbf{v}_{0:b}, \varphi_2, \tau)\right) \\ \min\left([\mathrm{R}]^{\mathsf{L}}(\mathbf{v}_{0:b}, \varphi_1, \tau), [\mathscr{R}]^{\oplus}(\mathbf{v}_{0:b}, \varphi_2, \tau)\right) \end{matrix}\right)$$

$$[\mathscr{R}]^{\oplus}(\mathbf{v}_{0:b}, \Box_I \varphi, \tau) := \sup_{t \in \tau + I}\left(\min\left([\mathscr{R}]^{\oplus}(\mathbf{v}_{0:b}, \varphi, t), [\mathrm{R}]^{\mathsf{L}}(\mathbf{v}_{0:b}, \Box_I \varphi, \tau)\right)\right)$$

$$[\mathscr{R}]^{\oplus}(\mathbf{v}_{0:b}, \varphi_1 \, \mathcal{U}_I \, \varphi_2, \tau) := \sup_{t \in \tau + I}\left(\max\left(\begin{matrix} \min\left(\begin{matrix} \sup_{t' \in [\tau, t)} [\mathscr{R}]^{\oplus}(\mathbf{v}_{0:b}, \varphi_1, t') \\ \inf_{t' \in [\tau, t)} [\mathrm{R}]^{\mathsf{L}}(\mathbf{v}_{0:b}, \varphi_1, t') \\ [\mathrm{R}]^{\mathsf{L}}(\mathbf{v}_{0:b}, \varphi_2, t) \end{matrix}\right) \\ \min\left(\begin{matrix} \inf_{t' \in [\tau, t)} [\mathrm{R}]^{\mathsf{L}}(\mathbf{v}_{0:b}, \varphi_1, t') \\ [\mathscr{R}]^{\oplus}(\mathbf{v}_{0:b}, \varphi_2, t) \end{matrix}\right) \end{matrix}\right)\right)$$

an epoch can be considered as an explanation of a violation or satisfaction; the formal definition of epoch can be found in [2, 15]. Given access to trace diagnostic results at each instant, causation monitoring classifies the current time step b of $\mathbf{v}_{0:b}$ as follows: if the current instant b of $\mathbf{v}_{0:b}$ is included in the violation epoch, it is considered as a *violation causation*; if b is included in the satisfaction epoch, it is considered as a *satisfaction causation*; otherwise, it is *irrelevant*.

The causality monitor proposed in [13] achieves this as follows: at each instant, it computes two quantitative metrics $[\mathscr{R}]^{\ominus}(\mathbf{v}_{0:b}, \varphi, \tau)$ and $[\mathscr{R}]^{\oplus}(\mathbf{v}_{0:b}, \varphi, \tau)$, that respectively represent the distances of the current instant b from being a violation causation and a satisfaction causation. The formal definition of causation distances $[\mathscr{R}]^{\ominus}(\mathbf{v}_{0:b}, \varphi, \tau)$ and $[\mathscr{R}]^{\oplus}(\mathbf{v}_{0:b}, \varphi, \tau)$ are presented in Definition 6.

Definition 6 (Online causality monitor [13]). Let $\mathbf{v}_{0:b}$ be a partial signal and φ be an STL formula. At an instant b, an online causality monitor returns a *violation causation distance* $[\mathscr{R}]^{\ominus}(\mathbf{v}_{0:b}, \varphi, \tau)$ and a *satisfaction causation distance* $[\mathscr{R}]^{\oplus}(\mathbf{v}_{0:b}, \varphi, \tau)$, as defined in Table 1.

The causation verdict, regarding whether b is a causation or not, can be inferred by the results of the online causality monitor in Definition 6, as follows:

- if $[\mathscr{R}]^{\ominus}(\mathbf{v}_{0:b}, \varphi, \tau) < 0$, then b is a violation causation;
- if $[\mathscr{R}]^{\oplus}(\mathbf{v}_{0:b}, \varphi, \tau) > 0$, then b is a satisfaction causation;
- otherwise, i.e., $[\mathscr{R}]^{\ominus}(\mathbf{v}_{0:b}, \varphi, \tau) > 0$ and $[\mathscr{R}]^{\oplus}(\mathbf{v}_{0:b}, \varphi, \tau) < 0$, b is irrelevant.

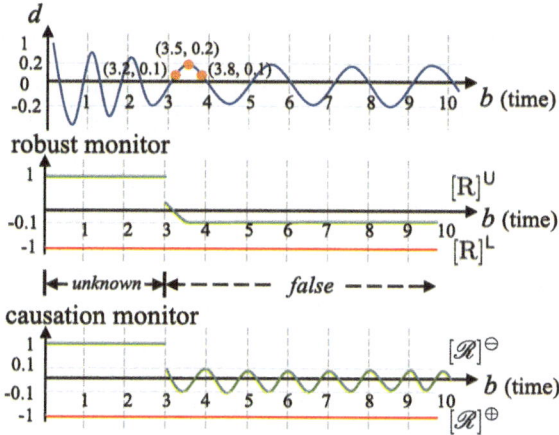

Fig. 6. Monitoring results for a trace of d and an STL formula φ : $\Box_{[0,10]}(\Diamond_{[0,1]}|d| < 0.1)$, by using robust monitoring and causality monitoring, respectively.

Below, we use an example to illustrate how online causality monitor works.

Example 2. We use the example in Fig. 6 to illustrate how causality monitoring works. According to the robust monitor, the specification is violated by the signal after $b = 3.2$. If we apply the trace diagnostics technique in [15] at the moment $b = 3.5$, we can find that the specification violation of the prefix of d till $b = 3.5$ is caused by the signal values of d during $[3.2, 3.5]$. Since $b = 3.5$ is included in this set, $b = 3.5$ should be considered as a causation of the violation. The causality monitor can directly report whether a moment is a causation of violation: by Fig. 6, we can see that the violation causation distance $[\mathscr{R}]^{\ominus}(\mathbf{v}_{0:b}, \varphi, 0) = -0.1 < 0$, which implies that $b = 3.5$ is indeed a violation causation.

Similarly, at $b = 4$, we obtain the causal interval for the violation of the specification, which is $[3.2, 3.8]$ that does not include $b = 4$, therefore, $b = 4$ is irrelevant to the violation. This is also shown by computing the causation distances $[\mathscr{R}]^{\ominus}(\mathbf{v}_{0:4}, \varphi, 0) = 0.1 > 0$ and $[\mathscr{R}]^{\oplus}(\mathbf{v}_{0:4}, \varphi, 0) = \mathrm{R}_{\min}^{\alpha} < 0$.

Relationship with Robust Monitors. By Example 2, we can see that the causality monitor is not monotonic, and does not suffer from the information masking problem. Lemma 1 states the relationship between causality monitoring and robust monitoring, i.e., for a given signal and specification, the monitoring results by using causality monitors in Definition 6 refine the results by using robust monitors in Definition 5. In other words, the information delivered by causality monitors is a superset of that can be delivered by classic robust monitors.

Lemma 1. The causality monitor in Definition 6 refines the classic online robust monitor in Definition 5, in the sense that the monitoring results of the robust monitor can be inferred from the results of the causation monitor, as follows:

$$[\mathrm{R}]^{\mathsf{U}}(\mathbf{v}_{0:b}, \varphi, \tau) = \inf_{t \in [0,b]} [\mathscr{R}]^{\ominus}(\mathbf{v}_{0:t}, \varphi, \tau), \ [\mathrm{R}]^{\mathsf{L}}(\mathbf{v}_{0:b}, \varphi, \tau) = \sup_{t \in [0,b]} [\mathscr{R}]^{\oplus}(\mathbf{v}_{0:t}, \varphi, \tau)$$

4   A Case Study

To demonstrate the modeling, simulation, and monitoring capabilities of MIMOS, we present a case study based on an autopilot for autonomous marine vessels.

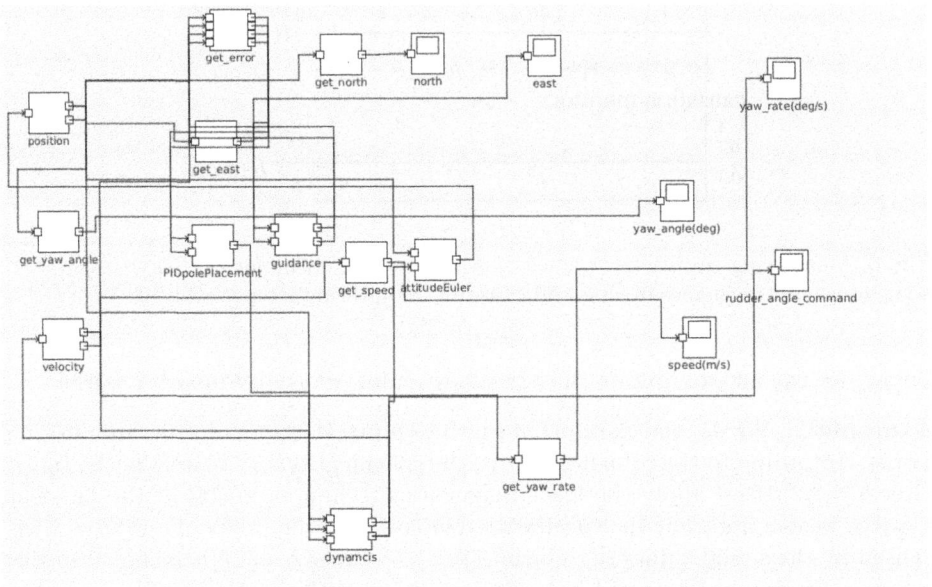

Fig. 7. Modeling a heading autopilot system in MIMOS.

4.1   A Heading Autopilot for Marine Vessels

An autopilot for autonomous marine vessels is a control system designed to regulate a vessel's position and orientation in three-dimensional space. This involves managing six degrees of freedom: three translational movements (surge, sway, heave) and three rotational movements (roll, pitch, yaw). While navigating a vessel from one point to another theoretically requires adjusting all six parameters, practical systems often prioritize specific aspects, such as position and heading, depending on the vessel and its operational needs. These autopilots are complex, integrating multiple control loops to account for dynamic and environmental conditions. In this study, we will examine the control system of a heading autopilot, which specifically adjusts the vessel's yaw angle, as detailed in [6].

Figure 7 illustrates the system model for the heading autopilot, designed to regulate the vessel's yaw angle through real-time Euler angle computations and yaw dynamics modeled using the Clarke approach. The vessel has specific parameters: a length of 70 m, a beam of 8 m, and a draft of 6 m. A PID controller, tuned using pole-placement, forms the system's core, adjusting the rudder to maintain the desired yaw angle. The system includes the following components:

- Guidance module: Generates smooth yaw angle and rate trajectories using a reference model.
- Position module: Extracts the current yaw angle from the vessel's state.
- Error calculation: Computes deviations between desired and actual yaw angle and rate.
- PID controller: Processes these errors to determine the rudder control input.
- Dynamics module: Simulates vessel motion using the Clarke model in response to rudder adjustments.
- Attitude module: Updates the Euler angles to reflect the vessel's orientation.

These components are interconnected, with feedback enabling continuous correction, forming a closed-loop control system that ensures precise and robust yaw control under varying dynamic and environmental conditions.

4.2   STL Specifications and Evaluation Results

Signal Temporal Logic (STL) is employed to formalize safety-critical requirements of the autopilot system. Specifically, the following STL specifications define the primary safety constraints:

$$\varphi_1 = \Box \left(|yaw\_rate(t)| < 5.0 \right)$$
$$\varphi_2 = \Box \left(|yaw\_angle(t)| < 12.0 \right)$$

Specification φ_1 ensures that the yaw rate remains within limits to prevent rapid rotational movements, mitigating severe rolling and protecting the crew, cargo, and vessel integrity. Similarly, specification φ_2 ensures that yaw angle deviations remain within acceptable boundaries, guaranteeing stable and controlled vessel maneuvering.

4.3   Simulation Results

Figure 8 presents comparative simulation results for two implementations of the autopilot system. The first implementation (top row, subfigures a-d) demonstrates stability with the yaw rate consistently staying below 5.0°/s, and the yaw angle remaining within ±12.0°. These results are reflected by the monitor outputs in Fig. 8c and 8d: in both subfigures, the violation causation distances (i.e., the yellow curves) are always positive, and the satisfaction causation distances (i.e., the orange curves) are negative constants (because the specifications are safety properties); these results deliver that the specifications are not violated during the simulation. This performance validates the PID controller's effectiveness in regulating rudder movements to achieve the target heading.

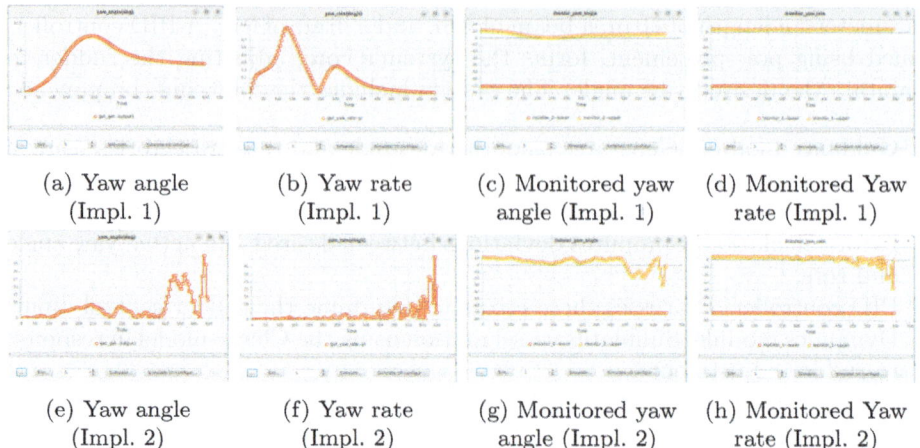

| (a) Yaw angle (Impl. 1) | (b) Yaw rate (Impl. 1) | (c) Monitored yaw angle (Impl. 1) | (d) Monitored Yaw rate (Impl. 1) |

| (e) Yaw angle (Impl. 2) | (f) Yaw rate (Impl. 2) | (g) Monitored yaw angle (Impl. 2) | (h) Monitored Yaw rate (Impl. 2) |

Fig. 8. Comparative simulation results for two implementations of the autopilot system. The top row (a-d) shows results for the first implementation, which maintains stability within safety thresholds. The bottom row (e-h) shows results for the second implementation, which exhibits pronounced oscillations and frequent violations of safety specifications.

In contrast, the second implementation (bottom row, subfigures e-h) exhibits pronounced oscillations and sharp transitions in both yaw angle and yaw rate. As demonstrated by the monitor outputs (see the yellow curves that depict the violation causation distances in Fig. 8g and 8h), this implementation frequently violates the safety thresholds specified by φ_1 and φ_2, indicating poor control stability that could compromise vessel safety and operational performance.

5   Conclusion

This paper presents the monitoring feature of MIMOS, an integrated tool environment for the design, analysis, and implementation of real-time systems. MIMOS is built on a deterministic and asynchronous data-flow model, supporting key functionalities such as modeling, simulation, verification, scheduling, timing analysis, and multi-core code generation. The tool enables a structured and modular approach to real-time system development, ensuring predictable behavior and facilitating dynamic updates without compromising system safety. We have introduced the causality monitoring plugin for Signal Temporal Logic, demonstrating its application within the MIMOS framework. A case study based on an autopilot control system for marine vessels is presented to illustrate the monitoring feature, which can be used to verify time-variant properties of MIMOS model at run-time.

References

1. Bartocci, E., Falcone, Y. (eds.): Lectures on Runtime Verification. LNCS, vol. 10457. Springer, Cham (2018). https://doi.org/10.1007/978-3-319-75632-5
2. Bartocci, E., Ferrère, T., Manjunath, N., Ničković, D.: Localizing faults in Simulink/Stateflow models with STL. In: Proceedings of the 21st International Conference on Hybrid Systems: Computation and Control (Part of CPS Week), HSCC 2018, pp. 197–206. Association for Computing Machinery, New York (2018). https://doi.org/10.1145/3178126.3178131
3. Deshmukh, J.V., Donzé, A., Ghosh, S., Jin, X., Juniwal, G., Seshia, S.A.: Robust online monitoring of signal temporal logic. Formal Methods Syst. Des. **51**(1), 5–30 (2017). https://doi.org/10.1007/s10703-017-0286-7
4. Donzé, A., Maler, O.: Robust satisfaction of temporal logic over real-valued signals. In: Chatterjee, K., Henzinger, T.A. (eds.) FORMATS 2010. LNCS, vol. 6246, pp. 92–106. Springer, Heidelberg (2010). https://doi.org/10.1007/978-3-642-15297-9_9
5. Fainekos, G.E., Pappas, G.J.: Robustness of temporal logic specifications for continuous-time signals. Theor. Comput. Sci. **410**(42), 4262–4291 (2009). https://doi.org/10.1016/j.tcs.2009.06.021
6. Fossen, T.I.: Handbook of Marine Craft Hydrodynamics and Motion Control. Wiley, Hoboken (2011)
7. Kahn, G.: The semantics of a simple language for parallel programming. In: Rosenfeld, J.L. (ed.) Information Processing, Proceedings of the 6th IFIP Congress 1974, Stockholm, Sweden, 5–10 August 1974, pp. 471–475. North-Holland (1974)
8. Liu, C.L., Layland, J.W.: Scheduling algorithms for multiprogramming in a hard-real-time environment. J. ACM (JACM) **20**(1), 46–61 (1973)
9. Maler, O., Nickovic, D.: Monitoring temporal properties of continuous signals. In: Lakhnech, Y., Yovine, S. (eds.) FORMATS/FTRTFT -2004. LNCS, vol. 3253, pp. 152–166. Springer, Heidelberg (2004). https://doi.org/10.1007/978-3-540-30206-3_12
10. Pnueli, A.: The temporal logic of programs. In: FOCS 1977, pp. 46–57. IEEE (1977). https://doi.org/10.1109/SFCS.1977.32
11. Selyunin, K., et al.: Runtime monitoring with recovery of the SENT communication protocol. In: Majumdar, R., Kunčak, V. (eds.) CAV 2017. LNCS, vol. 10426, pp. 336–355. Springer, Cham (2017). https://doi.org/10.1007/978-3-319-63387-9_17
12. Yi, W., Mohaqeqi, M., Graf, S.: MIMOS: a deterministic model for the design and update of real-time systems. In: ter Beek, M.H., Sirjani, M. (eds.) In Proceedings of 24th IFIP WG 6.1 International Conference on Coordination Models and Languages , COORDINATION 2022, Held as Part of the 17th International Federated Conference on Distributed Computing Techniques, DisCoTec 2022, Lucca, Italy, 13–17 June 2022. Lecture Notes in Computer Science, vol. 13271, pp. 17–34. Springer, Cham (2022). https://doi.org/10.1007/978-3-031-08143-9_2
13. Zhang, Z., An, J., Arcaini, P., Hasuo, I.: Online causation monitoring of signal temporal logic. In: Enea, C., Lal, A. (eds.) 35th International Conference on Computer Aided Verification, CAV 2023. Lecture Notes in Computer Science, vol. 13964, pp. 62–84. Springer, Cham (2023). https://doi.org/10.1007/978-3-031-37706-8_4

14. Zhang, Z., An, J., Arcaini, P., Hasuo, I.: Caumon: an informative online monitor for signal temporal logic. In: Platzer, A., Rozier, K.Y., Pradella, M., Rossi, M. (eds.) Formal Methods - 26th International Symposium, FM 2024, Milan, Italy, 9–13 September 2024, Proceedings, Part II. Lecture Notes in Computer Science, vol. 14934, pp. 286–304. Springer, Cham (2024). https://doi.org/10.1007/978-3-031-71177-0_18
15. Zhang, Z., Arcaini, P., Xie, X.: Online reset for signal temporal logic monitoring. IEEE Trans. Comput. Aided Des. Integr. Circuits Syst. **41**(11), 4421–4432 (2022). https://doi.org/10.1109/TCAD.2022.3197693

Beyond Concurrent Separation Logic: Who is Afraid of Completeness Proofs?

Frank S. de Boer$^{(\boxtimes)}$ and Hans-Dieter A. Hiep

Leiden Institute of Advanced Computer Science (LIACS), Leiden, The Netherlands
{f.s.de.boer,hdh}@cwi.nl

Abstract. This paper introduces a general semantic approach to the definition and formal justification (soundness and completeness) of the Owicki/Gries method for proving correctness of parallel programs.

Instantiating the general semantic approach allows to obtain sound and complete proof methods for both the standard model of shared variable concurrency and the concurrency model based on dynamically allocated variables, as described by separation logic.

1 Introduction

Christel Baier and Joost-Pieter Katoen provide a comprehensive overview of model-checking shared variable concurrency in [3]. An overview of techniques for combining deductive and model-checking methods to achieve greater scale, expressivity, and automation is presented in [15]. Here we discuss the seminal Owicki/Gries method [11] which extends Hoare logics for sequential programs to parallel programs that interact via shared variables. Following [3], Chap. 2 on modeling concurrent systems, we describe the semantics of parallel processes interacting via shared variables using the general format of transition systems. The general semantic analysis of the Owicki/Gries method, presented in this paper, provides a solid basis for further investigation into combining deductive and model-checking methods for the analysis of parallel systems interacting via different data-structures, for example, the heap structures in programming languages like Java.

This paper can also be read as a response to

"It is all too easy to get caught up in completeness and related issues for formal systems that turn out to be too complicated when humans try to apply them; it is more important first to get a sense for the extent to which simple reasoning is or is not supported."

as stated in [6] by the 2016 Gödel prize winners, Brookes and O'Hearn, for their invention of Concurrent Separation Logic.

Krzysztof Apt in [2] proved completeness of the Owicki/Gries method for parallel programs that compute over the integers. The basic idea of this proof is the construction of interference-free proof-outlines using *reachability* predicates. These predicates are used as annotations of those locations of the individual

N. Bertrand et al. (Eds.): Christel Baier Festschrift, LNCS 15760, pp. 301–319, 2026.
https://doi.org/10.1007/978-3-031-97439-7_15

sequential components which involve interleaving points. They describe the set of states that are reachable by a sequence of computation steps of the parallel program such that control (i.e., the *progam counter*) of the given component is at the associated location.

Model checking typically consists of an algorithm for computing these predicates as sets of states. In completeness proofs of program logics one typically has to show that these reachability predicates can be expressed in the logic (see for example [10] on *arithmetic* completeness for program logics). In general this requires a straightforward though tedious application of coding techniques.

The essential part of the completeness proof in [2] is then to define reachability predicates which are interference-free. Interference-free local proof-outlines require the introduction of *auxiliary* variables (in [2] this is formally proven for the correctness specification $\{x = 0\}\ x := x + 1 \parallel x := x + 1\ \{x = 2\}$). In the general case we need auxiliary variables which ensure the following: if a state can be reached by both a sequence of computation steps such that component C is at location l and a sequence of computation steps such that a (different) component C' is at location l' then this state can be reached by a *single* sequence of computation steps of the parallel program such that components C and C' are at location l and l', respectively. The basic observation underlying the completeness proof of [2] is that auxiliary variables which record the *state-changes* and as such fix the interleaving, trivially, ensure this property. Of course, this requires working out some details, but the basic idea is simple (nothing to be 'caught up in').

Instead of customizing the completeness proof of [2] to the application of separation logic to parallel programs that interact via *dynamically allocated variables*, we generalize the Owicki/Gries method by abstracting form the internal structure of states and underlying logics used to describe state properties. Given an abstract set of states, we abstract in the programming language from the syntax of the basic actions, that is, assignments and tests, which are directly represented by their semantics: Assignments are (partial) state transformations and tests are sets of states. Similarly, we use sets of states to annotate programs. One of the main technical challenges in this generalization is a semantic definition of the *auxiliary variables rule*. In the standard Owicki/Gries method this rule is formulated in terms of a syntactic condition which ensures that assignments to the auxiliary variables do not affect the flow of control of the given parallel program. We capture this condition semantically by a program-specific *congruence* relation on the abstract set of states.

Thus abstracting from the structure of states (and the underlying data structures) and any logics used to describe sets of states, allows to highlight the overall structure of the soundness and completeness proofs. Moreover, by the very nature of this generalization the proof method applies to any instantiation of the abstract states and corresponding logics.

In general, one can obtain a sound and complete instantiation by the specification of the underlying data structures and corresponding logics which allow for the formalization of

- the *weakest preconditions* of the basic actions, and
- the *reachability predicates* for programs.

For example, *separation logic* [14], which has been developed for modular reasoning about heap structures, satisfies these two requirements. As such, one can simply instantiate the general semantic definition of the Owicki/Gries method to obtain a sound and complete extension of the Owicki/Gries method based on separation logic.

The resulting proof method allows for arbitrary, non-deterministic interleaving of the (sequential) components which interact directly via a shared heap. In contrast, existing extensions of separation logic to parallel programs focus on restricting this interleaving to avoid *race conditions*. For example, in [4–6, 13] the parallel processes are required to operate only on *disjoint* local parts of the heap. Access to global heap locations is controlled by ownership of *resources* and corresponding conditional critical regions. This approach is an extension of the version of Owicki/Gries which restricts the use of shared variables, already anticipated in [12]. However, a simple *lock-free* example of a producer and consumer that interact via the heap (as described in Sect. 5, Fig. 1) cannot be proven correct in this version of concurrent separation logic without adaptation. It would require additional synchronization (with a global lock) between the producer and consumer. In general such additional synchronization decreases the performance (which benefits from loosely coupled process execution). The versions of concurrent separation logic described in [1, 13] uses *permissions* (in [13] additional to the use of resources) to restrict the interleaving by exclusive write access. However, it is worthwhile to emphasize here that the interference freedom test of the Owicki/Gries method itself already ensures that the scheduling of the parallel processes (and the resulting interleavings) does not affect its correctness (as specified by the global pre- and post-condition).

Further, we show in this paper how to instantiate the abstract Owicki/Gries method to the basic case of simple shared-variables itself and dynamic thread creation.

Completeness proofs, in our opinion, provide not only theoretical evidence that the proof method fully captures program correctness, but it also provides a formal basis for a general heuristic in proving program correctness by suitable abstractions (that is, abstractions of the auxiliary variables used in the completeness proof). Also it constrains the development of a plethora of ad-hoc separation logics (see for example [7]).

2   Abstract Programs and Specifications

Both abstract programs and correctness specifications are defined with respect to

- a set Σ of *states* with typical element σ,
- *assignments* which are modeled as partial functions $\Sigma \to \Sigma_\perp$ ($\Sigma_\perp = \Sigma \cup \{\perp\}$), with typical element a (note that $a(\sigma) = \perp$ indicates that $a(\sigma)$ is undefined),
- *predicates* which are modeled as sets of states, with typical element $p, q, \dots$.

A failing execution of an assignment a in a state σ is indicated by $a(\sigma) = \bot$. A (side-effect free) Boolean condition to be used in programs is denoted by $b \subseteq \Sigma$. *Statements* define the sequential flow of control and are specified by the following grammar.

$$
\begin{array}{lll}
S ::= & a & \text{(assignment)} \\
\mid & S_1; S_2 & \text{(sequential composition)} \\
\mid & \textbf{if } b\{S_1\}\textbf{else}\{S_2\} & \text{(choice)} \\
\mid & \textbf{while } b\{S\} & \text{(iteration)} \\
\mid & \textbf{await } b\{S\} & \text{(synchronization)}
\end{array}
$$

On the top-level we have a parallel composition $S_1 \parallel \cdots \parallel S_n$ where S_i are the (sequential) components. Such programs are typically denoted by $\pi, Q, \ldots$.

The local semantics of statements is described in terms of a small-step transition relation $\Rightarrow$ between configurations $\langle S, \sigma \rangle$: successful termination is indicated by the configuration $\langle \checkmark, \sigma \rangle$, and the symbol $\bot$ is also used in the place of a configuration to indicate failure.

Execution of assignments is described by the following transitions.

$$
\begin{array}{ll}
\langle a, \sigma \rangle \Rightarrow \langle \checkmark, a(\sigma) \rangle & \text{if } a(\sigma) \neq \bot \\
\langle a, \sigma \rangle \Rightarrow \bot & \text{otherwise}
\end{array}
$$

The following transition describes the execution of an await-statement: its body is executed atomically.

$$
\frac{\sigma \in b \text{ and } \langle S, \sigma \rangle \Rightarrow^* \langle \checkmark, \sigma' \rangle}{\langle \textbf{await } b\{S\}, \sigma \rangle \Rightarrow \langle \checkmark, \sigma' \rangle}
$$

Here $\Rightarrow^*$ denotes the reflexive, transitive closure of $\Rightarrow$. In case $\sigma \notin b$ execution of the await-statement suspends. The transitions for the other compound statements are standard (and therefore omitted). The global interleaving of parallel components is described by the following two rules.

$$
\frac{\langle S_i, \sigma \rangle \Rightarrow \langle S_i', \sigma' \rangle}{\langle S_1 \parallel \cdots \parallel S_i \parallel \cdots \parallel S_n, \sigma \rangle \Rightarrow \langle S_1 \parallel \cdots \parallel S_i' \parallel \cdots \parallel S_n, \sigma' \rangle}
$$

$$
\frac{\langle S_i, \sigma \rangle \Rightarrow \bot}{\langle S_1 \parallel \cdots \parallel S_i \parallel \cdots \parallel S_n, \sigma \rangle \Rightarrow \bot}
$$

Definition 1 (Semantics abstract programs and specifications). *The program semantics* $\mathcal{M}(\pi) : \Sigma \to \mathcal{P}(\Sigma_\bot)$ *is defined by*

$$
\mathcal{M}(\pi)(\sigma) = \{\sigma' \mid \langle \pi, \sigma \rangle \Rightarrow^* \langle \checkmark, \sigma' \rangle\} \cup \{\bot \mid \langle \pi, \sigma \rangle \Rightarrow^* \bot\}.
$$

Further, we identify the parallel composition $\checkmark \parallel \cdots \parallel \checkmark$ *of terminating statements with* $\checkmark$ *itself. A correctness specification is of the form* $\Sigma : \{p\}\ \pi\ \{q\}$, *where the 'type' Σ indicates that the predicates p and q, and program π are defined in terms of the state-space Σ. For notational convenience however, reference to the state-space is omitted in those cases where it is not relevant, or the context provides the information.*

The semantics of abstract specifications is defined by $\models \{p\} \; \pi \; \{q\}$ if and only if $\sigma \in p$ implies $\perp \notin \mathcal{M}(\pi)(\sigma)$ and $\sigma' \in q$, for all σ and $\sigma' \in \mathcal{M}(\pi)(\sigma)$.

Definition 2 (π-congruence). *Given a program π with state-space Σ, a π-congruence relation $\cong \; \subseteq \Sigma_\perp \times \Sigma_\perp$ satisfies*

- $\sigma \not\cong \perp$, *for any* $\sigma \in \Sigma$,
- $a(\sigma) \cong a(\sigma')$, *for any assignment* $a : \Sigma \to \Sigma_\perp$ *occurring in* π *and* $\sigma \cong \sigma'$,
- $\sigma \in b$ *if and only if* $\sigma' \in b$, *for any Boolean condition* $b \subseteq \Sigma$ *occurring in* π *and* $\sigma \cong \sigma'$.

The two conditions state that $\cong$ is a congruence relation with respect to the assignments $a : \Sigma \to \Sigma_\perp$ and Boolean conditions $b \subseteq \Sigma$ of π. The existential closure $\exists_\cong p$ of p is defined by $\sigma \in \exists_\cong p$ if and only if there exists $\sigma' \cong \sigma$ such that $\sigma' \in p$. A predicate $p \subseteq \Sigma$ is closed (under $\cong$) if $(\exists_\cong p) = p$, that is, if $\sigma \in p$ and $\sigma' \cong \sigma$ implies $\sigma' \in p$ (for any σ and σ').

Next we specify some basic semantics properties of (abstract) correctness specifications.

Proposition 1 (Isomorphism). *Let $f : \Sigma_\perp \to \Sigma'_\perp$ be a strict bijection (that is, $f(\perp) = \perp$) such that for any action $a : \Sigma \to \Sigma_\perp$, $f(a) : \Sigma' \to \Sigma'_\perp$ is defined by $f(a)(\sigma) = f(a(\sigma))$, and $f(p) \subseteq \Sigma'$, for any predicate $p \subseteq \Sigma$, is defined by $f(\sigma) \in f(p)$ if and only if $\sigma \in p$.*

Then $\models \Sigma : \{p\} \; \pi \; \{q\}$ if and only if $\models \Sigma' : \{f(p)\} \; f(\pi) \; \{f(q)\}$, where $f(\pi)$ transforms every assignment a and Boolean condition b of π to $f(a)$ and $f(b)$, respectively.

Proposition 2 (Congruence). *Given a π-congruence relation $\cong \; \subseteq \Sigma_\perp \times \Sigma_\perp$, let $\Sigma/\cong$ denote the set of equivalence classes $[\sigma]_\cong = \{\sigma' \mid \sigma' \cong \sigma\}$. Further, let $a_\cong([\sigma]) = [a(\sigma)]_\cong$, for any assignment $a : \Sigma \to \Sigma_\perp$ of π, and $r_\cong = \{[\sigma]_\cong : \sigma \in r\}$, for any $r \subseteq \Sigma$.*

Then for any predicates p and q which are closed (under $\cong$) it is the case that $\models \Sigma : \{p\} \; \pi \; \{q\}$ if and only if $\models \Sigma/\cong \; : \{p_\cong\} \; \pi_\cong \; \{q_\cong\}$, where $\pi_\cong$ results from replacing every assignment a and Boolean condition b of π by $a_\cong$ and $b_\cong$, respectively.

The above two propositions allows to identify isomorphic and congruent correctness specifications.

3 The Abstract Owicki/Gries Method

Let $WP(a,q) = \{\sigma \mid a(\sigma) \in q\}$ denote the weakest precondition of the assignment a with respect to post-condition q. Note that if $\sigma \in WP(a,q)$ then $a(\sigma) \neq \perp$ (because $q \subseteq \Sigma$).

In the following definition and throughout the paper, the symbol $\equiv$ is used to denote *syntactical* identity between statements. An occurrence of a sub-statement S of one of the components of a parallel program π represents an *interleaving point* if it does not occur as a sub-statement of an await-statement.

Definition 3 (Local proof-outlines). *A local proof-outline $PO(S)$ associates with each interleaving point R of S a precondition $pre(R)$ and a postcondition $post(R)$ such that*

- $pre(R) \subseteq WP(a, post(R))$, *for assignment a,*
- $pre(R) \subseteq pre(R_1)$, $post(R_1) \subseteq pre(R_2)$ *and* $post(R_2) \subseteq post(R)$, *where* $R \equiv R_1; R_2$
- $(pre(R) \cap b) \subseteq pre(R_1)$, $(pre(R) \cap \neg b) \subseteq pre(R_2)$ *and* $post(R_i) \subseteq post(R)$ *for* $i = 1, 2$, *where* $R \equiv$ **if** $b\{R_1\}$**else**$\{R_2\}$,
- $(pre(R) \cap b) \subseteq pre(R')$, $(pre(R) \cap \neg b) \subseteq post(R)$ *and* $post(R') \subseteq post(R)$, *where* $R \equiv$ **while** $b\{R'\}$,
- $(pre(R) \cap b) \subseteq pre(R')$ *and* $post(R') \subseteq post(R)$, *where* $R \equiv$ **await** $b\{R'\}$.

The existence of a proof-outline $PO(S)$ such that $p \subseteq pre(S)$ and $post(S) \subseteq q$ is denoted by $\vdash \{p\}\ S\ \{q\}$.

Definition 4 (Interference freedom test). *The proof-outlines $PO(S_i)$, $i = 1, \ldots, n$, are interference free, denoted by $IFT(PO(S_1), \ldots, PO(S_n))$, if for any interleaving point S of S_i and assignment or await-statement S' of S_j, $i \neq j$, it is the case that*

- $(p \cap pre(S')) \subseteq WP(S', p)$, *in case S' denotes an assignment,*
- $\vdash\ \{p \cap pre(S') \cap b\}\ R\ \{p\}$, *in case S' denotes an await-statement* **await** $b\{R\}$.

Here p denotes $pre(S)$ or $post(S)$.

The following rule allows to derive a program specification from interference free local proof-outlines of its components.

Parallel Composition Rule

$$\frac{IFT(PO(S_1), \ldots, PO(S_n))}{\{\bigcap_{i=1}^{n} pre(S_i)\}\ S_1 \parallel \cdots \parallel S_n\ \{\bigcap_{i=1}^{n} post(S_i)\}}$$

The following rule shows how to remove auxiliary assignments from a program specification.

Auxiliary Assignments Rule. Given a program π with state-space Σ and a π-congruence relation $\cong$, an *auxiliary* assignment $a : \Sigma \to \Sigma$ (note that a is thus required to be a total function) satisfies $a(\sigma) \cong \sigma$, for any $\sigma \in \Sigma$. Let π be obtained from π' by deleting all auxiliary assignments, and subsequently replacing every sub-statement **await true**$\{a\}$ by a.

Then

$$\frac{\{p\}\ \pi'\ \{q\}}{\{\exists_{\cong}p\}\ \pi\ \{\exists_{\cong}q\}}$$

The following consequence rule allows, as usual, to strengthen the precondition and to weaken the postcondition, semantically.

Consequence rule

$$\frac{p \subseteq p'\quad \{p'\}\ \pi\ \{q'\}\quad q' \subseteq q}{\{p\}\ \pi\ \{q\}}$$

By $\vdash \{p\}\ \pi\ \{q\}$ is denoted that $\{p\}\ \pi\ \{q\}$ is derivable by the parallel composition rule, the auxiliary assignments rule, and the consequence rule.

3.1   Soundness

Let $after(R, S)$ denote the remainder of S to be executed just after the execution of the sub-statement R of S has terminated. It is defined by a straightforward induction on the structure of S. The following lemma implies soundness of the parallel composition rule.

Lemma 1. *Let* $IFT(PO(S_1), \ldots, PO(S_n))$. *Then for any execution*

$$\langle S_1 \parallel \cdots \parallel S_n, \sigma \rangle \Rightarrow^* \langle S_1' \parallel \cdots \parallel S_n', \sigma' \rangle$$

such that $\sigma \in \bigcap_{i=1}^{n} pre(S_i)$, *it follows that*

- $\sigma' \in pre(R_i)$, *for* $S_i' \equiv R_i$; $after(R_i, S_i)$,
- $\sigma' \in post(R_i)$, *for* $S_i' \equiv after(R_i, S_i)$.

The proof proceeds by a straightforward, though tedious, induction on the length of the execution. The following lemma, which establishes a π-congruence as a simulation relation, implies soundness of the auxiliary assignments rule.

Lemma 2. *Let* π_0 *be a program and* $\cong$ *be a* π_0-*congruence relation. Let* π_0 *be obtained from* π_0' *by deleting all auxiliary assignments, and subsequently replacing every sub-statement* **await true**$\{a\}$ *by* a. *Then for every execution*

$$\langle \pi_0, \sigma_0 \rangle \Rightarrow^* \langle \pi_1, \sigma_1 \rangle$$

and state $\sigma_0' \cong \sigma_0$ *there exists a execution*

$$\langle \pi_0', \sigma_0' \rangle \Rightarrow^* \langle \pi_1', \sigma_1' \rangle$$

such that $\sigma_1' \cong \sigma_1$ *and* π_1 *can be obtained from* π_1' *by deleting all auxiliary assignments, and subsequently replacing every sub-statement* **await true**$\{a\}$ *by* a.

3.2   Completeness

Let π denote the parallel program $S_1 \parallel \cdots \parallel S_n$, defined with respect to state-space Σ. With each assignment that does not occur in an await-statement, and with each await-statement of π, we associate a unique label l, and introduce the set of extended states $L^* \times \Sigma$, where L denotes the set of labels of the given program π. The semantics of π is lifted to this extended set of states as follows (thus 'overloading' the notation of both assignments and Boolean conditions). Let a be an assignment of π. Then

$$a(\langle \tau, \sigma \rangle) = \begin{cases} \langle \tau, a(\sigma) \rangle & \text{if } a(\sigma) \neq \bot \\ \bot & \text{otherwise} \end{cases}$$

Any Boolean condition b of π is trivially extended as follows: $\langle \tau, \sigma \rangle \in b$ if and only if $\sigma \in b$.

Let a_l be defined by $a_l(\langle \tau, \sigma \rangle) = \langle \tau \cdot l, \sigma \rangle$ (here $\tau \cdot l$ denotes the result of appending l to τ). Next let $\pi' = S_1' \parallel \cdots \parallel S_n'$ denote the result of replacing

- every assignment a of π which does not occur in an await-statement by **await true**$\{a_l; a\}$, and
- every await-statement **await** $b\{R\}$ by **await** $b\{a_l; R\}$

where l denotes the label associated with the assignment and await-statement, respectively.

The generated sequence of labels thus records the scheduling of assignments and await-statements. It abstracts from the scheduling of the Boolean conditions. It is worthwhile to observe here that in fact the programming language does not allow to record the scheduling of the Boolean conditions. The following lemma states that the sequence of labels determines the result of a execution.

Lemma 3 (Determinism). *It is the case that*

$$\langle \pi', \langle \epsilon, \sigma \rangle \rangle \Rightarrow^* \langle Q_1, \langle \tau, \sigma_1 \rangle \rangle \quad and \quad \langle \pi', \langle \epsilon, \sigma \rangle \rangle \Rightarrow^* \langle Q_2, \langle \tau, \sigma_2 \rangle \rangle$$

implies $\sigma_1 = \sigma_2$.

Proof. The proof proceeds by a straightforward induction on the length of τ. The case that τ is empty is trivial because then both global executions consist only of the (side-effect free) evaluation of Boolean conditions which do not affect the initial state σ. Otherwise, there exists a label l such that the two global executions can decomposed as follows ($i = 1, 2$):

$$\langle \pi', \langle \epsilon, \sigma \rangle \rangle \Rightarrow^* \langle Q_i', \langle \tau, \sigma_i' \rangle \rangle \Rightarrow \langle Q_i'', \langle \tau \cdot l, \sigma_i \rangle \rangle \Rightarrow^* \langle Q_i, \langle \tau \cdot l, \sigma_i \rangle \rangle$$

where the intermediate configuration $\langle Q_i'', \langle \tau \cdot l, \sigma_i \rangle \rangle$ results in a single global execution step from the intermediate configuration $\langle Q_i', \langle \tau, \sigma_i' \rangle \rangle$ by the execution of the statement associated with l. Applying the induction hypothesis gives $\sigma_1' = \sigma_2'$, and so also $\sigma_1 = \sigma_2$. Since subsequent execution only involves the (side-effect free) evaluation of Boolean conditions, it is the case that $\sigma_1 = \sigma_2$.

It follows that the recorded sequence of labels fixes a global execution up to interleaving of the evaluation of the Boolean conditions. This allows to merge the executions of the individual components in the following sense.

Corollary 1 (Merging executions). *Let* $\langle \pi', \langle \epsilon, \sigma \rangle \rangle \Rightarrow^* \langle Q_i, \langle \tau, \sigma' \rangle \rangle$ *for* $i = 1, \ldots, n$. *Then* $\langle \pi', \langle \epsilon, \sigma \rangle \rangle \Rightarrow^* \langle Q, \langle \tau, \sigma' \rangle \rangle$ *where* Q *is the parallel program constructed from taking taking the i-th component of each Q_i for $i = 1, \ldots, n$.*

Proof. The proof follows closely that of the above lemma. As above, the proof proceeds by induction on the length of τ. Let R_i denote the i-th component of Q_i, for $i = 1, \ldots, n$. First, let τ be empty. It follows that, for $i = 1, \ldots, n$, the local execution

$$\langle S'_i, \langle \epsilon, \sigma \rangle \rangle \Rightarrow^* \langle R_i, \langle \epsilon, \sigma \rangle \rangle$$

consists only of the (side-effect free) evaluation of Boolean conditions which does not affect the initial state σ. Thus we obtain

$$\langle \pi', \langle \epsilon, \sigma \rangle \rangle \Rightarrow^* \langle Q, \langle \epsilon, \sigma \rangle \rangle$$

by an arbitrary interleaving of the local executions.

Otherwise, there exists a label l such that we can decompose the global executions $(i = 1, \ldots, n)$

$$\langle \pi', \langle \epsilon, \sigma \rangle \rangle \Rightarrow^* \langle Q'_i, \langle \tau, \sigma_i \rangle \rangle \Rightarrow \langle Q''_i, \langle \tau \cdot l, \sigma' \rangle \rangle \Rightarrow^* \langle Q_i, \langle \tau \cdot l, \sigma' \rangle \rangle$$

where the intermediate configuration $\langle Q''_i, \langle \tau \cdot l, \sigma' \rangle \rangle$ results from a single global execution step from the intermediate configuration $\langle Q'_i, \langle \tau, \sigma_i \rangle \rangle$ by the execution of the statement associated with l. The subsequent execution steps only involves the evaluation of Boolean conditions. By the above lemma it follows that $\sigma' = \sigma_i$, for some σ' and $i = 1, \ldots, n$. By the induction hypothesis it then follows that there exists a global execution

$$\langle \pi', \langle \epsilon, \sigma \rangle \rangle \Rightarrow^* \langle Q', \langle \tau, \sigma' \rangle \rangle$$

where the i-th component of Q' equals that of Q'_i, for $i = 1, \ldots, n$. Executing the l-labeled statement and subsequently arbitrarily interleaving the steps of the local executions (which only involve the evaluation of Boolean conditions), we obtain the desired result.

Next we introduce the set of extended states $\Sigma \times L^* \times \Sigma$ in the definition below of the abstract reachability predicates. For notational convenience we denote by $Q(i)$ the i-th component of the parallel composition Q.

Definition 5 (Abstract reachability predicates). *Let* $p \subseteq \Sigma$ *and* R *be an interleaving point of the i-th component of π'. Then*

- $\langle \sigma, \tau, \sigma' \rangle \in pre(R)$ *if and only if* $\sigma \in p$ *and* $\langle \pi', \langle \epsilon, \sigma \rangle \rangle \Rightarrow^* \langle Q, \langle \tau, \sigma' \rangle \rangle$, *for some* Q *such that* $Q(i) \equiv R; after(R, S'_i)$;
- $\langle \sigma, \tau, \sigma' \rangle \in post(R)$ *if and only if* $\sigma \in p$ *and* $\langle \pi', \langle \epsilon, \sigma \rangle \rangle \Rightarrow^* \langle Q, \langle \tau, \sigma' \rangle \rangle$, *for some* Q *such that* $Q(i) \equiv after(R, S'_i)$.

Note that the reachability predicate is expressed at the level of the extended states $\Sigma \times L^* \times \Sigma$, using the semantics of the program π' at the level of the extended states $L^* \times \Sigma$. To use the reachability predicates in the following theorem, we trivially lift the semantics of both π and π' to the extended state-space $\Sigma \times L^* \times \Sigma$ (again, 'overloading' the notation for assignments and Boolean conditions):

$$a_l(\langle \sigma, \tau, \sigma' \rangle) = \langle \sigma, \tau \cdot l, \sigma' \rangle$$

and

$$a(\langle \sigma, \tau, \sigma' \rangle) = \begin{cases} \langle \sigma, \tau, a(\sigma') \rangle & \text{if } a(\sigma') \neq \bot \\ \bot & \text{otherwise} \end{cases}$$

for any assignment a of π. For any Boolean condition b of π we define $\langle \sigma, \tau, \sigma' \rangle \in b$ if and only if $\sigma' \in b$.

Theorem 1 (Abstract completeness). $\models \Sigma : \{p\} \pi \{q\}$ *implies* $\vdash \Sigma : \{p\} \pi \{q\}$.

Proof. Let π' be defined as above. We lift the correctness specification $\{p\} \pi \{q\}$ to the extended state-space $\Sigma \times L^* \times \Sigma$, with $\langle \sigma, \tau, \sigma' \rangle \in p$ if and only if $\sigma \in p$ (and similarly for q). It is trivial to check that the relation $\langle \sigma_0, \tau, \sigma \rangle \cong \langle \sigma_0', \tau', \sigma' \rangle$ if and only if $\sigma = \sigma'$ is a π-congruence relation. By the above two Propositions 1 and 2 (note that both p and q are closed under $\cong$), it thus suffices to prove the extended interpretation of $\{p\} \pi \{q\}$ with respect to $\Sigma \times L^* \times \Sigma$.

It is straightforward, though tedious, to check that the annotation of the statements S_i', $i = 1, \ldots, n$, with the above abstract reachability predicates satisfies the verification conditions of a proof-outline. In order to apply the parallel composition rule to obtain

$$\{\bigcap_{i=1}^{n} pre(S_i')\} \pi' \{\bigcap_{i=1}^{n} post(S_i')\} \tag{1}$$

the proof-outlines $PO(S_i')$ must be interference-free:

- $\models \{pre(R) \cap pre(R') \cap b\} R' \{pre(R)\}$,
- $\models \{post(R) \cap pre(R) \cap b\} R' \{post(R)\}$,

for any interleaving point R of S_i', assignment or await-statement R' of S_j' ($i \neq j$), where b denotes **true** in case R' denotes an assignment, otherwise it denotes the Boolean condition of the await-statement R'. We prove

$$\models \{pre(R) \cap pre(a)\} a \{pre(R)\}$$

(the other cases are treated similarly). Let $\langle \sigma, \tau, \sigma' \rangle \in pre(R) \cap pre(a)$. Next observe that $\langle \sigma, \tau \cdot l, a(\sigma') \rangle \in pre(R)$, where l denotes the label uniquely associated with the assignment a (as a sub-statement of S_j'). This can be shown as follows. By definition of the reachability predicates there exist executions

$$\langle \pi', \langle \epsilon, \sigma \rangle \rangle \Rightarrow^* \langle Q_1, \langle \tau, \sigma' \rangle \rangle \quad \text{and} \quad \langle \pi', \langle \epsilon, \sigma \rangle \rangle \Rightarrow^* \langle Q_2, \langle \tau, \sigma' \rangle \rangle$$

such that $\sigma \in p$, $Q_1(i) = R; after(R, S_i')$ and $Q_2(j) = a; after(R, S_j')$. By the above corollary there exist a single execution

$$\langle \pi', \langle \epsilon, \sigma \rangle \rangle \Rightarrow^* \langle Q, \langle \tau, \sigma' \rangle \rangle$$

such that $Q(i) \equiv R; after(R, S_i')$ and $Q(j) \equiv a; after(R, S_j')$. By construction of π', $\langle \pi', \langle \epsilon, \sigma \rangle \rangle \Rightarrow^* \langle Q, \langle \tau, \sigma' \rangle \rangle$ implies $\langle \pi, \sigma \rangle \Rightarrow^* \langle Q, \sigma' \rangle$, and so $\models \{p\} \pi \{q\}$ implies that we can execute a. Executing assignment a then gives

$$\langle \pi', \langle \epsilon, \sigma \rangle \rangle \Rightarrow^* \langle Q', \langle \tau \cdot l, a(\sigma') \rangle \rangle$$

such that $Q'(i) = R; after(R, S_i')$ and $Q'(j) = after(R, S_j')$. By definition of $pre(R)$, it follows that $\langle \sigma, \tau \cdot l, a(\sigma') \rangle \in pre(R)$.

Next, apply the (abstract) auxiliary assignments rule to the correctness specification (1), using the above defined congruence relation $\cong$. We thus obtain by the auxiliary assignments rule

$$\{\exists_{\cong}(\bigcap_{i=1}^{n} pre(S_i'))\} \ \pi \ \{\exists_{\cong}(\bigcap_{i=1}^{n} post(S_i'))\}.$$

In order to apply next the consequence rule to derive $\{p\} \pi \{q\}$, defined with respect to $\Sigma \times L^* \times \Sigma$, we show that:

- $p \subseteq \exists_{\cong}(\bigcap_{i=1}^{n} pre(S_i'))$,
- $\exists_{\cong}(\bigcap_{i=1}^{n} post(S_i')) \subseteq q$.

To prove the first inclusion, let $\langle \sigma_0, \tau, \sigma \rangle \in p$. By definition of the abstract reachability predicates it trivially follows that $\langle \sigma, \epsilon, \sigma \rangle \in \bigcap_{i=1}^{n} pre(S_i')$, and so $\langle \sigma_0, \tau, \sigma \rangle \in \exists_{\cong}(\bigcap_{i=1}^{n} pre(S_i'))$.

Next, let $\langle \sigma_0, \tau, \sigma \rangle \in \exists_{\cong}(\bigcap_{i=1}^{n} post(S_i'))$. By definition there exists $\langle \sigma_0', \tau', \sigma \rangle \in \bigcap_{i=1}^{n} post(S_i')$. From the definition of $post(S_i')$ and the above corollary it follows that $\sigma_0 \in p$ and $\langle \pi', \langle \epsilon, \sigma_0' \rangle \rangle \Rightarrow^* \langle \checkmark, \langle \tau', \sigma \rangle \rangle$. By construction of π' we then also have that that $\langle \pi, \sigma_0 \rangle \Rightarrow^* \langle \checkmark, \sigma \rangle$, and so by the assumption $\models \Sigma : \{p\} \pi \{q\}$, we conclude that $\sigma \in q$ (and so $\langle \sigma_0, \tau, \sigma \rangle \in q$).

4   Shared Variable Concurrency

Here it is shown how to instantiate the above completeness proof to shared variable concurrency (SHV). The underlying data structure is the standard model of the natural numbers, and predicates $p, q, \ldots$ are arithmetic first-order formulas. Statements are constructed from assignments $x := e$ by the above standard sequential control structures.

Let V be an (infinite) set of program variables. We define $\Sigma_{shv} = V \to \mathbb{N}$. That the first-order formula p holds in σ is denoted by $\sigma \models p$ (which is defined in the standard manner). Further, $\sigma[\bar{x} := \bar{n}]$ denotes the simultaneous update of σ which assigns to the i-th variable x_i of $\bar{x}$ the corresponding value n_i of $\bar{n}$. The semantics of an assignment $x := e$ is defined by $(x := e)(\sigma) = \sigma[x := \sigma(e)]$, where $\sigma(e)$ denotes the value of e in state σ. For technical convenience only, it is here assumed that the evaluation of an expression does not give rise to failure, and thus that every assignment properly terminates (see the next section for an instantiation of assignments which may give rise to failure). The semantics of the compound statements and programs (as defined above) is given by instantiating the above general transition rules.

The definition of local proof-outlines includes the following standard verification condition for an assignment $x := e$: $pre(R) \rightarrow post(R)[e/x]$, where $[e/x]$ denotes the standard substitution operation in first-order logic. The auxiliary assignments rule is instantiated as follows. Let A be a set of (auxiliary) variables which do not appear free in p or q, and appear in π' only in assignments $x := e$, where $x \in A$. Let π be obtained from π' by deleting all auxiliary assignments, and subsequently replacing every sub-statement **await true**$\{x := e\}$ by $x := e$. Let $\cong \subseteq \Sigma_{shv} \times \Sigma_{shv}$ be the congruence relation defined by $\sigma \cong \sigma'$ if and only if $\sigma(x) = \sigma'(x)$, for every variable $x \in V \setminus A$. Let $\bar{x}$ be an enumeration of the auxiliary variables A in the following auxiliary assignments rule

$$\frac{\{p\} \ \pi' \ \{q\}}{\{\exists \bar{x} p\} \ \pi \ \{\exists \bar{x} q\}}$$

Identifying any first-order predicate r with its extensional meaning $\{\sigma : \sigma \in r\}$, note that $\exists_{\cong} r = \exists \bar{x} r$, for any first-order predicate r.

Let $\models \Sigma_{shv} : \{p\} \ \pi \ \{q\}$. As above, each assignment (which does not occur in an await-statement) and await-statement of π is associated with an unique label which is represented by a natural number. We introduce a variable u not appearing in π and denote by π' the result of replacing

- every assignment $x := e$ of π which does not occur in an await-statement by **await true**$\{u := u \cdot n; x := e\}$, and
- every await-statement **await** $b\{R\}$ by **await** $b\{u := u \cdot n; R\}$

where n denotes the number associated with the assignment and await-statement, respectively. Assuming some coding of sequences of natural numbers, $u \cdot n$ denotes the encoding the sequence obtained from appending n to the sequence encoded by u. As a special case, ϵ denotes the encoding of the empty sequence. Details of such an encoding are standard and therefore omitted.

For the formalization of the reachability predicates in the assertion language of first-order logic, a fresh 'freeze' variable z is introduced for each variable x of π, which is used to represent the initial value x. Further, $\bar{x}$ denotes an enumeration of the program variables of π, and $\bar{z}$ denotes the corresponding enumeration of the 'freeze' variables. Let A denote the set of auxiliary variables u and $\bar{z}$.

Definition 6 (Reachability predicates for SHV). *Let R be an interleaving point of the i-th component of π'. We define the following reachability predicates.*

- $\sigma \models pre(R)$ *if and only if* $\sigma[\bar{x} := \sigma(\bar{z})] \models p$ *and* $\langle \pi', \sigma[u, \bar{x} := \epsilon, \sigma(\bar{z})] \rangle \Rightarrow^*$ $\langle Q, \sigma \rangle$,
  for some Q such that $Q(i) \equiv R$; after(R, S_i'),

– $\sigma \models post(R)$ *if and only if* $\sigma[\bar{x} := \sigma(\bar{z})] \models p$ *and* $\langle \pi', \sigma[u, \bar{x} := \epsilon, \sigma(\bar{z})]\rangle \Rightarrow^*$
$\langle Q, \sigma \rangle\rangle,$
for some Q *such that* $Q(i) \equiv after(R, S_i')$.

By standard coding techniques one can formalize these predicates in the first-order assertion language. Lemma 3 and Corollary 1 are instantiated by setting Σ to $V' \to \mathbb{N}$, where $V' = V \setminus A$. Theorem 1 then can be instantiated by observing that

– $WP(x := e, post(R)) = post(R)[e/x]$,
– $\models \Sigma : \{p\}\ \pi\ \{q\}$ if and only if $\models \Sigma_{shv} : \{p\}\ \pi\ \{q\}$ (this follows from the Propositions 1 and 2, where the π-congruence on Σ_{shv} identifies states that coincide on $V \setminus A$), and
– every state $\sigma \in \Sigma_{shv}$ can be represented by $\langle \sigma'[\bar{x} := \sigma(\bar{z})], \sigma(u), \sigma' \rangle$, where σ' denotes σ restricted to the variables $V \setminus A$ (thus the above definitions of the reachability predicates coincide with the abstract versions in Definition 5).

5   Dynamically Allocated Variables

We follow the approach of separation logic as described in [14]. A heap h is represented by a finitely-based *partial* function $\mathbb{N} \rightharpoonup \mathbb{N}$.

As above, V denotes a set of program variables, ranging over natural numbers. A *store* s is a total function $V \to \mathbb{N}$. Abstracting from the syntax of arithmetic expressions, e denotes an expression, and Boolean expressions are denoted by b. The integer value of e in s is denoted by $s(e)$, and the Boolean value of b in s is denoted by $s(b)$. Following [14] expressions thus do not refer to the heap.

Basic instructions are for example $x := e$ (simple assignment), $x := [e]$ (look-up which updates the store by assigning $h(s(e))$ to x), $[x] := e$ (mutation which updates the heap by assigning $s(e)$ to the location $s(x)$), see [14] for the other instructions. Compound statements are constructed from these basic instructions using the sequential control structures introduced above. The successful execution of any basic instruction S is denoted by $\langle S, h, s\rangle \Rightarrow \langle \checkmark, h', s'\rangle$, whereas $\langle S, h, s\rangle \Rightarrow \bot$ denotes a failing execution (e.g. due to access of a 'dangling pointer'). See [14] for the details of their semantics. The semantics of a parallel program $\pi = S_1 \parallel \cdots \parallel S_n$ is defined by the above two general interleaving rules.

The truth relation of (classical) separation logic is denoted by $h, s \models p$. For example, $h, s \models (e \hookrightarrow e')$ iff $h(s(e)) = s(e')$ (which implies $s(e)$ is in the domain of h). For further details of the syntax and semantics of the assertion language of separation logic see [14]. The semantics of abstract program specifications is instantiated by $\models \Sigma_{sl} : \{p\}\ \pi\ \{q\}$, where Σ_{sl} denotes the set of all pairs (h, s): if $h, s \models p$, then $\langle \pi, h, s\rangle \not\Rightarrow^* \bot$ and $\langle \pi, h, s\rangle \Rightarrow^* \langle \checkmark, h', s'\rangle$ implies $h', s' \models q$, for all h', s'.

In the definition of local proof-outlines, the standard (complete) weakest precondition axiomatization of the basic instructions is assumed (again, see [14]). Figure 1 shows an example of a producer and consumer that interact via the heap. It allows for simple interference free local proof-outlines. Here $(e \hookrightarrow -)$ abbreviates $\exists z(e \hookrightarrow z)$. The consumer invariant I_c denotes the assertion $\forall i \in [0, y) : (i \hookrightarrow i)$, and the producer invariant I_p denotes $\forall i \in [0, n] : (i \hookrightarrow -)$. The interference freedom test boils down to showing that I_c is not affected by the assignments $[y] := y$ and $y := y + 1$, and that $x < y$ is not affected by $y := y + 1$ (details are straightforward). By the parallel composition rule and a trivial application of the consequence rule one obtains $\{x = y = 0 \wedge 0 \leq n \wedge I_p\}\ \pi\ \{sum = n(n-1)/2\}$ (where π denotes the program of Fig. 1). The interference freedom test becomes even simpler by replacing the inner while-statement by an await-statement (with the same body).

$\{x = 0 \wedge 0 \leq n \wedge I_c\}$
$sum := 0;$
$\{0 \leq x \leq n \wedge sum = \Sigma_{i=0}^{x-1} i \wedge I_c\}$
while $x < n$ **do**
  $\{0 \leq x < n \wedge sum = \Sigma_{i=0}^{x-1} i \wedge I_c\}$
  while $x < y$ **do**
    $\{x < y \wedge 0 \leq x < n \wedge sum = \Sigma_{i=0}^{x-1} i \wedge I_c\}$
    $z := [x];$
    $\{z = x \wedge 0 \leq x < n \wedge sum = \Sigma_{i=0}^{x-1} i \wedge I_c\}$
    $sum := sum + z; x := x + 1$
od od
$\{sum = n(n-1)/2\}$

$\|\ \{y = 0 \wedge 0 \leq n \wedge I_p\}$
$\{0 \leq y \leq n \wedge I_p\}$
while $y < n$ **do**
  $\{0 \leq y < n \wedge I_p\}$
  $[y] := y;$
  $\{(y \hookrightarrow y) \wedge 0 \leq y < n \wedge I_p\}$
  $y := y + 1$
  $\{0 \leq y < n \wedge I_p\}$
od
$\{\mathbf{true}\}$

Fig. 1. Example of a producer and consumer of a buffer that interact via the heap.

Note that the above example does not even need the introduction of auxiliary assignments. The logical instantiation of the auxiliary assignments rule coincides with the above rule for SHV. It is based on the following congruence relation: $(h, s) \cong (h', s')$ if and only if by $h = h'$ and $s(x) = s'(x)$, for every program variable $x \in V \setminus A$ (where A denotes the set of auxiliary variables).

Let $\models \Sigma_{sl} : \{p\}\ \pi\ \{q\}$. To instantiate the abstract completeness proof, as for SHV, each basic instruction (which does not occur in an await-statement) and await-statement of π is associated with an unique natural number. Let u be a variable not appearing in π. Then π' is defined as in the previous section: it denotes the result of replacing

- every basic instruction S of π which does not occur in an await-statement by **await** $\mathbf{true}\{u := u \cdot n; S\}$, and
- every await-statement **await** $b\{R\}$ by **await** $b\{u := u \cdot n; R\}$

where n denotes the number associated with the basic instruction and await-statement, respectively.

As above, for the formalization of the reachability predicates, a fresh 'freeze' variable z is introduced for each variable x of π, which is used to represent the initial value x. Let $\bar{x}$ be an enumeration of the program variables of π, and $\bar{z}$ the corresponding sequence of fresh 'freeze' variables. Additionally, let v be a fresh variable which is used to encode the initial heap, assuming some arithmetic encoding of heaps. For notational convenience, $s(v)$ is identified with the encoded heap. Let A denote the set of auxiliary variables u, v, and $\bar{z}$.

Definition 7 (Reachability predicates for separation logic). *Let R be an interleaving point of the i-th component of π'. The reachability predicates are defined by*

- *$h, s \models pre(R)$ if and only if $s(v), s[\bar{x} := s(\bar{z})] \models p$ and $\langle \pi', s(v), s[u, \bar{x} := \epsilon, s(\bar{z})]\rangle \Rightarrow^* \langle Q, h, s\rangle$, for some Q such that $Q(i) \equiv R$; $after(R, S_i')$,*
- *$h, s \models post(R)$ if and only if $s(v), s[\bar{x} := s(\bar{z})] \models p$ and $\langle \pi', s(v), s[u, \bar{x} := \epsilon, s(\bar{z})]\rangle \Rightarrow^* \langle Q, h, s\rangle$, for some Q such that $Q(i) \equiv after(R, S_i')$.*

By standard coding techniques one can formalize these predicates in the assertion language of separation logic (which subsumes first-order arithmetic). See [9] for more details.

Lemma 3 and Corollary 1 are instantiated by setting Σ to $H \times (V' \to \mathbb{N})$, where $H = \mathbb{N} \to \mathbb{N}$ and $V' = V \setminus A$. Theorem 1 then can be instantiated by observing that

- $WP(S, post(R))$, for any basic instruction S, denotes the standard weakest precondition in separation logic,
- $\models \Sigma : \{p\} \pi \{q\}$ if and only if $\models \Sigma_{sl} : \{p\} \pi \{q\}$ (this follows from the Propositions 1 and 2, where the π-congruence on Σ_{sl} is defined by $(h, s) \cong (h', s')$ if and only if $h = h'$ and $s(x) = s'(x)$, for $x \in V' = V \setminus A$), and
- every $(h, s) \in \Sigma_{sl}$ can be represented by $\langle (s(v), s'[\bar{x} := \sigma(\bar{z})]), s(u), (h, s')\rangle$, where s' denotes s restricted to the variables $V' = V \setminus A$ (thus the above definitions of the reachability predicates coincide with the abstract versions in Definition 5).

6 Dynamic Thread Creation

We extend the syntax of abstract statements (as introduced in Sect. 2) with a statement **spawn**(P), where P is a procedure identifier. Execution of this statement consists of the generation of a new process that starts executing the statement associated with P. Consequently, the statement associated with P can be executed by several processes in parallel. This requires an adaptation of the Owicki/Gries method so that one can distinguish between different processes originating from the same **spawn**(P) statement. This adaptation and corresponding adaptations of the soundness and completeness proofs are briefly sketched below.

Let in the sequel π, with state-space Σ, denote a program that consists of a set of process declarations $P_1 :: S_1, \ldots, P_n :: S_n$ and a main statement S_0. Without loss of generality, it is assumed that the main statement is executed atomically. Semantically, a parallel configuration is modeled as a *set* of indexed statements (k, S), where $k \in \mathbb{N}$ uniquely identifies the process executing S. To generate and reason about these process indices, let V denote a set of auxiliary variables, so that the set of (abstract) states is extended with a valuation $V \to \mathbb{N}$. Thus an abstract predicate p is a subset of $\Sigma^+ = \Sigma \times (V \to \mathbb{N})$. The set of free variables $var(p) \subseteq V$ of a predicate p is assumed to be finite, and satisfies the following closure property: $(\sigma, s) \in p$ if and only if $(\sigma, s') \in p$, where $s(x) = s'(x)$, $x \in var(p)$.

The local semantics for assignments and compound statements is specified as in Sect. 2. The semantics of a parallel program π is specified by a transition relation between configurations of the form $\langle X, \sigma, s \rangle$, where X is a set of local processes. A local process instance of P_i, $i = 1, \ldots, n$, is of the form $\langle k, S \rangle$, where k uniquely identifies the process.

Besides a straightforward adaptation of the two rules in Sect. 2 for lifting the local semantics of assignments and compound statements, the following global rule specifies the semantics of the spawn statement.

$$\langle X \cup \{(k, \mathbf{spawn}(P_i); S)\}, \sigma, s \rangle \Rightarrow \langle X \cup \{(k, S_i), (s(c_i), S)\}, \sigma, s[c_i := s(c_i) + 1] \rangle$$

This rule involves a distinguished built-in auxiliary variable c_i which is used to generate fresh indices for newly generated instances of P_i, $i = 1, \ldots, n$ ($s[x := n]$ denotes the result of assigning the value n to the variable x).

The semantics $M(\pi) : \Sigma \to P(\Sigma_\perp)$ is defined by

$$M(\pi)(\sigma) = \{\sigma' \mid \langle X_0, \sigma, s_0 \rangle \Rightarrow^* \langle X_\checkmark, \sigma', s' \rangle\} \cup \{\perp \mid \langle X_0, \sigma, s_0 \rangle \Rightarrow^* \perp\}$$

where $X_0 = \{\langle 0, S_0 \rangle\}$ (S_0 is the main statement), $s_0(c_i) = 0$, for $i = 1, \ldots, n$, and $X_\checkmark$ denotes a set of terminated processes, that is, $S \equiv \checkmark$, for any $\langle k, S \rangle \in X_\checkmark$.

The Owicki/Gries method is adapted as follows. First, the local proof-outlines for the statements $S_0, \ldots, S_n$ (statements $S_1, \ldots, S_n$ are associated with the procedure identifiers $P_1, \ldots, P_n$) are parameterized with respect to a distinguished built-in auxiliary variable $id \in V$ which identifies the current process. The local verification conditions for spawning a process can then be specified by

$$pre(\mathbf{spawn}(P_i)) \subseteq WP(c_i := c_i + 1, post(\mathbf{spawn}(P_i))) \tag{2}$$

and

$$pre(\mathbf{spawn}(P_i)) \subseteq WP(id, c_i := c_i, c_i + 1, pre(S_i)) \tag{3}$$

where $c_i := c_i + 1$, as expected, increments the value of c_i and $id, c_i := c_i, c_i + 1$ denotes a simultaneous update of the variables id and c_i.

The interference freedom test (Definition 4) is adapted as follows. Let p denote $pre(S)$ or $post(S)$, where S is an interleaving point of S_i, $i = 1, \ldots, n$ (thus excluding the main statement S_0), or p denotes $post(S_0)$ (the post-condition of the main statement). Further, let p' denote the predicate $WP(id := u, p)$, where

$u \in V$ is a fresh variable ($u \notin var(p) \cup var(pre(S'))$). This variable is introduced to distinguish between the process instances. Let S' be an assignment, spawn statement or an await-statement of S_j, $j = 1, \ldots, n$ (again, excluding the main statement S_0). Then

- $(p' \cap pre(S') \cap (i, id) \neq (j, u)) \subseteq WP(S', p')$, in case S' denotes an assignment,
- $(p' \cap pre(S') \cap (i, id) \neq (j, u)) \subseteq WP(c_k := c_k + 1, p')$, in case S' denotes a spawn statement **spawn**(P_k),
- $\vdash \{p' \cap pre(S') \cap b \cap (i, id) \neq (j, u)\}\ R\ \{p\}$, in case S' denotes an await-statement **await** $b\{R\}$.

Here $(i, id) \neq (j, u)$ denotes the predicate **true** in case $i \neq j$, otherwise it denotes the set $\{(\sigma, s) \mid s(id) \neq s(u)\}$. The symbol $\vdash$ denotes the existence of a local proof-outline, where spawn statements only need to satisfy the first local verification condition (2) above.

The following adaptation of the parallel composition rule allows to derive a program specification from interference free local proof-outlines of the main statement and the process definitions.

$$\frac{IFT(PO(S_0), \ldots, PO(S_n))}{\{pre(S_0) \cap \bigcap_{i=1}^{n} c_i = 0\}\ \pi\ \{post(S_0) \cap \bigcap_{i=1}^{n} \bigcap_{j=0}^{c_i - 1} WP(id := j, post(S_i))\}}$$

Soundness of the above parallel composition rule requires the following straightforward reformulation of the corresponding Lemma 1 in Sect. 3.1: Let $IFT(PO(S_0), \ldots, PO(S_n))$. Then for any execution $\langle X_0, \sigma, s \rangle \Rightarrow^* \langle X, \sigma', s' \rangle$ such that $(\sigma, s) \in pre(S_0) \cap \bigcap_{i=1}^{n} c_i = 0$, it follows that $(\sigma', s'[id := k]) \in pre(R)$, for $(k, R; after(R_i, S_i)) \in X$, and $(\sigma', s'[id := k]) \in post(R)$, for $(k, after(R, S_i)) \in X$. The main adaptation of the completeness proof consists of adding the process indices to the symbolic trace of the (static) location labels, that is, $\tau \in (\mathbb{N} \times L)^*$ additionally provides information about the process instance executing the statement associated with the label l. So, an update of τ consists of appending $(s(id), l)$.

7 Conclusion and Related Work

Cousot and Cousot [8] show that in general the verification conditions for proving program correctness can be obtained by a decomposition of a basic global induction principle. Soundness and completeness then follows from the particular decomposition. In this paper for the Owicki/Gries method abstract soundness and completeness proofs are given directly in terms of the verification conditions themselves. This allows for a more direct instantiation to different state-spaces (and corresponding logics).

Apart from the overall semantic approach the completeness proof presented in this paper differs from that of [2] in the use of symbolic traces, instead of recording the sequence of *state-changes*. Technically, recording such state-changes for programs that, for example, manipulate dynamically allocated variables requires heavy encoding machinery, whereas symbolic traces only require a simple labelling mechanism which is applicable to any abstract program.

What remains is a comprehensive and systematic evaluation of the Owicki/Gries proof method presented in this paper compared to other approaches using concurrent separation logic, e.g. in verifying the correctness of diverse case studies, such as Michael's algorithm for a non-blocking stack [13].

References

1. Amighi, A., Hurlin, C., Huisman, M., Haack, C.: Permission-based separation logic for multithreaded Java programs. Logical Methods Comput. Sci. **11** (2015)
2. Apt, K.R.: Recursive assertions and parallel programs. Acta Informatica **15**, 219–232 (1981)
3. Baier, C., Katoen, J.-P.: Principles of Model Checking (Representation and Mind Series). The MIT Press (2008)
4. Brookes, S.: A semantics for concurrent separation logic. Theor. Comput. Sci. **375**(1–3), 227–270 (2007)
5. Brookes, S.: Syntactic control of interference and concurrent separation logic. In: Berger, U., Mislove, M.W. (eds.) Proceedings of the 28th Conference on the Mathematical Foundations of Programming Semantics, MFPS 2012, Bath, UK, 6–9 June 2012. Electronic Notes in Theoretical Computer Science, vol. 286, pp. 87–102. Elsevier (2012)
6. Brookes, S., O'Hearn, P.W.: Concurrent separation logic. ACM SIGLOG News **3**(3), 47–65 (2016)
7. Cao, Q., Cuellar, S., Appel, A.W.: Bringing order to the separation logic jungle. In: Chang, B.-Y.E. (ed.) APLAS 2017. LNCS, vol. 10695, pp. 190–211. Springer, Cham (2017). https://doi.org/10.1007/978-3-319-71237-6_10
8. Cousot, P., Cousot, R.: Invariance proof methods and analysis techniques for parallel programs. In: Biermann, A.W., Guiho, G., Kodratoff, Y. (eds.) Automatic Program Construction Techniques, chapter 12, pp. 243–271. Macmillan, New York (1984)
9. Ameen, M., Tatsuta, M.: Completeness for recursive procedures in separation logic. Theor. Comput. Sci. **631**, 73–96 (2016)
10. Harel, D.: Arithmetical completeness in logics of programs. In: Ausiello, G., Böhm, C. (eds.) ICALP 1978. LNCS, vol. 62, pp. 268–288. Springer, Heidelberg (1978). https://doi.org/10.1007/3-540-08860-1_20
11. Owicki, S.S., Gries, D.: An axiomatic proof technique for parallel programs I. Acta Informatica **6**, 319–340 (1976)
12. Owicki, S.S., Gries, D.: Verifying properties of parallel programs: an axiomatic approach. Commun. ACM **19**(5), 279–285 (1976)
13. Parkinson, M., Bornat, R., O'Hearn, P.: Modular verification of a non-blocking stack. In: Proceedings of the 34th Annual ACM SIGPLAN-SIGACT Symposium on Principles of Programming Languages, pp. 297–302 (2007)

14. Reynolds, J.C.: Separation logic: a logic for shared mutable data structures. In: 17th IEEE Symposium on Logic in Computer Science (LICS 2002), 22–25 July 2002, Copenhagen, Denmark, Proceedings, pp. 55–74. IEEE Computer Society (2002)

15. Shankar, N.: Combining model checking and deduction. In: Clarke, E.M., Henzinger, T.A., Veith, H., Bloem, R. (eds.) Handbook of Model Checking, pp. 651–684. Springer, Cham (2018)

On the Verification of Quantum Circuits (Research Challenges and Opportunities)

Parosh Aziz Abdulla[1,2]([✉]), Yo-Ga Chen[3], Yu-Fang Chen[3], Kai-Min Chung[3], Lukáš Holík[6], Ondřej Lengál[4], Jyun-Ao Lin[5], Fang-Yi Lo[3], and Wei-Lun Tsai[3]

[1] Uppsala University, Uppsala, Sweden
parosh@it.uu.se
[2] Mälardalen University, Västerås, Sweden
[3] Academia Sinica, Taipei, Taiwan
[4] Brno University of Technology, Brno, Czechia
[5] National Taipei University of Technology, Taipei, Taiwan
[6] Aalborg University, Aalborg, Denmark

Abstract. Quantum technology is advancing at an exceptional pace and holds the potential to transform numerous sectors on both national and global scales. As quantum systems become more sophisticated and widespread, ensuring their correctness becomes critically important. This highlights the pressing need for rigorous tools capable of analyzing and verifying their behavior. However, developing such verification tools poses significant challenges. Fundamental quantum phenomena—most notably superposition and entanglement—lead to program behaviors that differ radically from those in classical computing. These characteristics give rise to inherently probabilistic models and result in exponentially large state spaces, even for systems of modest complexity.

In this paper, we outline initial steps toward addressing these challenges by drawing on insights gained from the verification of classical systems within our community. We then present a roadmap for designing novel verification frameworks that adapt the strengths of classical methods—such as succinct property specification, precise fault detection, automation, and scalability—to the quantum setting.

1 Introduction

Classical computing has advanced considerably over the decades and now forms the backbone of modern technological infrastructure. In contrast, quantum computing introduces a fundamentally new paradigm for information processing, grounded in the principles of quantum mechanics. This paradigm enables the solution of computational problems that are beyond the practical capabilities of classical methods.

The core distinction between classical and quantum computing lies in the way information is represented and manipulated. Classical computers operate on bits that exist in one of two binary states—0 or 1. In contrast, quantum computers use quantum bits, or *qubits*, which exhibit the quantum phenomena of *superposition* and *entanglement*. Superposition enables a qubit to exist in multiple states at once, thereby allowing parallel evaluation of many possibilities. Entanglement creates correlations between qubits such that the state of one qubit can instantaneously influence the state of another, regardless of the distance between them. These phenomena enable quantum systems to represent and manipulate exponentially many states simultaneously, and to encode complex correlations that classical systems cannot efficiently reproduce, resulting in powerful computational advantages.

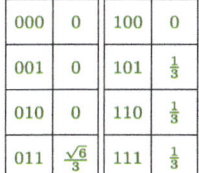
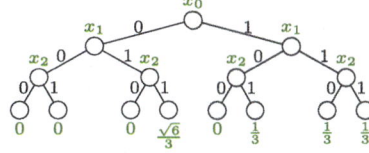

Fig. 1. A quantum state with three qubits and its tree representation. The system can be in any of the eight states, with the probability given in the table.

Figure 1 depicts a quantum state consisting of three qubits.[1] In a classical circuit, the system would occupy a single *basis* state from among the eight possible configurations: 000, 001, ..., 111. By contrast, a quantum system can exist in a superposition of all eight basis states simultaneously, each associated with a specific probability amplitude.

A quantum state can be interpreted as a distribution over basis states, where each state is assigned a complex amplitude. This distribution may be visualized as a tree, in which each path from the root to a leaf corresponds to a basis state, and each leaf stores a complex-valued amplitude. Unlike classical probabilistic models, quantum amplitudes are complex numbers, and the sum of the squares of their absolute values must equal one.[2]

A primary objective in the field of quantum computing is to achieve *quantum supremacy*—the point at which quantum computers can solve problems that are infeasible for even the most powerful classical supercomputers. Reaching this milestone would result in exponential speedups for specific computational tasks. Although the field is still in an early phase of practical realization, quantum computing applications are growing rapidly. In the near term, hybrid systems integrating classical and quantum processors are expected to gain prominence.

---

[1] Throughout the text, we illustrate the challenges encountered in verifying quantum circuits through examples. For clarity, these examples are simplified. Full technical details are beyond the scope of this document and can be found, for example, in [35].

[2] An *amplitude* generalizes the classical notion of probability. The square of the absolute value of a complex amplitude gives the corresponding probability. The use of complex numbers permits constructive and destructive interference, allowing for the cancellation of certain outcomes.

Looking ahead, quantum supremacy could revolutionize entire industries by solving problems that were previously considered intractable. Potential application areas for quantum technologies include medical diagnostics, cryptography, secure communications infrastructure, and autonomous systems, among others. These domains typically require stringent reliability and safety standards, as they cannot tolerate critical system failures. Consequently, they demand rigorous certification procedures and robust assurance mechanisms. Ensuring the reliability of quantum systems is therefore of critical importance. Nevertheless, verifying quantum systems presents formidable challenges. Their probabilistic behavior, combined with the exponential growth of state spaces as the number of qubits increases, introduces substantial complexity.

In this paper, we describe initial steps toward tackling these challenges by leveraging insights from our community's extensive experience in verifying classical systems. Building on this foundation, we will also propose a roadmap for the development of new verification frameworks that transfer the proven strengths of classical approaches—such as concise property specification, accurate fault detection, automation, and scalability—into the quantum domain.

We envision significant added value in bridging two complementary areas of computer science, especially when established techniques from a mature field are adapted to address complex problems in an emerging domain. This work embodies such interdisciplinary integration by applying well-developed methods from logic, automata theory, and symbolic verification to the problem of ensuring the correctness of quantum programs.

We present a novel application of automata theory—a foundational discipline in formal verification—to the verification of quantum systems. Our methodology integrates automata-based reasoning with symbolic representations tailored for the automated analysis of quantum behavior. The goal is to generalize core concepts from classical verification, such as state-space exploration and symbolic reasoning, into the quantum realm, thereby enabling the adaptation of robust classical paradigms to quantum computing, where formal assurances are essential. Interestingly, this approach also opens new avenues for research in automata theory itself, as the mathematical structures encountered in quantum systems introduce novel challenges and opportunities. It establishes a conceptual bridge between quantum program verification and automata theory, fostering potential for new theoretical developments and extending the expressive and analytical power of automata-based techniques into the context of quantum computation. We will give a high-level description of how to develop *symbolic representations*, based on *automata*, that serve as a theoretical and algorithmic foundation for efficiently modeling the state spaces of quantum circuits. These representations are grounded in the idea of modeling quantum states as binary trees, where each path from the root to a leaf corresponds to a computational basis state (see Fig. 1).

In the remainder of the paper, we first describe the modeling of quantum states, gates, and circuits, followed by a formalization of the verification problem. Subsequently, we outline a roadmap highlighting potential challenges and opportunities in the verification of quantum systems.

2   Probabilistic vs Quantum Systems

In probabilistic systems, there exists an inherent uncertainty in our knowledge of the physical state. Moreover, state transitions occur according to probabilistic laws. This implies that the evolution of such systems is governed by rules specifying the likelihood of transitions between states. A classical model capturing this behaviour is the *Markov chain*— an automaton whose transitions are labelled with *weights* [8,29]. A weight is a real number between 0 and 1, indicating the probability of moving from the source state to the target state. The probabilities of all transitions emanating from a given state sum to one.

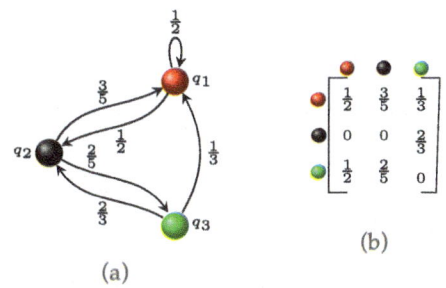

Fig. 2. (a) A Markov chain. (b) Its adjacency matrix.

We can represent a Markov chain using its adjacency matrix M, which is a square matrix with real-valued entries. Each row and column of M corresponds to a state, and the entry $M[q',q]$ denotes the weight of the transition from state q to state q'. By construction, the sum of the entries in each column of M is equal to 1. A probability distribution over the states of a Markov chain M is represented[3] by a column vector, for instance, $v = [1/2, 1/8, 3/8]^T$ (Fig. 2). This vector specifies that the system is in the states *red, black,* and *green* with probabilities 1/2, 1/8, and 3/8, respectively. To compute the distribution after one transition step, we multiply the transition matrix M by the column vector v, yielding a new distribution vector, here, $v' = [9/20, 1/4, 3/10]^T$.

The quantum setting operates over complex numbers rather than real-valued probabilities. In this context, a transition weight is not a real number $p \in [0, 1]$, but a complex number c such that the squared modulus $|c|^2$ lies in the interval $[0, 1]$. This seemingly subtle change has profound implications for system dynamics. Whereas real probabilities can only accumulate positively under addition, complex amplitudes can interfere—constructively or destructively. For example, if $p_1, p_2 \in [0, 1]$ are real numbers, then both $p_1 + p_2 \geq p_1$ and $p_1 + p_2 \geq p_2$ hold. In contrast, for complex numbers c_1, c_2, it is not necessarily the case that the squared modulus of their sum dominates the individual squared moduli:

$$|c_1 + c_2|^2 \leq \max\left(|c_1|^2, |c_2|^2\right)$$

This phenomenon, known as *interference*, is a hallmark of quantum systems. In particular, destructive interference allows for the cancellation of probability amplitudes, leading to outcomes with lower probabilities than suggested by the individual components. Such behaviours are central to the analysis and modelling

---

[3] To simplify notation, we may represent column vectors using their transposes.

of quantum processes, where unitary evolution replaces the stochastic matrices of classical probabilistic systems.

3   States

We present the two standard representations of quantum states used in quantum computing: the matrix (vector) representation and the Dirac notation. We then introduce our tree-based representation. We begin with the special case of classical states.

In classical computing, bits are the fundamental units of information. Each bit can exist in one of two possible states: 0 or 1. These may be conceptualized as switches that are either turned on or off. In contrast, quantum computing employs quantum bits, or *qubits*, to represent and process information. Qubits can exist in a *superposition*, meaning they may simultaneously be in state 0, state 1, or a combination of both. Additionally, a qubit can be subjected to a *measurement* operation. The state of a qubit is described by two complex numbers, called *amplitudes*, c_0 and c_1, where $|c_0|^2$ denotes the probability of measuring the state as 0, and $|c_1|^2$ the probability of measuring it as 1. These amplitudes must satisfy the normalization condition: $|c_0|^2 + |c_1|^2 = 1$.

$$\begin{bmatrix} 1 \\ 0 \end{bmatrix} \quad \begin{bmatrix} 0 \\ 1 \end{bmatrix} \qquad \begin{bmatrix} \sqrt{1/2} \\ \sqrt{1/2} \end{bmatrix} \qquad \begin{bmatrix} 1/3 \\ \sqrt{8}/3 \end{bmatrix} \qquad \begin{bmatrix} (2-i)/3 \\ 2i/3 \end{bmatrix}$$

$$|0\rangle \qquad |1\rangle \qquad \sqrt{1/2}\,|0\rangle + \sqrt{1/2}\,|1\rangle \qquad 1/3\,|0\rangle + \sqrt{8}/3\,|1\rangle \qquad (2-i)/3\,|0\rangle + 2i/3\,|1\rangle$$

(a)   (b)   (c)   (d)   (e)

Fig. 3. Different qubits and their vector and Dirac representations. (a, b) The basis states $|0\rangle$ and $|1\rangle$. (c, d, e) Quantum states in superposition. Measuring the state in (c) yields 0 or 1 with probability 0.5. Measuring the state in (d) yields 0 with probability 1/9 and 1 with probability 8/9.

There are two common representations used to describe a qubit state (cf. Fig. 3). The first is the vector notation, in which the state is written as a column vector of length 2, with complex entries corresponding to the amplitudes of the respective basis states. To simplify notation, we often use the transpose of this column vector. For example, we write $[1/3 \quad \sqrt{8}/3]^T$ to represent the state shown in Fig. 3(d). This results in a more compact row vector form. The second representation is the Dirac notation, where the state is expressed as a linear combination, e.g., $|0\rangle + |1\rangle$. Classical states are special cases where either $c_0 = 1$ and $c_1 = 0$, or $c_0 = 0$ and $c_1 = 1$. In quantum systems, these are referred to as the basis states $|0\rangle$ and $|1\rangle$.

We now generalize these concepts to systems with multiple qubits. For n qubits, there are 2^n possible basis states. In the case of two qubits, the system has four basis states. These can be represented as column vectors:

$$[1\ 0\ 0\ 0]^{\mathrm{T}} \quad [0\ 1\ 0\ 0]^{\mathrm{T}} \quad [0\ 0\ 1\ 0]^{\mathrm{T}} \quad [0\ 0\ 0\ 1]^{\mathrm{T}}$$

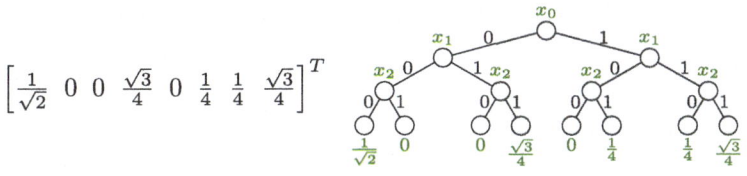

$$(1/\sqrt{2})\,|000\rangle + (\sqrt{3}/4)\,|011\rangle + (1/4)\,|101\rangle + (1/4)\,|110\rangle + (\sqrt{3}/4)\,|111\rangle$$

Fig. 4. A state involving three qubits. We provide the vector, Dirac, and tree-based representations. The tree spans all basis states, with the corresponding amplitudes shown as leaf labels. For instance, the left-most path represents $|000\rangle$ with amplitude $1/\sqrt{2}$.

A key feature of our framework is a symbolic representation of quantum states and sets of quantum states. We utilize *tree automata* as the underlying symbolic structure. Specifically, a quantum state over n qubits is encoded as a perfect binary tree of depth n. Each level of the tree corresponds to one qubit. We refer to the level associated with qubit x as the x-level, and the nodes at that level as x-nodes. For any x-node, its left and right subtrees represent the cases where qubit x has values 0 and 1, respectively. Each root-to-leaf path encodes a computational basis state. The leaf nodes hold the complex amplitudes associated with those basis states.

Consider Fig. 4. The left side shows a quantum state involving three qubits. The right side displays its corresponding tree representation.

4   Gates

We describe gate behaviors using matrices, and then describe the effect of a gate as a transformation of the tree representing the input state. We start by modeling classical digital gates in the quantum setting. Then, we introduce some quantum gates and show we can represent them using matrices.

Classical Gates. Logical gates, such as AND, OR, and NOT, are the fundamental components of digital circuits. They take binary inputs (0 s and 1 s) and produce a single binary output. Each type of gate implements a specific binary function, e.g., the output of an AND gate is the conjunction of its inputs. In order to prepare for quantum gates, we will represent the behavior of classical gates by matrices.

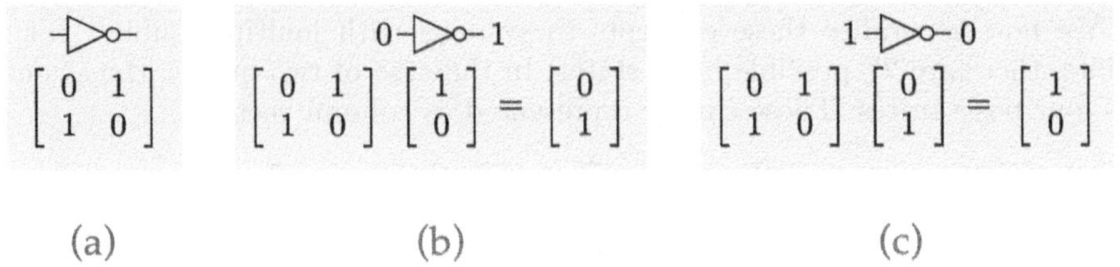

(a)　　　　　　　　(b)　　　　　　　　(c)

Fig. 5. (a) The NOT gate together with its matrix presentation. (b) Applying the gate to 0, represented by $[1\ 0]^T$, will give 1, i.e., $[0\ 1]^T$. (c) Applying the gate to 1 will give 0.

Let us take the NOT gate as an example (Fig. 5). For a given (basis) state, we compute the result of applying the gate by multiplying the matrix representing the gate with the input state. The gate has one input and one output bit, and hence its behavior is described by a 2×2-matrix. The input and output states are both column vectors.

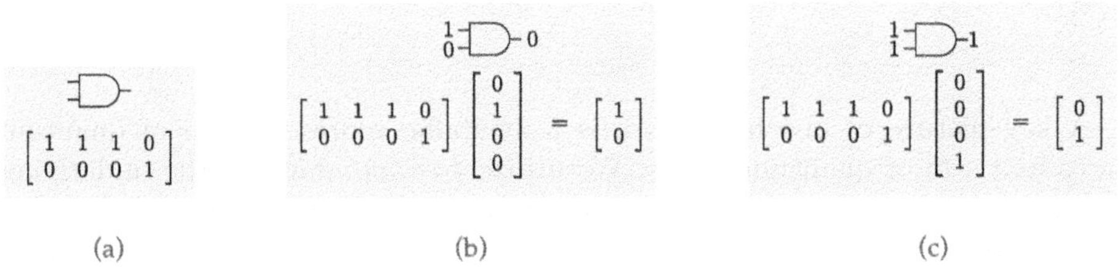

(a)　　　　　　　　(b)　　　　　　　　(c)

Fig. 6. (a) The AND gate together with its matrix presentation. (b) Applying the gate to 01, represented by $[0\ 1\ 0\ 0]^T$, will give 0. (c) Applying the gate to 11, represented by $[0\ 0\ 0\ 1]^T$, will give 1.

Figure 6 depicts the AND gate. The gate has two input bits and one output bit, and hence its behavior is described by a 2×4-matrix. The input state is of length 4 and the output state has length 2.

Quantum Gates. In a similar manner to classical gates, we will represent quantum gate behaviors by matrices. Quantum gates are *reversible*: if we run the gate from a state and then run the inverse of the gate, then we get the same state. Formally, the matrix representing the gate should be *unitary*, i.e., its conjugate transpose should be equal to its inverse. In Fig. 7 we show some examples of quantum gates and their behaviors. The gate I is the identity, and maps each input to an identical output. Next, we introduce the three Pauli gates.

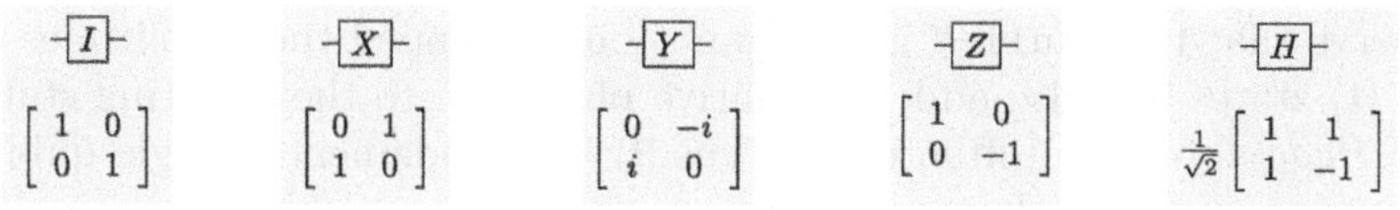

$$-\boxed{I}- \qquad -\boxed{X}- \qquad -\boxed{Y}- \qquad -\boxed{Z}- \qquad -\boxed{H}-$$

$$\begin{bmatrix} 1 & 0 \\ 0 & 1 \end{bmatrix} \qquad \begin{bmatrix} 0 & 1 \\ 1 & 0 \end{bmatrix} \qquad \begin{bmatrix} 0 & -i \\ i & 0 \end{bmatrix} \qquad \begin{bmatrix} 1 & 0 \\ 0 & -1 \end{bmatrix} \qquad \frac{1}{\sqrt{2}}\begin{bmatrix} 1 & 1 \\ 1 & -1 \end{bmatrix}$$

Fig. 7. The Identity gate (I), the Pauli gates (X, Y, Z), and the Hadamard gate (H).

The `Pauli-X` is the quantum analogue of the classical NOT gate, acting on a single qubit. It is often referred to as *bit-flip gate* because it inverts the value of the qubit. In classical computing, a NOT gate flips a binary value—changing 0 to 1 and vice versa. The `Pauli-X` gate performs the same operation in the quantum domain for the basis states.

$$X|0\rangle = |1\rangle \qquad X|1\rangle = |0\rangle$$

Moreover, unlike classical bits, qubits can exist in *superpositions* of $|0\rangle$ and $|1\rangle$. The `Pauli-X` gate also acts on such states by swapping the amplitudes of the two basis states. For example, consider a qubit in the superposition:

$$|\psi\rangle = a|0\rangle + b|1\rangle$$

where a and b are complex probability amplitudes. Applying the `Pauli-X` gate yields:

$$X|\psi\rangle = b|0\rangle + a|1\rangle$$

Thus, the `Pauli-X` gate exchanges the amplitudes of $|0\rangle$ and $|1\rangle$, effectively flipping the qubit's state within the superposition.

The `Pauli-Z` gate is also known as the phase-flip gate. Unlike the `Pauli-X` gate—which acts as a quantum analogue of the classical NOT gate by flipping the state between $|0\rangle$ and $|1\rangle$, the `Pauli-Z` gate leaves the computational basis states unchanged but inverts the phase of the $|1\rangle$ component. In other words, it modifies the relative phase between $|0\rangle$ and $|1\rangle$, a core concept in quantum mechanics. Thus, applying the `Pauli-Z` gate to a qubit in state $|0\rangle$ has no effect, while applying it to $|1\rangle$ multiplies the state by -1, effectively flipping its phase. Due to this nature, `Pauli-Z` is sometimes called phase-flip. This operation is essential in many quantum algorithms and plays a critical role in phase manipulation and interference, which are central to quantum computation.

The `Pauli-Y` gate is a combination of the `Pauli-X` and `Pauli-Z` gates, meaning it applies both a bit-flip and a phase-flip to the qubit. It changes both the state of the qubit (like the `Pauli-X` gate), and the relative phase between the states $|0\rangle$ and $|1\rangle$ (like the `Pauli-Z` gate).

For example, if you have a qubit in the state $|0\rangle$ and apply the `Pauli-Y` gate, the qubit's state changes to the state $i|1\rangle$. If you have a qubit in the state $|1\rangle$ and apply the `Pauli-Y` gate, the qubit's state changes to the state $-i|0\rangle$. Consider a qubit in the following superposition state: $|\psi\rangle = a|0\rangle + b|1\rangle$, where a and b are complex numbers representing the amplitudes of the qubit states. When the `Pauli-Y` gate is applied to this qubit, the result will be: $Y|\psi\rangle = -ib|0\rangle + ia|1\rangle$.

We observe that the `Pauli-Y` gate has not only swapped the amplitudes of the $|0\rangle$ and $|1\rangle$ states but also added a relative phase of i to the resulting states.

The *Hadamard* gate (often denoted as H) is a fundamental single-qubit gate in quantum computing. It creates superposition states, which are essential for quantum parallelism. It transforms the basis states as follows:

$$H|0\rangle = \frac{1}{\sqrt{2}}(|0\rangle + |1\rangle) \quad H|1\rangle = \frac{1}{\sqrt{2}}(|0\rangle - |1\rangle)$$

The H gate has three important properties:

- *Self-inverse*: $H^2 = I$, so applying it twice returns the original state.
- Creates superposition: Crucial for quantum algorithms like Deutsch-Jozsa, Grover's, and Shor's.
- Balanced output: Equal amplitude for both $|0\rangle$ and $|1\rangle$ (up to a sign), enabling interference.

Fig. 8. Applying the X gate on an input tree. We apply the gate to the qubit x_1. The application is modeled by swapping the left and right subtrees of all nodes at the level corresponding to the qubit x_1 (i.e., level 1 in this case).

Fig. 9. Applying the H gate on an input tree. The gate is applied to the qubit x_1. This transformation also targets all nodes at level 1 (corresponding to x_1), updating subtree structure and amplitudes accordingly.

Tree Transformations. In our setting, we encode gate applications as tree transformations. Given a tree representing a quantum state, a qubit x, and a gate, we compute a new tree that describes the quantum state resulting from applying the gate to the given qubit.

A major difference from the classical setting is that, due to quantum super-position, the effect of a gate application is generally *global*. That is, we must update all parts of the state in which the qubit x is involved.

In Figs. 8 and 9, we illustrate this by swapping all subtrees at nodes corresponding to x_1. As shown in the examples, a gate application can affect the amplitudes of exponentially many leaves—effectively updating an exponential number of classical states.

5 Circuits

A classical combinatorial circuit consists of a set of gates and acts as a Boolean function. It takes a sequence of bits as input and produces a sequence of bits as output. There are also internal bits that represent the connections between gates. In Fig. 10, the circuit has three input bits, one output bit, and three internal bits. Figure 11 shows a quantum circuit consisting of an H followed by a Pauli-Z gate.

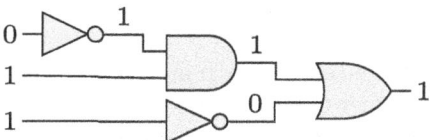

Fig. 10. A classical combinatorial circuit

We start with a qubit in a superposition state created using a Hadamard gate, and then apply a Pauli-Z gate to it. Here is a step-by-step description of the circuit behavior when the starting qubit is $|0\rangle$.

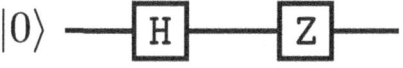

Fig. 11. A simple quantum circuit.

1. Start in $|0\rangle = [1 \ 0]^T$
2. Apply the Hadamard gate: $H|0\rangle = \frac{1}{\sqrt{2}}(|0\rangle + |1\rangle)$ Now the qubit is in an equal superposition.
3. Apply the Pauli-Z gate: $\texttt{Pauli-Z}\left(\frac{1}{\sqrt{2}}(|0\rangle + |1\rangle)\right) = \frac{1}{\sqrt{2}}(|0\rangle - |1\rangle)$. The $|1\rangle$ term's sign flips—this is phase inversion.

Intuitively, the Hadamard spreads the probability amplitude across both states. The Pauli-Z keeps the probability unchanged (as it only affects the phase), but introduces a phase shift to $|1\rangle$, which is crucial in interference patterns for quantum algorithms.

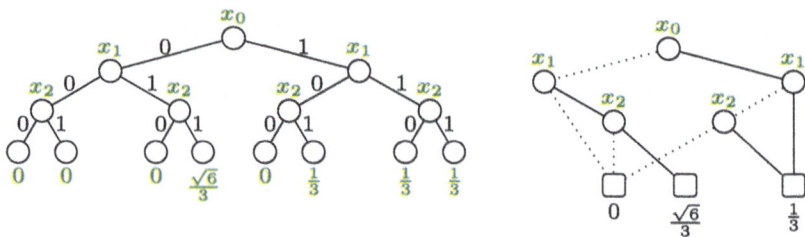

Fig. 12. The BDD representation of a quantum state.

6   Program Verification and Hoare Logic

Next, we turn our attention to the verification problem and consider the classical framework of *Hoare logic*. This formal verification method was initiated by Floyd [21] and further developed by Hoare [22]. The core concept of this logic is the *Hoare triple*, written in the form:

$$\{P\}\, C\, \{Q\}$$

where P is the *precondition*, Q is the *postcondition*, and C is a program. A Hoare triple is said to be *valid* if the following condition holds: If the program C is executed from an initial state σ satisfying the precondition P (i.e., $\sigma \models P$), then the execution terminates in a final state satisfying the postcondition Q. The initial obstacle in the application of the framework was the state space explosion problem that occurred due to the large size of the state space of C. A fundamental breakthrough in the verification of conventional computer systems was the invention of efficient data structures to represent sets of states. A case in point is the classical BDD (Binary Decision Diagram) data structure. Figure 12 depicts a BDD representation of a quantum state.

   In this paper, we take C to be a quantum circuit. The main challenge in using BDDs in quantum circuits is the fact that a BDD represents a single state. In conventional circuits, a BDD represents a (large) set of states. Thus, we need a framework in which we can handle sets of BDDs. Given that we represent quantum states by trees, a natural choice is to use tree automata to represent sets of quantum states. Figure 13(a) gives a tree automaton for accepting all basis states of size three. The set of rules is the following:

$$p \xrightarrow{x_0} (q_0, q_1) \qquad p \xrightarrow{x_0} (q_1, q_0)$$
$$q_0 \xrightarrow{x_1} (r_0, r_0) \qquad q_1 \xrightarrow{x_1} (r_1, r_0) \qquad q_1 \xrightarrow{x_1} (r_0, r_1)$$
$$r_0 \xrightarrow{x_2} (s_0, s_0) \qquad r_1 \xrightarrow{x_2} (s_1, s_0) \qquad r_1 \xrightarrow{x_2} (s_0, s_1)$$
$$s_0 \to 0 \qquad\qquad s_1 \to 1$$

Notice that we are characterizing 2^n basis states using only $3n + 1$ transitions. In our setting, a gate application corresponds to a tree transformation. For a circuit consisting of a number of gates, we provide an algorithm that transforms a tree automaton describing a set of input states to a new automaton

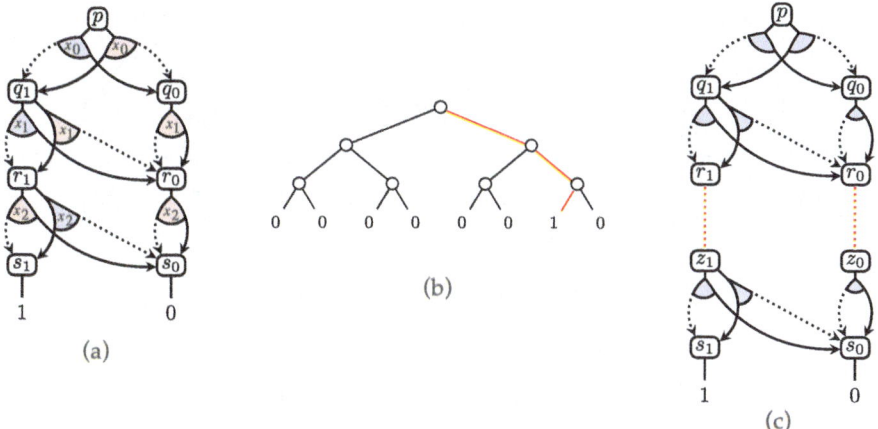

Fig. 13. (a) A tree automaton for generating all basis states of 3 qubits. The dashed lines represent the left-hand child, and the solid lines represent the right-hand child. (b) The tree corresponding to $[0\ 0\ 0\ 0\ 0\ 0\ 1\ 0]^T$. The rules, shadowed in pink, in the automaton are those used to generate the tree. (c) An automaton to generate all basis states of n qubits. (Color figure online)

describing the set of output states. We do this by constructing a sequence of automata that model the effects of each gate. Figure 14 gives a concrete example to demonstrate our approach. Assume we want to design a circuit constructing the Bell state, i.e., a 2-qubit circuit converting a basis state $|00\rangle$ to the maximally entangled state $\frac{1}{\sqrt{2}}(|00\rangle + |11\rangle)$. Given the automaton corresponding to the pre-condition (Fig. 14(a)), we derive the automaton corresponding to the post-condition (Fig. 14(c)). In this case, both automata use q as the root state and accept only one tree. We can observe the correspondence between quantum states and tree automata by traversing their structures.

7   Challenges

Research on the algorithmic verification of quantum systems remains in its early stages. Only recently have initial contributions emerged, including the first studies employing SAT solvers for quantum verification [17]. We have taken foundational steps in this direction through recent work on the symbolic verification of quantum circuits [1–3,14–16]. In contrast, the algorithmic verification of classical systems, particularly through model checking [8,18], has been an active and well-established area of research for nearly four decades. This field has yielded a substantial and mature body of work and continues to thrive, with ongoing research published regularly in premier venues such as POPL, CAV, PLDI, MICRO, ASPLOS, etc.

While extending verification techniques to quantum systems is a natural and necessary progression, it also introduces a host of novel and technically

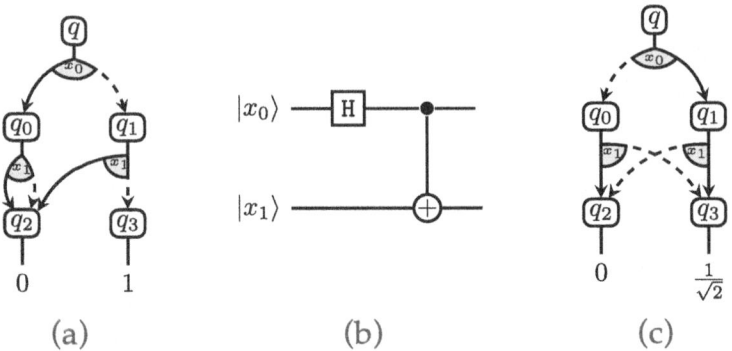

Fig. 14. The circuit for the Bell state.

challenging problems. In the remainder of this section, we briefly review the current state of the art and its limitations.

7.1   Symbolic Encodings

Symbolic state space encodings have been among the most influential techniques in the verification of conventional programs over the past three decades. By representing program semantics symbolically using various data structures and applying mathematical reasoning, one can efficiently verify whether a program satisfies its correctness properties. These data structures encode logical formulas, allowing for both compact modeling of large or infinite state spaces and the efficient execution of logical operations. Notable examples of symbolic representations that have significantly advanced the verification of classical systems include Binary Decision Diagrams (BDDs) for hardware verification and communication protocols, as well as automata over words and trees, for verifying parameterized systems and programs with dynamic heap structures.

Despite their success in conventional computing, the application of symbolic techniques remains limited in the quantum domain. This limitation stems from quantum computation's fundamentally different mathematical foundations, which involve state superposition, probabilistic measurement, and non-local entanglement—features that challenge conventional symbolic methods. Nevertheless, with techniques likely to be developed in the future, symbolic verification will eventually play an equally important role in advancing the scalability and precision of quantum program verification.

As quantum computing continues to evolve, there is a growing need to develop symbolic abstractions tailored to quantum systems, hybrid symbolic-numeric approaches, and novel verification frameworks. These directions represent promising and largely unexplored areas of research. In quantum circuit simulation, decision diagrams have been employed as compact representations of a single quantum state [10,13,24,27,31,38], and even of entire quantum circuits [11,30,32,33]. While their application to verification against formal specifications remains underexplored, encoding entire quantum circuits as decision diagrams has potential for such verification tasks. Some efforts have also investigated leveraging constraint solvers for quantum circuit simulation and equivalence checking [25, 26].

Aside from a few notable exceptions [7, 9, 12, 16, 20] and our own recent contributions [1–3, 14–16], there is a significant gap in research on symbolic encodings specifically designed for the automated verification of quantum circuits. By contrast, several studies have focused on adapting classical Hoare logic to quantum systems. In particular, the approach proposed in [19] and Ying's quantum Hoare logic [36] defines predicates as mappings from mixed states—i.e., probability distributions over pure quantum states—to real values in the interval $[0, 1]$, representing the probability that a given state satisfies a particular condition. However, these logics are inherently deductive and interactive, not algorithmic in nature. As such, they are not suited for automated verification, often requiring significant manual effort and domain-specific reasoning.

An alternative is to adopt a fundamentally different approach by proposing a set-based methodology, where predicates are defined as mappings from quantum states to the Boolean domain $\{0, 1\}$. Within this framework, automata serve as compact representations of such predicates. A tree representing a quantum state is accepted by an automaton if and only if the corresponding predicate maps that state to 1. One of the primary advantages of this set-based formulation is that it facilitates the construction of efficient and fully automated verification algorithms, making it significantly more scalable for practical use in the quantum domain.

Currently, there are no algorithms for minimization, simulation, and bisimulation over automata models designed specifically for the verification of quantum systems. Addressing this gap should be a central objective. We need to adapt and extend prior experience in classical automata theory to develop such algorithmic frameworks tailored to the specific characteristics of quantum systems.

7.2   Algorithmic Verification

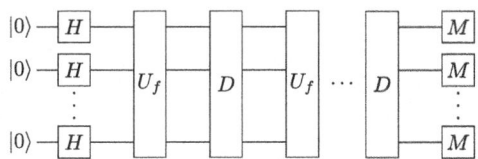

Fig. 15. The classical Grover's search algorithm.

As previously mentioned, algorithmic verification techniques such as *model checking* have been among the most influential methodologies for checking the correctness of classical programs over the past three decades. They offer an automated mechanism to determine whether a system satisfies a given specification.

Crucially, model checking is a complete verification method, meaning it can guarantee the absence of errors—unlike testing, which can only reveal the presence of bugs. Nevertheless, extending model checking techniques to quantum systems presents considerable challenges due to three primary reasons.

First, as previously discussed, quantum systems involve state spaces of infinite size, even when the system comprises only a finite number of qubits.

Second, quantum systems are intrinsically parameterized, typically along two dimensions. Consider, for instance, the circuit illustrated in Fig. 15, which implements the classical Grover's algorithm. This circuit accepts a Boolean function with n input variables and aims to find an input for which the function evaluates to 1. If such an input exists, the circuit identifies one with a runtime of $O(\sqrt{N})$, where $N = 2^n$. In contrast, any classical algorithm addressing the same problem would require $O(N)$ time. Since the algorithm is constructed to operate on an arbitrary number of input qubits, there is no fixed upper bound on the input size. Moreover, the system architecture comprises several stages, determined by $O(\sqrt{N})$. Intuitively, each algorithm stage carries out one so-called *Grover rotation*. This is an operation that increases the amplitudes of the basis states that satisfy the input function f. By performing an increasing number of iterations, the precision of the computation increases, and the probability that we will measure the correct answer approaches one. We reach the required precision after $O(\sqrt{N})$ iterations, which explains the number of stages[4]. The system is characterized by a two-dimensional parameterization: the number of qubits and the number of stages are unbounded. We need to verify the system's correctness across the entire space of possible values for these two parameters.

Third, unlike classical systems, quantum systems exhibit globally entangled transitions. In general, the application of a gate to a single qubit may affect an exponential number of classical basis states. For instance, in Fig. 1, negating the qubit x_1 would swap the leaves in positions 1, 3, 5, and 7 with the leaves in positions 2, 4, 6, and 8, respectively.

The following aspects are currently lacking in existing work:

- The paper [9] employs *symbolic execution* [23] to verify input-output relationships by formulating queries discharged to SMT solvers over the theory of real numbers. The SMT array theory approach [17] instantiated this framework by enabling a polynomial-sized encoding of quantum circuits. Nevertheless, it continues to suffer from scalability limitations. In [2,3], we demonstrate that automata-based techniques outperform these approaches by several orders of magnitude. Another scalable, fully automated method for analyzing quantum circuits is *quantum abstract interpretation* [28,37]. However, quantum abstract interpretation relies on over-approximation, which may limit its precision in bug detection.
- There are no methods supporting regular model checking or parameterized verification for quantum applications. Our recent paper [3] presents two examples of parameterized quantum circuits. However, it does not provide algo-

---

[4] Again, we give a high-level description here. We refer, e.g., to [35] for the algorithm details.

rithms for general parameterized verification. We need to bridge this gap by designing and integrating model checking algorithms with abstraction techniques tailored for quantum systems.

8 Research Opportunities

We outline some opportunities for research in quantum circuit verification.

8.1 Automata

We need to provide a comprehensive road map for designing efficient symbolic encodings based on automata to represent the state spaces of quantum systems. Furthermore, we should develop the first algorithms for minimization and checking language inclusion, language equivalence, simulation, and bisimulation relations for such automata.

Language Inclusion and Equivalence. A central challenge in formal verification, in general, and for quantum systems in particular, is *language inclusion*. We can use the operation to ensure that a system's behavior conforms to its specification and to formulate termination conditions within verification procedures. However, in many cases, checking language inclusion is computationally intensive or even undecidable. This problem is well known in classical automata and similarly manifests in the models for quantum systems in [3]. Therefore, we anticipate the models introduced in this project will exhibit the same computational difficulty. To address this, we may employ *simulation* relations as an efficient alternative and enhance them with subset-construction-like techniques to improve precision. We expect these simulation-based methods to be significantly more efficient than direct language inclusion checks when applied to quantum constraints. The disadvantage of using simulation relations is their incompleteness: a simulation preorder implies language inclusion, but the converse does not necessarily hold. To deal with incompleteness, we can, in parallel, develop methods based on the classical subset-construction for checking language inclusion. As in the classical case, we expect these algorithms to suffer from state space explosion, with an exponential increase in the number of states. To combine the strengths of both approaches, we need to adapt the antichain framework introduced in [4,34]. In this setting, the computed simulation relation will be used to prune unnecessary search paths during the execution of the subset-construction-based algorithm, thereby improving both performance and scalability.

Minimization. It is vital to design and implement algorithms to minimize the automata models developed in the project. Minimization techniques are essential for ensuring the scalability of our tools, as automata will serve as symbolic representations of state spaces. Consequently, the size of these automata directly impacts the computational complexity of language inclusion, simulation, and bisimulation problems addressed in Sect. 8.1, which underpin the verification procedures described in Sect. 8.2.

In general, computing a minimal automaton requires a determinization step, which may cause an intermediate exponential blow-up in the automaton's size. As a result, despite the potential compactness of the final minimized automaton, the computational cost of determinization may render the approach impractical, constituting a bottleneck in our verification pipeline.

We can adopt a pragmatic and computationally feasible strategy to circumvent this challenge: merge automaton states according to simulation and bisimulation relations. Although these relations are stricter than language equivalence, they can be computed efficiently (as discussed above). This yields a practical trade-off between the efficiency of the minimization algorithm and the degree of reduction it achieves.

8.2   Verification Techniques

We need to develop algorithms for parameterized verification of quantum circuits, and expand the regular model checking framework to account for the complexities of quantum systems.

Parameterized Verification. One can carry out parameterized verification of quantum circuits by employing *cut-off* techniques. Empirical evidence suggests that concurrent programs often exhibit a *small model property*, indicating that analyzing only a small number of threads can suffice to capture the reachability of bad configurations. In practice, if any problematic behavior exists, it typically manifests in relatively small instances of the system.

Building on this insight, we can develop methods that leverage the small model property by restricting the verification process to a limited set of fixed-size system instances. Inspired by the *view abstraction* approach of [6], one can design a procedure that involves a fixed-point iteration where we apply gate operations on fixed-size states that effectively sample the quantum state. If the results are inconclusive, we can refine the abstraction by increasing the size of the sampled states. We can repeat this process until we either uncover a specification violation or establish an invariant confirming the correctness of the circuit.

Regular Model Checking. A major current limitation in applying regular model checking to quantum circuits is the lack of suitable frameworks to represent the semantics of quantum gates using transducers. One promising approach is to introduce a new class of weighted transducers, where each transition is labeled not only with input and output symbols—as in classical models—but also with a unitary matrix that captures the behavior of a quantum gate.

In existing quantum computing frameworks, circuit semantics are defined via matrix operations such as composition (matrix multiplication) and the Kronecker (tensor) product. These operations can be naturally expressed through the composition and product of transducers, leveraging their inherent compositional semantics. A central technical challenge in this context is computing the transitive closure of transducer relations. In general, such closures are either not computable or involve prohibitive computational complexity. To mitigate this, we can enhance the framework with acceleration and abstraction techniques, drawing on methods developed in prior work [5], in order to approximate or accelerate the computation of transitive closures.

Because both the (potentially infinite) sets of initial and target states can be encoded as automata, the regular model checking framework will provide a powerful and expressive foundation for verifying both parameterized and fixed-parameter quantum circuits.

8.3   Applications

We can apply the frameworks defined in Sect. 8.1 and Sect. 8.2 to solve verification problems of particular interest in quantum computing. We mention two problems, circuit equivalence, and amplitude measurement, as examples, which we consider in two different tasks.

Circuit Equivalence. We need to develop a framework for checking circuit equivalence: given circuits C_1 and C_2, determine whether they define the same quantum function. The equivalence problem is relevant since we can use C_1 as a reference specification for a more complicated circuit C_2.

One can approach the solution to the equivalence problem based on the following observations. First, circuit equivalence can be verified by checking whether $C_1 C_2^\dagger$ implements the identity transformation, where $C_2^\dagger$ is obtained from C_2 by reversing the circuit (i.e., swapping inputs and outputs) and replacing each gate with its inverse. Since all quantum gates correspond to unitary matrices, their conjugate transposes give their inverses.

Second, to determine whether an n-qubit circuit is equivalent to the identity, we can test it on 2^n linearly independent *vectors*, verifying that each output vector matches its corresponding input. This criterion holds because the identity matrix is the only transformation that maps all vectors to themselves. This step is challenging and will require the techniques such as the ones proposed above.

Third, due to the linearity of quantum operations, it is sufficient to test a maximal set of linearly independent vectors rather than exhaustively considering the entire state space.

Amplitude Verification. Amplitude verification has emerged as an important objective in the verification of quantum circuits. Some initial approaches have been proposed in our recent work [3,14,15], and this remains an ongoing area of investigation.

In many quantum circuits, it is crucial to reason about *relational specifications* that impose numerical constraints on the amplitudes associated with the leaf nodes of tree representations of quantum states. These specifications can express probabilistic guarantees or amplitude transformations, such as: "Grover's circuit yields a correct answer with probability greater than 90%," or "the amplitude of a particular basis state increases after executing circuit C."

To formally capture such properties, we introduce *symbolic quantum states*, where each computational basis state is annotated with a symbolic numerical variable representing its amplitude. Relational specifications can then be encoded as arithmetic constraints over these variables.

Verification proceeds by performing symbolic execution of the quantum circuit, following the general paradigm established in [23]. This produces a symbolic representation of all possible executions, encapsulated as a tree automaton (TA). The correctness of the circuit with respect to a given specification is then reduced to checking whether all trees accepted by the TA satisfy the specified constraints.

To address the computational complexity of this task, we use a modified antichain-based algorithm for tree automata language inclusion [4]. These algorithms optimize the search by pruning the state space using simulation relations over automaton states. In our setting, we extend these simulation relations to incorporate the numerical constraints on symbolic amplitudes, thereby ensuring that the pruning strategy respects both the structural and quantitative aspects of quantum states.

9   Conclusions

In addition to its early practical promise, the approach opens new research directions in automata theory, where quantum structures introduce novel mathematical challenges and opportunities. It establishes a connection between quantum program verification and automata, promoting new possibilities to exploit the richness of automata theory and automata-based verification in quantum computing. It is worth noting that the methodology is not just about catching mistakes but also about building trust in quantum computing systems as we move toward an era in which they might solve problems classical systems cannot.

In conclusion, the paper represents a confluence of disciplines, opening pathways for collaboration between automata theorists, quantum physicists, and software engineers. Looking ahead, this line of research opens up exciting opportunities. Could similar automata-based techniques be adapted to other aspects of quantum software engineering? Could this approach handle quantum programming languages' more abstract and flexible constructs?

References

1. Abdulla, P.A.: A symbolic approach to verifying quantum systems. Commun. ACM (2025). https://doi.org/10.1145/3725725
2. Abdulla, P.A., et al.: An automata-based framework for quantum circuit verification. Presentation at PLanQC 2025, The Fifth International Workshop on Programming Languages for Quantum Computing (2025)

3. Abdulla, P.A., et al.: Verifying quantum circuits with level-synchronized tree automata. Proc. ACM Program. Lang. **9**(POPL), 923–953 (2025). https://doi.org/10.1145/3704868

4. Abdulla, P.A., Chen, Y.-F., Holík, L., Mayr, R., Vojnar, T.: When simulation meets antichains. In: Esparza, J., Majumdar, R. (eds.) TACAS 2010. LNCS, vol. 6015, pp. 158–174. Springer, Heidelberg (2010). https://doi.org/10.1007/978-3-642-12002-2_14

5. Abdulla, P.A., Delzanno, G., Henda, N.B., Rezine, A.: Regular model checking without transducers (on efficient verification of parameterized systems). In: Grumberg, O., Huth, M. (eds.) TACAS 2007. LNCS, vol. 4424, pp. 721–736. Springer, Heidelberg (2007). https://doi.org/10.1007/978-3-540-71209-1_56

6. Abdulla, P., Haziza, F., Holík, L.: Parameterized verification through view abstraction. Int. J. Softw. Tools Technol. Transfer **18**(5), 495–516 (2015). https://doi.org/10.1007/s10009-015-0406-x

7. Amy, M.: Towards large-scale functional verification of universal quantum circuits. In: Selinger, P., Chiribella, G. (eds.) Proceedings 15th International Conference on Quantum Physics and Logic, QPL 2018, Halifax, Canada, 3–7 June 2018. EPTCS, vol. 287, pp. 1–21 (2018). https://doi.org/10.4204/EPTCS.287.1

8. Baier, C., Katoen, J.: Principles of Model Checking. MIT Press, Cambridge (2008)

9. Bauer-Marquart, F., Leue, S., Schilling, C.: symQV: automated symbolic verification of quantum programs. In: Chechik, M., Katoen, J., Leucker, M. (eds.) Formal Methods - 25th International Symposium, FM 2023, Lübeck, Germany, 6–10 March 2023, Proceedings. Lecture Notes in Computer Science, vol. 14000, pp. 181–198. Springer, Cham (2023). https://doi.org/10.1007/978-3-031-27481-7_12

10. Burgholzer, L., Jimenez-Pastor, A., Larsen, K.G., Tribastone, M., Tschaikowski, M., Wille, R.: Forward and backward constrained bisimulations for quantum circuits using decision diagrams. ACM Trans. Quantum Comput. **6**(2) (2025). https://doi.org/10.1145/3712711

11. Burgholzer, L., Wille, R.: Advanced equivalence checking for quantum circuits. IEEE Trans. Comput.-Aided Des. Integr. Circuits Syst. **40**, 1810–1824 (2020)

12. Chareton, C., Bardin, S., Bobot, F., Perrelle, V., Valiron, B.: An automated deductive verification framework for circuit-building quantum programs. In: Yoshida, N. (ed.) ESOP. LNCS, vol. 12648, pp. 148–177. Springer, Cham (2021)

13. Chen, T.F., Chen, Y.F., Jiang, J.H.R., Jobranová, S., Lengál, O.: Accelerating quantum circuit simulation with symbolic execution and loop summarization. In: Proceedings of the IEEE/ACM International Conference on Computer-Aided Design (ICCAD). IEEE/ACM (2024)

14. Chen, Y.F., et al.: Autoq 2.0: from verification of quantum circuits to verification of quantum programs. In: Proceedings of the 31st International Conference on Tools and Algorithms for the Construction and Analysis of Systems (TACAS). ETAPS 2025. Springer, Cham (2025). https://doi.org/10.1007/978-3-031-90660-2_5

15. Chen, Y., Chung, K., Lengál, O., Lin, J., Tsai, W.: AutoQ: an automata-based quantum circuit verifier. In: Enea, C., Lal, A. (eds.) Computer Aided Verification - 35th International Conference, CAV 2023, Paris, France, 17–22 July 2023, Proceedings, Part III. Lecture Notes in Computer Science, vol. 13966, pp. 139–153. Springer, Cham (2023). https://doi.org/10.1007/978-3-031-37709-9_7

16. Chen, Y., Chung, K., Lengál, O., Lin, J., Tsai, W., Yen, D.: An automata-based framework for verification and bug hunting in quantum circuits. Proc. ACM Program. Lang. **7**(PLDI), 1218–1243 (2023). https://doi.org/10.1145/3591270

17. Chen, Y., Rümmer, P., Tsai, W.: A theory of cartesian arrays (with applications in quantum circuit verification). In: Pientka, B., Tinelli, C. (eds.) Automated Deduction - CADE 29 - 29th International Conference on Automated Deduction, Rome, Italy, 1–4 July 2023, Proceedings. Lecture Notes in Computer Science, vol. 14132, pp. 170–189. Springer, Cham (2023). https://doi.org/10.1007/978-3-031-38499-8_10

18. Clarke, E.M., Henzinger, T.A., Veith, H., Bloem, R. (eds.): Handbook of Model Checking. Springer, Cham (2018). https://doi.org/10.1007/978-3-319-10575-8

19. D'Hondt, E., Panangaden, P.: Quantum weakest preconditions. Math. Struct. Comput. Sci. **16**(3), 429–451 (2006)

20. Fang, W., Ying, M.: Symbolic execution for quantum error correction programs. Proc. ACM Program. Lang. **8**(PLDI), 1040–1065 (2024)

21. Floyd, R.W.: Assigning meanings to programs. **19**, 19–32 (1967)

22. Hoare, C.: An axiomatic basis for computer programming. Commun. ACM **12**(10), 576–580 (1969). https://doi.org/10.1145/363235.363259

23. King, J.C.: Symbolic execution and program testing. **19**(7) (1976). https://doi.org/10.1145/360248.360252

24. Kissinger, A., van de Wetering, J.: Simulating quantum circuits with ZX-calculus reduced stabiliser decompositions. Quantum Sci. Technol. **7**(4), 044001 (2022)

25. Mei, J., Bonsangue, M., Laarman, A.: Simulating quantum circuits by model counting. In: Computer Aided Verification. Springer, Cham (2024)

26. Mei, J., Coopmans, T., Bonsangue, M., Laarman, A.: Equivalence checking of quantum circuits by model counting. In: Automated Reasoning, pp. 401–421. Springer, Cham (2024)

27. Niemann, P., Wille, R., Miller, D.M., Thornton, M.A., Drechsler, R.: Qmdds: efficient quantum function representation and manipulation. IEEE Trans. Comput. Aided Des. Integr. Circuits Syst. **35**(1), 86–99 (2016)

28. Perdrix, S.: Quantum entanglement analysis based on abstract interpretation. In: International Static Analysis Symposium, pp. 270–282. Springer, Cham (2008)

29. Piribauer, J., Baier, C.: Partial and conditional expectations in Markov decision processes with integer weights. In: Bojańczyk, M., Simpson, A. (eds.) FoSSaCS 2019. LNCS, vol. 11425, pp. 436–452. Springer, Cham (2019). https://doi.org/10.1007/978-3-030-17127-8_25

30. Sistla, M., Chaudhuri, S., Reps, T.: Symbolic quantum simulation with quasimodo. In: Enea, C., Lal, A. (eds.) Computer Aided Verification, pp. 213–225. Springer, Cham (2023)

31. Tsai, Y.H., Jiang, J.H.R., Jhang, C.S.: Bit-slicing the hilbert space: scaling up accurate quantum circuit simulation. In: Proceedings of the 58th ACM/IEEE Design Automation Conference (DAC), pp. 439–444. IEEE (2021)

32. Vinkhuijzen, L., Coopmans, T., Elkouss, D., Dunjko, V., Laarman, A.: LIMDD: a decision diagram for simulation of quantum computing including stabilizer states. Quantum **7**, 1108 (2023). https://doi.org/10.22331/q-2023-09-11-1108

33. Wei, C.Y., Tsai, Y.H., Jhang, C.S., Jiang, J.H.R.: Accurate BDD-based unitary operator manipulation for scalable and robust quantum circuit verification. In: Proceedings of the 59th ACM/IEEE Design Automation Conference (DAC), pp. 523–528 (2022)

34. De Wulf, M., Doyen, L., Henzinger, T.A., Raskin, J.-F.: Antichains: a new algorithm for checking universality of finite automata. In: Ball, T., Jones, R.B. (eds.) CAV 2006. LNCS, vol. 4144, pp. 17–30. Springer, Heidelberg (2006). https://doi.org/10.1007/11817963_5

35. Yanofsky, N.S., Mannucci, M.A.: Quantum Computing for Computer Scientists, 1st edn. Cambridge University Press, Cambridge (2008)
36. Ying, M.: Floyd-Hoare logic for quantum programs. ACM Trans. Program. Lang. Syst. (TOPLAS) **33**(6), 1–49 (2012)
37. Yu, N., Palsberg, J.: Quantum abstract interpretation. In: Proceedings of the 42nd ACM SIGPLAN International Conference on Programming Language Design and Implementation, pp. 542–558 (2021)
38. Zulehner, A., Hillmich, S., Wille, R.: How to efficiently handle complex values? Implementing decision diagrams for quantum computing. In: 2019 IEEE/ACM International Conference on Computer-Aided Design (ICCAD), pp. 1–7 (2019)

Efficient Join Order for Constraint Automata Through LLM-Generated Heuristics

Ali Mehrani[1], Fatemeh Ghassemi[1(✉)], Marjan Sirjani[2], and Farhad Arbab[3]

[1] University of Tehran, Tehran, Iran
{ali.mehrani,fghassemi}@ut.ac.ir
[2] Mälardalen University, Västerås, Sweden
marjan.sirjani@mdu.se
[3] CWI, Amsterdam, The Netherlands
farhad@cwi.nl

Abstract. Reo is an exogenous coordination language designed for component-based systems based on channel-based connectors. Constraint automata is defined by Christel Baier et al. as the compositional operational semantics of Reo. Semantics of a Reo circuit is computed by joining the constraint automata of the connector elements. This computation can be costly when dealing with large connectors, making an improvement necessary. Improving this operation involves either improving the join algorithm or selecting a joining order that minimizes intermediate automata. While alternative algorithms for joining constraint automata have been proposed, identifying an efficient joining order remains a challenge. This paper proposes a heuristic-based approach for finding an efficient order of joining constraint automata. By feeding OpenAI's ChatGPT with data on the join algorithm and the structure of constraint automata, we ask it to generate diverse heuristics to identify the most efficient joining order and employ its suggestions. Our results demonstrate the impact of join order on the operation's performance. We analyze these results to identify the best heuristic for each set of CAs based on their characteristics. This highlights the potential of LLM-driven approaches in assisting the development of efficient solutions for computational tasks.

Keywords: Constraint Automata · Reo · Coordination languages · Component-based systems · Large Language Models

1 Introduction

Modern distributed systems are widely used as solutions for solving problems when multiple computers have to work together to achieve their goal. These

Dedication. This paper is dedicated to Christel Baier in celebration of her technical and leadership contributions to advancing the field of formal verification in computer science.

N. Bertrand et al. (Eds.): Christel Baier Festschrift, LNCS 15760, pp. 342–359, 2026.
https://doi.org/10.1007/978-3-031-97439-7_17

systems partition tasks among themselves and solve them separately, while also communicating with each other, sharing results, and processing tasks in parallel. Examples of such systems include file-sharing systems, large-scale databases, and service-oriented applications, which are widely adopted solutions for such scenarios. However, using and managing such systems has its own challenges, such as resilience to hardware, software, and network failures. One of the most important among them is managing coordination and concurrency due to the use of different communication protocols. Therefore, it's essential to understand the behavior of these component-based systems.

A good practice for countering this challenge is to separate the system's computation logic from its coordination logic. This allows for monitoring and modeling the coordination mechanism without addressing each node's computational complexity. To achieve this goal, coordination languages have emerged to model the coordination of component-based systems, including distributed and embedded systems. Reo is one of the most well-known coordination languages, introduced by [1] in 2001. The coordination among the components is realized by Reo connectors made of the built-in connectors, called channels, connected to each other through nodes. The data is later passed through the nodes and channels in this circuit. Reo can be used to model and analyze the interaction and communication of different components in a system, especially service-oriented applications. For example, in a work by [10], the interaction of different web services has been modeled using Reo.

The operational semantics of Reo circuits are defined by constraint automata (CAs), which capture the system's data flow. The constraint automaton of a Reo circuit is achieved by joining the constraint automata of its constituent elements, i.e., channels and nodes. There are some tools developed that convert a Reo connector to its constraint automata for verification purposes [4,7,8]. The result of the join operation is not sensitive to the join order, which means that if there were more than two automata to be joined, no matter which two are joined first, the result will always be the same. Different join orders lead to different intermediate automata, which might vary in size. The bigger the intermediate states get, the less efficient the entire operation will be. This is very similar to relational databases, where joining different tables does not affect the result, but can highly affect the operation's performance [2,5].

An inefficient implementation of the join algorithm can significantly reduce performance when dealing with large CAs, as mentioned in [4]. Therefore, it is crucial to improve the efficiency of the join operation. This can be achieved by improving the join algorithm while also identifying an efficient joining sequence that minimizes the size of intermediate automata, reducing the operation's computational load. While alternative algorithms for the former concern have been proposed [6,9], identifying an efficient join order remains a challenge.

Related Work. Automated conversion of Reo circuits to constraint automata has been introduced in some tools that employed different approaches for enhancing the operation's efficiency [5,6,9]. The tool introduced in [5] converts Reo circuits into their corresponding CA by employing a heuristic based on the number of transitions of CAs of adjacent elements to find the order of joining CAs. The employed heuristic examines all pairs of automata that share at least one transition label in common and selects the pair with the smallest transition product for joining. An alternative algorithm for constraint automata join was proposed by [9], being more efficient than the previously proposed algorithms by a constant factor. They also introduced a greedy algorithm for finding the selection order of CAs based on their shared transition labels, aiming to minimize the number of transitions in the intermediate CAs. Their proposed algorithm picks two automata with the largest set of common names in each iteration. An approach for grand composition (production) of constraint automata was introduced by [6]. The grand compositions are computed *state-by-state* instead of *iteratively*. Thus, only the reachable states of the final CA are computed, avoiding the unnecessary computation of intermediately reachable eventually leading to unreachable states.

In this paper, we present a heuristic-based approach for conducting the join operation on multiple constraint automata in such an order that the intermediate CAs become the smallest possible in size, which lowers the total operation's space consumption and boosts its performance. We provided [2] and [5] as context to OpenAI's ChatGPT, then prompted it to generate heuristics to be applied in the join operation. The heuristics are mainly compared by the operation's total computation time and the size of the intermediately generated CAs.

2   Preliminaries

In this section, we briefly introduce Reo connectors, Constraint Automata, and the join operation on Constraint Automata.

2.1   Reo

Reo is an exogenous coordination language used for the compositional reasoning and construction of component connectors, which allows to separate coordination from the behaviors of components. The Reo Connectors consist of channels and nodes that connect components. Each channel can be connected to at most a single component instance at any given time. A channel consists of two ends and a constraint that relates the timing and content of the data flow. Reo has two types of channel ends: *source* and *sink*, where the data is entered through the source end and distributed through the sink channel end. A node is a logical construct that consists of a set of channel ends. There are three types of nodes based on their containing channel ends: source, mixed, and sink. Data flows from source nodes to sink nodes [1, 2]. All channel ends in a source node are of source type. A write operation on this node is only possible when the data is accepted by all the source channel ends. Therefore, a source node acts as a *replicator*. A take operation from a mixed or sink node is possible if at least one of its sink channel ends offers a data item. If data is offered by more than one sink channel end, one is selected non-deterministically. Therefore, a sink (or mixed node with more than one sink channel end) node acts as a non-deterministic *merger*. Reo enables gluing channels of arbitrary types to build component connectors [1,3].

Reo provides a set of primitive channels that can be joined together to form a node in a Reo circuit, where channel ends coincide. Figure 1 shows basic Reo channels *Sync* (1a), *LossySync* (1b), *SyncDrain* (1c), *FIFO-1* (1d), and the merger node (mixed node) (1e) along with their corresponding constraint automata.

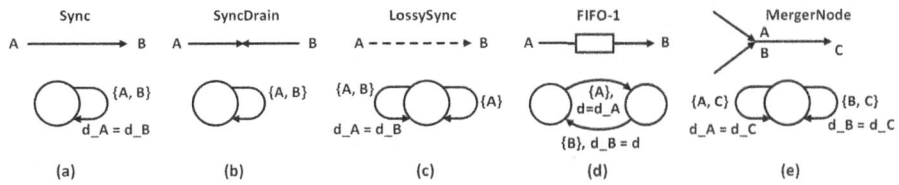

Fig. 1. Graphical representation for basic Reo channels Sync, SyncDrain, LossySync, FIFO-1, and the merger node along with their corresponding constraint automata

2.2   Constraint Automata

Constraint automata (CAs) was introduced in [2] to define the operational semantics of Reo circuits. In these automata, states represent the possible configurations, such as the contents of FIFO channels or other buffered channels, and transitions represent the possible data flow where their labels can be viewed as *sets of I/O-operations* that will be performed *in parallel* on channel ends.

Figure 1a shows a *Sync* channel with the source channel end A and the sink channel end B, along with its corresponding constraint automaton. The constraint automaton for the *Sync* channel has only one state, as this channel lacks a buffer or any other medium to store messages. The single transition in this CA represents the flow of data through the channel from A to B, while the data constraint ensures that the data item received at the source is identical to the one delivered at the sink. Similarly, Fig. 1d shows a *FIFO-1* channel alongside its constraint automaton, with the source end A and sink end B. Assuming that the data domain consists only of the data item d, the initial state of its CA represents an empty buffer, while the second state corresponds to the buffer filled with d. The transition from the initial state to the second state represents the data item d entering the buffer, while the reverse transition represents d leaving the buffer, returning it to an empty state.

Definition 1 *(Constraint Automaton).* *A constraint automaton (over the data domain Data) is a tuple* $\mathcal{A} = (Q, \mathcal{N}ames, \rightarrow, Q_0)$ *where*

- Q *is a set of states,*
- $\mathcal{N}ames$ *is a finite set of names,*
- $\rightarrow$ *is a subset of* $Q \times 2^{\mathcal{N}ames} \times DC \times Q$, *called the transition relation of* $\mathcal{A}$,
- $Q_0 \subseteq Q$ *is the set of initial states.*

We write $q \xrightarrow{N,g} p$ instead of $(q, N, g, p) \in \rightarrow$. We call N the name-set and g the guard of the transition. For every transition

$$q \xrightarrow{N,g} p$$

we require that: (1) $N \neq \emptyset$ and (2) $g \in DC(N, Data)$ is data constraint over N and $Data$. $\mathcal{A}$ is called finite iff Q, $\rightarrow$, and the underlying data domain $Data$ are finite.   □

The CA of a Reo circuit can be obtained by joining the CAs of its constituent elements, i.e., channels and nodes. This procedure combines each two nodes into a single node and creates a new CA from the two input CAs. The join algorithm for automata production is defined as follows:

Definition 2 (Product automaton). *The product automaton of the two constraint automata* $\mathcal{A}_1 = (Q_1, \mathcal{N}ames_1, \rightarrow_1, Q_{0,1})$ *and* $\mathcal{A}_2 = (Q_2, \mathcal{N}ames_2, \rightarrow_2, Q_{0,2})$ *is:*

$$\mathcal{A}_1 \bowtie \mathcal{A}_2 = (Q_1 \times Q_2, \mathcal{N}ames_1 \cup \mathcal{N}ames_2, \rightarrow, Q_{0,1} \times Q_{0,2})$$

where $\longrightarrow$ *is defined by the following rules:*

$$\frac{q_1 \xrightarrow{N_1,g_1}_1 p_1, \quad q_2 \xrightarrow{N_2,g_2}_2 p_2, \quad N_1 \cap \mathcal{N}ames_2 = N_2 \cap \mathcal{N}ames_1}{\langle q_1, q_2 \rangle \xrightarrow{N_1 \cup N_2, g_1 \wedge g_2} \langle p_1, p_2 \rangle}$$

and

$$\frac{q_1 \xrightarrow{N,g}_1 p_1, \quad N \cap \mathcal{N}ames_2 = \emptyset}{\langle q_1, q_2 \rangle \xrightarrow{N,g} \langle p_1, q_2 \rangle}$$

and the latter's symmetric rule.   □

In this definition, $\mathcal{N}$ of *names* is a finite set, e.g., $\mathcal{N} = \{A_1, \ldots, A_n\}$ where A_i stands for the i-th input/output port of a connector or component [2].

3   Heuristics for Joining Constraint Automata

To generate heuristics for joining CAs, we provide ChatGPT with data on the join algorithm and the structure of constraint automata, then ask it to create different heuristics for us. GPT suggested six heuristics. Among them, the transition-based heuristic, called *Min Transitions Heuristic* in Sect. 3.1, was similar to the heuristic proposed in [5], and the heuristic based on shared transition labels, called *Max Connectivity Heuristic* in Sect. 3.1, was used in both [5] and [9]. We applied these heuristics in various scenarios, each generating a specific sequence of CAs to be joined. Regardless of the heuristic used, the result of the join operation remains the same as expected. However, the applied heuristic may alter the join order and intermediate results, causing an improvement or, in rare cases, a decline in the operation's performance. The heuristics generally aim to avoid the generation of large intermediate CAs, also known as *intermediate compounds* [5,6,9].

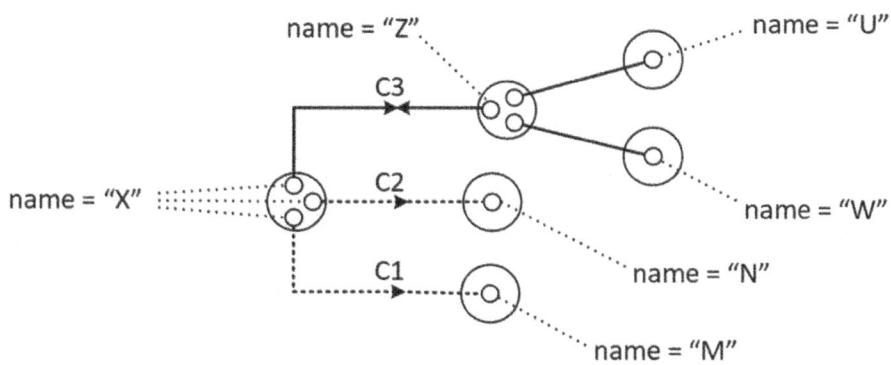

Fig. 2. The running example Reo circuit

We take advantage of a running example to show the applied joining order resulting from each heuristic. Figure 2 shows the Reo circuit for our running example containing two *LossySync* channels named C1 and C2, a *SyncDrain* channel named C3, a merger node, and a replicator node. Figure 3 demonstrates the corresponding CAs that should be joined, and their final product; two CAs for *LossySyncs* along with the CA for *SyncDrain*, a CA for the merger node.

We convert our Reo circuit to an undirected graph of squares shown in Fig. 4. In this graph, squares are connected by lines if their corresponding Reo elements are adjacent to each other in the Reo circuit. Elements in Reo are adjacent if they are connected directly or via a replicator node. Since replicator nodes have no constraint automata, the CA of the Reo circuit in Fig. 2, is achieved by joining the CAs of channels C1, C2, C3, and the merger node, which are named CA1 to CA4. According to this definition, CA1, CA2, and CA3 are adjacent in our square graph because they are adjacent in Fig. 2 via the replicator node, and CA3 and CA4 are adjacent in our square graph because they are connected directly in Fig. 2. We later use this square graph to visualize different joining sequences of constraint automata determined by our heuristics.

3.1   LLM-Generated Heuristics

We introduce the LLM-generated heuristics and explain their working procedures. Later, in Sect. 4, we elaborate on the effect of these heuristics on computation time and memory consumption, leveraging a set of experiments. If we do not use any heuristic for the join operation, the default join order would be $(((CA1 \bowtie CA2) \bowtie CA3) \bowtie CA4)$, which is shown in Fig. 5a. We also assume

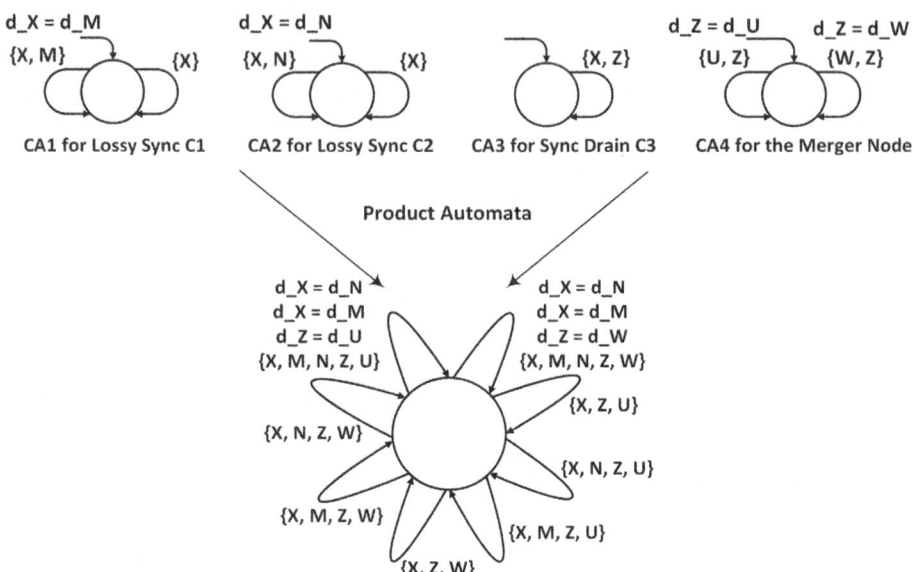

Fig. 3. The CAs of the Reo elements in the running example in Fig. 2 and the CA generated by their join.

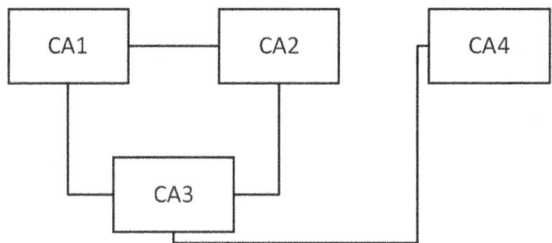

Fig. 4. The original square graph for the proposed running example, where each square represents a constraint automaton. CAs are connected by lines if their corresponding Reo elements are adjacent in the Reo circuit shown in Fig. 2.

that, when the heuristic values are equal, the two CAs with the lowest IDs are picked for the join, the same as the default approach.

Min Transitions Heuristic. This heuristic is similar to the one proposed in [5]. However, unlike that approach, it selects the two CAs with the fewest transitions among all available CAs and computes their product, instead of restricting the selection to adjacent ones (which share at least one transition label). The number of transitions in all CAs, including intermediate ones, is used as their heuristic score. It considers the number of transitions in a CA as an indicator of its size, prioritizing CAs with fewer transitions to keep intermediate CAs small. Figure 5b shows the joining sequence for our proposed example, generated using

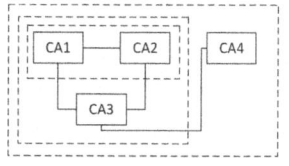 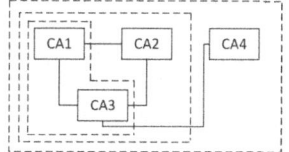 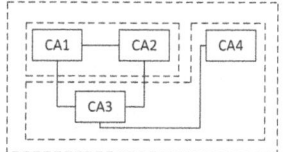

(a) The default join order

(b) Join order (((CA1 ⋈ CA3) ⋈ CA2) ⋈ CA4) selected by the *Min transitions* and *Transition Density* heuristics

(c) Join order ((CA1 ⋈ CA2) ⋈ (CA3 ⋈ CA4)) selected by the *Transition Disparity* heuristic

Fig. 5. The joining sequences generated by the LLM-suggested heuristics along with the default sequence. The dashed line between CAs denotes their join.

the *minimum-transitions* heuristic. It first selects CA3, as it has the fewest transitions (only 1). Among the remaining CAs, which all have 2 transitions, it picks CA1 (the one with the lowest ID) to be joined with CA3. The resulting CA from this join has 1 state and 2 transitions. The remaining CAs—(CA1 ⋈ CA3), CA2, and CA4—all have the same number of transitions, and the heuristic selects (CA1 ⋈ CA3) and CA2 for the next join iteration. The final join sequence generated by this heuristic is (((CA1 ⋈ CA3) ⋈ CA2) ⋈ CA4).

Min States Heuristic. This LLM-suggested heuristic is similar to the previous heuristic, except that it considers the automaton's number of states as its score value, selecting two CAs with the fewest states in each step. This criterion is also one of the main indicators of an automaton's size, making it useful for selecting the smallest CAs first during the join operation. For our proposed example, since the number of states in all CAs, including the intermediately generated ones, is equal to 1, this heuristic generates the same joining sequence as the default approach shown in Fig. 5a.

Transition Density Heuristic. This heuristic is calculated by dividing the total number of transitions by the total number of states in the automaton. It selects CAs with fewer transitions per state. In our running example, CA3 has 1 transition and 1 state, making it the CA with the lowest transition density value of 1, compared to other CAs, each of which has 2 transitions, 1 state, and a transition density value of 2. As a result, CA3 is selected first. Since the transition density values of the remaining CAs are the same, CA1, having the lowest ID, is chosen to be joined with CA3. The CA generated from this join, CA1 ⋈ CA3, has 2 transitions and 1 state, the same as the other remaining CAs. Thus, CA2 is selected to be joined with CA1 ⋈ CA3. The final joining sequence is (((CA1 ⋈ CA3) ⋈ CA2) ⋈ CA4) as shown in Fig. 5b.

Transitions Disparity Heuristic. The idea behind this heuristic is to avoid joining automata with significantly different transition counts. It selects a pair of CAs with the closest number of transitions and joins them, rather than joining CAs with a large difference in transition counts. Using this heuristic, CAs with the smallest difference in their number of transitions are joined first. In our proposed example, CA1, CA2, and CA4 each have 2 transitions, while CA3 has

1. Any pair among CA1, CA2, and CA4 has a transition disparity value of 0, the minimum possible for this heuristic, whereas pairing CA3 with any other CA results in a disparity value of 1. Among the possible pairs within CA1, CA2, and CA4, CA1 and CA2 are selected first, producing an automaton with 1 state and 4 transitions. After this join, CA1 $\bowtie$ CA2 has 4 transitions, while CA3 has 1, and CA4 has 2. The heuristic then selects CA3 and CA4 to be joined, as their transition disparity value (1) is the lowest among all remaining pairs. The final joining sequence is ((CA1 $\bowtie$ CA2) $\bowtie$ (CA3 $\bowtie$ CA4)), as shown in Fig. 5c.

States Disparity Heuristic. This heuristic works similarly to the previous one, except that it measures the difference in the number of states between automata. Thus, CAs with the smallest difference in their number of states are selected first for joining. In our running example, since all CAs—and any intermediate automata generated during the process—have exactly one state, the joining sequence remains the same as the default approach shown in Fig. 5a.

Max Connectivity Heuristic. This heuristic is the same as the selection criterion proposed in [9]. According to Definition 2, joining two CAs that have more transition labels in common generally results in a smaller CA compared to joining two CAs with the same number of transitions and states but fewer shared transition labels. This corresponds to the concept of connectivity in Reo circuits, in which CAs that have more transition labels in common are more connected in the original Reo circuit. Thus, this heuristic prioritizes joining automata that are highly connected in the original Reo circuit. In our proposed example, CA1, CA2, and CA3 share one transition label (X), while CA3 and CA4 share another label (Z). Since CA1 and CA2 have the lowest IDs, they are selected first for joining. The resulting automaton, CA1 $\bowtie$ CA2, also shares one transition label with CA3 and is then joined with it. The joining sequence is the same as the default joining sequence in Fig. 5a.

Each of these heuristics might alter the operation's performance based on the features of the input automata. We demonstrate the impact of different heuristics on the operation's efficiency through a set of test cases (experiments).

4   Evaluation Results

In this section, we evaluate the impact of the heuristics by analyzing the execution results of the join operation and their impact on performance. For this analysis, we consider two main factors: (1) the computation time of the operation and (2) memory usage, measured by the size of the intermediate automata (where the number of transitions in a CA indicates its size). We explain our developed framework for evaluating the operation's performance, later showing the operation's results along with a discussion. Implementation details can be accessed at: https://github.com/UT-ECE-FormalMethods/RtC.

4.1   Evaluation Framework

Our evaluation framework for joining CAs is developed in Java and assigns a heuristic score to each of them. The algorithm iteratively picks two CAs with the lowest or highest scores, depending on the selected heuristic, generates the product automaton, calculates its new heuristic score, and pushes it back to the list of CAs, ordered by their heuristic scores. The algorithm continues until one CA, which is the final result is remained in the list. We evaluate each heuristic across different test cases, each containing a set of CAs to be joined, which we elaborate on further in Sect. 4.2. During this process, we ensure that no optimization, such as caching, occurs and that the operation only uses heuristics to improve its performance. We remark that unreachable states are removed only from the final CA, as a state that is unreachable in an intermediate CA may become reachable in the next joins. Removing these states earlier could lead to an incorrect result.

Our framework is mainly divided into two parts, (1) the *SingleJoin*, which joins two constraint automata using the algorithm discussed in Sect. 2.2, and (2) the *MultiJoin* which takes a list of CAs and joins them in an order determined by the selected heuristic. The latter maintains a list of the CAs, picks them based on their heuristic values, and measures the time taken to join in the determined sequence. Regardless of the time complexity of the *SingleJoin*, a different joining sequence will affect the operation's total performance. In our framework, constraint automata are represented in JSON format. During this process, the ID of the resulting automaton in its corresponding JSON notation denotes the joining sequence used for its generation. For example, if the ID of the final CA is (CA3 ⋈ ((CA1 ⋈ CA2) ⋈ CA4)), it denotes that, to construct the final CA, CA1 and CA2 were first joined, then the intermediately generated CA was joined with CA4, and finally, the resulting CA was joined with CA3 to create the final CA. The framework will measure the total execution time, along with the number of transitions and states of intermediate CAs, to compare different joining sequences determined by the proposed heuristics. Each joining sequence determined by a heuristic is executed 10 times, and the average execution time is considered the total time spent on the operation.

4.2   Evaluation Results

In this part, we discuss the evaluation results of the heuristics on our test cases. We compare the efficiency of the heuristics based on two criteria: computation time and the size of intermediate automata (where the number of transitions in a CA indicates its size), which indicates memory usage. We evaluate the heuristics on a total of 20 test cases, where each test case contains a set of CAs to be joined. For better visualization of the operation's performance, we divide these test cases into four categories: Small, Medium, Large, and X-Large based on the average execution time for the join operation. Each category contains 5 test cases. The execution time for small test cases ranges from a few milliseconds to slightly more than a second, while medium test cases typically take between 1 and 25 s. Large test cases range from tens of seconds to several minutes, and the X-Large test cases have execution times ranging from a few minutes to up to 4 h. Table 1 provides more information about our experiments, where we introduce each experiment along with its characteristics, including the average number of states and transitions of the CAs, the average size of their transition label set (Names), the maximum number of transition labels they share, and the total number of CAs to be joined.

Table 1. Characteristics of the fifteen test cases used for evaluation in our framework. Each test case contains a set of CAs to be joined. Test cases are classified into four categories of Small, Medium, Large, and X-Large based on the average time taken for the join operation on the CAs in each test case.

		Test Case S-1	Test Case S-2	Test Case S-3	Test Case S-4	Test Case S-5
Small (S) Join Time ≤ 1 sec	Total CAs to be joined	3	3	3	4	4
	Avg # of states	3	4	7	5	5.5
	Avg # of transitions	3.3	5.3	8	6	7
	Avg # of CA names	3	4	3.3	4.75	4
	Max # of shared names	1	2	3	4	3.75
		Test Case M-1	Test Case M-2	Test Case M-3	Test Case M-4	Test Case M-5
Medium (M) 1 sec≤ Join Time ≤25 sec	Total CAs to be joined	3	4	5	4	4
	Avg # of states	9	8.5	5.6	7.25	6.5
	Avg # of transitions	15.6	18.75	6.6	11.75	9.25
	Avg # of CA names	5	8	4	5.75	4
	Max # of shared names	4	8	5	2	2
		Test Case L-1	Test Case L-2	Test Case L-3	Test Case L-4	Test Case L-5
Large (L) 15 sec≤ Join Time ≤37 min	Total CAs to be joined	5	4	4	4	4
	Avg # of states	7	10	9.5	10.5	13.25
	Avg # of transitions	12	21.25	12.5	16.25	23.25
	Avg # of CA names	4.6	10	5.5	9	9.5
	Max # of shared names	6	8	3	6	8
		Test Case XL-1	Test Case XL-2	Test Case XL-3	Test Case XL-4	Test Case XL-5
X-Large (XL) 3 min≤ Join Time ≤4 hrs	Total CAs to be joined	6	7	8	9	10
	Avg # of states	5.66	4.85	3.5	3	2.5
	Avg # of transitions	9.16	9.42	5.625	3.77	3.7
	Avg # of CA names	3.83	4.57	4.625	3	2.8
	Max # of shared names	4	6	6	2	4

The execution times were measured on a system equipped with an Intel Core i7-9750H CPU and 16 GB of memory. We expect the join operation to generally have a better performance when using heuristics compared to the default approach.

Figures 6, 7, 8, and 9 show the computation time and the average number of intermediate CAs transitions for small, medium, large, and X-large test cases, respectively, where heuristics are compared with the default approach in both of these criteria. To enhance the visualization of our plot, a small jitter value is added to the plotted values to prevent the lines from overlapping.

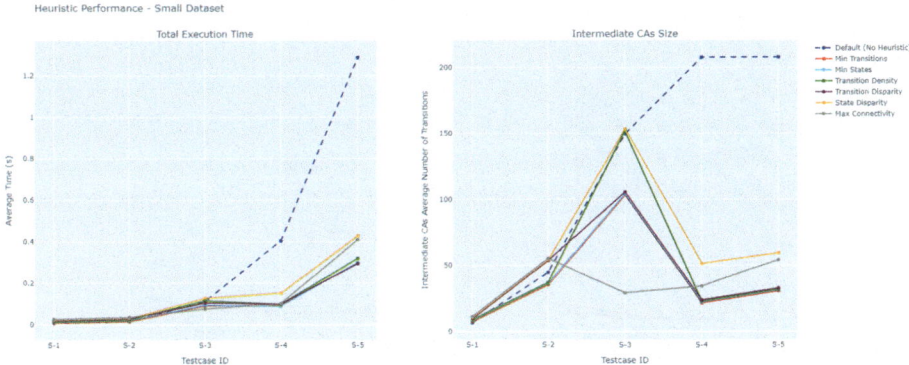

Fig. 6. Comparison of the default approach and six heuristic-based approaches on the Small test case set, based on execution time and the average number of transitions in intermediate CAs. The time unit is seconds.

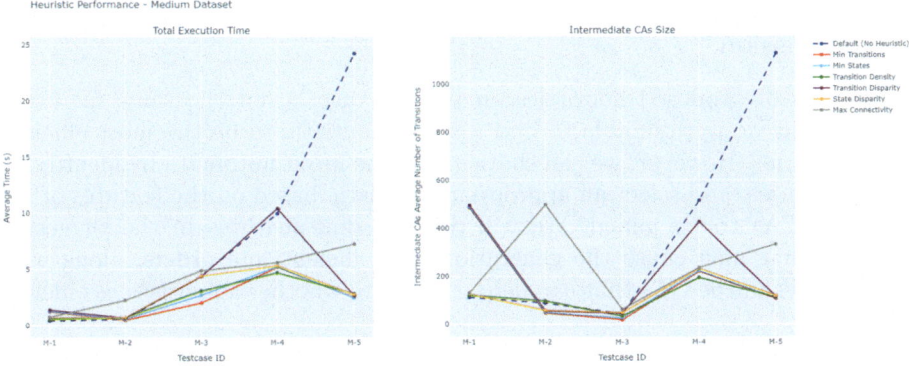

Fig. 7. Comparison of the default approach and six heuristic-based approaches on the Medium test case set, based on execution time and the average number of transitions in intermediate CAs. The time unit is seconds.

The results indicate that the performance difference between the default approach and the heuristic-based approach becomes more noticeable as the number of states and transitions in the input automata increases. A comparison between the size of intermediately generated CAs in each joining sequence and the total execution time confirms a clear connection between the size of intermediately generated CAs and the operation's performance. Minimizing the size of intermediate CAs improves the performance of the join operation.

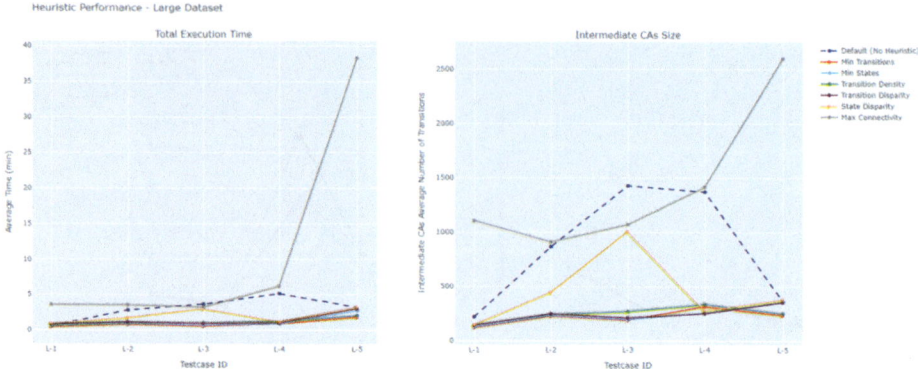

Fig. 8. Comparison of the default approach and six heuristic-based approaches on the Large test case set, based on execution time and the average number of transitions in intermediate CAs. The time unit is minutes.

4.3   Discussion

The impact of proposed heuristics on performance is not the same for every scenario, and we were unable to find a single heuristic to be the most efficient for all scenarios. However, we can characterize the input automata by identifying a set of features, and select an appropriate heuristic based on the features of the remaining CAs to be joined. We will discuss some features in the suggested heuristics that may cause the generation of inefficient join orders, along with some features in the CAs that can help us identify whether a heuristic is efficient or not.

Min Transitions Heuristic. As shown by our results, this heuristic is generally effective, since the number of transitions can be considered a good indicator to measure the size of the CAs. However, it may not always work well. Suppose that we have three CAs and we want to pick the two smallest automata to join. The heuristic first identifies a CA named CA1 (not to be confused with CA1 in Sect. 3) as the one with the fewest transitions. It then needs to decide between two CAs, CA2 and CA3, to join with CA1. We observe that CA2 has a total of n transitions and m states, while CA3 has slightly fewer transitions but significantly more states. Therefore, the heuristic selects CA3 over CA2 and joins it with CA1. However, CA2 would have been a better choice, as it has significantly fewer states than CA3. This leads to the concepts of *density* and *sparsity* in automata when modeled as graphs, where *dense* CAs have a high number of transitions per state compared to *sparse* CAs.

Min States Heuristic. Similar to the previous heuristic, this one considers the number of states in an automaton as its size indicator and may encounter the same issues. Specifically, it may choose a CA with fewer states but a large number of transitions over one with slightly more states but significantly fewer transitions. This heuristic is generally accurate when the numbers of transitions

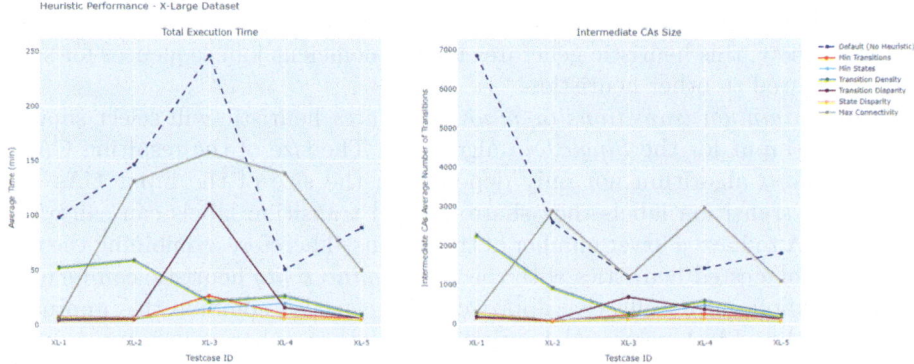

Fig. 9. Comparison of the default approach and six heuristic-based approaches on the X-Large test case set, based on execution time and the average number of transitions in intermediate CAs. The time unit is minutes.

and states in all CAs are close together, but it may not always generate the efficient join order. Considering additional heuristics, such as *max connectivity* or *transition density*, alongside this approach may improve the selection accuracy, especially when multiple CAs have similar state and transition counts.

Transition Density Heuristic. This heuristic first selects CAs with a low number of transitions per state. However, the density of a CA is not necessarily an indicator of its number of states and transitions, so this heuristic may fail to generate an efficient join sequence, particularly when dealing with CAs that vary in their state and transition counts. It would be more effective to combine this heuristic with others rather than using it as a standalone approach.

Transition Disparity Heuristic. This heuristic prioritizes joining CAs that are close to each other by their number of transitions to avoid the generation of large intermediate CAs. Since this heuristic only looks for a pair of CAs with the lowest disparity and does not analyze their own features, such as each CA's size, it might generate an inefficient sequence. It would be a better option to use this heuristic when all CAs are close to each other in size (number of transitions) or to combine it with other heuristics, such as the *minimum transitions* heuristic, to select a pair of CAs that are small and close to each other in size.

State Disparity Heuristic. Similar to the previous heuristic, this heuristic also does not consider the feature of each CA but only its relation to other CAs. Therefore, it would be better to combine it with other heuristics instead of using it as a standalone option to avoid generating inefficient join sequences.

Max Connectivity Heuristic. As previously mentioned, this heuristic was proposed to select CAs that are more connected to each other in the original Reo circuit to minimize the number of possible transitions in the resulting CA. Since this heuristic can greatly affect the size of the resulting CA, it would be a good idea to use it when the CAs are close to each other in size. However, when dealing with CAs diverse in size, we must not solely rely on it. As our test cases show, this heuristic generates highly inefficient join sequences for such cases compared to other heuristics.

Using *minimum transitions* or *minimum states* heuristic will select smaller CAs as the input for the *SingleJoin* algorithm. The size of the resulting CA of the *SingleJoin* algorithm not only depends on the size of the input CAs but also on the transition labels they share. Shared transition labels can cause the resulting CA to have a lower number of transitions. Therefore, combining the two previously suggested heuristics with the *Max Connectivity* heuristic can be used as the backbone of an effective approach. We can also combine this approach with the other LLM-suggested heuristics to further enhance our accuracy in selecting an efficient joining sequence.

4.3.1   Combining Heuristics

To reduce the shortage of an individual heuristic, different heuristics can be combined together as suggested in the previous section. We consider three combined heuristics using the *Min Transitions*, *Min States*, and *Max Connectivity* as the backbone of our combinations. The first two are chosen because they serve as the primary indicators of the size of an automaton and have performed well in test cases. *Max Connectivity* is also included as it is shown to be effective in our experiments when the CAs are close to each other in their number of states and transitions. Joining two CAs with more shared transition labels generally results in a smaller CA compared to joining CAs with the same numbers of states and transitions but fewer shared labels [9]. Our generated heuristics are as follows:

Transitions and States Product Heuristic. This heuristic combines the *States Count* and *Transitions Count* heuristics by computing the product of a CA's state and transition counts, treating both as indicators of its size. It treats these factors as equal contributors to an automaton's size, aiming to minimize their individual shortage. Since all CAs in our running example have the same number of states, this heuristic produces the same joining sequence as the *Minimum Transitions* heuristic, as shown in Fig. 5b.

Transitions and Connectivity Product Heuristic. This heuristic combines the *Min Transitions* and *Max Connectivity* heuristics. It first analyzes all possible pairs of CAs and computes the product of their transition counts. This value is then inverted and multiplied by the number of shared transition labels between the two CAs. The resulting value serves as the heuristic score for each pair. After computing scores for all pairs, the pair with the highest score is selected, joined, and the resulting CA is added to the list of remaining CAs. In our running example, CA1, CA2, and CA3 share a common transition label (X), while CA3 and CA4 share another label (Z). Among these pairs, (CA1, CA3), (CA2, CA3), and (CA3, CA4) have the lowest transition number product, each equal to 2. From these, CA1 and CA3 are selected for joining, resulting in a CA with one state and two transitions. This leaves three CAs, each with one state and two transitions. Next, CA1 ⋈ CA3 and CA2 are selected for joining. The final join sequence produced by this heuristic is $(((CA1 \bowtie CA3) \bowtie CA2) \bowtie CA4)$, the same as the sequence shown in Fig. 5b.

States and Connectivity Product Heuristic. This heuristic is similar to the previous one, but instead of transitions, it computes the product of the state counts for each pair of CAs, then inverts the result and multiplies it by the number of shared transition labels. In our running example, since all CAs (including the intermediate ones) have exactly one state, this heuristic produces the same joining sequence as the *Max Connectivity* heuristic, which matches the default sequence shown in Fig. 5a.

Figures 10 and 11, show the computation time and average number of intermediate CAs transitions for the Medium and X-Large datasets, respectively, where the *Min Transitions*, *Min States*, and the *Max Connectivity* heuristics are compared with these new combined heuristics and the default approach in both of these criteria. To enhance the visualization of our plot, a small jitter value is added to the plotted values to prevent the lines from overlapping. As seen in the results, the combined heuristics (particularly the *Transitions and Connectivity Product* and the *States and Connectivity Product* heuristics) have performed well and can be used as effective heuristics for joining constraint automata. By minimizing the individual shortage of single heuristics, these combined approaches can significantly improve the join operation's performance.

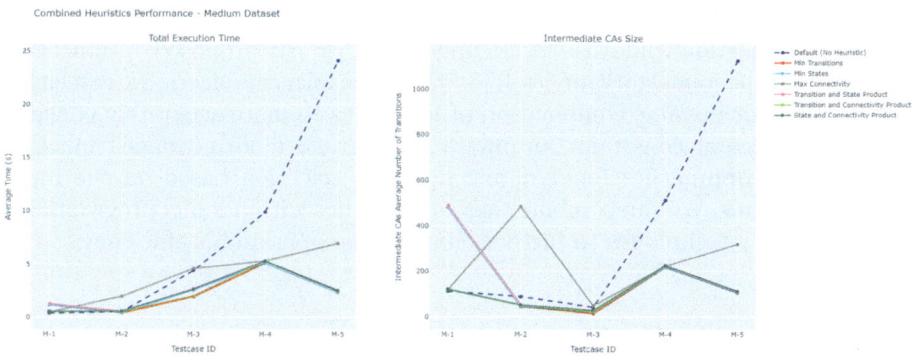

Fig. 10. Comparison of the default approach, *Min Transitions*, *Min States*, *Max Connectivity*, and the three combined heuristics on the Medium test case set, based on execution time and the average number of transitions in intermediate CAs. The time unit is seconds.

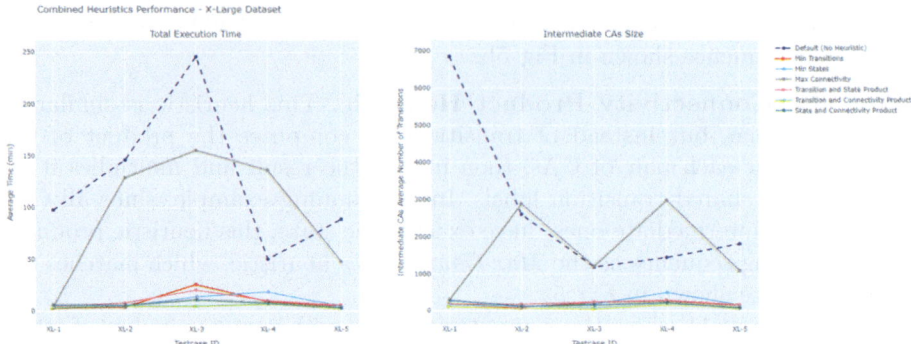

Fig. 11. Comparison of the default approach, *Min Transitions*, *Min States*, *Max Connectivity*, and the three combined heuristics on the X-Large test case set, based on execution time and the average number of transitions in intermediate CAs. The time unit is minutes.

5   Conclusion and Future Work

In this paper, we implemented and evaluated multiple LLM-suggested heuristics for improving the efficiency of the join operation on multiple constraint automata, reducing the size of intermediately generated compounds.

To further improve this operation, we can employ more advanced methods for the heuristic selection while also leveraging the previously proposed optimization techniques for the join algorithm itself [6,9]. For the former concern, we can either practice different possible combination of heuristics to more accurately consider different features of constraint automata or leverage more advanced machine learning-based approaches for dynamic heuristic selection based on the input automata features. We can combine these approaches with parallel programming and concurrency techniques to further enhance the operation's efficiency.

References

1. Arbab, F.: Reo: a channel-based coordination model for component composition. Math. Struct. Comput. Sci. **14**, 329–366 (2004)
2. Arbab, F., Baier, C., Rutten, J., Sirjani, M.: Modeling component connectors in reo by constraint automata. In: Proceedings of Second International Workshop on Foundations of Coordination Languages and Software Architectures (FLOCASA'03), vol. 97, pp. 25–46, July 2004
3. Baier, C.: Probabilistic models for reo connector circuits. J. Univ. Comput. Sci. **11**(10), 1718–1748 (2005)
4. Ghadiri, A.: A tool for constraint automata join, bs project. Technical report, ECE Department University of Tehran, 2004
5. Ghassemi, F., Tasharofi, S., Sirjani, M.: Automated mapping of reo circuits to constraint automata. In: FSEN 2005, volume 159 of ENTCS, pp. 99–115, 2006

6. Jongmans, S.S.T.Q., Kappé, T., Arbab, F.: Composing constraint automata, state-by-state. In: Braga, C., Ölveczky, P. (eds.) Formal Aspects of Component Software. FACS 2015. LNCS, vol. 9539, pp. 217–236. Springer, Cham (2016). https://doi.org/10.1007/978-3-319-28934-2_12
7. Mehta, N., Sirjani, M., Arbab, F.: Effective modeling of software architectural assemblies using constraint automata. Technical report SEN-R0309, CWI, Amsterdam, The Netherlands, 2003
8. Mehta, N., Medvidovic, N., Sirjani, M., Arbab, F.: Modeling behavior in compositions of software architectural primitives. In: 19th IEEE International Conference on Automated Software Engineering (ASE 2004), pp. 371–374, Linz, Austria, September 2004. IEEE Computer Society
9. Pourvatan, B., Rouhy, N.: An alternative algorithm for constraint automata product. In: Arbab, F., Sirjani, M. (eds.) FSEN 2007. LNCS, vol. 4767, pp. 412–422. Springer, Heidelberg (2007). https://doi.org/10.1007/978-3-540-75698-9_29
10. Tasharofi, S., Vakilian, M., Zilouchian Moghaddam, R., Sirjani, M.: Modeling web service interactions using the coordination language reo. In: Dumas, M., Heckel, R. (eds.) WS-FM 2007. LNCS, vol. 4937, pp. 108–123. Springer, Heidelberg (2008). https://doi.org/10.1007/978-3-540-79230-7_8

What-If Scenarios for the BedreFlyt Digital Twin

Åsmund Aqissiaq Arild Kløvstad[1] [ID], Paul Kobialka[1] [ID], Riccardo Sieve[1] [ID], Andrea Pferscher[1] [ID], Laura Slaughter[2], Silvia Lizeth Tapia Tarifa[1] [ID], and Einar Broch Johnsen[1] [(✉)] [ID]

[1] Department of Informatics, University of Oslo, Oslo, Norway
{aaklovst,paulkob,riccasi,andreapf,sltarifa,einarj}@uio.no
[2] dScience Center, University of Oslo, Oslo, Norway
l.a.slaughter@dscience.uio.no

Abstract. Digital twin technology is emerging as a valuable tool for both short-term decision-making and long-term strategic planning across domains such as process industry, energy, space, transport, and healthcare. This paper reports on ongoing work in designing a digital twin to enhance resource planning in hospitals, e.g., for in-patient needs. Our focus here is on a novel technique to express what-if scenarios in digital twins to improve strategic planning processes, spanning, e.g., average-case and worst-case resource needs, expected patient treatments, and ranging over variations in available resources such as bed bays in the hospital ward. Due to the modularity of our digital twin architecture, different what-if scenarios can be explored simply by configuring the digital twin's orchestrator, which triggers a formal methods analysis pipeline that combines executable formal models for simulation, optimization over constraints and a knowledge base that formalizes domain knowledge. We illustrate what-if scenario analysis in our digital twin architecture by considering the problem of bed bay allocation in a hospital ward.

1 Introduction

Predicting the future is easy: most likely, tomorrow will be exactly like today. However, sometimes we may wonder if events could have played out differently. Although people tend to blame destiny when things go wrong, Casanova claims that for the numerous bad turns in his life the blame was his alone (and that, if he could live again, he would do exactly the same) [8]. Thus, he lived his life according to a locally optimal strategy. In contrast to *predictive* analysis [22], which is concerned with what we expect to happen in the near future, *prescriptive* analysis [48] is concerned with so-called *what-if* scenarios, i.e., exploring and comparing the outcomes of strategies that decide between alternative choices. Humans have proven quite good at such prescriptive analysis; i.e., we routinely reason about and compare possible scenarios to derive appropriate strategies.

N. Bertrand et al. (Eds.): Christel Baier Festschrift, LNCS 15760, pp. 360–381, 2026.
https://doi.org/10.1007/978-3-031-97439-7_18

However, many of us struggle with our strategies when the problems get sufficiently complex. Here, computer-aided analyses can help to overcome human limitations and help us find and evaluate strategies. This problem of finding and exploring strategies touches on several aspects of Christel Baier's inspirational work, from the analysis of Markov decision processes (e.g., [4,5]) to learning strategies (e.g., [3,53]).[1]

Digital Twins (DTs) are virtual information constructs that capture the structure, context, and behavior of the "real" system they are twinning, are dynamically updated with data from the twinned system, have predictive capability, and inform decisions that realize value, according to a recent definition by the National Academies of Science, Engineering and Medicine (NASEM) [41]. Historically, DTs have been developed in engineering disciplines, where increasingly sophisticated "virtual replicas" have been used to simulate the behavior of a cyber-physical system and a closed feedback loop feeds control decisions back to the twinned system (e.g., [17]), however, the recent definition by NASEM is broader. Today, DTs can be found in many domains outside of cyber-physical systems, such as healthcare [51], manufacturing [6], and transportation [9].

In this paper, we are concerned with the use of DTs for the model-driven exploration of so-called what-if scenarios, moving from the predictive analysis of near-future events to the prescriptive analysis of hypothetical scenarios. We believe that DTs have a strong potential for applications in both short-term decision-making and long-term strategic planning in various domains. Seen from a formal methods perspective, DTs go beyond standard model-driven techniques by supporting the dynamic update of the model, leveraging a live data feed from the twinned system (known as the "physical twin"). Thus, the DT becomes an infrastructure for data-driven formal methods (e.g., [34]), in which the live data from the twinned system is used to configure a formal model. Similarly, the what-if scenario to be explored need not be fixed in advance, but may be requested on-the-fly by the user of the DT. This dynamically requested scenario may also determine aspects of the model's configuration, as well as the properties to be analyzed. In short, we may think of DT infrastructure as a self-adaptive system [52] for advanced model management, generating the different models and determining the analyses to be performed over these models.

Our focus here is on prescriptive analysis in BedreFlyt (/ˈbeːdrə flyːt/, Norwegian for "Better Flow") [47]. BedreFlyt is a DT for resource management in healthcare. The proper handling of resources at a hospital is crucial to efficient operations [40], e.g., to determine how trained staff, bed availability in the hospital ward, and necessary rooms and equipment match the needs of different activities at the hospital, such as the treatment of patients. The dynamic allocation of these resources is necessary to efficiently manage the workflow and adjust it to avoid bottlenecks in operations, and to improve the prioritization and utilization of available resources [56]. Simulations have been successfully

---

[1] In particular, the last author of this paper had the pleasure of collaborating with Christel in the EU project *CREDO* (including an unforgettable incident in Bonn, the further details of which shall not be unveiled).

used to improve resource allocation in a hospital [46]. By connecting simulation models to live data, the DT can ensure that the simulations more accurately reflect the actual resource allocation problems of the hospital. This way, a DT becomes a meeting point between static planning and dynamic optimization, allowing a better and more dynamic management of the workflow and its associated resources. By configuring the models to explore different scenarios, the DT further supports the comparison of resource management strategies under different assumptions concerning the resources as well as the incoming patients to the hospital.

The main contributions of this paper are (1) a technique to express what-if scenarios in DTs for prescriptive analysis that is parametric in risk tolerance, and (2) a simulation interface for such predictive analyses for human-in-the-loop decision making. We explore *worst-case scenarios* for the bed bay allocation problem at a hospital ward, as well as *sample* statistical information when assigning treatments to incoming patients, by enriching the domain knowledge of the BedreFlyt DT [47] with statistical distributions for patient treatments. We further seamlessly combine such sampling with worst-case scenarios to capture *risk tolerance* in long-term planning. The result is a wide variety of strategies for the long-term planning of bed allocation for patients, that minimizes the number of reallocations. We evaluate the design on a realistic patient diagnosis stream, based on a historical dataset for a hospital ward at the Norwegian hospital *Rikshospitalet.*[2]

2   The BedreFlyt Digital Twin

The BedreFlyt DT [47] aims to aid hospital staff with resource planning in a ward by solving the problem of room allocations for an incoming stream of patients. The complexity of this problem arises from the unknown, new patients arriving at every time step, creating a dynamic scheduling problem. Patients arrive at the hospital with a diagnosis and are assigned a treatment. Then, depending on the needs of the treatment, they will have different requirements in terms of monitoring and time over their stay. Additionally, the hospital wishes to separate patients by gender and to keep contagious patients isolated.

The DT takes a stream of patients with their diagnoses, genders, and contagiousness status as input and then outputs bed bay allocations for all patients so that all the requirements are met. The following describes the architecture and components of the BedreFlyt DT and introduces a simple running example. We start out with a high-level view and then discuss each component individually.

2.1   BedreFlyt DT Architecture

The BedreFlyt DT, depicted in Fig. 1, integrates several formal techniques into a tool chain for prescriptive analysis. Our DT combines formalized domain knowledge about patient treatments and hospital wards, an actor-based executable formal model to explore strategies for streams of incoming patients with

---

[2] https://www.oslo-universitetssykehus.no/steder/rikshospitalet/.

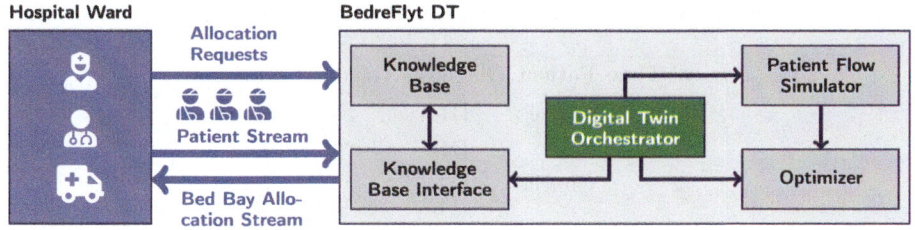

Fig. 1. BedreFlyt DT information flow.

associated treatments, and an optimizer to perform the actual bed bay allocation. Patient flow is expressed in the *abstract behavioral specification* language ABS [28,29], which specifies object-oriented control flow and flexible communication between actors with a timed semantics. The resulting model is compiled into Java and used as a patient flow simulator. We define the optimization problem for bed bay allocation as a constraint satisfaction problem. For this purpose, DT integrates the theorem prover Z3 [12], which is an established SMT solver implementation. The orchestration language SMOL [30,31] is used to connect the knowledge graph to the ABS and Z3 models. SMOL supports querying a knowledge base, which includes the reflection of the runtime state of the SMOL program itself, via SPARQL and SHACL queries (e.g., [24]). The ABS model transforms the stream of patient data into a stream of constraint problems that capture the bed bay allocation problem at different points in time. Together with a description of the ward, these are turned into optimization problems. The four key components of the architecture are: (1) a digital twin orchestrator, (2) a knowledge base and its interface, (3) a patient flow simulator, and (4) an optimizer.

Static domain-specific information is kept in the knowledge base, while dynamic information arrives in two input types. The first type of dynamic information is a stream of incoming patients on a daily basis. The second type of information is *allocation requests* that are received via a simulation interface. The user of the interface, presumed to be hospital staff, requests a bed bay allocation given the current patients, in response to which the interface proposes a possible allocation. Alternatively, the user may ask for a simulation of a stream of patients under different strategy assumptions. In the following, we describe the functionality of the individual components to output such an allocation. We provide details on the different strategies in Sect. 3. We note that this architecture is generally applicable when considering similar problems.

Communication between the orchestrator, patient flow simulator and optimizer components is via *discrete timed streams*. For a set X, a timed stream over X is a sequence of elements of X tagged with monotonically increasing timestamps in $\mathbb{N}$. We write $t_0{:}x_0$, $t_0{:}x_1$, $t_1{:}x_2$ to denote a timed stream where x_0 and x_1 occur at time t_0 and x_2 occurs at time t_1.

Table 1. Example of timed patient input stream.

Arrival Time	Patient	Diagnosis	Gender	Contagious
1	Alice	D1	♀	True
	Bob	D1	♂	False
2	Charlie	D2	♀	False

2.2 The Digital Twin Orchestrator

The digital twin orchestrator is the interface to the twin and coordinates information flow to the twin's other components. It receives a stream of patients and allocation requests, and creates a timed stream of so-called *packages* detailing patient and treatment information, that serves as input for the patient flow simulator.

Let P be the set of patients. A patient $p \in P$ is a tuple $\langle id, g, q, d \rangle$, where id is a unique identifier, $g \in \{♀, ♂\}$ their gender as distinguished by the hospital, $q \in \mathbb{B}$ a Boolean value indicating if the patient is contagious, and d their diagnosis.

The digital twin orchestrator receives at a time $t \in \mathbb{N}$ a set of patients $\{p_1, \ldots, p_n\}$, which all arrive at t. It then constructs a timed stream of *packages* for the patient flow simulator by selecting treatments for each patient's diagnosis based on a strategy. A package is a triple $\langle t, p, \phi \rangle$ consisting of the time of arrival t, patient tuple $p \in P$, and a sequence of tasks ϕ associated with a treatment $tr \in Tr$. The strategy for selecting a treatment tr for a given diagnosis can be based on cost functions or probabilities, we detail the strategies currently supported in the BedreFlyt DT in Sect. 3. After providing the packages, the twin orchestrator receives the stream of bed bay allocations from the optimizer, formats the data and returns it to the user.

Example 1. Table 1 depicts a stream of three patients arriving over two time steps. In the first step, two patients with diagnosis D1 arrive, and in the second step a single patient with diagnosis D2. Note that the patient identified by Alice is contagious and should therefore be isolated. From this patient stream, the digital twin orchestrator may generate the package stream

$$\langle 1, p_{\text{Alice}}, \langle 1, 3 \rangle \langle 2, 2 \rangle \rangle, \langle 1, p_{\text{Bob}}, \langle 1, 3 \rangle \langle 2, 2 \rangle \rangle, \langle 2, p_{\text{Charlie}}, \langle 1, 3 \rangle \langle 1, 2 \rangle \rangle,$$

by picking the tasks the most frequent treatment (see Table 2); here p_{Alice} is the patient tuple for Alice, etc.

2.3 The Knowledge Base and Its Interface

The knowledge database contains static information about the rooms in the hospital ward and about the considered diagnoses and their associated treatments.

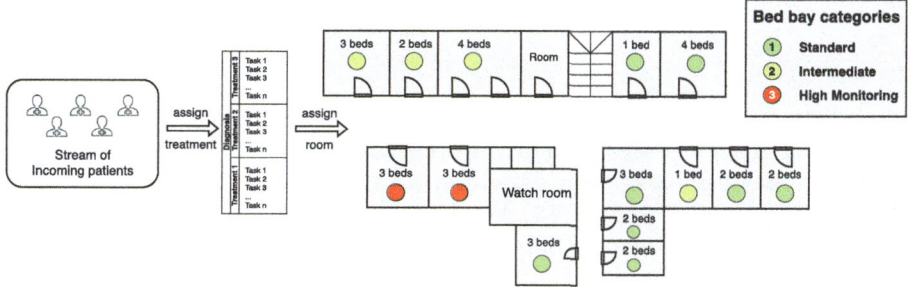

Fig. 2. Example of a hospital ward in BedreFlyt DT.

For a well-structured representation of this knowledge, we use ontologies [45]. The BedreFlyt DT ontology was modeled based on existing available healthcare ontologies and standards, e.g., [11,42,54].

To interface the knowledge base, we use SMOL [30], a small imperative object-oriented programming language that leverages ontologies to develop DTs through semantic reflection. SMOL allows an easy integration of ontologies in DT architectures and provides access to the knowledge modeled in the ontology for DT orchestration [32]. Currently, SMOL is only used to query the knowledge base. In the future, SMOL could orchestrate components of the DT represented in a knowledge base, where queries to the knowledge base consider the semantic reflection of the runtime state [30].

One part of the ontology covers the modeling of assets, i.e. rooms, in the hospital ward. Our example hospital ward of *Rikshospitalet* is depicted in Fig. 2. We describe a room from the set of rooms R as a triplet $\langle r, b, c \rangle$, where r is a room identifier, $b \in \mathbb{N}$ denotes the number of beds in the room, and $c \in C$ defines the monitoring category. In the following, we refer to a room in R also by its identifier r. We distinguish between three levels of monitoring categories $C = \{1, 2, 3\}$, which describe the amount of monitoring capabilities a room provides. Categories are arranged in ascending order, where category 1 maps to standard monitoring efforts, 2 to intermediate, and 3 to high monitoring efforts.

The second part of the ontology models diagnoses and their subsequent treatments, where each treatment consists of a sequence of tasks. Let Ta be the set of task, where a task $ta \in Ta$ is a pair $\langle d, c \rangle \in \mathbb{N} \times C$, with d the task duration, and c the minimal required monitoring category.

We describe a treatment as a pair $tr = \langle \phi, \omega \rangle$, where $\phi = \langle d_1, c_1 \rangle, \ldots,$ $\langle d_n, c_n \rangle$ is a sequence of n tasks and $\omega \in \mathbb{R}_{\geq 0}$ is a real-valued weight. The weight allows the assignment of a real-valued effort to a treatment, where the higher the weight the higher the effort for the treatment. The assignment of the weight is domain-specific, and relies on expert knowledge from the hospital staff. For our DT setup, we used for a treatment $\langle \phi, \omega \rangle$ a simple approximation for ω based on the sequence of tasks ϕ using

$$\omega = \sum_{\langle d_i, c_i \rangle \in \phi}^{n} d_i \cdot c_i. \tag{1}$$

A *distribution* over a finite set X is a function $\mu\colon X \to [0,1]$, where $\sum_{x \in X} \mu(x) = 1$. We write $Dist(X)$ for the set of distributions over X. Let $\mathcal{D}$ be the set of identifiers of possible diagnoses; in the data from *Rikshospitalet* they are alphanumeric codes like "D320", "I60" and "C713". We define a probabilistic treatment function $R\colon \mathcal{D} \to Dist(Tr)$ that maps a diagnosis to a distribution of treatments.

Table 2. Example ontology of diagnoses.

Diagnosis	Treatment	Frequency	Weight	Tasks
D1	T1	0.8	7	$\langle 1,3 \rangle, \langle 2,2 \rangle$
	T2	0.2	8	$\langle 2,3 \rangle, \langle 1,2 \rangle$
D2	T3	0.6	5	$\langle 1,3 \rangle, \langle 1,2 \rangle$
	T4	0.4	3	$\langle 1,2 \rangle, \langle 1,1 \rangle$

Example 2. Table 2 depicts a treatment knowledge base consisting of two diagnoses, each with two treatments. The knowledge base includes the calculated weight and a distribution over the treatments for each diagnosis. For example, the diagnosis D2 has two possible treatments (T3 and T4) and the treatment T3 consists of two length 1 tasks with respective monitoring categories 3 and 2. The probabilistic treatment function R maps diagnosis D2 to the distribution defined by the frequencies of treatments T3 and T4, thus, $R(D2)(T3) = 0.6$ and $R(D2)(T4) = 0.4$. Additionally, the knowledge base describes the hospital ward depicted in Fig. 2, i.e., a hospital ward with 14 rooms, two of which have the highest monitoring category.

2.4   The Patient Flow Simulator

We simulate the workflow by connecting the static structure in the knowledge base, see Sect. 2.3, with the dynamic patient/treatment stream. The workflow simulator takes a timed stream of packages, see Sect. 2.2, including patient information with associated treatments, as input and produces a timed stream of bed bay requirements as output, each of which captures the bed bay allocation problem to be solved at a particular point in time.

The simulator, implemented in ABS [28], receives[3] an input stream of data from the digital twin orchestrator. The simulator retrieves new data from the

---

[3] Technically, the digital twin orchestrator stores the data locally in an embedded SQLite database (see https://sqlite.org), that is queried from the simulator.

digital twin orchestrator at different points in time and reuses the notion of packages internally to capture the active patient treatments; a package consists of patient information and the remaining tasks in the patient's treatment at a certain point of time.

We define a bed bay requirement as a tuple $\beta = \langle id, g, q, c \rangle$ where id is a patient identifier, $g \in \{\female, \male\}$ is a gender, $q \in \mathbb{B}$ indicates contagiousness, and $c \in \mathcal{C}$ is the minimum monitoring category. The bed bay requirements are calculated from the sequence of tasks ϕ in the treatment tr. The ABS model keeps track of active packages and their remaining tasks. At time t, the simulator

1. checks for new packages (i.e., the incoming packages for time t);
2. for each active package $\langle t', p, \langle d_1, c_1 \rangle \ldots \langle d_n, c_n \rangle \rangle$, with $t' \leq t$:
   (a) output the bed bay requirement $\langle id, g, q, c_1 \rangle$ for time t where id, g, q are the identifier, gender and contagiousness status of patient p and c_1 is the minimum monitoring category of the current task, and
   (b) decrement d_1 and remove the associated task if it reaches 0; and
3. remove any packages that have no more tasks

The ABS simulator runs as long as there are active packages, generating a stream of bed bay requirements for different points in time. This output is used to generate a stream of optimization problems for the optimizer component. Remark that the simulation model in ABS is more general than our current case study, because the architecture of the simulator can handle tasks that occur at the same time and have different resource needs; e.g., a laboratory test can occur while the patient occupies a bed bay during recovery. Furthermore, dynamic, unforeseen variations in task duration can be simulated by exploiting the timed semantics of ABS [29].

2.5 The Optimizer

We use the theorem prover Z3 [12] to compute a bed bay allocation for a given stream of packages (patients with treatments); i.e., for each simulated step $t \in T$, we compute an assignment from patients to beds such that all constraints on gender, monitoring categories, contagiousness status, and room capacities are satisfied. We then introduce a target function to minimize the number of required bed bay reallocations, i.e. the number of bed bay changes of single patients over their stay, by minimizing the number of required bed bay reallocations, i.e. the number of bed bay changes of single patients over their stay.

Recall that room $\langle r, b, c \rangle \in R$ is a tuple over the number of beds b and the monitoring category c, and is referenced by r. Further, a bed bay requirement for a patient p is a tuple $\beta = \langle id, g, q, c \rangle$. The input to the optimizer is a timed stream of bed bay requirements $S = t_0{:}\beta_1^0, \ldots, t_0{:}\beta_{m_{t_0}}^0, \ldots, t_n{:}\beta_1^n, \ldots, t_n{:}\beta_{m_{t_n}}^n$ where m_t denotes the number of patients arriving at time t. To shorten notation, let $P^t = \{id \mid t : \langle id, g, q, c \rangle \in S\}$ be the set of patient ids for a time t, and g_{id}^t, q_{id}^t, and c_{id}^t the gender, contagiousness, and minimum room requirement of a patient with identifier id at time t. Note that gender and contagiousness are

$$\varphi^t_{patient} := \bigwedge_{id \in P^t} \sum_{r \in R} a^t_{id,r} = 1,$$
$$\varphi^t_{room} := \bigwedge_{r \in R} \sum_{id \in P^t} a^t_{id,r} \leq b,$$
$$\varphi^t_{gender} := \bigwedge_{id \in P^t, r \in R} a^t_{id,r} \Longrightarrow g^t_{id} = g^t_r,$$
$$\varphi^t_{contagious} := \bigwedge_{id \in P^t, r \in R} a^t_{id,r} \wedge q^t_{id} \Longrightarrow \bigwedge_{id' \in P^t \setminus \{id\}} \neg a^t_{id',r},$$
$$\varphi^t_{category} := \bigwedge_{id \in P^t, r \in R} a^t_{id,r} \Longrightarrow c^t_{id} \leq c$$

Fig. 3. The sub-formulas of the bed bay allocation constraint problem.

constant over time. To compute a valid bed bay allocation, we reformulate the entire problem into a quantifier-free linear real arithmetic formula.

To encode the constraint problem, we introduce two types of variables: (1) variable $a^t_{id,r} \in \{0,1\}$ encodes that at time t, a patient with identifier id is assigned to room r, and (2) variable g^t_r specifies the gender of the room r at that time step. For each time t, the assignment problem is decomposed into the following sub-formulas:

- $\varphi^t_{patient}$ assigns each patient to exactly one room,
- φ^t_{room} limits the number of patients in a room by the room's capacity,
- φ^t_{gender} ensures that patients sharing a room have the same gender by enforcing that all patients in a room have the gender assigned to that room,
- $\varphi^t_{contagious}$ ensures that contagious patients are alone in their room, and
- $\varphi^t_{category}$ restricts the bed bay assignable to a patient based on the monitoring category.

The formulas are detailed in Fig. 3. Then, the formula $\varphi^t := \varphi^t_{patient} \wedge \varphi^t_{room} \wedge \varphi^t_{gender} \wedge \varphi^t_{contagious} \wedge \varphi^t_{category}$ ensures that there exists an assignment of patients to bed bays at time t if and only if that assignment is sound.

We further constrain φ^t to respect the previous bed bay allocation by minimizing the number of required reallocations, avoiding patients being moved around when they stay at the hospital. To this aim, we introduce variable δ^t_{id} indicating whether the patient with identifier $id \in P^t \cap P^{t-1}$ had to move beds between time $t - 1$ and t. Minimizing $\sum_{t \in T} \sum_{id \in P^t} \delta^t_{id}$ under $\bigwedge_{t \in T} \varphi^t \wedge \varphi^t_{changes}$ constructs bed bay allocations minimizing the aggregated patient moves for all time steps, where

$$\varphi^t_{changes} := \begin{cases} \bigwedge_{id \in P^0} a^0_{id,r_{id}}, & \text{if } t = 0, \\ \bigwedge_{id \in P^t \cap P^{t-1}} (a^{t-1}_{id,r} \Longrightarrow a^t_{id,r}) \vee \delta^t_{id}, & \text{otherwise.} \end{cases}$$

Let $a^0_{id,r_{id}}$ denote the initial bed bay allocation for patients $p \in P^0$. If no such allocation is given, the first case in $\varphi^0_{changes}$ defaults to $\top$.

Further, we note that by encoding all time steps, patients and rooms into a single problem, a significant number of variables is introduced. However, our experiments in Sect. 4 reveal that the problem remains computationally feasible in the context of *optimization modulo theories* as implemented in Z3 [7]. If the constructed optimization problem is satisfiable, an optimal allocation of rooms is returned along with the patients that need to be moved to a different room.

3   What-If Scenarios

To reason about potential futures, the BedreFlyt DT employs what-if scenarios. In particular, there might be several treatments for the same diagnosis, depending on the availability of equipment, the patient's preference and underlying health conditions; and a choice of treatment changes the scenario.

Adopting the terminology of *Scenic* [19], a *scenario* is a distribution over configurations, while a *scene* is one such configuration. In our setting, scenes are timed streams of patients with assigned treatments—the inputs to the patient flow simulator. Scenarios are distributions over scenes, given by a patient stream and a (potentially stochastic) *strategy* for selecting treatments—the inputs to the DT orchestrator. Since a strategy determines a scenario if the patient stream is fixed, we conflate the two if the patient stream is clear from the context or does not matter. For example, "worst-case scenario" means the worst-case strategy applied to an understood patient stream.

The BedreFlyt DT implements one deterministic and one stochastic strategy, where the stochastic strategy can be used in simulations to compute expected outcomes. We first explain strategies considered for the hospital ward planning problem in the BedreFlyt DT and how they are used to explore what-if scenarios, before a brief discussion of implementation.

3.1   Different Strategies for the BedreFlyt DT

The existing BedreFlyt DT framework [47] for bed bay allocation is useful for understanding the current state of the hospital ward in relation to the incoming patients by solving the bed bay allocation of incoming patients with given treatments. To analyze the hospital's ability to accommodate patients under different treatments, we now develop a what-if analysis by considering different *strategies* for assigning treatments to patients. We implement (1) worst-case and (2) sampled-case strategies for assigning treatments to patients. When performing simulations, a so-called *risk tolerance* parameter additionally determines the probability of defaulting to the worst-case strategy. By sampling and varying this risk tolerance, we can construct a wide range of what-if scenarios and simulate their expected outcomes.

1. **Worst-case strategy.** The worst-case strategy always picks the treatment with the highest weight. Since the choice of treatment may depend on many factors outside our control (i.e., not-modeled resource requirements, patient health, patient preferences, etc.) this strategy allows us to simulate a scenario in which the hospital is maximally unlucky. This strategy is *deterministic* in that the same diagnosis always results in the same treatment. In the case of a tie we assume a total order on the treatments and pick the first.
2. **Sampled-case strategy.** The sampled-case strategy stochastically picks a treatment for a given diagnosis $\mathcal{D}$ based on the frequencies provided by the probabilistic treatment function $R(\mathcal{D})$. Thus, a treatment with a frequency of 0.2 is expected to be picked one in five times, for patients arriving with that diagnosis.

The treatment strategy is used by the DT to map each patient to a treatment in each *run*, where a run comprises the patient flow simulation and optimizer allocation for a scene. Each run consists of three steps, (1) the DT orchestrator receives a scenario and a strategy, and generates a scene using the given strategy, (2) then the patient flow simulator turns the scene into a stream of optimization problems, (3) finally the optimizer computes a stream of bed allocations.

The non-deterministic sampled-case strategy and the Monte Carlo method allow us to calculate expected values by aggregating a number of runs into a *simulation*. We compute the expected amount of time taken, the expected proportion of time steps without a feasible bed allocation, and the expected number of patients that need to be moved between bed bays in the hospital ward.

We parameterize each simulation by a so-called *risk tolerance* $\tau \in [0, 1]$ that indicates the degree to which we account for a worst-case assignment of treatments. The sampled-case strategy is used with a probability of τ in each run, otherwise the worst-case strategy is used. Thus, with a risk tolerance of 0, the twin will use the worst-case strategy for all patients in all runs. Note that a choice of risk tolerance and a patient stream constitute a scenario—it defines a distribution of treatment assignments.

Example 3. Consider again the setting from Examples 2 and 1. Using what-if scenarios, there are now multiple ways to assign the patients to their bed bays.

Under the *worst-case strategy*, Alice and Bob will both be assigned the treatment T2, and Charlie will be assigned T3. At time 1, all is well as Alice and Bob are assigned to the two high monitoring rooms. At time 2, there is a problem: Charlie needs to be kept in high monitoring, but she cannot be assigned to Bob's room because of their different genders, nor can she be assigned to Alice's room because Alice is contagious. At time 3, Alice and Bob are moved to standard rooms, and there is space for Charlie in an intermediate room (assuming she got the first step of her treatment elsewhere).

Under the *sampled-case strategy*, there are multiple solutions. If Alice is assigned T1, there will be a free room for Charlie because the female high monitoring room is no longer contagious, and if Bob is assigned T1 the similar situation applies since Charlie can move into the now unoccupied male room and turn it female. On the other hand, Charlie could be assigned T4, and not need a high monitoring room at all.

Performing a *simulation* with the simulation tolerance set to 0.8 over 1000 runs, we find that the scenario always takes 4 time steps, the expected number of unsatisfiable allocation problems is ≈ 0.24, and the expected number of room changes per satisfied time step is ≈ 0.11.

If we take the step size to be days, this means the hospital should expect to be unable to accommodate the patients on $\frac{24\%}{4} = 6\%$ of days, and have to move a patient on 11% of days in this scenario.

3.2   Implementation of Strategies in the BedreFlyt DT

Exploiting the modular nature of the BedreFlyt DT, strategies are implemented entirely in the DT orchestrator—directly reusing the patient flow simulator and optimizer components of the existing digital twin architecture [47].

Allocation requests may contain a strategy flag indicating the strategy to be used, and the choice of strategy determines how the DT orchestrator constructs scenes for the patient flow simulator.

Alternatively, a user may send a simulation request to estimate expectations for a given stream of patients. This request is parameterized by a simulation tolerance as described above, a number of runs to perform, and the stream of patients. The DT orchestrator performs the requested number of runs and reports the expected amount of time treatments for all patients will take, the expected number of unsatisfiable time steps (that is, the number of steps where there are not enough bed bays of the right categories), and the expected number of room changes. Since the worst-case strategy is deterministic, its results are cached and reused.

Compared to the previous work [47], we have decoupled the patients diagnosis from their treatments in the BedreFlyt DT. By enriching the knowledge base with multiple treatments for a diagnosis—as well as the additional information concerning their respective weights and frequencies—we can use this information to implement the strategies described above.

4   Evaluation

We evaluate the resulting extension of the BedreFlyt DT along two axes: (1) the use of simulation to analyze what-if scenarios, and (2) the optimality of the solution with respect to the number of bed bay changes to which patients are subjected. Using the simulation requests described in Sect. 3, we investigate the expected number of satisfiable time steps and number of bed changes for a fixed stream of patients across different what-if scenarios. We note that the BedreFlyt DT implements an *online* approach, i.e., one that considers only one step at the time. Alternatively, if the full patient stream is known in advance, we may compute an optimal offline solution as described in Sect. 2.5, we evaluate the quality of the online solution in comparison to that optimal solution.

Comparing Strategies. To investigate the impact of the strategy choice, we create different simulation experiments—sets of simulations with different tolerance levels and bed bay availability. We fix the diagnosis-treatment information in the knowledge base and the stream of patients for each experiment, and vary the tolerance and the number of available high-category bed bays. This answers the question: "Given a tolerance for risk, how many bed bays do we need to upgrade to a higher monitoring category for the incoming stream of patients?". Note that we do not add or remove rooms, but upgrade existing rooms by adjusting their monitoring category. Furthermore, the number of steps it takes to treat all the patients in a stream may vary in each run since the selection of treatments for

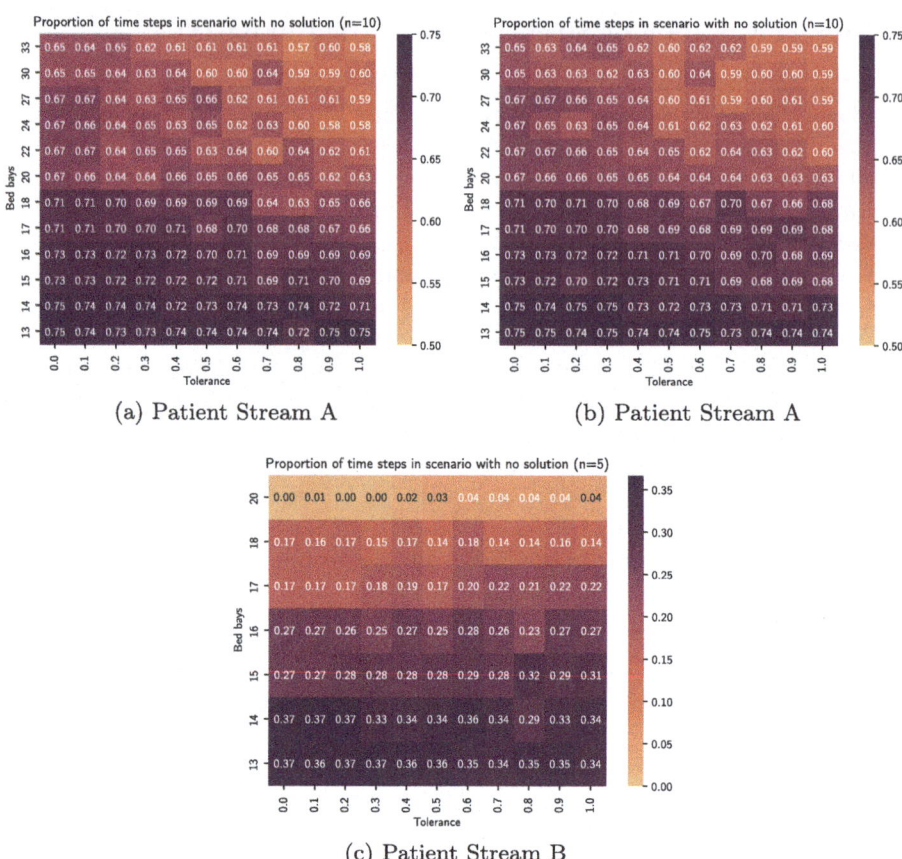

(a) Patient Stream A   (b) Patient Stream A

(c) Patient Stream B

Fig. 4. Unsatisfiable time steps by tolerance and available category-3 bed bays for two different patient streams using simulations with n runs. Plots (a) and (b) show different simulations of the what-if scenario for Patient Stream A.

the same diagnosis is stochastic in general, and so, the DT outputs the average *proportion* of unsatisfiable steps. For our experiments, we randomly generate incoming patient streams, using the anonymized patient identifiers and diagnoses from given historical hospital data. Stream A has 350 unique patients arriving over the course of 35 time steps and stream B has 75 patients over 7 steps.

Having fixed a stream of patients, we execute simulations with n runs with the sampled-case strategy and a tolerance $\tau \in [0, 1]$. If there exists a time step for which no feasible assignment was found, we find the room with the smallest number of bed bays and a less than maximal category. We upgrade this room to the maximal category, and run again n simulations. We continue this process until there are no more rooms to upgrade or no steps are unsatisfiable.

Figure 4 depicts our obtained results. The expectation before the experiments was to confirm that higher tolerance levels will *decrease* the number of unsat-

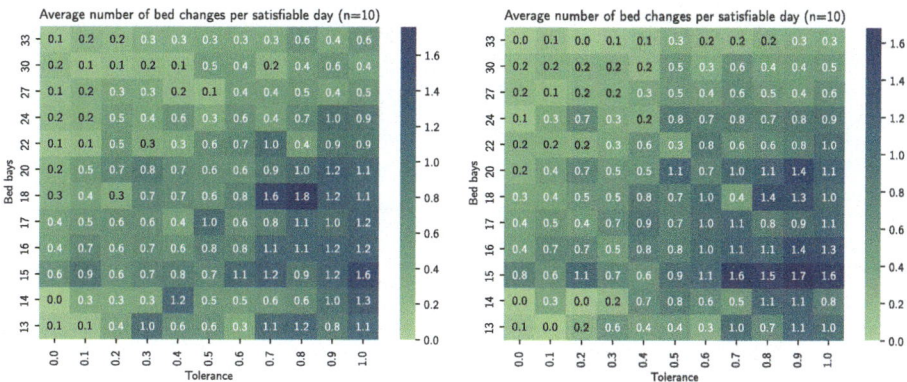

Fig. 5. Average number of bed bay changes per time step, for Patient Stream A; the plots show different simulations of the same what-if scenario.

isfiable problems, since fewer patients receive the worst-case treatment. Consequently, the number of unsatisfiable problems should be inversely proportional to the number of available category-3 bed bays, as there are simply more available slots. Remember that a category-3 room can host a patient of any category, as discussed in Sect. 2.5. Our experiment results approximately align with our expected result—the proportion of unsatisfiable problems generally decreases going up and right, where high tolerance is captured. However, we also observe some noise, which is caused by the complex interactions of patients' requirements. For example, one patient being assigned a treatment with less weight may result in moving them to a higher category room sooner, thus conflicting with other patients already there. Note that the results are relatively stable across simulations. The plots (a) and (b) depict two different simulations of the same patient stream and are very similar—though not exactly the same.

If there exist multiple possible allocations of patients to bed bays such that all constraints are satisfied, the hospital would like to pick the one that requires the fewest number of patients to be moved. To this end, we provide the optimizer with the allocation for the previous time step and compute the number of room changes as explained in Sect. 2.5. The optimizer then computes an allocation that minimizes the number of changes. Figure 5 reports the number of changes per satisfiable time step in two simulations of the same patient stream. As before, we vary the tolerance and number of available category-3 bed bays. Note that more bed bays and lower tolerance levels lead to a larger number of satisfiable problems and, hence, a larger number of possible moves. For this reason, Fig. 5 shows the average number of bed bay changes per satisfiable time step.

Quality of the Online Solution. Due to the limited predictability of bed bay allocations in hospitals in real time, BedreFlyt DT assigns bed bays in an online fashion. Specifically, BedreFlyt DT implements a greedy algorithm that minimizes the number of bed bay changes for the current time step. Note that we

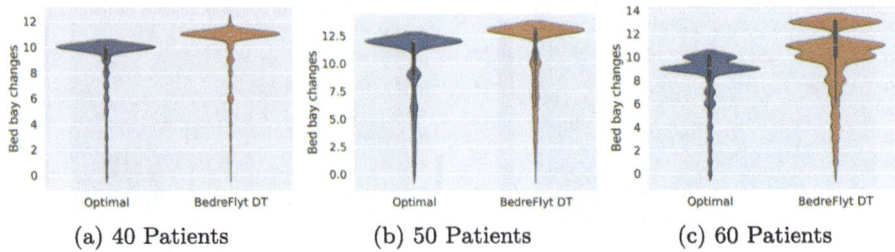

(a) 40 Patients   (b) 50 Patients   (c) 60 Patients

Fig. 6. Comparison of optimal and BedreFlyt DT solutions

can easily construct a situation where moving one patient now will prevent two forced moves in the next step, thus the greedy solution cannot be optimal. In fact, the greedy solution for the very similar *k-server problem* can be arbitrarily bad, but competitive algorithms exist [35].

To investigate the gap between online and optimal allocation solutions, we implemented a *t*-indexed version of the optimization problem, computing a optimal offline solution and comparing it with the greedy approach in the BedreFlyt DT. We employ a meta-heuristic search for synthetic package streams that maximize the number of bed bay changes of the optimal solution, and bypass the simulation component to directly generate problem instances for the optimizer. Specifically, we search for length 5 streams for the same hospital ward as before. This hospital ward has a total of 37 beds—13, 11, and 13 of respectively category 1, 2, and 3—distributed across 17 rooms. We vary the total number of patients and then compare the optimal solution, i.e., the minimum possible number of bed changes when all arrivals are known, to the online solution (with partial information) employed by BedreFlyt DT. For each choice of patent number, we generate in total 30 000 instances, and report the number of changes for the optimal (left) and the online solution (right). The results are displayed in Fig. 6. For 40 and 50 patients, both approaches produce similar distributions of results— the median BedreFlyt DT solution makes one more move than optimal. For 60 patients, BedreFlyt DT skews further than the optimal solution, requiring up to 14 moves where optimal does not reach above 11 moves.

These results indicate that while theoretically, the greedy solution can be arbitrarily bad, it performs similar to the optimal solution on realistic scenes with a moderate number of patient moves required. The gap grows slightly larger with larger numbers of patients, but stays within reasonable bounds.

5   Related Work

Digital Twins in healthcare is an emerging topic [1,57], especially with the impact that COVID-19 had on our lives [23], where emergent DT technologies, e.g. [37], can be integrated with different devices to provide a more comprehensive view of the patients' health. Many solutions explore the application of AI to improve

DTs for healthcare, as done in [33], which combined with context-sensitive applications, can lead to a comprehensive and scalable health system, e.g. [13]. Furthermore, existing work reports on the use of DTs in conjunction with AI to monitor the health of patients in real time [39], management of computational resources [26], etc. In contrast, our work explores digital twins for operational resource analysis in healthcare. Although the use of AI for resource allocation in hospitals has also been explored, e.g. [36], our work focuses on the use of formal methods for resource analysis.

Digital twins have proven to be effective in resource management and resource allocation, e.g. [43,49], but they have not been extensively explored in the context of healthcare [14,25,51]. Resource analysis in hospitals is a critical issue that has been extensively addressed, especially in the Emergency Department (ED) [18], where crucial and trustworthy decisions must be made effectively under uncertainties; in this domain, having a tool to make informed decisions can help drastically. Simulation tools for resource management and decision making are a well-established technique in healthcare [2,44], where they can be used for e.g., resource allocation [16], demographic trends forecasting [21], etc. However, such simulation models have not been integrated into digital twins, as done in this paper. As explored in this paper, the notion of a digital twin goes beyond simulation to include tighter integration between models, data, and decisions [55].

In the context of healthcare and resource management, a data driven performance measurement technique has been used to evaluate the efficiency of hospitals [10] to align the resource allocation needs, while meta-heuristic methods [50] have been assessed for patient allocation. Toward data-driven applications, the integration of different models for resource and capacity allocation in hospitals have been explored [58,59], in both cases, strategies were used to leverage the big volume of data incoming from the hospitals to create a more efficient system to improve operational performance. Closer to the analysis of live incoming streams, as done in digital twins, existing work reports on the use of an adaptive method from near real-time data to predict future bed occupancy levels during a pandemic wave [20], but not for concrete allocation of bed bays at the hospital ward.

In contrast to all the work reported in this section, our work explores digital twin architectures that take advantage of domain knowledge, captured in a knowledge base, for the online and offline analysis of incoming streams of data for decision making support. Furthermore, our work also explores the orchestration of formal models for digital twins, with analysis strategies that are parametric in risk tolerance and consider a human in the loop, touching on various open challenges for digital twins [55].

6   Conclusion

We report on the development of the BedreFlyt digital twin (DT), a tool for resource allocation planning in healthcare. The BedreFlyt DT combines a patient flow simulator and an optimizer to allocate bed bays in a hospital ward according to a stream of incoming patients and their treatments. The proposed DT architecture orchestrates a knowledge base, a patient flow simulator, and an optimizer. In this paper, we consider how the capabilities of the BedreFlyt DT can be extended to analyze hypothetical scenarios over patient streams and strategies for selecting treatments, so called what-if scenarios. We showed how the Monte Carlo method allows the DT to explore expected outcomes for a given knowledge base and patient stream, parameterized by a tolerance factor determining how many of the patient treatments we should expect to be worst case.

Finally, we evaluate the DT's proposed bed bay allocations along two axes: a *qualitative* look at what-if simulations of the same patient stream under different tolerance levels and bed bay availabilities, and a *comparison* of online (day to day) bed bay allocation results with the optimal allocation. The simulation results confirm our assumption that a lower number of bed bays and a lower risk tolerance lead to more unsatisfactory time steps. However, results can vary and be non-monotone due to complex interacting factors. In terms of optimizing the necessary bed bay moves during a patient's stay, we find that the online solution can be arbitrarily sub-optimal, but that for realistic problems the deficit is relatively small.

There are several interesting directions for future work. First, our analysis of what-if scenarios currently work over fixed patient streams, and we would like to sample traffic patterns in a similar way as we now sample treatments. Further, the greedy online bed bay allocation has proven reasonably effective, but does not utilize all the available data. By integrating more knowledge about future steps in treatment sequences, we may further reduce the number of necessary bed bay changes. In addition, the assumption of static information in the knowledge base could be too restrictive in practice. By using techniques to self-adapt the knowledge base, we may ensure that DT better reflects the hospital's reality. Furthermore, online learning could be used to make what-if scenarios more realistic by reinforcing observed behavior in the selected strategies. We note that our assumptions on treatment frequencies and weights are artificial; Using real data from the hospital instead would allow for more realistic modeling and prediction, where it will be interesting to explore advanced stochastic techniques for resource allocation to address, e.g., uncertainties, as explored in [15,38]. Such techniques may also include probabilistic sampling over distributions of patient streams to enable even greater variation in scenario modeling. Finally, we would like to explore the use of Markov decision processes (MDPs) in the healthcare setting. As a modeling and analysis tool for systems with both probabilistic and non-deterministic elements, MDPs are well suited to the world of hospitals where the flow of patients is unknown and staff makes non-deterministic decisions.

Acknowledgment. The work was partly funded by the South-Eastern Norway Regional Health Authority (Helse Sør-Øst) through the project *BedreFlyt*, and the Research Council of Norway through the *Smart Journey Mining* project (grant no. 312198). We thank the other project participants, especially Céline Cunen, Ingrid Konstanse Ledel Solem, Frode Strisland and Manuela Zucknik who helped collect and organize data about patient treatments and patient streams at Oslo University Hospital - Rikshospitalet. We thank the anonymous reviewers for their helpful feedback.

References

1. Alazab, M., et al.: Digital twins for Healthcare 4.0 - recent advances, architecture, and open challenges. IEEE Consum. Electron. Mag. **12**(6), 29–37 (2023). https://doi.org/10.1109/MCE.2022.3208986
2. Almagooshi, S.: Simulation modelling in healthcare: challenges and trends. Procedia Manuf. **3**, 301–307 (2015). https://doi.org/10.1016/j.promfg.2015.07.155, 6th Int. Conf. on Applied Human Factors and Ergonomics and the Affiliated Conferences, AHFE 2015
3. Baier, C., Dubslaff, C., Wienhöft, P., Kiebel, S.J.: Strategy synthesis in Markov decision processes under limited sampling access. In: Rozier, K.Y., Chaudhuri, S. (eds.) Proceedings of the 15th International Symposium of NASA Formal Methods - (NFM 2023). LNCS, vol. 13903, pp. 86–103. Springer, Cham (2023). https://doi.org/10.1007/978-3-031-33170-1_6
4. Baier, C., Katoen, J.: Principles of Model Checking. MIT Press, Cambridge (2008)
5. Baier, C., Piribauer, J., Ziemek, R.: Foundations of probability-raising causality in Markov decision processes. Log. Methods Comput. Sci. **20**(1) (2024). https://doi.org/10.46298/lmcs-20(1:4)2024
6. Billey, A., Wuest, T.: Energy digital twins in smart manufacturing systems: a case study. Robot. Comput. Integr. Manuf. **88**, 102729 (2024). https://doi.org/10.1016/j.rcim.2024.102729
7. Bjørner, N.S., Phan, A., Fleckenstein, L.: νz - an optimizing SMT solver. In: Baier, C., Tinelli, C. (eds.) Proceedings of the 21st International Conference on Tools and Algorithms for the Construction and Analysis of Systems (TACAS 2015). LNCS, vol. 9035, pp. 194–199. Springer, Cham (2015). https://doi.org/10.1007/978-3-662-46681-0_14
8. Casanova, G.: Histoire de ma vie. Robert Laffont (1993). manuscript originally written 1789–1798
9. Chang, X., Zhang, R., Mao, J., Fu, Y.: Digital twins in transportation infrastructure: an investigation of the key enabling technologies, applications, and challenges. IEEE Trans. Intell. Transp. Syst. **25**(7), 6449–6471 (2024). https://doi.org/10.1109/TITS.2024.3401716
10. Chu, J., Li, X., Yuan, Z.: Emergency medical resource allocation among hospitals with non-regressive production technology: a DEA-based approach. Comput. Ind. Eng. **171**, 108491 (2022). https://doi.org/10.1016/J.CIE.2022.108491
11. Dang, J., Hedayati, A., Hampel, K., Toklu, C.: An ontological knowledge framework for adaptive medical workflow. J. Biomed. Inform. **41**(5), 829–836 (2008). https://doi.org/10.1016/j.jbi.2008.05.012
12. De Moura, L., Bjørner, N.: Z3: An efficient SMT solver. In: Proceedings of the 14th International Conference on Tools and Algorithms for the Construction and Analysis of Systems. LNCS, vol. 4963, pp. 337–340. Springer, Cham (2008). https://doi.org/10.1007/978-3-540-78800-3_24
13. Elayan, H., Aloqaily, M., Guizani, M.: Digital twin for intelligent context-aware IoT healthcare systems. IEEE Internet Things J. **8**(23), 16749–16757 (2021). https://doi.org/10.1109/JIOT.2021.3051158

14. Elkefi, S., Asan, O.: Digital twins for managing health care systems: rapid literature review. J. Med. Internet Res. **24**(8) (2022). https://doi.org/10.2196/37641
15. Fan, G., Huang, H.: Scenario-based stochastic resource allocation with uncertain probability parameters. J. Syst. Sci. Complex. **30**(2), 357–377 (2017). https://doi.org/10.1007/S11424-017-6178-5
16. Feng, Y.-Y., Wu, I.-C., Chen, T.-L.: Stochastic resource allocation in emergency departments with a multi-objective simulation optimization algorithm. Health Care Manag. Sci. **20**(1), 55–75 (2015). https://doi.org/10.1007/s10729-015-9335-1
17. Fitzgerald, J., Gomes, C., Larsen, P.G. (eds.): The Engineering of Digital Twins. Springer, Cham (2024). https://doi.org/10.1007/978-3-031-66719-0
18. Florencia, J., Moyaux, T., Trilling, L., Bouleux, G., Cheutet, V.: Toward improving dynamic resource scheduling in the context of digital twin of emergency department. IEEE Trans. Autom. Sci. Eng. 1–13 (2024). https://doi.org/10.1109/TASE.2024.3463489
19. Fremont, D.J., Dreossi, T., Ghosh, S., Yue, X., Sangiovanni-Vincentelli, A.L., Seshia, S.A.: Scenic: a language for scenario specification and scene generation. In: Proceedings of the 40th ACM SIGPLAN Conference on Programming Language Design and Implementation, pp. 63–78. PLDI '19, ACM (June 2019). https://doi.org/10.1145/3314221.3314633
20. Garcia-Vicuña, D., López-Cheda, A., Jácome, M.A., Mallor, F.: Estimation of patient flow in hospitals using up-to-date data. Application to bed demand prediction during pandemic waves. PLOS ONE **18**(2) (2023). https://doi.org/10.1371/journal.pone.0282331
21. Hajłasz, M., Mielczarek, B.: Simulation modeling for predicting hospital admissions and bed utilisation. Oper. Res. Decis. **30**(3) (2020). https://doi.org/10.37190/ord200301
22. Hasan, A., Widyotriatmo, A., Fagerhaug, E., Osen, O.: Predictive digital twins for autonomous ships. In: Proceedings of the Conference on Control Technology and Applications (CCTA 2023), pp. 1128–1133. IEEE (2023). https://doi.org/10.1109/CCTA54093.2023.10252433
23. Hassani, H., Huang, X., MacFeely, S.: Impactful digital twin in the healthcare revolution. Big Data Cogn. Comput. **6**(3), 83 (2022). https://doi.org/10.3390/BDCC6030083
24. Hitzler, P., Krötzsch, M., Rudolph, S.: Foundations of Semantic Web Technologies. Chapman and Hall/CRC Press, Boca Raton (2010). https://doi.org/10.1201/9781420090512
25. Hu, X., Cao, H., Shi, J., Dai, Y., Dai, W.: Study of hospital emergency resource scheduling based on digital twin technology. In: 2021 IEEE 2nd International Conference on Information Technology, Big Data and Artificial Intelligence (ICIBA), vol. 2, pp. 1059–1063 (2021). https://doi.org/10.1109/ICIBA52610.2021.9688239
26. Jameil, A.K., Al-Raweshidy, H.S.: AI-enabled healthcare and enhanced computational resource management with digital twins into task offloading strategies. IEEE Access **12**, 90353–90370 (2024). https://doi.org/10.1109/ACCESS.2024.3420741
27. Ji, S., Pan, S., Cambria, E., Marttinen, P., Yu, P.S.: A survey on knowledge graphs: representation, acquisition, and applications. IEEE Trans. Neural Netw. Learn. Syst. **33**(2), 494–514 (2022). https://doi.org/10.1109/TNNLS.2021.3070843
28. Johnsen, E.B., Hähnle, R., Schäfer, J., Schlatte, R., Steffen, M.: ABS: a core language for abstract behavioral specification. In: Aichernig, B.K., de Boer, F.S., Bonsangue, M.M. (eds.) Formal Methods for Components and Objects. FMCO 2010. LNCS, vol. 6957, pp. 142–164. Springer, Berlin, Heidelberg (2010). https://doi.org/10.1007/978-3-642-25271-6_8

29. Johnsen, E.B., Schlatte, R., Tapia Tarifa, S.L.: Integrating deployment architectures and resource consumption in timed object-oriented models. J. Log. Algebraic Methods Program. **84**(1), 67–91 (2015). https://doi.org/10.1016/J.JLAMP.2014.07.001

30. Kamburjan, E., Klungre, V.N., Schlatte, R., Johnsen, E.B., Giese, M.: Programming and debugging with semantically lifted states. In: Verborgh, R., et al. (eds.) The Semantic Web. ESWC 2021. LNCS, vol. 12731, pp. 126–142. Springer, Cham (2021). https://doi.org/10.1007/978-3-030-77385-4_8

31. Kamburjan, E., Klungre, V.N., Schlatte, R., Tapia Tarifa, S.L., Cameron, D., Johnsen, E.B.: Digital twin reconfiguration using asset models. In: Margaria, T., Steffen, B. (eds.) Leveraging Applications of Formal Methods, Verification and Validation. Practice. ISoLA 2022. LNCS, vol. 13704, pp. 71–88. Springer, Cham (2022). https://doi.org/10.1007/978-3-031-19762-8_6

32. Kamburjan, E., Pferscher, A., Schlatte, R., Sieve, R., Tapia Tarifa, S.L., Johnsen, E.B.: Semantic reflection and digital twins: a comprehensive overview. In: Hinchey, M., Steffen, B. (eds.) The Combined Power of Research, Education, and Dissemination. LNCS, vol. 15240, pp. 129–145. Springer, Cham (2025). https://doi.org/10.1007/978-3-031-73887-6_11

33. Kaul, R., et al.: The role of AI for developing digital twins in healthcare: the case of cancer care. WIREs Data Min. Knowl. Discov. **13**(1) (2023). https://doi.org/10.1002/widm.1480

34. Kobialka, P., Pferscher, A., Bergersen, G.R., Johnsen, E.B., Tapia Tarifa, S.L.: Stochastic games for user journeys. In: Platzer, A., Rozier, K.Y., Pradella, M., Rossi, M. (eds.) Formal Methods. FM 2024. LNCS, vol. 14934, pp. 167–186. Springer, Cham (2024). https://doi.org/10.1007/978-3-031-71177-0_12

35. Koutsoupias, E.: The k-server problem. Comput. Sci. Rev. **3**(2), 105–118 (2009). https://doi.org/10.1016/j.cosrev.2009.04.002

36. Lazebnik, T.: Data-driven hospitals staff and resources allocation using agent-based simulation and deep reinforcement learning. Eng. Appl. Artif. Intell. **126**, 106783 (2023). https://doi.org/10.1016/J.ENGAPPAI.2023.106783

37. Liu, Y., et al.: A novel cloud-based framework for the elderly healthcare services using digital twin. IEEE Access **7**, 49088–49101 (2019). https://doi.org/10.1109/ACCESS.2019.2909828

38. Melouk, S.H., Fontem, B.A., Waymire, E., Hall, S.N.: Stochastic resource allocation using a predictor-based heuristic for optimization via simulation. Comput. Oper. Res. **46**, 38–48 (2014). https://doi.org/10.1016/J.COR.2013.12.010

39. Mohamed, N., Al-Jaroodi, J., Jawhar, I., Kesserwan, N.: Leveraging digital twins for healthcare systems engineering. IEEE Access **11**, 69841–69853 (2023). https://doi.org/10.1109/ACCESS.2023.3292119

40. Mukamel, D., Zwanziger, J., Bamezai, A.: Hospital competition, resource allocation and quality of care. BMC Health Serv. Res. **2**, 10 (2002). https://doi.org/10.1186/1472-6963-2-10

41. National Academies of Sciences, Engineering, and Medicine (NASEM): Foundational Research Gaps and Future Directions for Digital Twins. The National Academies Press (2024). https://doi.org/10.17226/26894

42. Neumann, J., et al.: Ontology-based surgical workflow recognition and prediction. J. Biomed. Inform. **136**, 104240 (2022). https://doi.org/10.1016/j.jbi.2022.104240

43. Peng, K., Huang, H., Bilal, M., Xu, X.: Distributed incentives for intelligent offloading and resource allocation in digital twin driven smart industry. IEEE Trans. Ind. Inform. **19**(3), 3133–3143 (2023). https://doi.org/10.1109/TII.2022.3184070

44. Pitt, M.: A generalised simulation system to support strategic resource planning in healthcare. In: Proceedings of the 29th Conference on Winter Simulation (WSC'97), pp. 1155–1162. ACM (1997). https://doi.org/10.1109/WSC.1997.641004
45. Shadbolt, N., Berners-Lee, T., Hall, W.: The semantic web revisited. IEEE Intell. Syst. **21**(3), 96–101 (2006). https://doi.org/10.1109/MIS.2006.62
46. Sharma, V., Abel, J., Al-Hussein, M., Lennerts, K., Pfründer, U.: Simulation application for resource allocation in facility management processes in hospitals. Facilities **25**, 493–506 (2007). https://doi.org/10.1108/02632770710822599
47. Sieve, R., Kobialka, P., Slaughter, L., Schlatte, R., Johnsen, E.B., Tapia Tarifa, S.L.: BedreFlyt: improving patient flows through hospital wards with digital twins. In: Proc. 1st Intl. Workshop on Autonomous System Quality Assurance and Prediction with Digital Twins (ASQAP 2025). Electronic Proceedings in Theoretical Computer Science, vol. 418, pp. 1–15. Open Publishing Association (2025). https://doi.org/10.4204/EPTCS.418.1
48. Steward, J., et al.: Unleashing the power of what if: cloud-enabled high performance computing workflows in digital twins for scenario exploration. In: Proceedings of the International Geoscience and Remote Sensing Symposium (IGARSS 2024), pp. 2315–2318. IEEE (2024). https://doi.org/10.1109/IGARSS53475.2024.10642587
49. Sun, W., Wang, P., Xu, N., Wang, G., Zhang, Y.: Dynamic digital twin and distributed incentives for resource allocation in aerial-assisted internet of vehicles. IEEE Internet Things J. **9**(8), 5839–5852 (2022). https://doi.org/10.1109/JIOT.2021.3058213
50. Taieb, C., Tlili, T., Nouaouri, I., Krichen, S.: Towards an efficient hospital allocation to patients with resource constraints. In: 10th International Conference on Control, Decision and Information Technologies (CoDIT 2024), pp. 2284–2289. IEEE (2024). https://doi.org/10.1109/CODIT62066.2024.10708224
51. Vallée, A.: Digital twin for healthcare systems. Front. Digit. Health **5**, 1253050 (2023). https://doi.org/10.3389/fdgth.2023.1253050
52. Weyns, D.: An Introduction to Self-Adaptive Systems: A Contemporary Software Engineering Perspective. Wiley-IEEE Computer Society Press, Hoboken (February 2021). https://doi.org/10.1002/9781119574910
53. Wienhöft, P., Suilen, M., Simão, T.D., Dubslaff, C., Baier, C., Jansen, N.: More for less: safe policy improvement with stronger performance guarantees. In: Proceedings of the 32nd International Joint Conference on Artificial Intelligence (IJCAI 2023), pp. 4406–4415. ijcai.org (2023). https://doi.org/10.24963/IJCAI.2023/490
54. Wilk, S., et al.: An ontology-driven framework to support the dynamic formation of an interdisciplinary healthcare team. Int. J. Med. Inform. **136** (2020). https://doi.org/10.1016/j.ijmedinf.2020.104075
55. Willcox, K., Segundo, B.: The role of computational science in digital twins. Nat. Comput. Sci. **4**(3), 147–149 (2024). https://doi.org/10.1038/S43588-024-00609-4
56. Withanachchi, N., Uchida, Y., Nanayakkara, S., Samaranayake, D., Okitsu, A.: Resource allocation in public hospitals: Is it effective? Health Policy **80**, 308–313 (2007). https://doi.org/10.1016/j.healthpol.2006.03.014
57. Xames, M.D., Topcu, T.G.: A systematic literature review of digital twin research for healthcare systems: research trends, gaps, and realization challenges. IEEE Access **12**, 4099–4126 (2024). https://doi.org/10.1109/ACCESS.2023.3349379

58. Yu, W., Liu, Q., Zhao, G., Song, Y.: Exploring the effects of data-driven hospital operations on operational performance from the resource orchestration theory perspective. IEEE Trans. Eng. Manag. **70**(8), 2747–2759 (2023). https://doi.org/10.1109/TEM.2021.3098541

59. Zhu, T., Liao, P., Luo, L., Ye, H.Q.: Data-driven models for capacity allocation of inpatient beds in a Chinese public hospital. Comput. Math. Methods Med. **2020**(1) (2020). https://doi.org/10.1155/2020/8740457

RUNNING.CHRISTEL: A Stochastic Hybrid Case-Study Optimizing Battery Pack Usage

Lisa Willemsen[1]([✉])(ID), Anne Remke[1,2](ID), Boudewijn R. Haverkort[1](ID),
and Johann L. Hurink[1](ID)

[1] University of Twente, Enschede, The Netherlands
{l.c.willemsen,b.r.h.m.haverkort,j.l.hurink}@utwente.nl
[2] University of Münster, Münster, Germany
anne.remke@uni-muenster.de

Abstract. Over the last two decades, batteries have become essential components in many high-tech systems, enabling the storage of (electrical) energy for later use. However, the high costs and scarcity of the materials used for high-end batteries make their efficient use and dimensioning a crucial aspect in system design. Enabling such dimensioning requires accurate modeling of the battery behavior under realistic workload conditions. In this paper we use the Kinetic Battery Model (KiBaM) to effectively capture key behavioral aspects of batteries at reasonable modeling costs. Also an accurate battery workload model is needed that describes the demand of energy over time; such workload models often include stochastic elements. Timed automata models have been used to evaluate battery lifetimes under (primarily) *deterministic* workloads for only small battery configurations. The previously proposed approach required a discretization that led to a computational error that could not be quantified in general. Instead, this paper adopts a stochastic hybrid modeling (SHM) approach to better capture battery dynamics as well as *stochastic* workloads. This paper, hence, presents an exploration of state-of-the-art SHM methods and tools for the analysis of battery systems under stochastic workloads, and an investigation of the advantages and disadvantages of these techniques.

Keywords: Batteries · Stochastic Hybrid Models · Optimization

1 Introduction

Batteries play an important role in a wide range of wireless high-tech systems and applications. They also enable significant sustainability contributions, by their ability to temporarily store energy, hence their capability to decouple energy generation from energy use. In modern households, for example, solar panels combined with stationary batteries allow for the storage and later use of self-generated solar electricity, particularly during periods when sunlight is not

© The Author(s), under exclusive license to Springer Nature Switzerland AG 2026
N. Bertrand et al. (Eds.): Christel Baier Festschrift, LNCS 15760, pp. 382–407, 2026.
https://doi.org/10.1007/978-3-031-97439-7_19

available. In addition, batteries also support grid stability by enabling energy balancing, which improves the overall safety and reliability of the electricity grid [26]. Despite these advantages, the relatively high cost and limited availability of battery materials necessitate a careful and efficient use of batteries to maximize their utility. Accurate modeling of battery behavior is crucial for optimizing performance in various systems, such as smart home networks and space exploration technologies. In particular, understanding battery aging and optimizing their lifetime are essential to ensure that tasks can be completed reliably and timely within the constraints of a given battery configuration. A battery model that accounts for key non-linear battery-internal phenomena, such as the rate-capacity effect and the recovery effect (to be described below) is therefore essential. The so-called Kinetic Battery Model (KiBaM) [35] has proven to be effective in representing such non-linear behavior, making it a valuable tool in battery system modeling.

In addition to the battery model itself, an effective and realistic workload model is required to capture the stochastic nature of energy demand and use. Previous studies, cf. [27,28], have employed a deterministic approach to model the combination of two batteries and evaluate their lifetimes under various fixed workload patterns using model checking techniques for timed automata, thereby using the tool UPPAAL. This approach relies on a discretization of the battery storage levels and can predict a system's life-

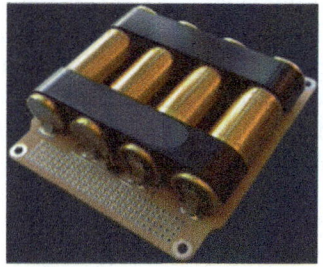

Fig. 1. 4-cell battery pack.

time for deterministic workloads. However, it is less suitable for stochastic workloads due to the inherent variability. Furthermore, the scalability of this approach is limited by state-space explosion, constraining evaluations to just two batteries. To address these limitations, this paper proposes a stochastic hybrid modeling (SHM) approach, aiming to capture the complex interactions between the battery systems and less predictable stochastic workloads. We explore different simulation and analysis tools capable of extracting meaningful insights from SHM models, with the goal of improving battery system design and management under real-world conditions.

This paper is dedicated to the 60th birthday of Christel Baier. Its topic is a deliberate choice, as it connects us to Christel Baier. Since the beginning of our cooperation, we have been working on stochastic models, bringing together expertise from the formal methods field within theoretical computer science, and from operations research, including stochastic processes, Markov chains, Markov decision processes and computer performance and reliability. We are thankful to Christel for her open attitude and admire her willingness to really get to the bottom of interesting research questions. Also, we appreciate the many things we learned while cooperating, as well as the many fruitful and pleasant days of cooperation, in Bonn, Dresden, Aachen and Twente, as well as during other events around the globe. Anne is still grateful for her time as visiting scholar in Dresden with Christel's group in 2014.

Outline. In Sect. 2 we discuss related work, before we address battery models in Sect. 3. In Sect. 4 we briefly present stochastic hybrid models and in Sect. 5 we discuss tools that support their analysis. Some of the considered tools support reinforcement learning, which we touch upon in Sect. 6. In Sect. 7 we introduce a case-study addressing the lifetime of a system with N batteries. The implementation of the case-study is described in Sect. 8 and results are discussed in Sect. 9. Finally, we conclude the paper in Sect. 10.

2   Related Work

Over the past decades many lines of research focused on quantifying safety properties for safety-critical systems. To provide a quantitative analysis, multiple formalisms for stochastic hybrid models have been proposed, which extend hybrid automata [4,23] with stochastic components. For example, [7] introduces the class of stochastic timed automata, which resolves the inherent non-determinism in timed automata stochastically. A more general definition of stochastic hybrid models named CTSHA is given in [34], where continuous variables evolve according to a stochastic differential equation and two stochastic kernels are used to determine the jump times and destinations within the execution of the hybrid automaton. Time-discrete versions of CTSHA are discussed in [1]. Further, [14,46] extend subclasses of HA with random clocks, which model a stochastic resolution of continuous non-determinism. While the above formalisms allow to accurately model the interplay of discrete, continuous and stochastic behavior, the formal analysis of such models is a challenging task due to their inherent complexity. This issue has been tackled via other lines of work which focus on the simulation of stochastic hybrid models. For instance, in [39,40] hybrid Petri net models of systems are simulated to train a reinforcement learning agent to take optimal decisions in the given system.

The tool UPPAAL [6] is widely used and offers a range of tools and verification techniques for networks of timed automata. Stochastic extensions for more expressive models can be evaluated with *statistical model checking* (SMC) and optimal strategies can be identified and constructed [5,13]. Other tools for the analytical evaluation of stochastic hybrid models focus on *discrete-time* stochastic systems [11,47]. Algorithms for non-linear continuous dynamics are provided in [45]. The tools applied in this paper are introduced in more depth in Sect. 5.

Next to these research-focused tools, the simulation tool Simulink is well-established in both industry and research [36]. The combination of deductive verification and SMC-based learning have been considered in [2,3] to obtain safe and resilient systems in Simulink. A formal semantics is provided for a subclass of Simulink models using a transformation in [9].

The lifetime of a system with two attached batteries is analyzed in [27,28], which corresponds to the time in which at least one battery is not empty. It is assumed that jobs may have different service times and energy demands. During the system execution, jobs are served sequentially, where idle periods may occur. To determine a schedule maximizing the system's lifetime, [27, 28] model the two batteries via a discretized version of the KiBaM and implement the system as a priced-timed automata (PTA). This PTA is then analyzed using UPPAAL Cora resulting in a control schedule which determines which battery should be used to serve the job in order to achieve a maximal overall system lifetime. If a battery runs empty while serving, it is marked empty and the other battery, if still non-empty, continues the job. Note that a battery that has been marked empty cannot be chosen again. Furthermore, the work presented in [27, 28] compares the schedule maximizing the lifetime of the battery against other scheduling techniques such as round robin or sequential discharging. To avoid state-space explosion in the model-checking process, the number of scheduling decisions has to be kept low in [27, 28]. Thus, the maximum battery capacity and the number of batteries needs to be restricted. Furthermore, the discretization of the KiBaM leads to small deviations to the behavior of the original KiBaM. The above case-study is extended in [8] by considering a stochastic extension of the KiBaM [25]. This extension is applied to a nano-sattelite of type GOMX-3, which is also considered in [56]. Here, the satellite is considered to have one battery and is modeled by a stochastic timed automaton in UPPAAL Stratego. The work then uses reinforcement learning to obtain strategies maximizing the longevity of a battery.

3   Battery Modeling

Formally speaking, a battery consists of multiple electrochemical cells, however, a single cell is also often called a battery. Within a cell, an electrochemical reaction converts the stored chemical energy into electrical energy. For a more detailed overview, we refer to [29]. Depending on the chemical reaction type, some batteries can be recharged, while others cannot. In general, the *maximal capacity* C_{max} of a battery indicates how much energy can be stored in the battery. Furthermore, the *effective capacity* states how much energy can still be used during the current discharge-cycle. Due to the electrochemical processes, modeling the behavior of batteries is complex. In particular, during discharge two main non-linear effects occur [30]:

– If a battery (cell) is discharged with a high current, the voltage drops slowly and the resulting effective capacity of the battery is lower, which is called the *rate-capacity effect.*
– The effective capacity can recover in an idle period after periods of discharge. This *recovery effect* may again increase the perceived battery lifetime.

Clearly, a linear battery model is unable to capture these effects but instead models only an ideal case, where voltage remains constant during discharge, and the effective capacity drops to zero as soon as the battery is empty (and not earlier than that). To better capture the above two physical effects, a large variety of models have been proposed ranging from simple analytical models, e.g., based on Peukert's law [42], to detailed electrochemical models, as summarized in [12,30]. Detailed electrochemical models allow for more precise modeling but are also more complex and more difficult to use in practice [32].

In the remainder of this section we discuss an intuitive analytical model, the so-called *Kinetic Battery Model* (KiBaM) [35], which offers a good balance between accuracy and complexity. The KiBaM, as depicted in Fig. 2, assumes that the charge of the battery is distributed over two wells, called available (*a*) and bounded (*b*) charge.

The proportional width of the available charge is given by a parameter $c \in (0,1)$ and the proportional width of the bounded charge well is $1 - c$. When applying a load s to the battery, the energy is taken (consumed) from the available charge. Meanwhile, the available charge well is refilled from the bounded charge, via a pipe with fixed conductance k. The change of charge within each well is described by differential equations with parameters c, k, and C_{max}, as follows:

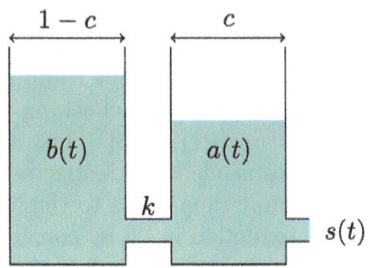

Fig. 2. Graphical representation of the KiBaM.

$$\dot{a}(t) = k \cdot \left(\frac{b(t)}{1-c} - \frac{a(t)}{c} \right) + s(t),$$

$$\dot{b}(t) = k \cdot \left(\frac{a(t)}{c} - \frac{b(t)}{1-c} \right), \tag{1}$$

where $a(t)$ and $b(t)$ denote the available and bounded charge at time t and $s(t)$ is the load that is applied to the battery at time t. If the available charge is at its upper (lower) boundary and $s(t)$ is larger (smaller) than $-k \cdot \left(\frac{b(t)}{1-c} - \frac{a(t)}{c} \right)$, $\dot{a}(t)$ is defined to be 0.

In the following we assume that the initial charge is distributed over the wells according to the proportional width, i.e., $a(0) = c \cdot C_{max}$ and $b(0) = (1-c) \cdot C_{max}$ and we assume that the workload $s(t)$ only changes at given discrete time points $t_1, t_2, \ldots, t_n$. Note, that the KiBaM model indeed captures certain non-linear behavior, like the *rate-capacity effect* and the *recovery effect*.

4   Stochastic Hybrid Automata

Hybrid automata (HA) [24] are a well-established formalism to model complex real-world applications, as they can capture the interplay of discrete and continuous behavior. Naturally, HA include inherent non-determinism via multiple possible discrete or continuous actions and can be used to perform a qualitative analysis, however, in many cases a quantitative analysis is desired. We give a formal definition below.

Definition 1. (Hybrid Automata: Syntax [4]). *A HA $\mathcal{H}$ is a tuple (Loc, Var, Flow, Inv, Edge, Init), with the following components:*

- *Loc denotes a finite set of locations.*
- *Var is a finite set of real-valued variables. A function $v : Var \rightarrow \mathbb{R}$ assigns a real valuation to each variable. The set of all valuations is denoted V. The state of a HA is given by the tuple (q, v), which consists of a location $q \in Loc$ and a valuation $v \in V$. The location of the current state is called the control location.*
- *Flow is a labeling function that assigns a set of flows to each location. Each flow $f \in Flow(q)$ is given by a function $f : \mathbb{R}_{\geq 0} \rightarrow V$ and specifies the evolution of the continuous state. It is required that flows are time-invariant. Thus, for all locations $q \in Loc$, all flows $f \in Flow(q)$ and $t \in \mathbb{R}_{\geq 0}$ also $(f + t) \in Flow(q)$, with $(f + t)(t') = f(t + t')$, for all $t' \in \mathbb{R}_{\geq 0}$.*
- *Inv is a labeling function which assigns an invariant $Inv(q) \subseteq V$ to each location $q \in Loc$. The invariant indicates the domain of the continuous state within the current location.*
- *Edge $\subseteq Loc \times 2^V \times (V \rightarrow V) \times Loc$ is a finite set of discrete transitions or jumps. For a jump $(\ell, g, r, \ell') \in Edge$, ℓ and ℓ' are its source resp. target locations, a its label, g its guard, and r its reset.*
- *Init is a non-empty set of initial states.*

Note that compared to [4], we omit labels from the above definition. Adding labels would allow us to specify the parallel compositions of HA which, however, is not required for the course of this paper.

The execution of an HA distinguishes between discrete and timed steps, where discrete steps correspond to taking a transition and timed steps correspond to the passage of time within the control location. A discrete step can only be taken if the continuous state fulfills the specified guard and invariant at the target location. Similarly, a timed step is only possible if a passage of time would not violate the invariant of the current location.

Definition 2. (Hybrid Automata: Semantics [4]). *For a HA (Loc, Var, Flow, Inv, Edge, Init), the semantics of a discrete step is:*

$$\frac{(q, \mu, q') \in Edge \quad (v, v') \in \mu \quad v \in Inv(q) \quad v' \in Inv(q')}{(q, v) \rightarrow (q', v')} Rule_{discrete}.$$

The timed step semantics of the automaton is given by:

$$\frac{f \in Flow(q) \quad f(0) = v \quad f(t) = v' \quad \forall 0 \leq t' \leq t : f(t') \in Inv(q)}{(q, v) \xrightarrow{t} (q, v')} Rule_{time}.$$

A path $\varrho_{\mathcal{H}}$ of an HA $\mathcal{H}$ is a finite or infinite sequence of states, where changes in the state happen according to either a timed step or a discrete step.

HA have been extended with stochastic components [1,7,15,21,34] which (partially) resolve the HA's inherent non-determinism. Several approaches to include stochastic behavior have been generalized as well as classified w.r.t. their expressivity in [53,54]. In particular, they extend HA with stochastic kernels in various ways to allow for stochastic decisions on the discrete or continuous action, as summarized in the following definition.

Definition 3. (Stochastic Hybrid Automaton: Syntax). *A stochastic hybrid automaton is a tuple* $\mathcal{S} = (\mathcal{H}, \Psi)$, *where:*

- $\mathcal{H} = (Loc, Var, Flow, Inv, Edge, Init)$ *is a hybrid automaton,*
- Ψ *is a set of stochastic kernels [31].*

Depending on how the stochastic kernels are used in the semantics, five different subclasses of stochastic hybrid automata can be identified. To keep this paper focused, we refer the interested reader to [53,54] and do not further discuss these differences here.

5   Tools for SHA

In the following we present the tools for the analysis of SHA that are used to evaluate the case-study in this paper.

MODEST TOOLSET. The MODEST TOOLSET [21,22] provides a *single-formalism, multiple solution* approach. It contains a collection of tools suitable for the analysis of different subclasses of networks of stochastic hybrid automata (SHAs) [21]. In general, SHAs allow to use continuous time and to specify the evolution of variables according to differential (in)equations. Moreover, choices within the execution of the model can be resolved non-deterministically or probabilistically. While the MODEST TOOLSET supports various input formats, models are typically specified using the MODEST language, which is inspired by process algebras. The MODEST TOOLSET includes the `modes` [10] statistical model checker which supports SHM with linear dynamics as well as lightweight scheduler sampling [33] for non-hybrid models [17,18]. Further, the MODEST TOOLSET includes the tool `prohver` [21], which relies on a modification of PHAVer [19] and computes safe upper bounds for maximum reachability properties.

`RealySt`. The tool `RealySt`[1] [14,15] is capable of computing maximal reachability probabilities for the class of rectangular hybrid automata extended with random clocks. It uses a combination of forward reachability analysis with backward refinement to optimize the reachability probabilities. For the flowpipe-based reachability analysis `RealySt` uses the library `HyPro` [44]. Moreover, the GNU Scientific Library (GSL) [20] provides Monte Carlo integration methods, which are used during the analysis performed in `RealySt`.

`Simulink` *Toolbox*. `Simulink` [36] is an industrially well established, graphical modeling language for hybrid systems. `Simulink` models consist of blocks with discrete or continuous behavior that are connected by so-called signals. The `Simulink` block library provides a large set of predefined blocks, from arithmetic over control flow blocks to integrators and complex transformations. The

---

[1] Available at https://zivgitlab.uni-muenster.de/ag-sks/tools/realyst.

reinforcement learning Toolbox [50] provides an RL agent block, which enables executing RL algorithms within Simulink. It periodically samples observations, provides rewards and chooses actions.

6   Reinforcement Learning

As some of the above-mentioned tools apply Reinforcement Learning (RL), we give a brief overview on the topic. RL is a class of machine learning methods in which an agent learns via trial and error to take actions in a dynamic environment in order to maximize some given reward [49]. Often the environment is expressed as a Markov decision process (MDP). An MDP [43] is defined by a tuple (S, A, R, P), where S is a set of states, A is a set of actions, $R \subset \mathbb{R}$ is a set of rewards, where $R_a(s, s')$ indicates the reward for a transition from state s to state s' according to action a. Finally, P specifies the transition probabilities of the MDP and $P_a(s, s')$ gives the probability of moving from state s to s' in case action a has been chosen. The agent interacts with the environment at discrete time steps. Each step t, the agent observes the current state $s_t \in S$ of the environment and chooses an action $a_t \in A$, which may depend on the current state s_t. The environment then responds with the state $s_{t+1} \in S$ that was reached as well as a numerical reward $R_{a_t}(s_t, s_{t+1}) \in R$ which is associated with the transition from state s_t to state s_{t+1} via the action a_t. A so-called *policy* $\pi(a|s)$ of an RL agent indicates the action a that is taken when observing state s. The aim of RL is that the agent learns a policy $\pi(a|s)$, which maximizes the cumulative reward.

Two well-established algorithms for RL are *Q-Learning* [51] and *Deep Q-Learning* (DQN) [38]. The first maintains a discrete *Q-Table*, where each state-action pair is explicitly stored. Thus, the (i, j)-th entry of the table indicates the reward for action j, when being in state s_i. During the training, the entries of the Q-Table are updated according to the observed rewards and actions. After training, in state s_i, the action a is chosen for which the Q-Table indicates the highest reward. Q-Learning requires discrete and finite state-spaces, as well as discrete and finite action spaces. DQN replaces that Q-Table by a deep neuronal network. This allows to handle high-dimensional and continuous state-spaces.

7   Case-Study: An Optimal Route for RUNNING.CHRISTEL

As a case-study, we address a new research project of the Algebraic and Logic Foundations of Computer Science group at the TU Dresden, which aims to find the most efficient route to run around the world in an interactive way. To achieve this, a stochastic model of a runner is developed, denoted RUNNING.CHRISTEL. Next to that, a nano-satellite is being employed, as previously modeled by her friends and long-time co-authors Hermanns and Haverkort [41,48].

The nano-satellite is equipped with multiple batteries and an integrated camera. It orbits the earth and carries out three tasks: (1) its camera takes a batch of pictures of the earth's surface, (2) the images of the batch are compressed to reduce their overall size, and (3) the compressed images are sent to a base station in Dresden, where they are used to find the optimal route for RUN-NING.CHRISTEL. After the compressed images have been sent, the procedure repeats as long as at least one battery in the satellite has enough charge to operate the system.

To take as many images as possible, the system's lifetime should be maximized, i.e., the time in which at least one of the on-system batteries has some charge left. This can be achieved by switching between the batteries to make use of the non-linear recovery effect of batteries. The case-study at hand uses the approach of [27,28] and aims at predicting the best switching strategy (between the batteries) such that the overall lifetime of the nano-satellite reaches at least a certain threshold L.

7.1   Modeling RUNNING.CHRISTEL

Within the satellite a controller decides which of the batteries is used as power source for an incoming job, based on a given workload model that characterizes the arrival time of jobs, their power demand and required service time. We formalize the workload model as follows.

Definition 4. (Workload Model). *A workload model W for horizon h is a sequence of tuples*

$$W = [(I_i, arr_i, dur_i, dem_i)]_{1 \leq i \leq h},$$

where the i-th tuple represents a job i with job ID I_i, arrival time arr_i, required service time dur_i and power demand dem_i.

In the following, we denote the workload specification of a job k for a given workload model W as $W(k)$. While serving the jobs according to the given workload model W, the controller can only choose a non-empty battery. If a battery runs empty while serving, it is marked as empty and a non-empty battery is chosen immediately to continue the job. The lifetime of the system corresponds to the time in which at least one battery is still not marked as empty. We aim to identify the lower bound of the system's maximal lifetime, hence we only consider one charging cycle of the batteries and assume that the batteries are not recharged. Depending on the decisions of the controller, different lifetimes can be reached. To evaluate which lifetimes are possible for the workload model W representing jobs 1 to h, the system can be modeled as an hybrid automaton $\mathcal{H}_W$. An example for an HA modeling a 2-battery system can be found in Fig. 3. Table 1 indicates the initial values for each variable as well as their flow that holds unless given otherwise in Fig. 3. Here, batteries are modeled by the KiBaM which requires three continuous variables for each battery: a_i and b_i to keep track of the available resp. bounded charge of the i-th battery and e_i to track whether the i-th battery is marked as empty (in which case we set $e_i = 1$; initially, all $e_i = 0$). A battery is marked as empty as soon as its available charge runs empty for the first time. Additionally, five continuous variables are needed to monitor the state of the system: the stopwatch L tracks the lifetime of the system and is

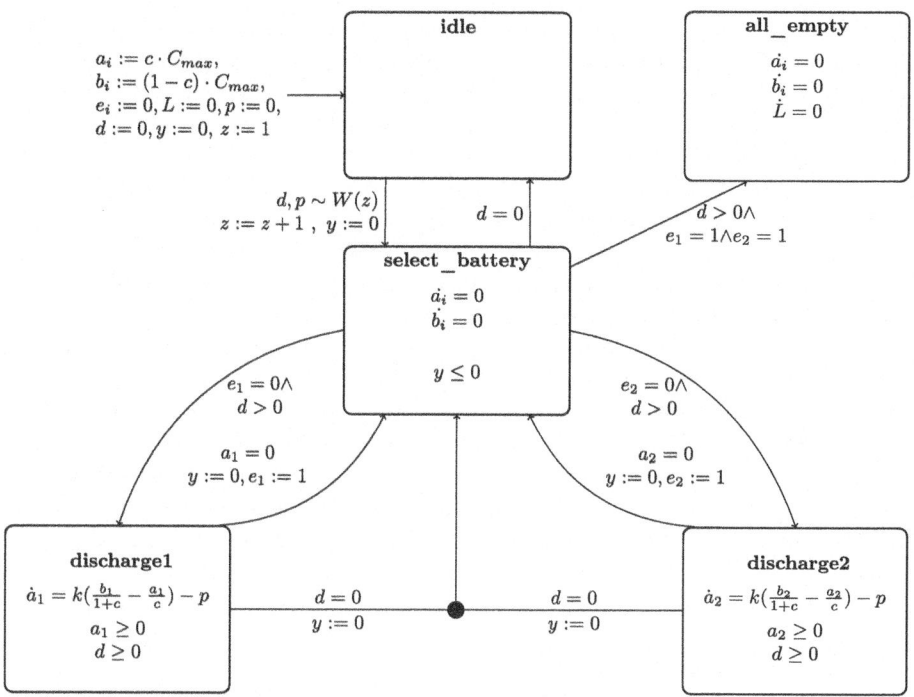

Fig. 3. HA model for a system with 2 batteries. If not stated otherwise in the location, variables evolve according to Table 1.

Table 1. Overview of continuous variables, their derivatives and initial conditions.

Variable	Flow	Initial condition
a_i	$\dot{a}_i = k(\frac{b_i}{1-c} - \frac{a_i}{c})$	$c \cdot C_{max}$
b_i	$\dot{b}_i = -k(\frac{b_i}{1-c} - \frac{a_i}{c})$	$(1-c) \cdot C_{max}$
e_i	$\dot{e}_i = 0$	0
d	$\dot{d} = -1$	0
p	$\dot{p} = 0$	0
L	$\dot{L} = 1$	0
y	$\dot{y} = 1$	0
z	$\dot{z} = 0$	1

paused as soon as all batteries are marked as empty. The variable z stores the id of the job that is currently served, p indicates the power demand of the job that is currently served and d the remaining serving time of that job. Finally, y is an auxiliary variable.

If we have N batteries there are $3N + 5$ continuous variables in the hybrid automaton. The HA has $N + 3$ locations and initially, the control location of the

HA is *idle*, where it remains until the arrival of $W(1)$. Afterwards the system moves from location *idle* to location *select_battery* and the values for p and d are set to the specifications of $W(1)$. As long as there exists an $i \in \{1, \ldots, N\}$ with $e_i = 0$, the systems moves immediately from *select_battery* to a location *discharge_j*, where $e_j = 0$ with $j \in \{1, \ldots, N\}$ holds. This corresponds to choosing battery j to serve the current job. Location *discharge_j* models the discharging of the chosen battery. From location *discharge_j* we move to location *select_battery* as soon as either $d = 0$ holds, which corresponds to successfully completing the job, or as soon as $a_j = 0$. In the latter case, the variable e_j is set to 1, meaning that battery j can no longer be chosen for future jobs. If $d = 0$ location *select_battery* is left immediately to location *idle*. If $d > 0$ we either move immediately to location *all_empty* if for all $i \in \{1, \ldots, N\}$ we have $e_i = 1$, or to location *discharge_k* with $e_k = 0$ otherwise.

After moving to location *idle*, the process repeats with the next job according to the workload profile. Note that the horizon of the applied workload model is chosen in such a way that regardless of the scheduler which resolves the inherent non-determinism between choosing a discharge location, location *all_empty* is eventually reached.

7.2   Workload Model

In the following we parametrize the workload model. Note that the chosen values are primarily chosen to highlight the stochastic impact and do not necessarily reflect reality. During its lifetime, the satellite orbits the earth and takes images of its surface for exactly two hours and has a power demand of 0.2 W. After two hours of taking images, the satellite starts compressing the images and moves to a position above Dresden from where it can send the images back to earth.

The size of the compression and with that the time until the full batch is sent to earth is not fixed, but strongly depends on the content of the images. Thus, the time taken for compression and sending may vary, where we assume it to be uniformly distributed between 1 and 2 h. Furthermore, the power demand for compressing and sending is 0.9 W.

After the images have been sent, the satellite immediately continues with the task of taking images. Thus, our workload model of horizon h is given by

$$W_s = [(1, 0, dur_1, dur_1, dem_1), (2, dur_1, dur_2, dem_2),$$
$$\ldots, (k, arr_{k-1} + dur_{k-1}, dur_k, dem_k),$$
$$\ldots, (n, arr_{h-1} + dur_{h-1}, dur_h, dem_h)], \text{where}$$

$$dem_i = \begin{cases} 0.2\text{W}, & \text{if } i \bmod 2 = 0, \\ 0.9\text{W}, & \text{otherwise.} \end{cases} \text{ and } dur_i = \begin{cases} 2\text{h}, & \text{if } i \bmod 2 = 0, \\ \text{Unif}(1,2)\text{h}, & \text{otherwise.} \end{cases}$$

We also consider a deterministic workload model taken from [27,28], where all jobs have a required service time of 1h, a power demand of 0.25W and no idling time occurs between jobs. Thus, for an $h \in \mathbb{N}$:

$$W_d = [(1, 0, 1, 0.25W), (2, 1, 0.25W), \ldots, (h, h, 1, 0.25W)].$$

Since W_d is purely deterministic, $\mathcal{H}_{W_d}$ is a hybrid automaton as described in Sect. 4. In contrast, the workload model $\mathcal{H}_{W_s}$ includes stochasticity and $\mathcal{H}_{W_s}$ is a SHA. If $W_s(z)$ corresponds to the job which compresses images and sends them to earth, the required service time of the sending and compression process is set according to a uniform distribution.

The maximal lifetime of the system can possibly be improved by choosing only batteries with sufficient effective capacity for the current job. However, precomputing the power demand requires solving the differential equations of the KiBaM. In general, this is only possible for a deterministic workload model, but not if the required service time is sampled from a probability distribution.

Numerical Parameters. We assume that the nano-satellite is equipped with a battery system similar to Fig. 1 and has four attached batteries, where each has a nominal voltage of 3.6 V and range in capacity from 1800 mAh to 3600 mAh. As this corresponds to a range from 6.48 Wh to 12.96 Wh, we fixed the capacities of the batteries in this case-study such that the first battery (B_1) has a capacity of 7 Wh, the second battery (B_2) has a capacity of 8.5 Wh, the third one (B_3) a capacity of 10 Wh and the last battery (B_4) a capacity of 12 Wh.

Since each battery is modeled by the KiBaM, the systems lifetime strongly depends on the choice of the parameters c and k, as they govern the recovery effect of each battery (see Sect. 3). It follows directly from Equation (1) that the flow from the bounded charge to the available charge is maximal if the height difference between the two wells is maximal, i.e., if $a(t) = 0$ and $b(t) = (1 - c) \cdot C_{max}$. In this case, the flow would be $k \cdot C_{max}$. During our evaluation we set $c = 0.5$ to ensure that enough energy is stored in the available charge to increase the number of batteries from which the scheduler can choose, while also ensuring that the bounded charge stores enough energy, to showcase the recovery effect. According to [27,28], we set $k = 0.017$.

8 Modeling the N-Battery System in Different Tools

We implemented the N-battery system in each of the considered tools given in Sect. 5. The corresponding source files are available online[2]. The Simulink model is explained in Sect. 8.1 followed by a specification of the case-study in the input language of the MODEST TOOLSET in Sect. 8.2 and a discussion of the implementation in RealySt in Sect. 8.3. Even though the case-study of [27,28] was performed in UPPAAL Cora, for this Festschrift we used academic tools developed at the University of Münster and the University of Twente.

---

[2] https://zivgitlab.uni-muenster.de/ag-sks/tools/misc/execution-scripts/2025-04-optimal-battery-pack-usage.

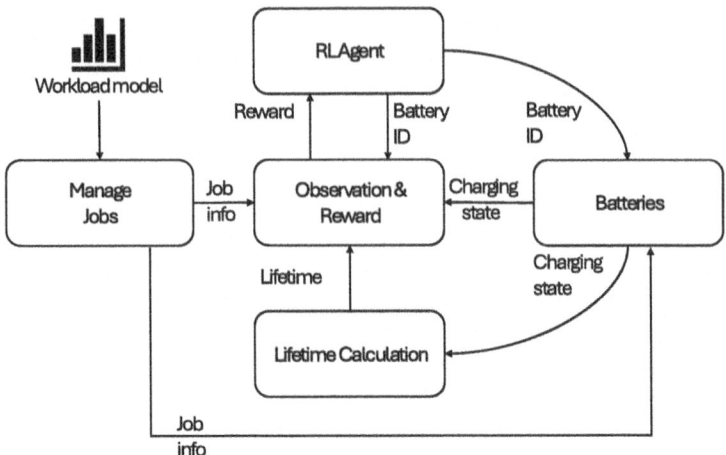

Fig. 4. Schematic overview on the setup of the Simulink Model.

8.1 Setup Simulink

The Simulink model shown in Fig. 4 consists of five subsystems: ManageJobs, ObservationAndReward, Batteries, RL agent and LifetimeCalculation. Note that this model can easily be adapted for a different number of batteries and for different workload models. In the following we describe the tasks assigned to the different subsystems. ManageJobs processes the given workload model (either deterministic or stochastic) and forwards information on the current job $W(z)$, i.e., its demand and its remaining service time, to ObservationAndReward and Batteries. ObservationAndReward collects and processes information from all other subsystems and provides the system state and a reward to the RL agent.

Since Simulink performs discrete-time simulation, the RL agent makes a decision after each time-step. This decision is based on its observations (as explained below) and identifies a battery to be used until the next decision. Note that in case a battery runs empty or the job is completed in-between two time steps, the model evolution is paused, hence, the lifetime and the available and bounded charge of the batteries do not evolve further until the next decision of the agent. The decision of the RL agent is forwarded to ObservationAndReward and to Batteries. The latter uses the input from the RL agent and the received job information, to initiate the discharging of the corresponding battery. Recall from Sect. 7 that each battery is modeled as a KiBaM in its own subsystem (see Fig. 5b). Finally, LifetimeCalculation computes the current lifetime of the system and performs the statistical evaluation. Thus, in case the stochastic workload model is applied, subsystem Lifetime Calculation checks whether the lifetime surpasses a given threshold. For the stochastic workload model, we compute the Wilson-Score Confidence Interval [55].

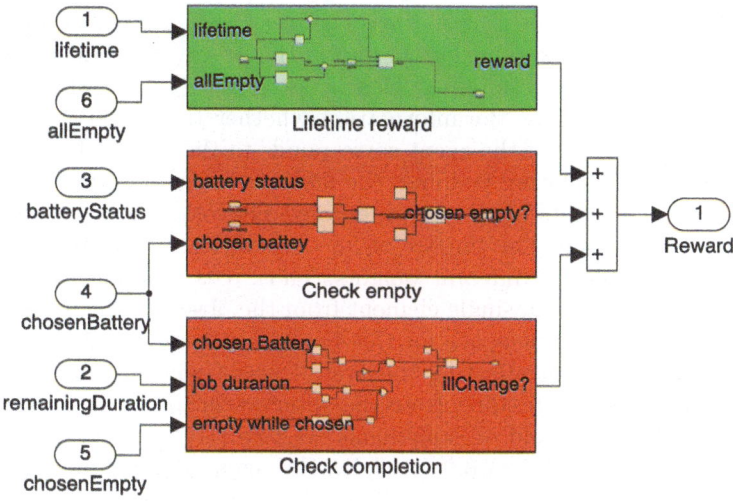

(a) `Simulink` subsystem for reward calculation.

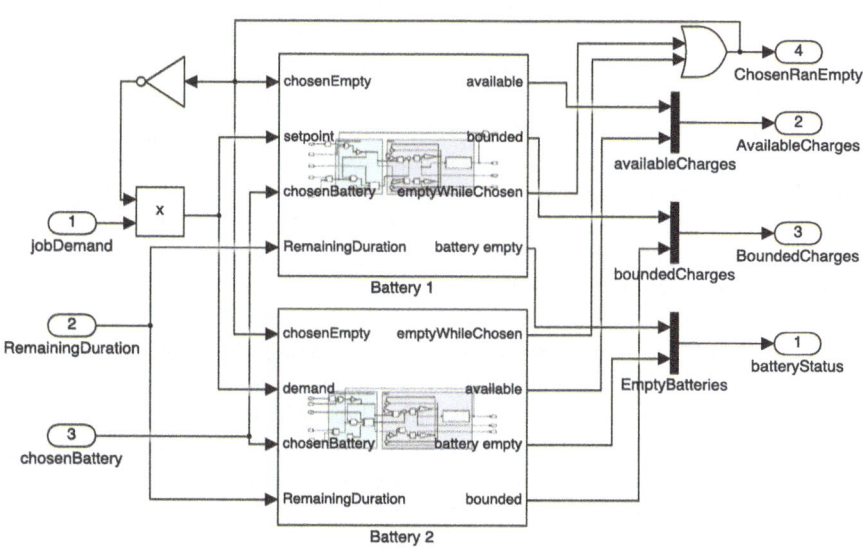

(b) `Simulink` subsystem for batteries.

Fig. 5. Two subsystems in a `Simulink` model of a 2-battery system.

We run `Simulink` with a variable-step solver with a maximum step size of 0.025. Differential equations are solved with the Dormand-Prince method (ode45) [16]. We use the parallel training and simulation option of `Simulink`.

The RL agent is implemented as a Deep Q-Network agent and receives a set of observations on the state of the system, which include: the previously chosen battery, the current number of battery changes, the current lifetime as well as the current job information. Furthermore, the observation includes features based on information on the state of each battery, i.e., the bounded and the available charge, as well as the information whether the battery is marked as empty. The action space of the agent corresponds to the IDs of the N batteries, ranging from 1 to N. During training, the features in the observation space are used to learn the weights of the deep neuronal network, which determines the action of the controller. The impact of an observation to the result is driven by a complex interplay of different weights. Thus, it is generally challenging to predict the influence that a single element from the observation space has. Note that the chosen observation space for this paper is not necessarily minimal, but could be reduced without affecting the overall performance of the DQN agent. The importance of single features from the observation space could be evaluated with methods from the field of *explainable AI* (see [37] for a survey). The first layer of the deep neural network used by the agent is an input layer which processes the observations. This is followed by two alternating layers of fully connected units (256 each) and ReLU activation functions. The final layer is a fully connected output layer, which generates a Q-value for each possible action for the given observation. As introduced in [38], `Simulink` uses a target critic and memory replay. We use a learning rate of $5 \cdot 10^{-3}$ for the critic and adapt the standard settings of the agent to `ExperienceBufferLength`$=25 \cdot 10^{-3}$. To enhance learning, we normalize the observations.

The aim is to train the agent, such that its decisions maximize the overall lifetime of the system. For this, a reward function calculates for each action of the agent a reward within a subsystem that is depicted in Fig. 5a and integrated into `Observation & Reward`. The reward is composed out of three different components, where the first is the `Lifetime reward` (indicated in green), which rewards the agent with a value from the interval $[0, 1]$, corresponding to the increase in lifetime since the last decision of the agent. Furthermore, we use *reward shaping* to ensure that the agent only takes *valid actions*, i.e., no empty battery is chosen and a new battery is only chosen if either the previous job is completed or the previous battery ran empty. The subsystems `Check empty` and `Check completion` (indicated in red) provide the agent with a high negative reward of -2 in case the last action was not valid.

8.2   Setup MODEST TOOLSET

We implemented the HA of Fig. 3 within the Modest Toolset. Hereby, each location of the HA is described as a separate process and the changes between the processes are specified according to the guard conditions. Listing 1.1 shows the specification of location *discharge1* of Fig. 3 in Modest. The invariant of the process specifies the flows for each continuous variable, as well as the conditions on the continuous state. The transitions leading to location *select_battery* are encoded into the `alt` statement. If a process is changed, Modest allows to reset variables to samples of a given probability distribution. Thus, the specifications for an HA with W_d or W_s in Modest only differ in the specification of the conditions when the process *Idle* is left. Note that for the analysis with modes, guard conditions cannot be specified as an open interval. To avoid Zeno behavior, we therefore replace the guard condition $d > 0$ with $d \geq 0.000001$.

Listing 1.1. Implementation of location *discharge1* from Fig. 3 in Modest.

```
process discharge1()
{
    invariant(der(a1) == (k * (b1 / (1 - c) - a1 / c)) - p
    && der(b1) == k * (a1 / c - b1 / (1 - c))
    && der(a2) == k * (b2 / (1 - c) - a2 / c)
    && der(b2) == k * (a2 / c - b2 / (1 - c))
    && der(p) == 0 && der(d) == (-1) && der(L) == 1
    && der(h) == 1 && der(T) == 1  && a1 >= 0 && d >= 0)
    alt {
  :: when ((d >= eps) && (a1 <= 0))
        tau {= e1 = 1,h = 0 =}; Select_Battery()
  :: when (d <= 0) tau {= h = 0 =}; Select_Battery()
  }
}
```

We use the simulation tool modes with its Q-Learning extension to analyze whether the lifetime of the system can exceed a given threshold L. The agent aims at resolving the discrete non-determinism arising in the HA of Fig. 3 such that the probability to exceed the given threshold is maximized. Unless stated otherwise, we use the Q-Learning extension with the following settings:

rate = 0.1, disc-factor = 10, final-rate = 0.05, discount = 1.0, epsilon = 0.15, final-epsilon = 0.05.

For the statistical evaluation, again a Wilson-Score Confidence Interval is computed. Note that the agent observes the current sampling of the random variable and the remaining duration of stochastic delays is part of the observation space. Hence, we obtain a prophetic scheduler if a stochastic workload model is applied, which corresponds to the computation of a best case that can be achieved in this setting. For deterministic workload models, we also apply a uniform scheduler to all non-deterministic decisions instead of Q-Learning. In that case each decision is resolved probabilistically, where each choice has the same probability. We then evaluate the maximal threshold L for which the probability to eventually reach a lifetime of L is larger than 0. Note that the uniform scheduler returns a probability for reaching a predefined lifetime L, but is not able to return a schedule.

8.3   Setup RealySt

For a proper analytical evaluation of the model in RealySt, the support of continuous dynamics in the form of Linear-Time-Invariant (LTI) systems [52] is required. However, this kind of analysis is not yet supported in RealySt, nor in prohver. However, by adapting the HA shown in Fig. 3 to continuous variables with constant rates, the tool can be used to determine an over-approximation of the system lifetime. For this, we construct a singular automaton [24] as follows:

The lifetime of the model strongly depends on the KiBaM parameters, as discussed in Sect. 7. To make the model analyzable for the analytical tools, we have to assume a maximal flow between the two wells of the KiBaM and we set $\dot{a}_i = k \cdot C_{max}$, when the battery i is idling and $\dot{a}_i = k \cdot C_{max} - p$, otherwise. This over-approximates the lifetime of the model as explained in Sect. 7.2. Moreover, we ensure that $0 \leq a_i \leq c \cdot C_{max}$ holds at each time. The resulting singular automaton is specified in the input format of RealySt. To improve computation time, we replace guards of the form $d > 0$ by $d \geq 0.000001$.

9   Evaluation

In this section the lifetime of the nano-satellite is evaluated by checking whether it reaches a certain threshold L. In Sect. 9.1, we introduce the specifications of our experiments. Section 9.2 considers a two-battery system with a deterministic workload model which is analyzed with analytical tools. In Sect. 9.3 we consider various settings which are analyzed with simulation-based tools.

9.1   Specifications

We consider different settings to showcase important aspects in the analysis of our proposed hybrid models with the tools discussed in Sect. 5. We identify the different settings via a tuple of the form (N, C_{max}, c), where N indicates the number of batteries used, C_{max} the maximal capacity of each battery and c the chosen value for the parameter c of the KiBaM. Note that in each setting, we set $k = 0.017$, therefore, this parameter is not included in the tuple.

All experiments using Simulink or a tool from the MODEST TOOLSET have been conducted on a machine equipped with two AMD EPYC 9354 processors with 32 kernels running at 3.25 GHz and a total of 512 GB of RAM. The experiments using RealySt have been conducted on a machine equipped with an Intel ® Core™ i7 processor running at 10×1.7 GHz with 64 GB of RAM. Unless stated otherwise we train the DQN-agent in Simulink for 600 and the Q-Learning agent in modes for 10000 episodes, as these two training techniques require a different number of training episodes. In both tools one simulation run is performed with the learned strategy. To simulate modes with the uniform scheduler, we use 10000 simulation runs.

Table 2. Lifetime obtained for a $(2, 5.5\ \text{Wh}, 0.166)$ system with deterministic workload and jobs with power demand 0.4 W and service time 1 h. Per setting, 10 simulation runs are performed. The lifetime (LT) is indicated by the maximum and the training time (TT) by the average.

RealySt	prohver	modes		Simulink	
		Q-Learning	*Uniform Scheduler*		
9.2	9.2	5.75	5.75	5.73	LT [h]
58.82	1.36	2.28	6.64	94.09	TT [s]

9.2 Results Obtained by Analytical Tools

As mentioned in Sect. 8.3, both analytical tools `prohver` and `RealySt` do not support the analysis of systems with linear dynamics such as the KiBaM (yet). However, by over-approximating the recovery effect modeled in the KiBaM via a singular automaton, an upper bound for the lifetime can be determined. Table 2 compares the results achieved by the different tools for the setting $(2, 5.5\ \text{Wh}, 0.166)$ when applying a deterministic workload model, where jobs with demand 0.4 W and a required service time of $1h$ arrive consecutively. While the analytical models over-approximate the recovery effect as explained above, Modes and `Simulink` analyze the lifetime taking into account the non-linear effects modeled by the KiBaM.

The over-approximations provided by `RealySt` and `prohver` both lead to a lifetime of $9.2h$. Hereby, the computation time of `prohver` is much smaller than `RealySt`. Also the results obtained by the simulation-based tools coincide, where `modes` is considerably faster than `Simulink`. Due to the deterministic workload model no confidence intervals are provided. The over-approximation results in a lifetime that is approximately 60% larger than the simulated lifetimes. To obtain a meaningful upper bound it could be interesting to investigate a tighter over-approximation of the recovery effect in the KiBaM in future work. Because the over-approximation of the recovery effect increases the lifetime significantly, also the number of scheduling decisions increases drastically. Similar to [27,28], this leads to a state-space explosion resulting in very high computation times. Hence, in the following, we only consider simulation-based tools which are able to deal with larger state-spaces and with the nonlinear dynamic induced by the KiBaM.

9.3 Results Obtained by Simulation

2- or 3-Battery Systems. We first consider the setting $(2, 5.5\ Wh, 0.166)$ which corresponds to one of the settings in [27,28]. To highlight the scalability of our model we further extend this setting towards higher values of c and C_{max}, as well as towards 3 batteries. Each of these changes leads to a larger amount of scheduling decisions, increasing the complexity of the problem.

Table 3. Lifetimes and computation times for settings (i) (2, 5.5 Wh, 0.166), (ii) (2, 5.5 Wh, 0.5), (iii) (2, 11 Wh, 0.5) (iv) (2, 30 Wh, 0.5) and (v) (3, 5.5 Wh, 0.166), for a deterministic workload model W_d. The computation time (TT) is the average of the 10 training runs performed, and the lifetime (LT) is their maximum.

	Simulink	modes		
		Q-Learning	**Uniform**	
(i)	11.89	11.89	11.93	LT [h]
	78.36	6.83	12.57	TT [s]
(ii)	30.97	30.75	30.98	LT [h]
	114.87	11.42	27.87	TT [s]
(iii)	73.23	73.2	73.24	LT [h]
	202.10	24.00	59.02	TT [s]
(iv)	224.97	225.1	225.15	LT [h]
	602.37	64.75	155.47	TT [s]
(v)	23.58	15.8	24.05	LT [h]
	101.75	7.57	22.90	TT [s]

Table 3 shows results for *setting (i)* (2, 5.5 Wh, 0.166), *(ii)* (2, 5.5 Wh, 0.5), *(iii)* (2, 11 Wh, 0.5), *(iv)* (2, 30 Wh, 0.5) and *(iv)* (3, 5.5 Wh, 0.166) in combination with the deterministic workload model W_d. For each case, we use the uniform scheduler of `modes`, to obtain the maximal possible lifetime for this scenario. For a system with two batteries, both Deep Q-Learning in `Simulink` and Q-Learning in `modes` obtain schedules which maximize the system's lifetime. The computation time of `modes` with Q-Learning outperform `Simulink` in all cases, while the resulting lifetime is similar in all cases, except for *setting*(v), which is discussed together with two other interesting settings below.

For *setting (i)* an optimal schedule is provided in [27,28] which is computed for a discretized version of the KiBaM and results in a lifetime of 12.04h. As the source code of [27,28] is not available, we could not analyze the computation time of UPPAAL on a modern machine. The results of our models are within the 1% error induced by the discretization of the KiBaM for the considered workload model, as indicated by the authors.

For *setting (iv)* the discretization factor which determines the discretization, had to be reduced to 1 in `modes` to allow its Q-Learning agent to learn optimal decisions. This reduces the amount of possible states drastically and implies that it is more challenging to find an optimal solution using Q-Learning for settings with a large state-space. Both, the number of batteries and their maximal capacity increase the state-space, hence, the lifetime obtained via Q-Learning for *setting (v)*, which is a setting with three batteries, is significantly lower than the one obtained via Deep Q-Learning in `Simulink`.

Table 4. Probability to reach a lifetime of at least threshold L for *setting* (i) and a stochastic workload model. Confidence intervals (CI) are Wilson-score with confidence level 0.95 for 5000 simulation runs. Computation time is specified for training (TT) and simulation (ST).

L	≥ 7	≥ 8	≥ 9	
modes	[1.00,1.00]	[0.44,0.47]	[0.00,0.00]	CI
	2.58	4.82	6.01	TT [s]
	1.01	2.09	2.88	ST[s]
Simulink	[1.00, 1.00]	[0.44, 0.47]	[0.00,0.00]	CI
	100.22	100.22	100.22	TT[s]
	112.67	112.67	112.67	ST [s]

For each setting, the uniform scheduler in `modes` outperforms the results obtained by Q-Learning. We suppose that this is caused by the high number of possible schedules, where finding the optimal one is very rare and therefore hard to learn for RL agents. Recall that, simulation with a uniform scheduler, do not lead to schedules which maximizes the lifetime. In contrast, applying `modes` with Q-learning or `Simulink` with Deep Q-learning results in a trained agent that provides a scheduling strategy.

When a stochastic workload model, as in Table 4 is applied, the lifetime strongly depends on the samples of the service times. Hence, a maximal lifetime cannot be determined as for the previous case in which a deterministic workload model was applied. Instead, statistical model checking is used to indicate the probability that a certain lifetime is reached. Table 4 shows the results obtained for a stochastic workload model where every second job has a service time that is uniformly distributed over $(0, 1)$ and a demand of 0.4W. This workload model is applied to the setting $(2, 5.5 \text{ Wh}, 0.166)$, where results obtained with `Simulink` and `modes` match. Even though `Simulink` is able to compute results for all three considered lifetimes in one run, i.e., the indicated computation time is only needed once, `modes` again has a lower execution time than `Simulink`.

4-battery system for RUNNING.CHRISTEL The simulation tools are able to evaluate a system with four batteries, which was not possible in [28], and is also not possible for the analytical tools considered in this paper. Tables 6 and 5 summarize the simulation results for the 4-battery system with a deterministic and a stochastic workload, respectively.

First, Table 5 shows the lifetimes in a setting with four batteries and a workload model in which the required service time of every second job is fixed to either $1, 1.5$ or 2 h. The values in column *Max.* indicate the lifetime that can be achieved if for each battery the available charge and the bounded charge are fully depleted. Similarly, column *Min.* indicates the smallest lifetime achieved by a scheduler that only discharges the available charge of each battery. Note that *Min.* and *Max.* provide a theoretical range for the lifetime, where *Max.* cannot be reached in practice, as it is usually not possible to completely drain a battery.

Table 5. Lifetime in hours for a the deterministic workload model $[(0, 0, 2h, 200W), (1, 2, x_1, 900W), (3, 2 + x, 2h, 200W), (4, 4 + x, xh, 900W), \ldots]$ with $x = \{1, 1.5, 2\}$, applied to the setting $(4, [7 \text{ Wh}, 8.5 \text{ Wh}, 10Wh, 12 \text{ Wh}], 0.5)$. Trainings have been run 10 times, the time (TT) is the average of these runs and the lifetime the maximum obtained from these runs.

	Min.	Max.	Simulink	modes		
				Q-Learning	Uniform Scheduler	
$x = 1$	43.26	86.53	71.08	57.6	71.2	LT [h]
			222.18	17.10	56.39	TT[s]
$x = 1.5$	37.5	75	59.01	48.8	59.2	LT [h]
			223.51	13.12	41.20	TT[s]
$x = 2$	34.09	68.18	51.28	43.3	52	LT [h]
			216.58 s	11.29	33.89	TT [s]

Table 6. Probability to reach a life-time of at least L for the stochastic workload model W_s in settings $(4, [7 \text{ Wh}, 8.5 \text{ Wh}, 10 \text{ Wh}, 12 \text{ Wh}], 0.5)$. Confidence intervals (CI) are computed using the Wilson-score interval method with confidence level 0.95 and 5000 simulation runs. Computation time is specified for training (TT) and simulation (ST).

L	≥ 52	≥ 54	≥ 55	≥ 56	≥ 58	
modes	[0.53, 0.55]	[0.26, 0.29]	[0.14, 0.16]	[0.07, 0.08]	[0.01, 0.02]	CI
	14.23	15.54	16.21	16.64	16.97	TT[s]
	7.06	7.62	7.95	8.09	8.18	ST[s]
Simulink	[0.99, 1.00]	[0.97, 0.98]	[0.97, 094]	[0.84, 0.84]	[0.46, 0.49]	CI
	232.26	232.26	232.26	232.26	232.26	TT [s]
	130.49	130.49	130.49	130.49	130.49	ST[s]

Table 5 indicates that the DQN-agent in `Simulink` is able to learn schedules which lead to lifetimes close to the lifetimes obtained by the uniform scheduler with `modes`. The best results learned by `modes` are achieved for a discretization factor of 1000, which however still leads to significantly lower lifetimes. This can be explained by the large state-space for the system with four batteries.

If the stochastic workload model W_s is applied, in which every second job has a service time that is uniformly distributed on $[1, 2]$, the lifetime strongly depends on the sampled delay. As expected, results for the stochastic workload model lie between the results for $x = 1$ and $x = 2$, i.e., between 52 and 71.2. As higher lifetimes can only be reached, when small service times are sampled, the probability to reach higher lifetime decreases. This can be seen in Table 6, where the confidence intervals for the probability to reach a certain threshold in the lifetime are given. Within `Simulink` it is much more likely to reach a lifetime that exceeds certain threshold than with `modes`. This matches the results of

Table 5, where Q-Learning in `modes` resulted in significantly lower lifetimes for the model with four batteries. This is due to the increasing size of the Q-Table, which makes efficient Q-Learning infeasible. Finally, the results of Table 6 show that the probability that the satellite can be used for at least 56 h is 84%.

10   Conclusion

This paper presented how the maximal lifetime of a system equipped with N batteries can be determined in the presence of stochastic or deterministic workload models. To achieve this, we proposed a hybrid automaton modeling the system and we specified it in four state-of the art tools. In particular for this paper, we focused on two simulation-based tools in which a reinforcement learning agent is trained to resolve the decisions on which battery to use. In both tools, the RL agent is trained in such a way that it aims to make decisions leading to a long lifetime of the system. Both tools train with a different RL algorithm. The achieved results are compared to results obtained by two tools providing a formal analysis of hybrid systems.

Our results show that for small systems with a relatively small state-space, both the DQN-agent included in the `Simulink` and the Q-Learning agent trained with `modes`, learn schedules that maximize the system's lifetime even for stochastic workload models. This shows that the discretization of the state-space required for the Q-Learning approach in `modes` has no negative impact on the result for small systems. For such systems `modes` shows a lower runtime and it is guaranteed that only valid actions (i.e. no empty battery is chosen and the used battery is not changed within a job) are carried out. In the `Simulink` model, we use reward shaping which gives a high incentive for the agent to avoid invalid actions. However, it is not guaranteed that they do not occur. Our evaluations also showed that for systems with more than two batteries, the schedule induced by the Q-agent results in significantly lower lifetimes than the schedule induced by the DQN-agent. This shows, that adding more batteries results in a higher dimensional state-space which cannot be efficiently handled by Q-Learning. Adding a pre- or post shield to the RL-agent of the `Simulink` model guarantees that the agent only takes valid actions. In future work we plan to evaluate how a shield can be added and how it influences the learning behavior. Further, we plan to investigate a tighter over-approximation for the KiBaM.

References

1. Abate, A., Prandini, M., Lygeros, J., Sastry, S.: Probabilistic reachability and safety for controlled discrete time stochastic hybrid systems. Automatica **44**(11), 2724–2734 (2008). https://doi.org/10.1016/j.automatica.2008.03.027
2. Adelt, J., Bruch, S., Herber, P., Niehage, M., Remke, A.: Shielded learning for resilience and performance based on statistical model checking in Simulink. In: Proceedings of the first Int. Conf. on Bridging the Gap Between AI and Reality, pp. 94–118. Springer Nature Switzerland (2024). https://doi.org/10.1007/978-3-031-46002-9_6

3. Adelt, J., Herber, P., Niehage, M., Remke, A.: Towards safe and resilient hybrid systems in the presence of learning and uncertainty. In: Proceedings of the 11th Int. Symposium on Leveraging Applications of Formal Methods, Verification and Validation. Verification Principles, pp. 299–319. Springer (2022). https://doi.org/10.1007/978-3-031-19849-6_18

4. Alur, R., et al.: The algorithmic analysis of hybrid systems. Theor. Comput. Sci. **138**(1), 3–34 (1995). https://doi.org/10.1016/0304-3975(94)00202-T

5. Behrmann, G., Cougnard, A., David, A., Fleury, E., Larsen, K.G., Lime, D.: UPPAAL-Tiga: time for playing games! In: Proceedings of the 19th Int. Conf. on Computer Aided Verification, pp. 121–125. Springer (2007).https://doi.org/10.1007/978-3-540-73368-3_14

6. Bengtsson, J., Larsen, K., Larsson, F., Pettersson, P., Yi, W.: UPPAAL — a tool suite for automatic verification of real-time systems. In: Alur, R., Henzinger, T.A., Sontag, E.D. (eds.) HS 1995. LNCS, vol. 1066, pp. 232–243. Springer, Heidelberg (1996). https://doi.org/10.1007/BFb0020949

7. Bertrand, N., Bouyer, P., Brihaye, T., Menet, Q., Baier, C., Größer, M., Jurdzinski, M.: Stochastic timed automata. Logical Methods in Computer Science **10**(4) (2014). https://doi.org/10.2168/LMCS-10(4:6)2014

8. Bisgaard, M., Gerhardt, D., Hermanns, H., Krčál, J., Nies, G., Stenger, M.: Battery-aware scheduling in low orbit: the GomX–3 case. Formal Aspects Comput. **31**(2), 261–285 (2018). https://doi.org/10.1007/s00165-018-0458-2

9. Blohm, P., Herber, P., Remke, A.: Towards quantitative analysis of simulink models using stochastic hybrid automata. In: Proceedigs of the 19th Int. Conf. on Integrated Formal Methods, pp. 172–193. Springer Nature Switzerland (2025). https://doi.org/10.1007/978-3-031-76554-4_10

10. Budde, C.E., D'Argenio, P.R., Hartmanns, A., Sedwards, S.: An efficient statistical model checker for nondeterminism and rare events. Int. J. Softw. Tools Technol. Transf. **22**(6), 759–780 (2020). https://doi.org/10.1007/s10009-020-00563-2

11. Cauchi, N., Abate, A.: Stochy - automated verification and synthesis of stochastic processes: poster abstract. In: Proceedings of the 22nd ACM Int. Conf. on Hybrid Systems: Computation and Control, pp. 258–259. Association for Computing Machinery (2019). https://doi.org/10.1145/3302504.3313349

12. Cloth, L., Jongerden, M.R., Haverkort, B.R.: Computing battery lifetime distributions. In: Proceedings of the 37th Annual IEEE/IFIP Int. Conf. on Dependable Systems and Networks, pp. 780–789 (2007).https://doi.org/10.1109/DSN.2007.26

13. David, A., Jensen, P.G., Larsen, K.G., Mikučionis, M., Taankvist, J.H.: UPPAAL Stratego. In: Proceedings of the 21st Int. Conf. on Tools and Algorithms for the Construction and Analysis of Systems, pp. 206–211. Springer (2015). https://doi.org/10.1007/978-3-662-46681-0_16

14. Delicaris, J., Schupp, S., Ábrahám, E., Remke, A.: Maximizing reachability probabilities in rectangular automata with random clocks. In: Proceedings of the 17th Int. Symposium on Theoretical Aspects of Software Engineering, pp. 164–182. Springer (2023). https://doi.org/10.1007/978-3-031-35257-7_10

15. Delicaris, J., Stübbe, J., Schupp, S., Remke, A.: RealySt: A C++ tool for optimizing reachability probabilities in stochastic hybrid systems. In: Proceedings of the 16th EAI Int. Conf. on Performance Evaluation Methodologies and Tools, pp. 170–182. Springer Nature Switzerland (2024). https://doi.org/10.1007/978-3-031-48885-6_11

16. Dormand, J.R.: Numerical Methods for Differential Equations: A Computational Approach. CRC Press, 1st edn. (1996). https://doi.org/10.1201/9781351075107

17. D'Argenio, P., Legay, A., Sedwards, S., Traonouez, L.-M.: Smart sampling for lightweight verification of Markov decision processes. Int. J. Softw. Tools Technol. Transfer **17**(4), 469–484 (2015). https://doi.org/10.1007/s10009-015-0383-0

18. D'Argenio, P.R., Hartmanns, A., Sedwards, S.: Lightweight statistical model checking in nondeterministic continuous time. In: Margaria, T., Steffen, B. (eds.) ISoLA 2018. LNCS, vol. 11245, pp. 336–353. Springer, Cham (2018). https://doi.org/10.1007/978-3-030-03421-4_22

19. Frehse, G.: PHAVer: algorithmic verification of hybrid systems past HyTech. In: Proceedings of the 8th Int. Workshop on Hybrid Systems: Computation and Control, pp. 258–273. Springer (2005). https://doi.org/10.1007/978-3-540-31954-2_17

20. Galassi, M., Davies, J., Theiler, J., Gough, B., Jungman, G.: GNU Scientific Library - Reference Manual, Third Edition, for GSL Version 1.12. Network Theory Ltd (2009)

21. Hahn, E., Hartmanns, A., Hermanns, H., Katoen, J.P.: A compositional modelling and analysis framework for stochastic hybrid systems. Formal Methods Syst. Des. **43** (2013). https://doi.org/10.1007/s10703-012-0167-z

22. Hartmanns, A., Hermanns, H.: The modest toolset: an integrated environment for quantitative modelling and verification. In: Proceedings of the 20th Int. Conf. on Tools and Algorithms for the Construction and Analysis of Systems, pp. 593–598. Springer (2014). https://doi.org/10.1007/978-3-642-54862-8_51

23. Henzinger, T.A.: The theory of hybrid automata. In: Proceedings of the 11th Annual IEEE Symposium on Logic in Computer Science, pp. 278–292. IEEE Computer Society (1996). https://doi.org/10.1109/LICS.1996.561342

24. Henzinger, T.A., Kopke, P.W., Puri, A., Varaiya, P.: What's decidable about hybrid automata? J. Comput. Syst. Sci. **57**(1), 94–124 (1998). https://doi.org/10.1006/jcss.1998.1581

25. Hermanns, H., Krčál, J., Nies, G.: How is your satellite doing? Battery kinetics with recharging and uncertainty. Leibniz Trans. Embed. Syst. **4**(1), 1–28 (2017).https://doi.org/10.4230/LITES-v004-i001-a004

26. Hoogsteen, G., Molderink, A., Smit, G., Hurink, J., Kootstra, B., Schuring, F.: Charging electric vehicles, baking pizzas, and melting a fuse in Lochem. Int. Conf. Exhib. Electr. Distrib. **2017**(1), 1629–1633 (2017). https://doi.org/10.1049/oap-cired.2017.0340

27. Jongerden, M., Haverkort, B., Bohnenkamp, H., Katoen, J.P.: Maximizing system lifetime by battery scheduling. In: Proceedings of the 2009 IEEE/IFIP Int. Conf. on Dependable Systems & Networks, pp. 63–72 (2009).https://doi.org/10.1109/DSN.2009.5270351

28. Jongerden, M., Mereacre, A., Bohnenkamp, H., Haverkort, B., Katoen, J.P.: Computing optimal schedules of battery usage in embedded systems. IEEE Trans. Industr. Inf. **6**(3), 276–286 (2010). https://doi.org/10.1109/TII.2010.2051813

29. Jongerden, M.: Model-based energy analysis of battery powered systems. Phd thesis - research UT, graduation UT, University of Twente (2010).https://doi.org/10.3990/1.9789036531146

30. Jongerden, M., Haverkort, B.: Which battery model to use? IET Softw. **3**(6), 445–457 (2009). https://doi.org/10.1049/iet-sen.2009.0001

31. Klenke, A.: Probability Theory: A Comprehensive Course. Springer, Cham (2014). https://doi.org/10.1007/978-1-4471-5361-0_1

32. Lahiri, K., Raghunathan, A., Dey, S., Panigrahi, D.: Battery-driven system design: a new frontier in low power design. In: Proceedings the 7th Asia and South Pacific Design Automation Conf. and 15h Int. Conf. on VLSI Design, pp. 261–267 (2002). https://doi.org/10.1109/ASPDAC.2002.994932

33. Legay, A., Sedwards, S., Traonouez, L.-M.: Scalable verification of Markov decision processes. In: Canal, C., Idani, A. (eds.) SEFM 2014. LNCS, vol. 8938, pp. 350–362. Springer, Cham (2015). https://doi.org/10.1007/978-3-319-15201-1_23

34. Lygeros, J., Prandini, M.: Stochastic hybrid systems: a powerful framework for complex, large scale applications. Eur. J. Control **16**(6), 583–594 (2010). https://doi.org/10.3166/ejc.16.583-594

35. Manwell, J.F., McGowan, J.G.: Lead acid battery storage model for hybrid energy systems. Sol. Energy **50**(5), 399–405 (1993). https://doi.org/10.1016/0038-092X(93)90060-2

36. MathWorks: MATLAB Simulink. www.mathworks.com/products/simulink.html

37. Milani, S., Topin, N., Veloso, M., Fang, F.: Explainable reinforcement learning: a survey and comparative review. ACM Comput. Surv. **56**(7) (2024). https://doi.org/10.1145/3616864

38. Mnih, V., et al.: Human-level control through deep reinforcement learning. Nature **518**(7540), 529–533 (2015). https://doi.org/10.1038/nature14236

39. Niehage, M., Hartmanns, A., Remke, A.: Learning optimal decisions for stochastic hybrid systems. In: Proceedings of the 19th ACM-IEEE Int. Conf. on Formal Methods and Models for System Design, p. 44–55. Association for Computing Machinery (2021). https://doi.org/10.1145/3487212.3487339

40. Niehage, M., Remke, A.: Learning that grid-convenience does not hurt resilience in the presence of uncertainty. In: Proceedings of the 20th Int. Conf. on Formal Modeling and Analysis of Timed Systems, pp. 298–306. Springer (2022). https://doi.org/10.1007/978-3-031-15839-1_17

41. Nies, G., et al..: Mastering operational limitations of LEO satellites – the GOMX-3 approach. Acta Astronautica **151**, 726–735 (2018). https://doi.org/10.1016/j.actaastro.2018.04.040

42. Rakhmatov, D., Vrudhula, S.: An analytical high-level battery model for use in energy management of portable electronic systems. In: Proceedings of the 2001 Int. Conf. on Computer Aided Design, pp. 488–493 (2001).https://doi.org/10.1109/ICCAD.2001.968687

43. Schneider, S., Wagner, D.H.: Error detection in redundant systems. In: Proceedings of the Western Joint Computer Conf, pp. 115–121. Association for Computing Machinery (1957). https://doi.org/10.1145/1455567.1455587

44. Schupp, S., Ábrahám, E., Makhlouf, I.B., Kowalewski, S.: HyPro: a C++ library of state set representations for hybrid systems reachability analysis. In: Proceedings of the 9th Int. Symposium on NASA Formal Methods, pp. 288–294. Springer (2017). https://doi.org/10.1007/978-3-319-57288-8_20

45. Shmarov, F., Zuliani, P.: ProbReach: verified probabilistic δ-reachability for stochastic hybrid systems. In: Proceedings of the 18th Int. Conf. on Hybrid Systems: Computation and Control, pp. 134–139. Association for Computing Machinery (2015). https://doi.org/10.1145/2728606.2728625

46. da Silva, C., Schupp, S., Remke, A.: Optimizing reachability probabilities for a restricted class of stochastic hybrid automata via Flowpipe-construction. ACM Trans. Model. Comput. Simul. **33**(4) (2023). https://doi.org/10.1145/3607197

47. Soudjani, S.E.Z., Gevaerts, C., Abate, A.: FAUST^2 : formal abstractions of uncountable-state stochastic processes. In: Proceedings of the 21st Int. Conf. on Tools and Algorithms for the Construction and Analysis of Systems, pp. 272–286. Springer (2015).https://doi.org/10.1007/978-3-662-46681-0_23

48. Stock, G., Fraire, J.A., Mömke, T., Hermanns, H., Babayev, F., Cruz, E.: Managing fleets of LEO satellites: nonlinear, optimal, efficient, scalable, usable, and robust.

IEEE Trans. Comput. Aided Des. Integr. Circuits Syst. **39**(11), 3762–3773 (2020). https://doi.org/10.1109/TCAD.2020.3012751

49. Sutton, R.S., Barto, A.G.: Reinforcement Learning: An Introduction, 2nd edn. The MIT Press (2018)

50. The MathWorks: Reinforcement Learning Toolbox. https://www.mathworks.com/products/reinforcement-learning.html

51. Watkins, C., Dayan, P.: Q-learning. Mach. Learn. **8**(3), 279–292 (1992). https://doi.org/10.1007/BF00992698

52. Willems, J.C.: From time series to linear system – part i. finite dimensional linear time invariant systems. Automatica **22**(5), 561–580 (1986). https://doi.org/10.1016/0005-1098(86)90066-X

53. Willemsen, L., Remke, A., Ábrahám, E.: Comparing two approaches to include stochasticity in hybrid automata. In: Proceedings of the 20th Int. Conf. on Quantitative Evaluation of Systems, pp. 238–254. Springer Nature (2023). https://doi.org/10.1007/978-3-031-43835-6_17

54. Willemsen, L., Remke, A., Ábrahám, E.: (de-)composed and more: Eager and lazy specifications (camels) for stochastic hybrid systems. In: Principles of Verification: Cycling the Probabilistic Landscape : Essays Dedicated to Joost-Pieter Katoen on the Occasion of His 60th Birthday, Part III. Springer Nature (2025). https://doi.org/10.1007/978-3-031-75778-5_15

55. Wilson, E.B.: Probable inference, the law of succession, and statistical inference. J. Am. Stat. Assoc. **22**(158), 209–212 (1927). https://doi.org/10.1080/01621459.1927.10502953

56. Wognsen, E.R., Haverkort, B.R., Jongerden, M., Hansen, R.R., Larsen, K.G.: A score function for optimizing the cycle-life of battery-powered embedded systems. In: Sankaranarayanan, S., Vicario, E. (eds.) FORMATS 2015. LNCS, vol. 9268, pp. 305–320. Springer, Cham (2015). https://doi.org/10.1007/978-3-319-22975-1_20

Termination in Extended Probabilistic Threshold Automata

Mouhammad Sakr$^{(\boxtimes)}$ and Marcus Völp

SnT, Luxembourg University, Esch-sur-Alzette, Luxembourg
{mouhammad.sakr,marcus.voelp}@uni.lu

Abstract. Scaling up distributed systems increases also the chance that some node ceases to operate correctly, which turns fault tolerance into an essential trait. To achieve fault tolerance, numerous distributed algorithms, including reliable broadcast and consensus, depend on threshold guards. A threshold guard can, for example, ensure that a process waits for a majority of its peers to acknowledge that they reached a certain state in the distributed algorithm, before the process makes any progress. Threshold automata are computational models that allow fully automated parameterized verification of single- and multi-round threshold-based distributed algorithms (FTDA), where often the number of processes and the proportion of faulty processes are parameters. However, due to the fact that such algorithms have to cope with faulty processes not answering or, more generally, behaving in an arbitrary potentially malicious manner, liveness must only depend on a subset of processes, while ideally all correct processes should be considered in the termination properties of such algorithm. In this paper, we present a novel reasoning technique for proving almost-sure termination in extended probablistic threshold automata, by detecting strongly connected components (SCCs) in extended probabilistic and in ordinary threshold automata to reduce almost-sure termination to reachability in a finite abstract system.

Keywords: Formal Verification · Fault-tolerance · Probabilistic Systems · Termination · Strongly Connected Components

A personal note by Marcus

Dear Christel, remember when we first met. Me coming from operating systems, it took us a while before we understood each other. Model checkers were always suspicious to me, in particular after you confirmed that variable declaration order can make a difference between getting results and waiting for millenia. Thanks for taking away this skepticism and for making me cherish the prospect of obtaining the highest assurance guarantees known today, just by pressing a button. Remember also, when we stumbled over a strongly connected component (SCC) in our characterization of conditional probabilities on the long run of our model of locks in coherent caches [1]. What started as a toy example, with a number of cores that was already outpaced by reality, turned into a scalable solution for core counts that we barely see today. In the present paper, Mouhammad

N. Bertrand et al. (Eds.): Christel Baier Festschrift, LNCS 15760, pp. 408–424, 2026.
https://doi.org/10.1007/978-3-031-97439-7_20

and I return to SCCs to prove almost-sure termination in extended probablis-tic threshold automata. Thanks for leading the paths of so many researchers, including myself, to not shy away from formal verification and model checking in particular.

I wish you all the best,

Marcus

1   Introduction

In large scale distributed systems, faults are inevitable and may lead to arbitrary and possibly intentionally malicious, that is, Byzantine behavior of nodes and the processes they execute. Consequently, the distributed algorithms that govern these systems must tolerate faults, whether they are accidental or caused by cyberattacks. Prominent examples of such fault-tolerant distributed algorithms (FTDAs) are reliable broadcast [9], Byzantine fault tolerant agreement [19], and the various variants of the two that currently form the dissemination and consensus layers of modern permissioned and permissionless blockchains [22,23]. Unfortunately, FTDAs are hard to get correct, so we need formal verification and tools to assist developers constructing them.

Threshold automata (TAs) are formal models of FTDAs that avoid the need to enumerate faulty behavior and that are able to characterize systems in a parametric manner. While the parameterized verification problem is generally undecidable [21], it is decidable for certain classes of systems [10,11], including threshold automata [2,12].

FTDAs, like the ones above, must on the one side be able to detect and recover from or, better, mask any behavior of faulty processes, including no response to requests at all, but also possibly long correct behavior patterns. Consequently, given a fault tolerance threshold t, where the actual number of faulty processes satisfies $f \leq t$, the algorithm must ensure progress with only $n - t$ processes, as it cannot depend on responses from potentially faulty ones. In addition, FTDAs must also ensure that no conflicting decisions are taken, be that binary consensus (e.g., whether to deliver a message) by following a simple majority (e.g., more than $n/2$), or agreement on a value by requiring a correct process in the intersection of any two quorums [18] that can take such a decision (e.g., $2Q - n \leq t + 1$ where Q denotes the quorum size). Threshold automata characterize such algorithms, by demanding that a threshold of processes are in a given state (e.g., of agreement), before progress can be made.

In this work, we deal with randomized systems, so we use probabilistic thresh-old automata (PTAs). One randomized binary consensus algorithm is Ben-Or's algorithm [4], see Algorithm 1 and its corresponding TA in Fig. 1. It proceeds in rounds, where $n - t$ processes execute rounds in lock-step in an asynchronous manner. Each round is comprised of two phases. In the first, each process tries to identify a value $v \in \{0, 1\}$ that is supported by a majority. In the second, decision or ratification phase, processes finalize their decision if the value is pro-posed by at least $t + 1$ processes among the $n = 2t + 1$ processes. In case no such simple majority can be found, Ben-Or's algorithm causes some nodes to change their votes for the next round, repeating the agreement until eventually probabilistically the system converges to a consensus value in one of the future rounds.

In this work, we lift two restrictions that were imposed on threshold automata during their verification, namely that cycles are not allowed and that coin tosses may only appear at the end of the automaton. With these restrictions in place, algorithms like Ben-Or's could only be verified under round-rigid adversarial schedules [6], which require all processes to complete the previous round r before any process can start round $r+1$. The algorithm as verified, would have to either wait for faulty processes to acknowledge this fact, or would have to prevent them from behaving correctly and push other correct processes beyond this boundary.

We do so by introducing a novel algorithm for detecting strongly connected components and apply it to prove termination in a system model for a network of an arbitrary number of deterministic threshold automata. The algorithm further detects strongly connected components and almost-sure termination (i.e., termination with probability one) in a system model for a network of an arbitrary number of probabilistic threshold automata.

We start by introducing our system model and the extension of probabilistic threshold automata (PTAs) that allows cycles in Sect. 2. In Sect. 3, we introduce a finite abstract domain for the shared variables of TAs and an abstraction based on parametric interval abstraction. Section 4 demonstrates how this new abstraction allows dropping the resilience condition, the function N that determines the number of processes to be modeled, and the exact number of processes in a configuration. Section 5 introduces our algorithm for detecting almost-sure termination and strongly connected components. Section 6 relates our work to the works of others, before we draw conclusions and highlight future work in Sect. 7.

2   System Model

In this section, we build upon the existing concept of probabilistic threshold automata (PTA) [6], extending the definition to allow shared variables to be reset, similar to [3]. We then show how to transform a PTA into a nondeterministic threshold automaton (TA), where non-determinism arises from multiple rules (transitions) being enabled simultaneously. We also define the semantics of an unbounded number of non-deterministic TAs running in parallel.

Definition 1. *A* probabilistic threshold automaton *(PTA) is a tuple* $PA = (L, \mathcal{I}, \Gamma, \Pi, \mathcal{R}, RC)$ *where:*

- L *is a finite set of* locations.
- $\mathcal{I} \subseteq L$ *is the set of* initial locations.
- $\Gamma = \{x_0, \ldots, x_m\}$ *is a finite set of* shared variables *over* $\mathbb{N}_0$.

Algorithm 1. Ben-Or's Algorithm for Byzantine Faults [4]

```
 1: bool v ← input_value({0, 1})
 2: int r ← 1
 3: while true do
 4:     send (R, r, v) to all
 5:     wait for n − t messages (R, r, *)
 6:     if received at least (n + t)/2 messages (R, r, w) then
 7:         send (P, r, w, D) to all
 8:     else
 9:         send (P, r, ?) to all
10:     wait for n − t messages (P, r, *)
11:     if received at least t + 1 messages (P, r, w, D) then
12:         v ← w
13:         if received at least (n + t)/2 messages (P, r, w, D) then
14:             decide w
15:     else
16:         v ← random({0, 1})
17:     r ← r + 1
```

- Π is a finite set of parameter variables over $\mathbb{N}_0$. Usually, $\Pi = \{n, t, f\}$, where n is the total number of processes, t is a bound on the number of tolerated faulty processes, and f is the actual number of faulty processes.
- RC, the resilience condition, is a linear integer arithmetic formula over parameter variables. E.g.: for Ben-Or, $RC = n > 2t \wedge t \geq f$.
  For a vector $\mathbf{p} \in \mathbb{N}_0^{|\Pi|}$, we write $\mathbf{p} \models RC$ if RC holds after substituting parameter variables with values according to $\mathbf{p}$. Then the set of admissible parameters is $\mathbf{P}_{RC} = \{\mathbf{p} \in \mathbb{N}_0^{|\Pi|} : \mathbf{p} \models RC\}$.
- $\mathcal{R}$ is a set of rules where a rule is a tuple $r = (from, \delta_{to}, \varphi, \mathbf{uv}, \tau)$ such that:
  - $from \in L$ is a location.
  - δ_{to} is a probability distribution over the target locations.
  - φ is a conjunction of lower guards and upper guards. A lower guard has the form: $a_0 + \sum_{i=1}^{|\Pi|} a_i \cdot p_i \leq x$; An upper guard has the form: $a_0 + \sum_{i=1}^{|\Pi|} a_i \cdot p_i > x$, with $x \in \Gamma$, $a_0, \ldots, a_{|\Pi|} \in \mathbb{Q}$, $p_1, \ldots, p_{|\Pi|} \in \Pi$. We denote these inequalities as a lower guard on x and an upper guard on x respectively.
    The left-hand side of a lower or upper guard is called a threshold.
  - $\mathbf{uv} \in |\mathbb{N}_0|^{|\Gamma|}$ is an update vector for shared variables.
  - $\tau \subseteq \Gamma$ is the set of shared variables to be reset to 0.

Remark 1. For any tuple-based structure $X = (A, B, C, \ldots)$ with named components, we use dot notation to refer to individual components. For instance, $PA = (L, \mathcal{I}, \Gamma, \Pi, \mathcal{R}, RC)$ we use $PA.L, PA.\mathcal{I}, PA.\Gamma, PA.\Pi, PA.\mathcal{R}, PA.RC$ to refer to the components $L, \mathcal{I}, \Gamma, \Pi, \mathcal{R}$, and RC respectively.

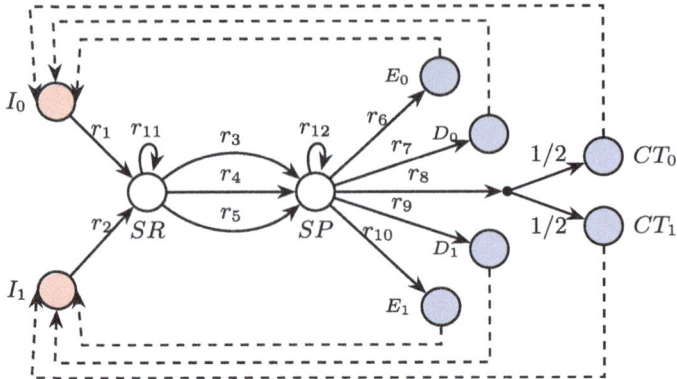

Fig. 1. Ben-Or's algorithm as a probabilistic threshold automaton with resilience condition $n > 2t \wedge t \geq f \geq 0 \wedge t > 0$. Check Table 1 for the rules' notations.

From a Probabilistic Threshold Automaton (PTA), one can derive a non-probabilistic threshold automaton simply by replacing all probabilities with non-determinism. That is, every probabilistic rule $r = (from, \delta_{to}, \varphi, \mathbf{uv}, \tau)$ is replaced by non-deterministic rules of the form $r_{to} = (from, to, \varphi, \mathbf{uv}, \tau)$ for every location to with $\delta_{to} > 0$. While we can specify shared variables in τ to be reset, our algorithm cannot handle resets, yet.

Definition 2. *Given a PTA, $PA = (L, \mathcal{I}, \Gamma, \Pi, \mathcal{R}, RC)$, its corresponding non-probabilistic threshold automaton is $A_{np} = (L, \mathcal{I}, \Gamma, \Pi, \mathcal{R}_{np}, RC)$ where the set of rules $\mathcal{R}_{np}$ is defined as follows:* $\{ (from, to, \varphi, \mathbf{uv}, \tau) \mid (from, \delta_{to}, \varphi, \mathbf{uv}, \tau) \in \mathcal{R} \wedge to \in L \wedge \delta_{to} > 0 \}$.

Example 1. Figure 2 illustrates a second, more simpler threshold automaton with the following components: $\mathcal{I} = \{V_0, V_1\}$, $L = \{V_0, V_1, \text{Wait}, D_0, D_1\}$, $\Gamma = \{x_0, x_1\}$, and $\Pi = \{n, t, f\}$. In this model, $n > 3t$ and $t \leq f$ holds. A process in V_0 has a vote of 0, while a process in V_1 has a vote of 1. The decision will be 0 (or 1) if more than $\frac{n-t}{2}$ processes vote 0 (or 1, respectively). This outcome is modeled by all processes transitioning to state D_0 (if the decision is 0) or D_1 (if the decision is 1).

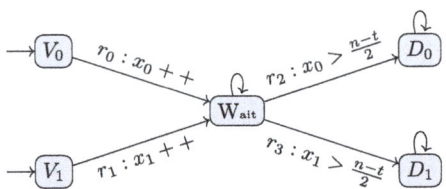

Fig. 2. A threshold automaton for simple voting.

2.1   Semantics of a Threshold Automaton

Given a TA $A = (L, \mathcal{I}, \Gamma, \Pi, \mathcal{R}, RC)$, let the function $N : \mathbf{P}_{RC} \to \mathbb{N}_0$ determine the number of processes to be modelled, typically defined as $N(n, t, f) = n -$

Table 1. Rules of the probabilistic threshold automaton for Ben-Or's algorithm in Fig. 1.

Rule	Guard	Update
r_1	$true$	$x_0 + +$
r_2	$true$	$x_1 + +$
r_3	$x_0 + x_1 \geq n - t - f \ \wedge \ x_0 \geq (n+t)/2 - f$	$y_0 + +$
r_4	$x_0 + x_1 \geq n - t - f \ \wedge \ x_1 \geq (n+t)/2 - f$	$y_1 + +$
r_5	$x_0 + x_1 \geq n - t - f \ \wedge \ x_0 \geq (n-3t)/2 - f \ \wedge \ x_1 \geq (n-3t)/2 - f$	$y? + +$
r_6	$y_0 + y_1 + y? \geq n - t - f \ \wedge \ y? \geq (n-3t)/2 - f \ \wedge \ y_0 \geq t + 1 - f$	$-$
r_7	$y_0 + y_1 + y? \geq n - t - f \ \wedge \ y_0 \geq (n+t)/2 - f$	$-$
r_8	$y_0 + y_1 + y? \geq n - t - f \ \wedge \ y? \geq (n-3t)/2 - f \ \wedge \ y? > n - 2t - f - 1$	$-$
r_9	$y_0 + y_1 + y? \geq n - t - f \ \wedge \ y_1 \geq (n+t)/2 - f$	$-$
r_{10}	$y_0 + y_1 + y? \geq n - t - f \ \wedge \ y? \geq (n-3t)/2 - f \ \wedge \ y_1 \geq t + 1 - f$	$-$
r_{11}	$true$	$-$
r_{12}	$true$	$-$

f, where n represents the total number of processes and f denotes the actual number of faulty processes. The concrete semantics of a system consisting of $N(\mathbf{p})$ threshold automata running in parallel are captured by a *counter system*.

Formally, a counter system is an abstraction of $N(\mathbf{p})$ instances of a given TA running in parallel, as it only keeps track of *how many* TA are in which location, but not exactly which of the processes. However, this abstraction is well-known to be sound and complete for distributed systems with identical/anonymous processes.

Definition 3. *A counter system* (CS) *of a non-probabilistic threshold automaton* $A = (L, \mathcal{I}, \Gamma, \Pi, \mathcal{R}, RC)$ *is a transition system* $\mathsf{CS}(A) = (\Sigma, \Sigma_0, \mathcal{T})$ *where*

- *Σ is the set of all configurations. A configuration is a tuple $\sigma = (\mathbf{k}, \mathbf{g}, \mathbf{p})$ where:*
  - *$\mathbf{k} \in \mathbb{N}_0^{|L|}$ is a vector representing the counter values at each location. Specifically, $\mathbf{k}[i]$ indicates the number of processes present in location i. We refer to locations by their indices in L.*
  - *$\mathbf{g} \in \mathbb{N}_0^{|\Gamma|}$ is a vector of values for the shared variables, where $\mathbf{g}[i]$ is the value of variable $x_i \in \Gamma$.*
  - *$\mathbf{p} \in \mathbf{P}_{RC}$ is an admissible vector of parameter values.*
- *The set Σ_0 consists of all* initial *configurations, which satisfy the following conditions:*
  - *$\forall x_i \in \Gamma : \ \sigma.\mathbf{g}[i] = 0$.*
  - *$\sum_{i \in \mathcal{I}} \sigma.\mathbf{k}[i] = N(\mathbf{p})$.*
  - *$\sum_{i \notin \mathcal{I}} \sigma.\mathbf{k}[i] = 0$.*
- *$\mathcal{T} \subseteq \Sigma \times \mathcal{R} \times \Sigma$ is the set of transitions, where $(\sigma, r, \sigma') \in \mathcal{T}$ if and only if the following conditions hold:*

- *The parameter values are unchanged:* $\sigma'.\mathbf{p} = \sigma.\mathbf{p}$.
- *The counter value at the target location of the rule is incremented:* $\sigma'.\mathbf{k}[r.to] = \sigma.\mathbf{k}[r.to] + 1$ *(one process moves to r.to).*
- *The counter value at the source location of the rule is decremented:* $\sigma'.\mathbf{k}[r.from] = \sigma.\mathbf{k}[r.from] - 1$ *(one process moves out of r.from).*
- *The guard condition $r.\varphi$ holds in the current configuration:* $\sigma.\mathbf{g} \models r.\varphi$ *(i.e., φ holds after replacing shared variables with values $\sigma.\mathbf{g}$).*
- *The shared variable values are updated according to the rule:* $\sigma'.\mathbf{g} = \sigma.\mathbf{g} + r.\mathbf{uv}$.
- *Shared variables in τ are reset to 0:* $\forall x_i \in \tau\ \sigma'.\mathbf{g}[i] = 0$.

Instead of $(\sigma, r, \sigma') \in \mathcal{T}$ we also write $\sigma \xrightarrow{r} \sigma'$. If $(\sigma, r, \sigma') \in \mathcal{T}$, we say that r is enabled *in configuration σ; otherwise, r is* disabled.

Paths of CS. A sequence $\sigma_0, r_0, \sigma_1, \ldots, \sigma_{k-1}, r_{k-1}, \sigma_k$ of alternating configurations and rules, is called a *path* of a counter system $\mathsf{CS}(A) = (\Sigma, \Sigma_0, \mathcal{T})$ if and only if the following conditions hold:

- σ_0 is an initial configuration, i.e., $\sigma_0 \in \Sigma_0$.
- For each $0 \leq i < k$, the transition $(\sigma_i, r_i, \sigma_{i+1}) \in \mathcal{T}$ holds.

In this case, we also write $\sigma_0 \rightarrow^* \sigma_k$ to denote the existence of this path. The set of all such paths in $\mathsf{CS}(A)$ is denoted by $Paths(\mathsf{CS}(A))$.

Example 2. Let $N(n, t, f) = n - f$ and $RC = n > 3t \wedge t \geq f$. For the case where $n = 5$, $t = 1$, and $f = 1$, the following sequence represents a valid path in the counter system of the threshold automaton shown in Fig. 2:
$[(4, 0, 0, 0, 0)(0, 0)], \mathbf{r_0}, [(3, 0, 1, 0, 0)(1, 0)], \mathbf{r_0}, [(2, 0, 2, 0, 0)(2, 0)], \mathbf{r_0},$
$[(1, 0, 3, 0, 0)(3, 0)], \mathbf{r_0}, [(0, 0, 4, 0, 0)(4, 0)], \mathbf{r_2}, [(0, 0, 3, 1, 0)(4, 0)].$

3   Abstract Threshold Automata

A TA is an infinite state automaton due to the infinite domains of its shared variables and parameters. Therefore, to enable the use of parameterized verification techniques for finite-state processes, we introduce a finite abstract domain for the shared variables and we introduce an abstraction of TAs based on parametric interval abstraction [3].

Abstract Domain for Shared Variables. The key idea is that along a run of an automaton, we are not interested in the exact values of shared variables. Instead, we only care whether they satisfy a guard condition. Given a threshold automaton A, we define the set of thresholds as

$$\mathcal{TH} = \{d_0, d_1, \ldots, d_k\}$$

where $d_0 = 0$, $d_1 = 1$, and for all $i > 1$, d_i is a threshold in A. We assume that for all i, j, if $i < j$, then $d_i < d_j$. This ordering is always feasible for a fixed $\mathbf{p} \in \mathbf{P}_{RC}$. If different values of $\mathbf{p} \in \mathbf{P}_{RC}$ result in different orderings of the d_i, we consider each of the finitely many such orderings separately. Based on this, we define the finite set of intervals

$$\mathcal{D} = \{I_0, I_1, \ldots, I_k\}$$

where $I_i = [d_i, d_{i+1}[$ for $i < k$, and $I_k = [d_k, \infty[$.

Definition 4. Abstract Threshold Automata [3]. Given a threshold automaton $A = (L, \mathcal{I}, \Gamma, \Pi, \mathcal{R}, RC)$, we define the *abstract threshold automaton* (or $\overline{TA}$) as $\overline{A} = (L, \mathcal{I}, \overline{\Gamma}, \Pi, \overline{\mathcal{R}})$, where:

- A and $\overline{A}$ share the components $L, \mathcal{I}, \Pi$.
- Let $\Gamma = \{x_0, \ldots, x_m\}$, then $\overline{\Gamma} = \{\overline{x}_0, \ldots, \overline{x}_m\}$, where each $\overline{x}_i$ takes values in the domain $\mathcal{D} = \{I_0, I_1, \ldots, I_k\}$.
- $\overline{\mathcal{R}}$ is the set of abstract rules. An *abstract rule* is a tuple $\overline{r} = (from, to, \overline{\varphi}, \mathbf{uv}, \tau)$, where *from*, *to*, $\mathbf{uv}, \tau$ remain unchanged from A, and the *abstract guard* $\overline{\varphi}$ is a Boolean expression over equalities between shared variables and abstract values.
  Formally, let $\varphi = \varphi_0 \wedge \ldots \wedge \varphi_k$. Then, $\overline{\varphi} = \overline{\varphi}_0 \wedge \ldots \wedge \overline{\varphi}_k$, where:
  - If $\varphi_i = (d_j \leq x)$, then

$$\overline{\varphi}_i = \bigvee_{c=j}^{k-1} (\overline{x} = [d_c, d_{c+1}[) \vee (\overline{x} = [d_k, \infty[).$$

  - If $\varphi_i = (d_j > x)$, then

$$\overline{\varphi}_i = \bigvee_{c=0}^{j-1} (\overline{x} = [d_c, d_{c+1}[).$$

Example 3. Revisit the simple threshold automaton in Fig. 2 with $N(n, t, f) = n - f$ and $RC = n > 3t \wedge t \geq f > 1$. We define $\mathcal{TH} = \{0, 1, t, \frac{n-t}{2}\}$ and $\mathcal{D} = \{[0, 1[, [1, t[, [t, \frac{n-t}{2}[, [\frac{n-t}{2}, \infty[\}$, where the order is induced by the condition RC. The thresholds 0 and 1 are always included to allow detection of whether a shared variable is equal to 0.

Moreover, we have the following abstract rules:

- $\overline{r}_0 = r_0$, $\overline{r}_1 = r_1$ (due to the absence of a guard),
- $\overline{r}_2.\overline{\varphi} = (\overline{x}_0 = [\frac{n-t}{2}, \infty[)$ since the concrete guard is $x_0 > \frac{n-t}{2}$,
- $\overline{r}_3.\overline{\varphi} = (\overline{x}_1 = [\frac{n-t}{2}, \infty[)$ since the concrete guard is $x_1 > \frac{n-t}{2}$.

3.1 (0, 1)-Abstraction

Semantics of $\overline{TA}$. In this section the semantics of a system composed of an arbitrary number of $\overline{TA}$s is over-approximated by a $(0, 1)$-*counter system* (or ZCS). The main component in a ZCS is the $(0, 1)$-configuration. The key idea is that, for the specifications of interest here (reachability and termination), knowing the exact number of processes at a location is unnecessary; it is sufficient to determine whether a location contains any processes. Such an abstraction transforms our system into a finite-state one, enabling the use of symbolic techniques to implement our algorithm.

A $(0, 1)$-*configuration* is a tuple $\overline{\sigma} = (\overline{\mathbf{k}}, \overline{\mathbf{g}})$, where $\overline{\mathbf{k}} \in \mathbb{B}^{|L|}$ and $\overline{\mathbf{g}} \in \mathcal{D}^{|\overline{\Gamma}|}$. Here, $\overline{\mathbf{k}}[i]$ indicates the presence (1) or absence (0) of at least one process at location i, and $\overline{\mathbf{g}}$ is a vector of shared variable values, where $\overline{\mathbf{g}}[i]$ is the parametric interval currently assigned to $\overline{x}_i$. In a $(0, 1)$-configuration $\overline{\sigma} = (\overline{\mathbf{k}}, \overline{\mathbf{g}})$, we denote $\overline{\mathbf{k}}$ as the 01-counter-valuation and $\overline{\mathbf{g}}$ as the 01-var-valuation. If $\overline{\mathbf{k}}[i] = 1$, we say that location i is covered in $\overline{\sigma}$.

Definition 5. *A $(0, 1)$-counter system (or* ZCS*) [3] of an abstract threshold automaton $\overline{A} = (L, \mathcal{I}, \overline{\Gamma}, \Pi, \mathcal{R})$ is a transition system* $\mathsf{ZCS}(\overline{A}) = (\overline{\Sigma}, \overline{\Sigma}_0, \mathcal{T})$, *where:*

- $\overline{\Sigma} = \mathbb{B}^{|L|} \times \mathcal{D}^{|\overline{\Gamma}|}$ *is the set of $(0, 1)$-configurations. Each configuration $\overline{\sigma} = (\overline{\mathbf{k}}, \overline{\mathbf{g}})$ consists of:*
  - $\overline{\mathbf{k}} \in \mathbb{B}^{|L|}$, *where $\overline{\mathbf{k}}[i] = 1$ indicates that location i contains at least one process, and $\overline{\mathbf{k}}[i] = 0$ otherwise.*
  - $\overline{\mathbf{g}} \in \mathcal{D}^{|\overline{\Gamma}|}$, *where $\overline{\mathbf{g}}[i]$ represents the parametric interval assigned to $\overline{x}_i$.*
- $\overline{\Sigma}_0 \subseteq \overline{\Sigma}$ *is the set of initial $(0, 1)$-configurations satisfying:*
  - $\forall i \in \overline{\Gamma} : \overline{\sigma}.\overline{\mathbf{g}}[i] = I_0$.
  - $\forall i \in L : \overline{\sigma}.\overline{\mathbf{k}}[i] = 1 \Leftrightarrow i \in \mathcal{I}$.
- *The transition relation $\mathcal{T}$ consists of transitions $(\overline{\sigma}, \overline{r}, \overline{\sigma}')$, where:*
  - $\overline{r} = \{from, to, \overline{\varphi}, \mathbf{uv}\} \in \mathcal{R}$.
  - *The condition $\overline{\sigma}.\overline{\mathbf{g}} \models \overline{r}.\overline{\varphi}$ holds.*
  - *The source location is occupied: $\overline{\sigma}.\overline{\mathbf{k}}[\overline{r}.from] = 1$.*
  - *The transition updates $\overline{\mathbf{k}}$: either $\overline{\sigma}'.\overline{\mathbf{k}}[\overline{r}.from] = 0$ or it remains 1.*
  - *The target location is covered after the transition: $\overline{\sigma}'.\overline{\mathbf{k}}[\overline{r}.to] = 1$.*
  - *Shared variables are updated: $\overline{\sigma}'.\overline{\mathbf{g}} = \overline{\sigma}.\overline{\mathbf{g}} \dotplus \mathbf{uv}$, defined as follows:*
    1. $\overline{\sigma}'.\overline{\mathbf{g}}[i] = \overline{\sigma}.\overline{\mathbf{g}}[i]$, *if $\overline{r}.\mathbf{uv}[i] = 0$*
    2. $(\overline{\sigma}'.\overline{\mathbf{g}}[i] = \overline{\sigma}.\overline{\mathbf{g}}[i]) \vee (\overline{\sigma}'.\overline{\mathbf{g}}[i] = \overline{\sigma}.\overline{\mathbf{g}}[i].next)$, *if $\overline{r}.\mathbf{uv}[i] = 1$. The first disjunct is omitted if $\overline{\sigma}.\overline{\mathbf{g}}[i] = I_0$.*
  - *Reset variables are reinitialized: $\forall x_i \in \overline{r}.\tau, \overline{\sigma}'.\overline{\mathbf{g}}[i] = I_0$.*

Paths. A *path* of $\mathsf{ZCS}(\overline{A})$ is a sequence of alternating $(0, 1)$-configurations and abstract rules, given by $\overline{\sigma}_0, \overline{r}_0, \overline{\sigma}_1, \ldots, \overline{\sigma}_{k-1}, \overline{r}_{k-1}, \overline{\sigma}_k$, such that for all $i < k$, the transition $(\overline{\sigma}_i, \overline{r}_i, \overline{\sigma}_{i+1})$ belongs to $\mathcal{T}$. The set of all paths of $\mathsf{ZCS}(\overline{A})$ is denoted by $Paths(\mathsf{ZCS}(\overline{A}))$.

Example 4. Consider again the TA from Fig. 2, and $I_0 = [0, 1[, I_1 = [1, \frac{n-t}{2}[, I_2 = [\frac{n-t}{2}, \infty[$. The following is a valid path of its $(0, 1)$-counter system:
$[(1, 0, 0, 0, 0)(I_0, I_0)], \overline{r}_0, [(1, 0, 1, 0, 0)(I_1, I_0)], \overline{r}_0, [(1, 0, 1, 0, 0)(I_2, I_0)], \overline{r}_2,$
$[(0, 0, 1, 1, 0)(I_2, I_0)]$.

Specifications. We say that a 01-configuration $\overline{\sigma}$ satisfies a reachability specification $L_{spec} = (L_{=0}, L_{>0})$, denoted $\overline{\sigma} \models L_{spec}$, if for all $i \in L_{=0}$, $\overline{\sigma}.\overline{\mathbf{k}}[i] = 0$, and for all $i \in L_{>0}$, $\overline{\sigma}.\overline{\mathbf{k}}[i] > 0$.

Monotonicity. Since global variables in a ZCS (or a CS) are initialized to 0 and never decrease, the following property holds:

Property 1. *(Monotonicity) Given a TA A, the monotonicity property states that in any execution of* ZCS($\overline{A}$) *(CS(A)):*

- *Once a lower guard becomes enabled, it remains enabled forever.*
- *Once an upper guard becomes disabled, it remains disabled forever.*

4   CS Vs ZCS

In comparison to CS, in ZCS we drop the resilience condition, the function N that determines the number of processes to be modeled, as well as the exact number of processes in a global configuration. Moreover, a transition in ZCS may jump from one interval to the next too early and may stay in the same interval although it had to move. In our previous work [3], we showed that the abstraction from CS to ZCS is complete with respect to reachability specifications.

In the following, we show how to detect whether a behavior of the $(0, 1)$-counter system corresponds to a concrete behavior of a counter system.

A path $\overline{\pi} = \overline{\sigma}_0, \overline{r}_0, \dots, \overline{r}_{m-1}, \overline{\sigma}_m$ in ZCS($\overline{A}$) = $(\overline{\Sigma}, \overline{\Sigma}_0, \mathcal{T})$ *corresponds* to the paths $\pi = \sigma_0, r_0^{c_0}, \dots, r_{m-1}^{c_{m-1}}, \sigma_m$ (where $r_i^{c_i}$ simulates applying r_i c_i times) of CS(A) = $(\Sigma, \Sigma_0, \mathcal{T})$ that satisfy the following conditions:

- $RC \wedge (\sum_{j \in \mathcal{I}} \sigma_0.\mathbf{k}[j] = N(n, t, f))$
- $\forall i < m \; \sigma_i.\mathbf{k}[r_i.from] = c_i + \sigma_{i+1}.\mathbf{k}[r_i.from] \wedge \sigma_{i+1}.\mathbf{k}[r_i.to] = c_i + \sigma_i.\mathbf{k}[r_i.to]$
- $\forall i < m \; \forall x_j \in \Gamma \; x_j \notin r_i.\tau \implies \sigma_{i+1}.\mathbf{g}[j] = \sigma_i.\mathbf{g}[j] + c_i \cdot r_i.\mathbf{uv}[j]$
- $\forall i < m \; \forall x_j \in r_i.\tau \; \sigma_{i+1}.\mathbf{g}[j] = 0$
- $\forall i < m \; \forall x_j \in \Gamma \; \sigma_i.\mathbf{g}[j] \in \overline{\sigma}_i.\overline{\mathbf{g}}[j] \wedge \sigma_{i+1}.\mathbf{g}[j] \in \overline{\sigma}_{i+1}.\overline{\mathbf{g}}[j]$
- $\forall i < m \; c_i > 1 \implies ((\sigma_{i+1}.\mathbf{g} - r_i.\mathbf{uv}) \models r_i.\varphi)^1$

Let $Concretize(\overline{\pi})$ denote the conjunction of the constraints above, where quantified formulas are instantiated as a finite conjunction of quantifier-free formulas. Since $Concretize(\overline{\pi})$ is a quantifier-free formula in linear integer arithmetic, a satisfying assignment (which can be computed by an SMT solver) represents a path of CS(A) corresponding to $\overline{\pi}$. A path $\overline{\pi} \in Paths(\text{ZCS}(\overline{A}))$ is said to be *spurious* if $Concretize(\overline{\pi})$ is unsatisfiable.

5   Almost-Sure Termination and SCC Detection

In our previous work [3], we introduced a reachability algorithm for ZCS. The algorithm begins with the set of target states and performs a backward traversal

---

[1] This is needed only in cases where an update affects any of the guards of $r_i.\varphi$.

of the state space until it reaches a fixed point. Within this fixed point, all paths originating from the initial state are examined to determine whether they are spurious, as described in Sect. 4. If at least one such path is not spurious, the algorithm concludes that the target states are reachable. This reachability algorithm serves as two subprocedures, *IsReachable* and *ComputeFixedPoint*, in Algorithm 2.

In this section, we present our algorithm (Algorithm 2) for proving almost-sure termination and for detecting strongly connected components (SCCs) in probabilistic threshold automata, as well as for proving termination and detecting SCCs in ordinary threshold automata. Then we prove that our algorithm is sound. We assume that the input probabilistic automaton is deterministic, reset-free, and that its corresponding $(0,1)$-counter system is free of both local and global deadlocks.

Remark 2. A fault-tolerant system terminates if and only if all correct processes (modeled by a TA) reach a final state.

Before presenting the algorithm, we first provide the following definitions of the key terms and concepts used throughout the algorithm.

- For a given set S, let $\mathcal{P}_\emptyset(S)$ denote the powerset of S excluding the empty set.
- A set $X \subset V$ is strongly connected if, for every two elements $v, u \in X$, there is a path from v to u. A *strongly connected component* (SCC) is a maximal strongly connected set $S \subseteq V$. We denote a strongly connected component over L by local SCC, and a strongly connected component over $\overline{\Sigma}$ by simply SCC.
- A *strongly connected component* (SCC) of a directed graph is a maximal subgraph where for any two vertices u and v, there exists a path from u to v and vice versa. We refer to an SCC within a TA as a *local SCC* and an SCC within ZCS as an *01-SCC*. Given a local SCC $\mathcal{C}$, we denote by $\mathcal{C}.Locations$ the set of locations in $\mathcal{C}$, and by $\mathcal{C}.Rules$ the set of rules in $\mathcal{C}$. We use the same notations for a 01-SCC.
- A local SCC $\mathcal{C}$ is *valid* if there exists at least one $(0,1)$-*configuration* $\overline{\sigma} = (\overline{\mathbf{k}}, \overline{\mathbf{g}})$ where $\forall \overline{r}_i \in \mathcal{C}.Rules : \overline{\mathbf{g}} \models \overline{r}_i.\overline{\varphi}$. Otherwise, $\mathcal{C}$ is called *invalid*.
- A local SCC $\mathcal{C}$ is *finite-traverse* if there exist indices i, j and a variable k such that $\overline{r}_i, \overline{r}_j \in \mathcal{C}.Rules$ where $\overline{r}_i.\mathbf{uv}[k] > 0$ and $\overline{r}_j.\overline{\varphi}$ includes an upper guard on k. Otherwise, $\mathcal{C}$ is *infinite-traverse*. A local SCC $\mathcal{C}$ is valid-infinite if $\mathcal{C}$ is valid and infinite-traverse.
- We extend the latter definition to subsets of local SCCs as follows: A *finite-traverse subset* is a subset of local SCCs $S^\mathcal{C} = \{\mathcal{C}_1, \ldots, \mathcal{C}_c\}$ for which there exist indices i, j, rules $\overline{r}_i, \overline{r}_j$, and a variable k such that $\mathcal{C}_i, \mathcal{C}_j \in S^\mathcal{C}$ where $\overline{r}_i \in \mathcal{C}_i.Rules \wedge \overline{r}_i.\mathbf{uv}[k] > 0$, and $\overline{r}_j \in \mathcal{C}_j.Rules$ with $\overline{r}_j.\overline{\varphi}$ includes an upper guard on k. Otherwise, $S^\mathcal{C}$ is called an *infinite-traverse subset*.

Algorithm 2. Termination and SCC detection. If the given automaton is probabilistic, it is converted first to its non-deterministic version (see Definition 2).

1: **Input:** Abstract TA ($\overline{TA}$), final locations, and a probabilistic flag
2: **Output:** termination flag.
3: **procedure** TERMINATIONCHECK($ATA, final\_locs, isProbabilistic$)
4:     $localSCCs \leftarrow ExtractAllSCCs(ata)$ // Extract all local SCCs from the ata.
5:     // keep only infinite-traverse local SCCs.
6:     $localSCCs \leftarrow \{\mathcal{C} \in localSCCs \mid \mathcal{C}.IsInfiniteTraverse\}$
7:     $sccValidIntrvls \leftarrow \emptyset$
8:     // keep only valid local SCCs.
9:     **for all** $\mathcal{C} \in localSCCs$ **do**
10:        // $varGuards$ maps a variable to its guards on the local SCC.
11:        $varGuards \leftarrow GetVarGuards(\mathcal{C})$
12:        **for all** $(var, guards) \in varGuards$ **do**
13:            // $varValidIntrvls$ maps a var to intervals satisfying $guards$.
14:            $varValidIntrvls \leftarrow GetVarEnablingIntrvls(var, guards)$
15:            **if** $\exists var \in varValidIntrvls.Keys : varValidIntrvls[var] == \emptyset$ **then**
16:                $localSCCs \leftarrow localSCCs \setminus \{\mathcal{C}\}$ // remove invalid local SCCs
17:            **else**
18:                $sccValidIntrvls[\mathcal{C}] \leftarrow \bigotimes_{V \in varValidIntrvls.Values} V$
19:                // $sccValidIntrvls[\mathcal{C}] = \{\overline{\mathbf{g}} \mid \forall \overline{r} \in \mathcal{C}.Rules : \overline{\mathbf{g}} \models \overline{r}.\overline{\varphi}\}$
20:     $subsets \leftarrow \mathcal{P}_\emptyset(localSCCs)$ // generate all subsets.
21:     // keep only infinite-traverse subsets.
22:     $subsets \leftarrow \{subset \in subsets \mid subset.isInfiniteTraverse\}$
23:     **if** $isProbabilistic$ **then**
24:        // 01-configurations in which only final locations are covered.
25:        $final01Configs \leftarrow ConvertTo01Configs(\mathcal{P}_\emptyset(final\_locs))$
26:     $scc01Configs \leftarrow \emptyset$
27:     **for all** $subset \in subsets$ **do**
28:        // $subsetValidIntrvls[subset] = \{\overline{\mathbf{g}} \mid \forall \overline{r} \in subset : \overline{\mathbf{g}} \models \overline{r}.\overline{\varphi}\}$
29:        $subsetValidIntrvls[subset] \leftarrow \bigcap_{\mathcal{C} \in subset} sccValidIntrvls[\mathcal{C}]$
30:        **if** $subsetValidIntrvls[subset] == \emptyset$ **then** continue
31:        $locsSubsets \leftarrow \mathcal{P}_\emptyset(subset.Locations)$
32:        // convert 01-counter-valuations and 01-var-valuations into 01-configuration
33:        $01Configs \leftarrow locsSubsets \otimes subsetValidIntrvls[subset]$
34:        // check if traversing $subset$ infinitly has probability 0.
35:        **if** $isProbabilistic$ **then**
36:            $nonDetImage \leftarrow ComputeNonDetImage(01Configs)$
37:            $fixedPoint \leftarrow ComputeFixedPoint(final01Configs)$
38:            **if** $fixedPoint \cap nonDetImage \neq \emptyset$ **then** $01Configs \leftarrow \emptyset$
39:        // collcet all configurations that are on an SCC
40:        $scc01Configs \leftarrow scc01Configs \cup 01Configs$
41:     **Return** $IsReachable(scc01Configs)$

5.1   Detailed Description for Algorithm 2

Algorithm 2 starts in Line 4 by extracting all local SCCs (*localSCCs*) in $\overline{TA}$. This can be computed efficiently using either Tarjan's algorithm, which runs in linear time, or the symbolic CHAIN algorithm [15], which is also linear. Line 6 removes all *finite-traverse* local SCCs from *localSCCs*. Lines 9 to 19 filter out invalid local SCCs and compute, for each *valid infinite-traverse* local SCC, the set of 01-var-valuations that enable all its rules. The cross-product operation in Line 18 can be efficiently implemented using a BDD manager by computing the conjunction of disjunctions over the intervals of each set in *varValidIntrvls.Values*. At this stage all local SCCs in *localSCCs* are guaranteed to be valid-infinite due to Monotonicity property (see Property 1). Lines 20 to 22 compute all subsets of *infinite-traverse* local SCCs. Line 25 computes all $(0, 1)$-configurations in which only final locations (*final_locs*) are covered. Line 29 computes the set of 01-var-valuations that enable all rules in every cycle of a given *subset*. If and only if *subset* is valid (Line 30), its corresponding 01-counter-valuations and 01-var-valuations are converted into $(0, 1)$-configurations (Lines 31–33). Lines 35 to 38 check whether traversing the local SCC set (*subset*) infinitely has probability 0. The key idea is that if infinite travese local SCC $\mathcal{C} \in subset$ contains a non-deterministic transition, one remaining within $\mathcal{C}$ and another leading outside and reaches a final location, then the probability of infinitely choosing the intra-SCC transition is 0. $ComputeNonDetImage(01Configs)$ in Line 36 is a procedure that computes the image of $01Configs$ as follows:

- Select all $(0, 1)$-configurations σ from $01Configs$ that satisfy the following condition: If a location l belongs to *locsSubsets* and is covered in σ, then l has a non-deterministic transition to a location outside *locsSubsets*. We refer to this set as NDT.
- For each $(0, 1)$-configuration $\sigma \in NDT$, compute a new $(0, 1)$-configuration σ'. The configuration σ' is obtained by executing all non-deterministic transitions that originate from locations in *locsSubsets* (covered in σ) and lead to locations outside *locsSubsets*.

Line 40, collects all $(0, 1)$-configurations that belong to a 01-SCC, and checks if any configuration in $scc01Configs$ is reachable. The procedures in Lines 37 and 41 follow the approach we presented in [3] which also takes care of spurious paths.

Remark 3. To simplify the presentation of the algorithm, we assumed that a local SCC is either *valid-infinite* or not. However, in general, a local SCC that is not *valid-infinite* can be decomposed into multiple *valid-infinite* local SCCs.

5.2   Correctness

Algorithm 2 is sound. Soundness follows directly from the fact that $scc01Configs$ contains all $(0, 1)$-configurations that may traverse a 01-SCC inifintely often. Also since, for probabilistic systems, SCCs with a probability of 0 of being traversed infinitely often are excluded (see Lines 35– 38),

Theorem 1. (Soundness SCC Detection). *Algorithm 2 is sound for SCC detection. That is, if the algorithm computes a non-empty scc01Configs, then every* $(0,1)$ − *configuration in scc01Configs belongs to a* 01-*SCC.*

The below corollary is obtained since SCC detection is sound and since the reachability algorithm in [3] is proven to be correct.

Theorem 2. (Soundness (A.S.) Termination). *Algorithm 2 is sound for termination and almost-sure termination. Specifically, if the algorithm returns true, it ensures that a system consisting of a network of TAs does not terminate, regardless of its size.*

We will leave completness and the handling of resets for future work.

6   Related Work

Recently, there have been many works [2, 3, 12–14] that target the parameteried verification of non-probabilistic threshold automata. In [12,13], Konnov et al. proposed an approach for detecting TA traces that violate reachability specifications, as well as lasso-shaped TA traces that violate a given liveness property [13]. These methods have been implemented in the ByMC tool [14]. The decidability and complexity of verification and synthesis for threshold automata were also addressed in [2]. Their decision procedure relies on an SMT encoding of potential error paths, where, in general, the size of the SMT formula increases exponentially with the length of the paths. In [3], the authors extended the threshold automaton with resets and variable decrements, and introduced a new algorithm to check reachability and coverability. Additionally, their approach removed the cycle absence restriction that was required in previous works.

In [6], the authors introduced the probabilistic threshold automaton, a threshold automaton extended with coin tosses, and proposed a new approach for verifying them and checking almost-sure termination. This approach works under two key restrictions, which we lift in this paper: 1) it does not allow cycles inside the automaton, and 2) coin tosses may only appear at the end of a round. An extension of the PTA was presented in [8], where the authors incorporated common coins into PTAs. They reduced the formal verification of the extended PTA to single-round queries on non-probabilistic threshold automata, which are then verified using ByMC [12].

Unfortunately, there are few works [5, 7, 16, 17, 20, 24] that address the automatic verification of probabilistic parameterized systems. Unlike this work (and [6,8]), these approaches rely on process templates with a finite state space and use a single parameter, the number of process template instances. In [5], the authors use a probabilistic single-clock timed automaton with a broadcasting communication primitive. They verify whether a configuration in which one process reaches a target state almost surely. Their approach is based on well-structured transition systems. In [17], a method was introduced to prove liveness for randomized parameterized systems under arbitrary schedulers, while [16] presented a fully automatic verification method for proving almost-sure termination of probabilistic parameterized concurrent systems. Both approaches [5, 16, 17] are based on regular model checking and do not support arithmetic resilience conditions or shared variables over infinite domains. The seminal work by Pnueli and Zuck [20] requires shared variables to be over finite domains and restricts the use of thresholds to only 1 and n. In [24], authors presented a novel approach for checking liveness in probabilistic parameterized protocols, by abstracting a parameterized Markov Decision Process (MDP) to a finite MDP.

7   Conclusion

In this paper, we introduced an extension of probabilistic threshold automata to support modelling resets of shared variables. We further presented a sound algorithm for detecting almost-sure termination in randomized fault-tolerant algorithms, modelled with the help of probabilistic threshold automata, and for verifying termination in ordinary fault-tolerant algorithms, albeit so far only for reset-free automata. Our approach enables the detection of strongly connected components and termination in system models consisting of an arbitrary number of threshold automata. Additionally, it identifies strongly connected components and verifies almost-sure termination in system models with an arbitrary number of probabilistic threshold automata. Furthermore, we lifted two key restrictions previously imposed on threshold automata: (1) cycles were not allowed, and (2) coin tosses could only appear at the end of the automaton. For future work, we plan to investigate the completeness of our approach, handle resets of shared variables, relax the deadlock restriction by incorporating a fairness notion, and develop an implementation of the proposed algorithm.

Acknowledgments. This research was funded in whole or in part by the Luxembourg National Research Fund (FNR) grant C22/IS/17432184 (project FM-CReST). For the purpose of open access, and in fulfilment of the obligations arising from the grant agreement, the author has applied a Creative Commons Attribution 4.0 International (CC BY 4.0) license to any Author Accepted Manuscript version arising from this submission.

References

1. Baier, C., et al.: Chiefly symmetric: results on the scalability of probabilistic model checking for operating-system code. In: Prof. of the 7th Conference on Systems Software Verification (SSV'12). Electronic Proceedings in Theoretical Computer Science, vol. 102, pp. 156–166 (2012). https://doi.org/10.4204/EPTCS.102.14
2. Balasubramanian, A., Esparza, J., Lazić, M.: Complexity of verification and synthesis of threshold automata. In: International Symposium on Automated Technology for Verification and Analysis, pp. 144–160. Springer (2020)

3. Baumeister, T., Eichler, P., Jacobs, S., Sakr, M., Völp, M.: Parameterized verification of round-based distributed algorithms via extended threshold automata. In: International Symposium on Formal Methods, pp. 638–657. Springer (2024)

4. Ben-Or, M.: Another advantage of free choice (extended abstract): completely asynchronous agreement protocols. In: Proceedings of the Second Annual ACM Symposium on Principles of Distributed Computing, PODC '83, pp. 27–30. Association for Computing Machinery, New York (1983).https://doi.org/10.1145/800221.806707

5. Bertrand, N., Fournier, P.: Parameterized verification of many identical probabilistic timed processes. In: IARCS Annual Conference on Foundations of Software Technology and Theoretical Computer Science (FSTTCS 2013), pp. 501–513. Schloss Dagstuhl–Leibniz-Zentrum für Informatik (2013)

6. Bertrand, N., Konnov, I., Lazić, M., Widder, J.: Verification of randomized consensus algorithms under round-rigid adversaries. Int. J. Softw. Tools Technol. Transfer , 1–25 (2021). https://doi.org/10.1007/s10009-020-00603-x

7. Esparza, J., Gaiser, A., Kiefer, S.: Proving termination of probabilistic programs using patterns. In: Madhusudan, P., Seshia, S.A. (eds.) CAV 2012. LNCS, vol. 7358, pp. 123–138. Springer, Heidelberg (2012). https://doi.org/10.1007/978-3-642-31424-7_14

8. Gao, S., Zhan, B., Wu, Z., Zhang, L.: Verifying randomized consensus protocols with common coins. In: 2024 54th Annual IEEE/IFIP International Conference on Dependable Systems and Networks (DSN), pp. 403–415 (2024).https://doi.org/10.1109/DSN58291.2024.00047

9. Guerraoui, R., Kuznetsov, P., Monti, M., Pavlovic, M., Seredinschi, D.: Scalable byzantine reliable broadcast. In: Suomela, J. (ed.) 33rd International Symposium on Distributed Computing, DISC 2019, October 14-18, 2019, Budapest, Hungary. LIPIcs, vol. 146, pp. 1–16. Schloss Dagstuhl - Leibniz-Zentrum für Informatik (2019).https://doi.org/10.4230/LIPICS.DISC.2019.22

10. Jacobs, S., Sakr, M.: Analyzing guarded protocols: better cutoffs, more systems, more expressivity. In: International Conference on Verification, Model Checking, and Abstract Interpretation, pp. 247–268. Springer (2018)

11. Jacobs, S., Sakr, M., Zimmermann, M.: Promptness and bounded fairness in concurrent and parameterized systems. In: International Conference on Verification, Model Checking, and Abstract Interpretation, pp. 337–359. Springer (2020)

12. Konnov, I., Lazić, M., Veith, H., Widder, J.: Para 2: parameterized path reduction, acceleration, and SMT for reachability in threshold-guarded distributed algorithms. Formal Methods Syst. Des. **51**(2), 270–307 (2017)

13. Konnov, I., Lazić, M., Veith, H., Widder, J.: A short counterexample property for safety and liveness verification of fault-tolerant distributed algorithms. In: Proceedings of the 44th ACM SIGPLAN Symposium on Principles of Programming Languages, pp. 719–734 (2017)

14. Konnov, I., Widder, J.: BYMC: byzantine model checker. In: International Symposium on Leveraging Applications of Formal Methods, pp. 327–342. Springer (2018)

15. Larsen, C.A., Schmidt, S.M., Steensgaard, J., Jakobsen, A.B., de Pol, J.V., Pavlogiannis, A.: A truly symbolic linear-time algorithm for SCC decomposition. In: International Conference on Tools and Algorithms for the Construction and Analysis of Systems, pp. 353–371. Springer (2023)

16. Lengál, O., Lin, A.W., Majumdar, R., Rümmer, P.: Fair termination for parameterized probabilistic concurrent systems. In: Tools and Algorithms for the Construction and Analysis of Systems: 23rd International Conference, TACAS 2017,

Held as Part of the European Joint Conferences on Theory and Practice of Software, ETAPS 2017, Uppsala, Sweden, April 22–29, 2017, Proceedings, Part I 23, pp. 499–517. Springer (2017)

17. Lin, A.W., Rümmer, P.: Liveness of randomised parameterised systems under arbitrary schedulers. In: International Conference on Computer Aided Verification, pp. 112–133. Springer (2016)

18. Malkhi, D., Reiter, M.: Byzantine quorum systems. In: Proceedings of the Twenty-Ninth Annual ACM Symposium on Theory of Computing, STOC '97, pp. 569–578. Association for Computing Machinery, New York (1997). https://doi.org/10.1145/258533.258650

19. Neiheiser, R., Matos, M., Rodrigues, L.: Kauri: scalable BFT consensus with pipelined tree-based dissemination and aggregation. In: Proceedings of the ACM SIGOPS 28th Symposium on Operating Systems Principles, pp. 35–48 (2021)

20. Pnueli, A., Zuck, L.: Verification of multiprocess probabilistic protocols. In: Proceedings of the third annual ACM Symposium on Principles of Distributed Computing, pp. 12–27 (1984)

21. Suzuki, I.: Proving properties of a ring of finite-state machines. Inf. Process. Lett. **28**(4), 213–214 (1988)

22. Zamani, M., Movahedi, M., Raykova, M.: RapidChain: scaling blockchain via full sharding. In: Lie, D., Mannan, M., Backes, M., Wang, X. (eds.) Proceedings of the 2018 ACM SIGSAC Conference on Computer and Communications Security, CCS 2018, Toronto, ON, Canada, October 15–19, 2018, pp. 931–948. ACM (2018). https://doi.org/10.1145/3243734.3243853

23. Zarbafian, P., Gramoli, V.: Lyra: fast and scalable resilience to reordering attacks in blockchains. In: 2023 IEEE International Parallel and Distributed Processing Symposium (IPDPS), pp. 929–939. IEEE (2023)

24. Zuck, L.D., McMillan, K.L., Torf, J.: Planner-less proofs of probabilistic parameterized protocols. In: International Conference on Verification, Model Checking, and Abstract Interpretation, pp. 336–357. Springer (2017)

Author Index

N. Bertrand et al. (Eds.): Christel Baier Festschrift, LNCS 15760, pp. 425–426, 2026.
https://doi.org/10.1007/978-3-031-97439-7

The manufacturer's authorised representative in the EU is Springer
Nature Customer Service Centre GmbH, Europaplatz 3, 69115 Heidelberg,
Germany. If you have any concerns regarding our products, please
contact ProductSafety@springernature.com

Printed and bound by CPI Group (UK) Ltd, Croydon, CR0 4YY
28/04/2026
02098528-0002